# Renewable Energy Systems for Building Designers

*Renewable Energy Systems for Building Designers* presents a comprehensive introduction to the latest resources and technologies used in high performance and net zero energy buildings, with a practical focus on the design and integration of these systems. This textbook and convenient reference offers a single-source guide to renewable technologies, balancing broad knowledge with the details of implementation crucial for successful sustainable design. It equips students and professionals with foundations and critical information needed to confidently plan for and meet the highest standards of energy efficiency in new construction and retrofitted buildings.

Part I of the book establishes key principles of renewable systems, power production, and design for climate, introducing energy modeling and measurements of performance. Part II focuses in more depth on renewable energy systems, including photovoltaics, heat pumps, solar thermal, and more. Dedicated chapters break down the fundamental concepts behind each renewable technology and present guidelines for configuration and installation including system requirements, equipment specification, sizing, and location of components. Part III discusses topics relevant across renewable systems, including energy storage, control and monitoring, and cost/payback calculation. Part IV comprises case studies of exemplary renewable energy projects.

Features:

- Covers resources and technologies including photovoltaics, solar thermal hot water, heat pumps, biomass, wind and microhydro turbines, marine renewable energy, deep cycle rechargeable batteries, and system controllers.
- Compiles up-to-date, essential information on designing with renewable systems in one location, organized by technology for easy reference.
- Presents clear explanations of all concepts and system aspects, using US/SI units and full-color diagrams and illustrations throughout.
- Features case studies of renewable energy systems in completed projects, demonstrating a range of climate-specific applications.
- Includes study questions, a comprehensive guide to terminology and acronyms, spreadsheets for calculations, system sizing worksheets, and additional online resources.

*Renewable Energy Systems for Building Designers: Fundamentals of Net Zero and High Performance Design* will serve as an essential introduction and enduring reference for students of architecture, engineering, construction, and building science. Equally valuable as a professional resource, it will quickly become the go-to guide for energy efficient design for practitioners in these areas.

**Dorothy Gerring** has been a registered architect since 1989 and an Associate Professor of Architecture and Sustainability at the Pennsylvania College of Technology for 32 years. She holds certifications for LEED AP BD+C and Passive House US Certified Passive House Consultant (CPHC). She has presented at conferences and created media promoting sustainability and universal design, as well as leading multiple student teams in the DOE Solar Decathlon Design competition. She served on the National Green Building Standard 2021 revision Task Group 5 for Chapter 7, Energy Efficiency.

# Renewable Energy Systems for Building Designers

## Fundamentals of Net Zero and High Performance Design

Dorothy Gerring

Routledge
Taylor & Francis Group
NEW YORK AND LONDON

Cover image: Dorothy Gerring

First published 2023
by Routledge
605 Third Avenue, New York, NY 10158

and by Routledge
4 Park Square, Milton Park, Abingdon, Oxon, OX14 4RN

*Routledge is an imprint of the Taylor & Francis Group, an informa business*



*Library of Congress Cataloging-in-Publication Data*
Names: Gerring, Dorothy, author.
Title: Renewable energy systems for building designers : fundamentals of net zero and high performance design / Dorothy Gerring.
Description: First edition. | New York, NY : Routledge, 2023. |
Includes bibliographical references and index.
Identifiers: LCCN 2022011293 (print) | LCCN 2022011294 (ebook)
Subjects: LCSH: Sustainable buildings–Design and construction. |
Buildings–Energy conservation. | Buildings–Mechanical equipment. |
Solar buildings–Design and construction.
Classification: LCC TH880 .G47 2023 (print) |
LCC TH880 (ebook) | DDC 690.028/6–dc23/eng/20220527
LC record available at https://lccn.loc.gov/2022011293
LC ebook record available at https://lccn.loc.gov/2022011294

ISBN: 978-1-032-28648-8 (hbk)
ISBN: 978-1-032-28647-1 (pbk)
ISBN: 978-1-003-29781-9 (ebk)

DOI: 10.1201/9781003297819

Typeset in Univers
by Newgen Publishing UK

Access the Support Material: www.routledge.com/9781032286471

# Dedication

This book is dedicated to everyone who would make the earth more fruitful and habitable for all its inhabitants.

# Contents

# Disclaimer

The author and publisher of this book have put forth their best effort to ensure that the content contained herein is accurate and up to date. The purpose of this content is to assist the reader to understand the scope of work and the issues that need to be addressed when designing a building with a renewable energy system. Professional aid should be obtained when designing the specifics of any renewable energy system. The author and publisher shall not be held liable for incidental or consequential damages connected to, or arising from, the use of this book's content.

# Acknowledgements

This book is in your hands because of my students: thank you for wanting to know how to create net zero sustainable buildings and be the awesome creators of a better future for our built environment. I couldn't have finished this project without the support of my dear husband, William Geyer, and my wonderful illustrator, Alexander Geyer. Thank you so much for helping me create this book.

I'm very grateful to everyone who got excited every time I started talking about my book and to those who volunteered their projects for the case studies: Bailey Thompson and the design team from ZGF Architects LLP for their time and efforts providing insight and information about RMI's Innovation Center; Jim Hammons from the City of Libby and Brett Bauer of Canyon Hydro for information about the City of Libby's municipal microhydro project; Jim Maskrey from the University of Hawaii at Manoa for writing up the classroom case study and sharing my book project; Dawn Linn and Dave Rickard from Northcentral PA YWCA for allowing me to take photos and regularly traipse my students through their building; Jörg Rügemer from Atelier Jörg Rügemer for sharing 125 Haus and being a fellow sustainability guru; and Clarke Snell for writing such a lovely preface.

Many thanks to Carrie Mayer, Georgia Laudenslager, and my editors Chloe Layman and Lisa Mosier at Routledge, Taylor & Francis Group for helping me prepare my manuscript and publish this book.

There are many groups who gave permission to use their images in this book and I'd like to thank all of them and encourage you to visit their websites in the citations – especially check out the great information that is regularly updated and published by IEA and REN21.

Finally, I'd like to thank the Pennsylvania College of Technology for supporting my growth through the years and for offering our fabulous sustainable architecture degree.

# Preface

Are you ever frustrated because you can't quickly and easily put your hands on the information you need? Are you unable to determine if the information you are finding is actually up to date and accurate? What happens to the validity of your work if you don't know what you are missing – where is that darn consultant anyway?

These problems are the reasons I created this book. Students, young professionals, and interested people can have a difficult time piecing together clear and accurate information about renewable energy systems and how they work with a building's design. There are many variables to juggle. A designer can't design something without asking questions – and you have to have enough information to start asking questions. Students don't get to have a consultant (not withstanding that, in my experience, professionals are quite willing to help students when asked).

This book is intended for college-aged students, undergraduate and graduate, in the fields of architecture, construction, mechanical, or architectural engineering. Young professionals, as well as those studying the potential of developing new building products related to energy efficiency or renewable energy systems, can use it as a guide for deepening their understanding.

In a world where net zero performance is needed now (and required in some places), we are responsible for creating a world filled with fabulous buildings that provide comfort and performance while improving the quality of life for everything on the planet.

I could not find a book that covered these issues with both breadth and depth, and my students needed a better system for tracking what they learned about renewable energy systems so that they could fully integrate the systems into their projects.

This book is intended to be a resource that can be referred to many times: the consultant that is available 24/7. The chapters are intended to stand on their own as just-in-time training, while forming a cohesive whole. The concepts are laid out clearly, with lots of helpful illustrations and explanations.

Please let me know how the book works for you. If there's an online resource that you can't live without, and I didn't mention it, tell me about it. Above all, share your successes through the wonderful projects you've created that you *know* are going to work because you asked the right questions to integrate performance and renewable energy in innovative ways.

# Foreword

Let's review. For thousands of years, we designed and built buildings that created an inside only loosely separated from the outside. We kept rain out while letting water vapor and air move freely through the envelope. If we wanted heat, we brought our campfire indoors. Cooling was passive, delivered through shading and/or a thermal mass flywheel. If this sounds primitive, think again. It represents the full canon of immortal buildings covered in your architectural history class from the Pantheon to Ronchamp. To generalize, you can assume that pretty much every hut, house, office, cathedral, and castle built before 1900 met this description. Many of these buildings were beautiful, most were durable and functional, but all lacked substantial environmental control.

As we started demanding more nuanced control over the indoor environment, we considerably complicated the picture. Desire for large and pinpointed temperature differentials between inside and out triggered the need to control air and water vapor migration which led to concerns about fresh air and the broader concept of **indoor air quality (IAQ)**. This brave new world splintered architectural practice, creating roles for a variety of engineers to design the complicated mechanical, electrical, and plumbing systems required to deliver this new level of control. In the process the etymological role of "architect" (master-builder) was downgraded to "form maker". This set up a design process that compartmentalized form and performance with decidedly mixed results. Environmental control was increased but so were problems associated with all manner of durability and IAQ issues. Counterintuitively, building energy efficiency did not generally improve because oversized mechanical systems worked overtime to make up for huge energy loads mandated by form imagined by form makers alone. Still, everyone settled into their roles and this disjointed approach might have continued indefinitely.

Enter climate change, a complex existential problem that mandates a very simple solution, or at least a fundamental first step. Insert here the thousands of reports, articles, position papers, books, and book intros that make the same basic point: we have to drastically cut our combustion of fossil fuels to lower atmospheric carbon dioxide levels to a baseline accepted by scientific consensus. Since buildings are responsible for a big chunk of our collective carbon footprint, building design professionals have a huge role to play. And since most of building related carbon is in operational energy (heating, cooling, hot water, lighting, and plug loads), then a clear strategy is obvious: reduce operational energy loads so that low-carbon renewable energy systems can be sized to fulfill that demand.

Building operational demand can be considerably lowered through passive envelope design. Architects and architecture schools are beginning to accept that role, reclaiming as their own the orphaned zone between outer façade and interior finish. Classes or modules in building science, energy modeling, and sustainable design are becoming commonplace. Poche is out, construction details are in. This is an important step but only a beginning. What is needed is a complete merging of form and performance, in other words, a return to our roots as architects. In the age of environmental control, this means that a building's form is a matrix of all its systems, including those that heat, cool, illuminate, and **power** it. This is not to say that architects should become engineers, only that we need to educate ourselves to understand the full entity we are forming.

There are a lot of informational resources of all types for architects on load reduction through low-energy envelope design in the context of the architect's workflow. Building information modeling (BIM) platforms offer a design infrastructure and performance standards like Passive House provide solid conceptual design and modeling rubrics; they incorporate mechanical systems, air exchange rates, and quantify energy use projections allowing performance to take shape simultaneously with form. In other words, the topic of load reduction is well covered.

But energy production is the second half of the puzzle and therefore a component of architecture in the age of climate change. The low hanging fruit are building integrated energy systems because they are clearly part and parcel of the architectural form. Since they require physical access to sun, wind, and/or water, these systems are as important in schematic design as any other element. Architects who ignore this fact will find their vague plans to incorporate renewables in design development or later will be a waste of time. However, on-site production is only part of the picture. Every act of architecture needs to take full responsibility for the energy it will require through its life cycle. This expands the scope of our work to include off-site production, not designing the infrastructure of course but planning how our projects will interface with commercial renewable production. The point is that architects can no longer design form and designate "by others" to power it.

Hence this book. It is written for architecture students, their professors, and any design or construction professional who needs an introduction to renewable energy system design as a piece of a holistic architectural puzzle. It is conceived to bridge the gap between energy efficiency and power system design. The book begins by setting renewables in our current energy production context, then lays out the basics of how renewable systems are located, sized, and the energy delivered. It covers both on-site (project designed) and off-site (project interfaced) approaches and includes basic primers on related topics like electricity and solar insolation. Load reduction is covered with sections on designing for climate, building envelope performance, and IAQ. Loads and production intersect through energy modeling, so a primer on that topic rounds out the picture. Practical considerations of costs and payback periods are clearly introduced. Finally, a series of case studies puts the full scope into context. Written in an approachable way, with non-technical language and clear graphics throughout, this book is an important introduction to a missing link in the architectural workflow for projects in the age of climate change.

*F. Clarke Snell*
Architect
Associate Professor
Stevens Institute of Technology

PART I

# Introduction to Renewable Energy and Building Performance

CHAPTER 1

# Introduction

> **Key Terms**
>
> Greenhouse gasses (GHG): any gas that is able to absorb infrared radiation, therefore trapping heat in the earth's atmosphere.
>
> Renewable energy (RE): any energy source that comes from naturally renewing sources or processes.

## Why This Book Now?

Architecture is the synthesis of art and technology. The ancient Roman architect Vitruvius said architecture is about the relationships between firmness (structural integrity), commodity (functioning of spaces and systems), and delight (joy and excitement in discovery and beauty). His ideas have stood up through the centuries but today we need to include respect and support for our communities and the world's living systems. We are experiencing unprecedented times where our quick action is needed to produce a dramatic reduction in **greenhouse gasses (GHG)** which are changing earth's climate. By paying attention to how we design, build, and remodel our built environment, we have the unique opportunity to make a big difference in the future of our planet, and improve the comfort and safety of building users.

This book introduces you to **renewable energy (RE)** systems and explains why we must drastically improve the energy efficiency of our buildings. Renewable energy systems on inefficient buildings have to be quite large if you want to produce as much energy as your building uses. To achieve sustainable building performance, we must address the complexities of designing for the climate, detailing the building envelope, using only electricity to power the building, and selecting appropriate renewable energy systems.

Today it seems like every day there is another climate driven disaster causing damage to neighborhoods and cities and creating both human and economic stress and loss. Buildings are responsible for as much as 40% of all **energy** use and 76% of electricity use in the USA[1] as well as 20% of all energy use worldwide.[2] Much of that energy is from non-renewable sources. We can, and are, doing better at lowering our carbon footprint through better buildings. We need to keep improving efficiency and comfort while taking advantage of appropriate climate strategies and renewable energy systems. This book provides an overall view of all these issues in an easy-to-understand format, providing you a single resource for fundamental concepts.

The chapters in this book each present a particular concept which is fairly independent of other chapters. The earlier chapters have information about the configuration of renewable energy systems, how to site them properly, understanding electricity and wiring, climate design, building performance and energy modeling, and the importance of indoor air quality. These are followed by chapters on the design and specific parts of different renewable energy systems, costs and payback of systems, issues of resiliency, and case studies of successful renewable energy projects.

This book is intended to get you started so you know what issues to address in order to apply energy efficient design strategies and renewable energy systems into your particular building. We can only provide thoughtful integration of systems, aesthetics, and functions if we have asked lots of good questions about them and how they are working together. We can only achieve the synergy of cost savings and great architecture by understanding how to maximize performance and truly integrate all of our building systems.

## Concepts for Success

Renewable energy is an energy source that is naturally abundant and regularly renews itself. Renewable energy

DOI: 10.1201/9781003297819-2

is also called clean energy because it is not adding carbon atoms to the current carbon cycle. Carbon dioxide, methane, nitrous oxide, and fluorinated gasses are greenhouse gasses that are changing how our atmosphere heats and cools. The increase of GHG in the atmosphere is happening because of burning fossil fuels and is causing global climate change.

People use energy in agricultural processes, industry, transportation, and buildings. There are over 7.76 billion people in the world. In 2019 there were 759 million people who did not have access to electricity.[3] The highest percentage of energy use is in industry, with production centers growing in China and India, the two most populous countries in the world.

Different fuels have varying levels of energy available in them (energy output) and they are often measured in different units depending on the fuel type. It is important to compare unit costs and outputs by converting them all to similar units. When including fossil fuels and **biomass** in comparisons with electricity from renewable energy, comparisons are made either in **British thermal units (Btu)** or **joules (J)**.

The use and design of buildings have a measurable impact on their energy demand. Newer buildings tend to be more energy efficient than older ones, especially in countries with building codes that keep updating and requiring higher performance. Designing and retrofitting buildings to be energy efficient and use renewable energy is very important. In developing countries energy demand is rising rapidly, so providing efficient buildings helps to keep up with energy demand and provide energy security.

The basic types of renewable energy that create electrical power include hydropower, wind, biomass, solar photovoltaics, and concentrated solar power, while solar thermal hot water systems heat water.

## Carbon Cycle

You probably remember studying about circular cycles of natural systems in science class. One that you are very familiar with is the water cycle: all the water on our planet has always been here. It evaporates, it rains (snows, sleets, etc.), it absorbs into the ground, and it runs off. Then it evaporates again. Much of the water is located in the oceans, frozen in icecaps, and underground aquifers. Just over 1% is accessible and potable.

The carbon cycle is similar to the water cycle, and just as fundamental to life. Pretty much everything on the planet contains carbon. Carbon is stored in rocks, ocean, atmosphere, living things, soil, and fossil fuels (which used to be plants and animals). As shown in figure 1.1, some carbon cycles are short, such as the seasonal uptake of carbon atoms into plants through photosynthesis and its subsequent release through plant respiration and decomposition. **Carbon dioxide ($CO_2$)** is released through respiration, decomposition, and emissions, while being taken up and stored in plants.

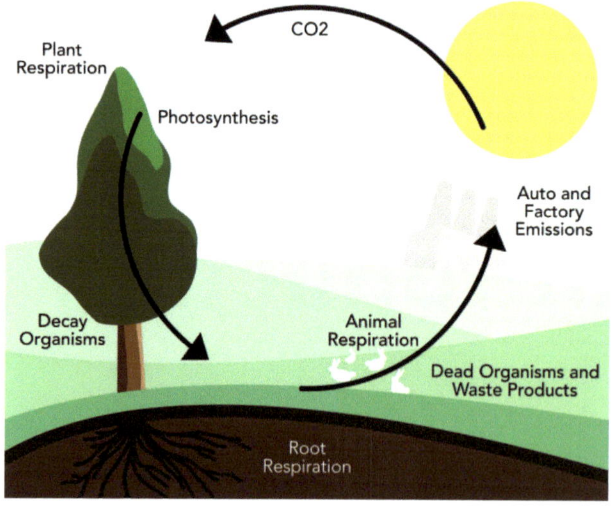


**Figure 1.1** Seasonal carbon cycle showing the addition and uptake of $CO_2$.

These cycles can be measured annually through the counting of carbon molecules in our atmosphere. Since more of the world's forests exist in the northern hemisphere, every fall there is an increase in global atmospheric carbon and every spring there is a decrease. There is also a carbon cycle for the oceans where carbon molecules are used in the life-cycles of water-based life forms. These quick cycles of carbon exchange are considered to be current carbon, carbon that is always moving within our living systems from one form to another.

Slower carbon cycles include the movement of carbon between rocks, soil, ocean, and the atmosphere. This can take many millions of years. This cycle includes the creation of rocks that are made from organic carbon, such as the ancient forests and animals that, over the course of time, got compressed and heated and became sedimentary rocks such as shale or fossil fuels such as oil, coal or natural gas. These carbon atoms made from organic materials are said to be sequestered carbon because the organic material has been taken out of the current, short carbon cycle.

Using materials for energy that contain sequestered carbon causes problems because it adds more carbon to the atmosphere than can be taken up in the normal carbon cycle. This changes the balance of the different gasses that make up our atmosphere which changes how the atmosphere works while it protects our planet from the nothingness of space.

## Greenhouse Gasses

The earth's atmosphere creates a boundary between space and the planet's surface. It filters out some of the sun's harmful radiation, such as **ultraviolet (UV)** rays, and creates an insulative barrier that acts to balance day and night temperatures. It also provides us with air to breathe.

GHG are a problem because they are being added to our atmosphere at a high enough rate that they have changed, and continue to change, its chemical composition. The sun's radiation shines through the atmosphere, passing through air and clouds to the earth's surface, where it is absorbed and later radiated back towards space, cooling off the planet at night. GHG interrupt this cycle by absorbing almost 90% of this infrared heat and then radiating it back to the earth's surface. Figure 1.2 illustrates how the changing composition of the atmosphere reflects radiant heat back toward the earth, rather than allowing it to escape to space.

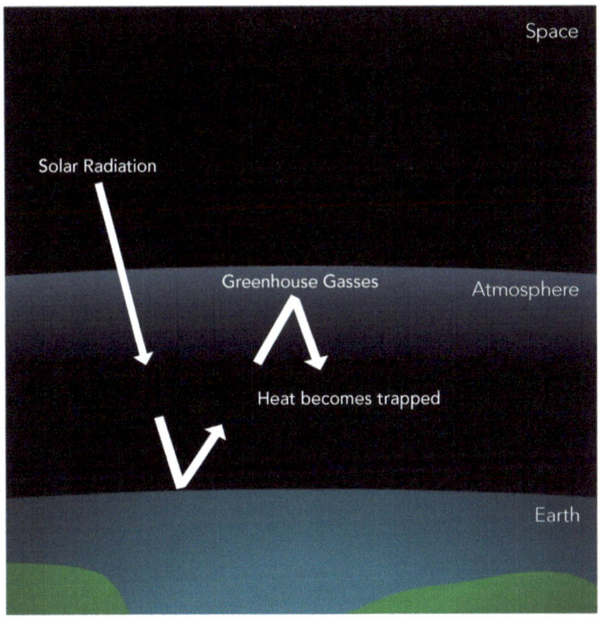


Figure 1.2 Greenhouse gasses reflect heat back to earth rather than allowing the atmosphere to cool off to space.

The inability of our atmosphere to release this heat has caused a gradual warming of the earth over time and has led to global climate change. The trapping of heat is similar to how a greenhouse can stay warm in cold weather so farmers can start plants earlier than in the field. You have probably experienced this trapped heat phenomenon if you've gotten into a car that has been parked in the sun, even on a cold day, to discover it is quite warm inside.

The four most important GHG tracked by United States Environmental Protection Agency (EPA) are carbon dioxide ($CO_2$), **methane ($CH_4$)**, **nitrous oxide ($N_2O$)**, and fluorinated gasses. The largest emissions globally are of carbon dioxide. Climate Trace[4] on the web is a great interactive tool for looking at independent data for real-time global GHG emissions. Figure 1.3 shows the breakdown of global greenhouse gas emissions by the type of gas. You can see that the pie chart shows $CO_2$ accounts for 76% of these emissions, followed by $CH_4$ at 16%, $N_2O$ at 6%, and fluorinated gasses at 2%.

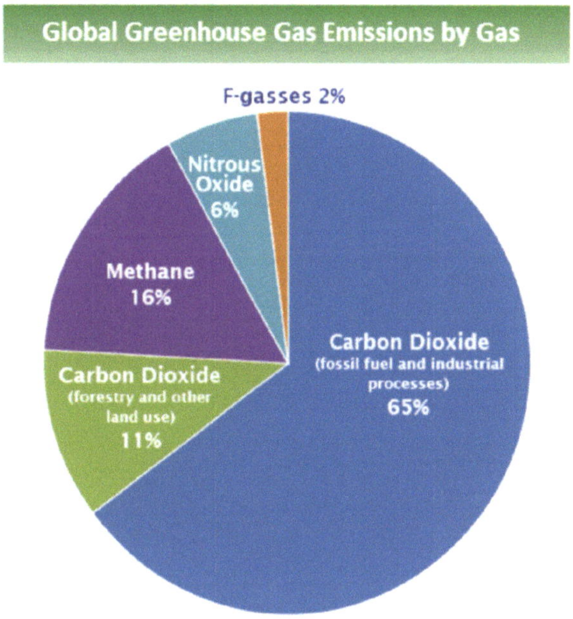


Figure 1.3 Global greenhouse gas emissions by gas.[5]

Looking at figure 1.4, you can see GHG emissions by economic sector in the USA in 2019, with the total emissions equaling 6,558 million metric tons of $CO_2$ equivalent ($CO_2$-eq).[6] The transportation sector was responsible for 29% of the emissions, where over 90% of all fuel used is gasoline and diesel fossil fuels for cars, trucks, ships, trains, and planes. Electricity production accounted for 25% of emissions because the USA creates about 62% of all electricity from burning fossil fuels (coal and natural gas). Buildings use 76% of the electrical power. Industries producing goods from raw materials accounted for 23% of emissions. The commercial and residential sector of the pie chart shows 13% for heating, cooling, and power (except electricity) used for buildings. The 10% agriculture number accounts for emissions from livestock (such as cows), agricultural soils, and rice production. Buildings account for all the 13% emissions in the commercial and residential part of the chart. Transportation impacts carry into the building sector because of transportation of construction materials and commuting habits of workers. Some portion of the industrial emissions can also be attributed to the manufacture of building materials and systems. As you can see, buildings have a significant impact on GHG emissions, so improving building performance and changing to renewable energy sources will help slow climate change.

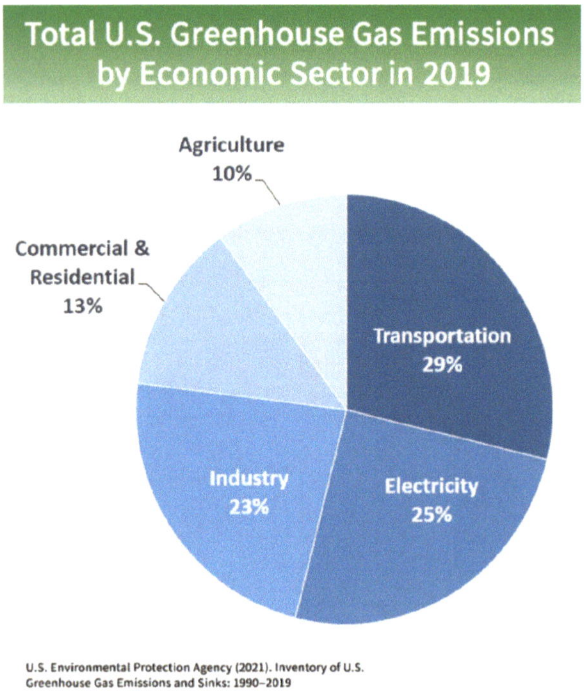


Figure 1.4 2019 USA greenhouse gas emissions by economic sector.[7]

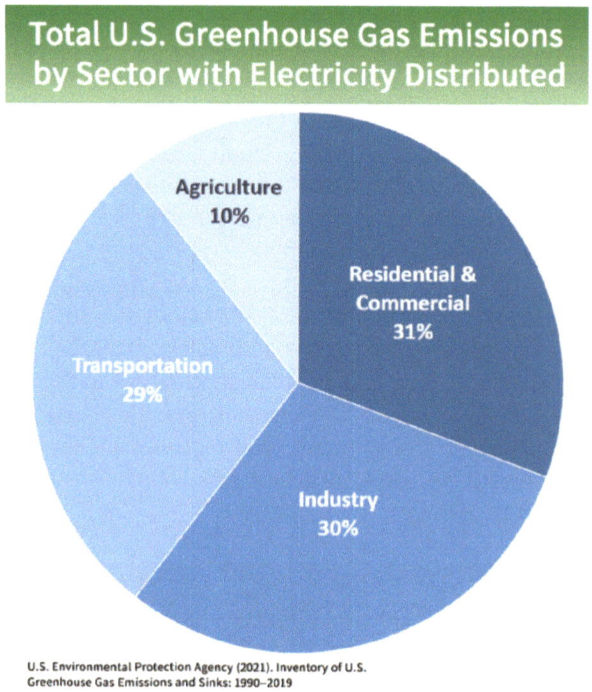


Figure 1.5 2019 USA greenhouse gas emissions by economic sector.[7]

Figure 1.5 shows the change in the pie chart when electricity is distributed among the various economic sectors. Residential and commercial buildings end up with the largest GHG emissions at 31%, with industry close behind at 30%. Transportation and agriculture remain unchanged at 29% and 10% respectively. We can surmise from this chart that creating more efficient buildings running on renewable energy can have a big effect on reducing GHG emissions.

Buildings use energy for lighting, heating and cooling, water heating, and running appliances. Figure 1.6 shows data for the energy usage for homes in the USA from 2015. This chart shows that heating and cooling demands together use the most energy, no matter what the housing type is, followed by hot water. The average for all homes' heating and cooling is 51% – over half the energy demand. The water heating average is 19%. These three things – heating, cooling, and hot

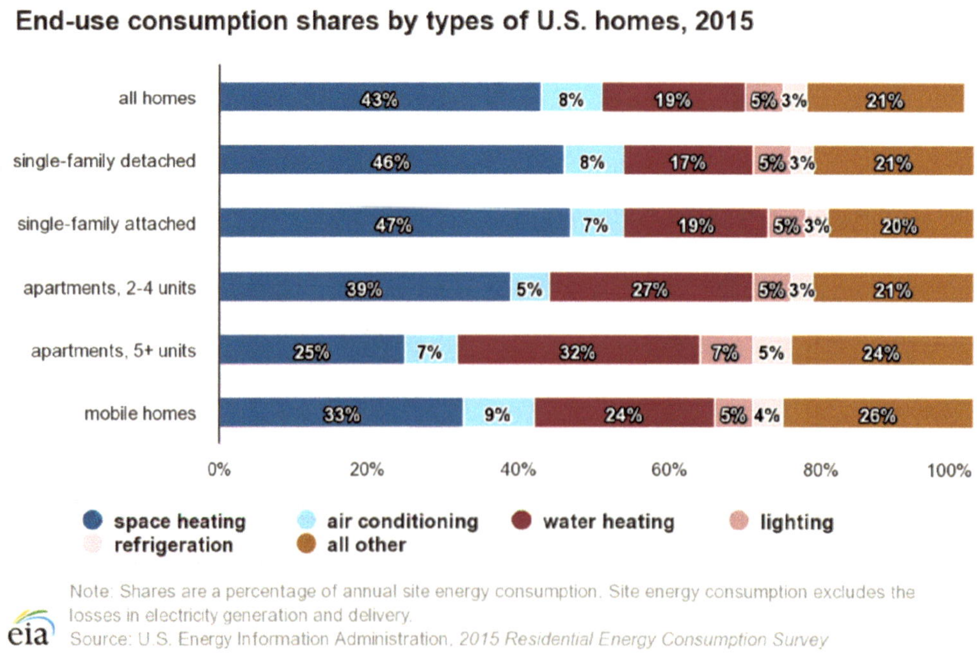


Figure 1.6 2015 End-use energy consumption by shares for homes in USA.[8]

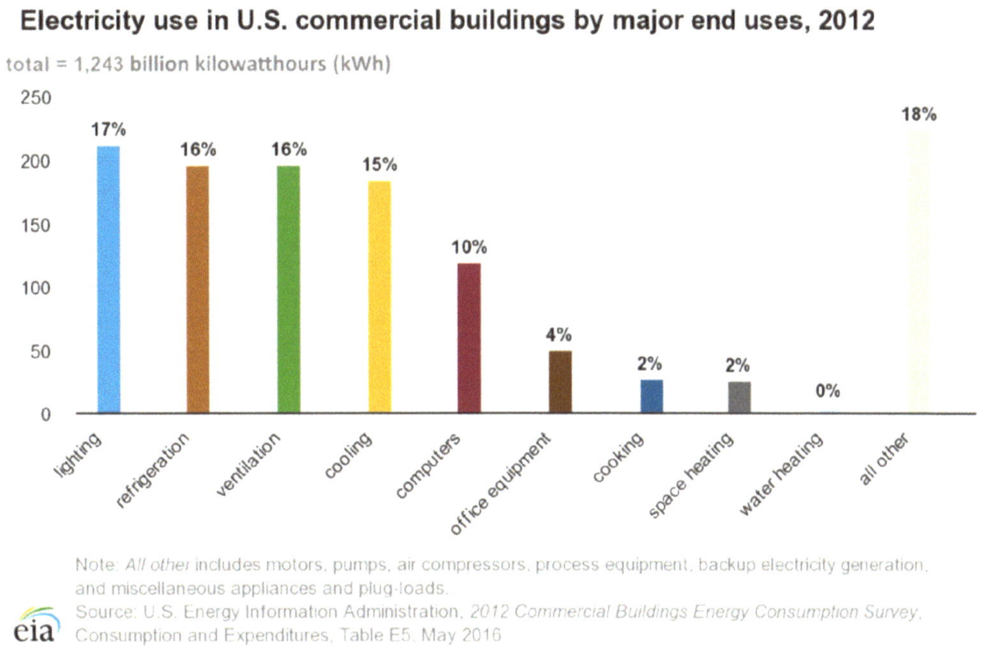


**Figure 1.7** 2012 electricity use in US commercial buildings.[9]

water – account for a whopping 70% of all energy used. This shows how important it is to address these issues in the design/renovation of the building.

Figure 1.7 shows data for electrical usage in commercial buildings in the USA from 2012. The highest demand is for electrical lighting. The advent of **light emitting diode (LED)** lights has presented a way to dramatically reduce the amount of energy needed for lighting in both residential and commercial buildings. Heating in commercial buildings is typically done using fossil fuels, so it doesn't show up as a big demand on this chart. As buildings change over to electricity for meeting space heating demands, this number will increase. LED lights run on direct current – the type of electricity most renewables produce – so there is a potential for compounded energy savings in countries where alternating current is the standard. Comparing these usage charts speaks to the importance of designing buildings for the climate and appropriately detailing the building's envelope so heating, cooling, and lighting demands can be lowered.

## World Population

As of July 1, 2021, there are over 7.76 billion people in the world.[10] The top three most populous countries in the world are China with 1.4 billion people, India with 1.3 billion people, and the United States of America (USA) with 332 million people. Not all the people in the world have equal access to healthy buildings and clean energy. Even in countries that have a high standard of living, there are many people who live in substandard conditions. Over half the world's population is located in Asia, with the next largest aggregate population numbers in Africa. Many of these people have no access to electricity.

There is a clear call at the present time to pay attention to equity across the world and within cultures. We can no longer choose to squander resources while others have little access to healthy, comfortable living conditions. We cannot continue to alter the world's climate without concern for the many who end up experiencing loss because of brutal climate impacts. Estimates range on the number of climate refugees there will be by the year 2050, however, **UNHCR** (the **United Nations Refugee Agency**) believes the number of people needing aid for climate related disasters could be over 200 million people every year.[11]

## Energy

Everything you interact with every day uses energy or needed energy to produce it. This is true of both natural items and systems as well as engineered products. Energy, as defined in physics, is the capacity to do work. You might remember (from a physics class) that it can be in forms such as thermal (heat), chemical, electrical, kinetic, and potential.

While energy is never lost – as energy is converted from one form to another – we don't get 100% in the new form. Often some is converted to other forms or is unusable for the desired task. For example, burning natural gas to create electricity is about 45% efficient, while coal is about 32%. Further losses happen as the electricity is sent over (distributed through) utility grid power lines. This means that more than 60% of energy used for generating electricity can be lost in its conversion and distribution. There is a lot of room to improve this system!

But energy isn't just electricity. There are lots of forms and measurements for energy. In order to compare all forms, many charts convert to British thermal units (Btu) or joules (J); 1 Btu will heat 1 pound of water 1 degree Fahrenheit; 1Btu = 1055.06J, often rounded to 1055J. If you are looking at charts for energy use in the whole world, that's a lot of energy, so the units are in quadrillion Btu (quads), which is $10^{15}$Btu (approximately $1.055 \times 10^{18}$J).

Globally, most energy is used for industrial manufacturing. The next biggest category is transportation. As seen in figure 1.8, China uses much more energy than any other nation. Based on population forecasts and the continuing shift of manufacturing to China and India, it is anticipated that by 2050 China and India will use more energy than the rest of the world combined. In the chart, OECD stands for Organization for Economic Cooperation and Development.

Figure 1.9 shows that the greatest electrical use by sector is in the combined total of residential and commercial buildings, both now and in the future prediction. In 2050, as more of the world's population has electricity and electrical appliances, residential consumption will almost double when compared to 2018. Therefore, buildings will have the greatest growth in demand in any sector.

Figure 1.10 illustrates renewables will displace petroleum as the most used primary energy source, moving from below 100 quads ($1.06 \times 10^{20}$J) in 2018 to over 250 quads ($2.64 \times 10^{20}$J) in 2050. Electricity as an end-use consumption will grow faster than any other fuel as it more than doubles in demand between 2018 and 2050 while other sources remain fairly flat in their growth.

## Renewable Energy

Renewable energy is any energy source from naturally renewing sources or processes. Typically, you don't have to pay for the energy. This means that while you have to pay for the equipment to produce the energy, the actual energy source is free. Examples of these energy types include sunlight, wind, waves, and biomass. In 2020, renewables provided about 30% of global electricity and it is estimated that they will continue to grow at a rapid rate into the future. Figure 1.11 shows estimated demand for renewables in the year 2050 will more

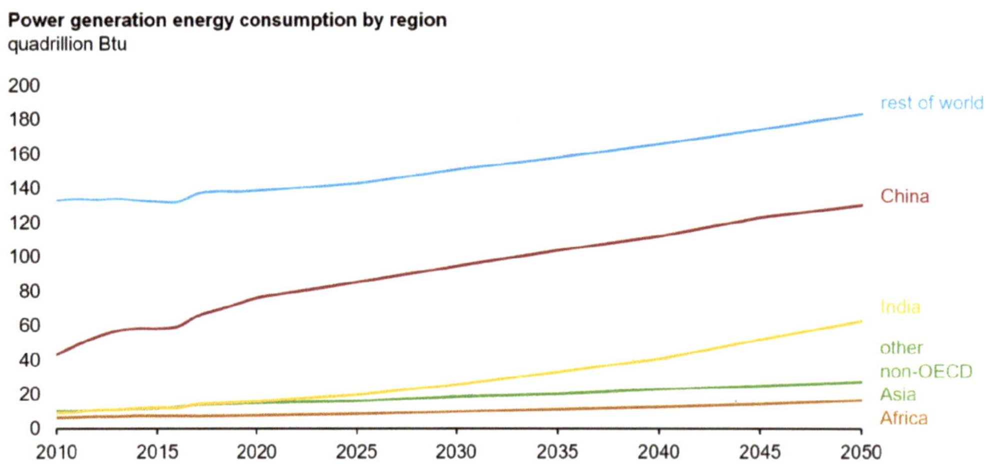


Figure 1.8 Power generation energy consumption by world region.[12]

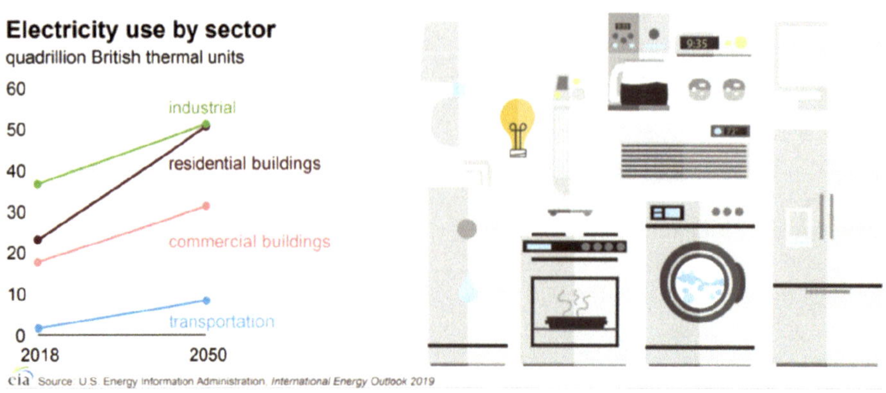


Figure 1.9 Global increase in electricity use by sector from 2018–2050.[12]

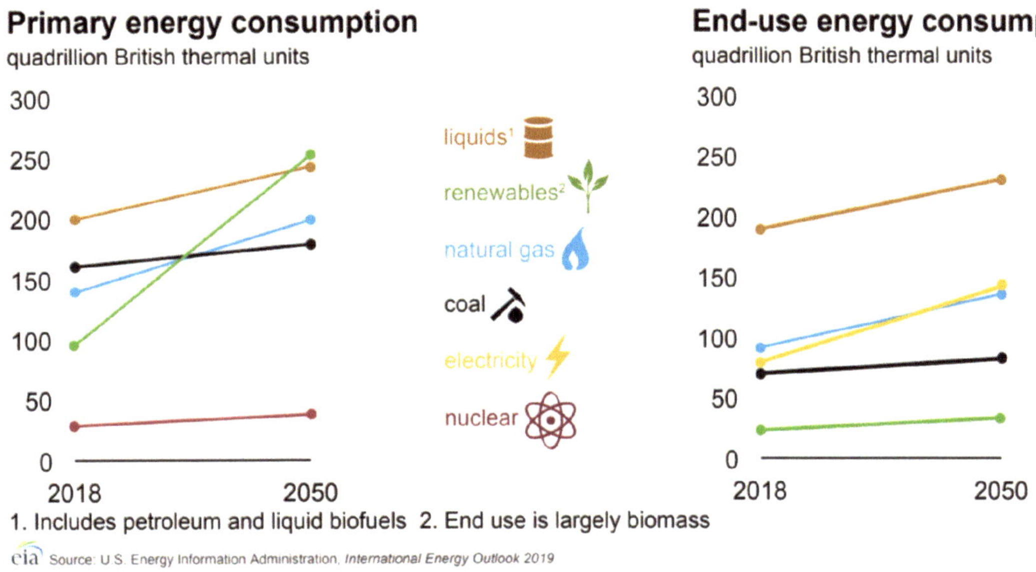


Figure 1.10 Primary energy growth is in renewables and end use consumption is shifting to electricity.[12]

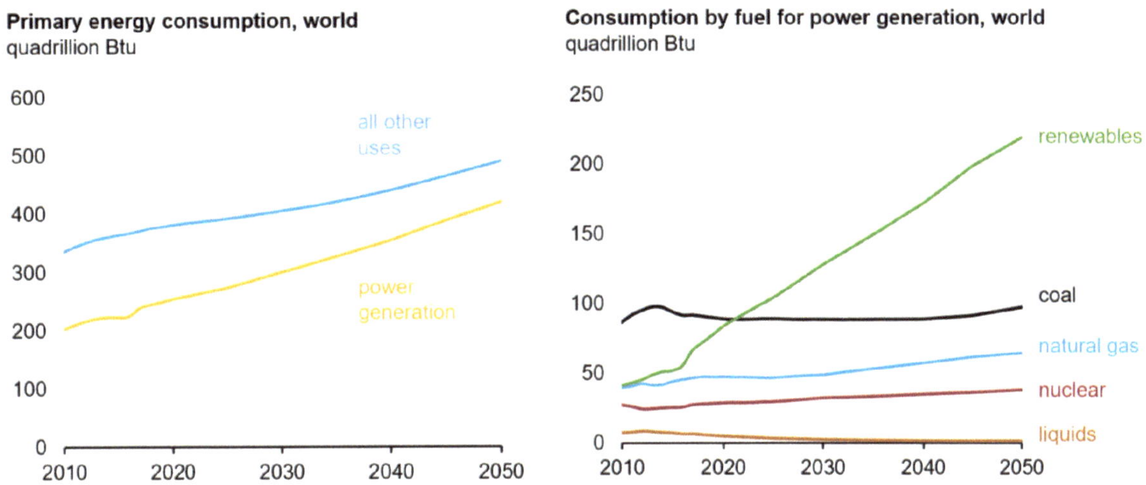


Figure 1.11 Global energy consumption and sources of power generation from 2010–2050.[11]

than double compared to today's demand, while other sources stay about the same.

In the USA, renewable energy accounted for 12% of all energy consumed in 2020 (a 1% increase from 2019). Figure 1.12[13] shows that various forms of biomass accounted for 39% of the total energy consumed, followed by wind at 26%, hydroelectric at 22%, solar at 11%, and geothermal at 2%. Wind and solar both saw production gains while biomass decreased. The pie chart also shows that petroleum (35%) is barely ahead of natural gas (34%). These, added to 10% for coal, make up 79% of the energy consumed in the USA, all of which are sequestered carbon adding to GHG emissions.

The **International Energy Agency (IEA)**, an international association comprising 30 countries, is "committed to shaping a secure and sustainable energy future for all"[14] and annually creates various global forecasts in their Renewables Reports. This data is used by many other worldwide organizations. Figure 1.13 shows the forecast for growth for different types of renewable energy in the next few years (to 2024) both as a base case and an accelerated case. This graph shows that photovoltaics (PV) have the highest increase each year. It is interesting to note that other renewable energy technologies are remaining flat or decreasing in installed gigawatts. One **gigawatt (GW)** is equal to one billion ($10^9$) watts.

We saw that heating and cooling buildings accounts for the largest share of energy used in buildings. Figure 1.14 shows that globally, in 2018, it was estimated by the IEA that renewables (traditional biomass and modern renewables) provided 38.1% of the demand. Fossil fuels still accounted for over half of the demand at 53.9%. On the renewables breakdown, you'll notice that the majority of the energy was used for heating and was provided through burning biomass. Renewable

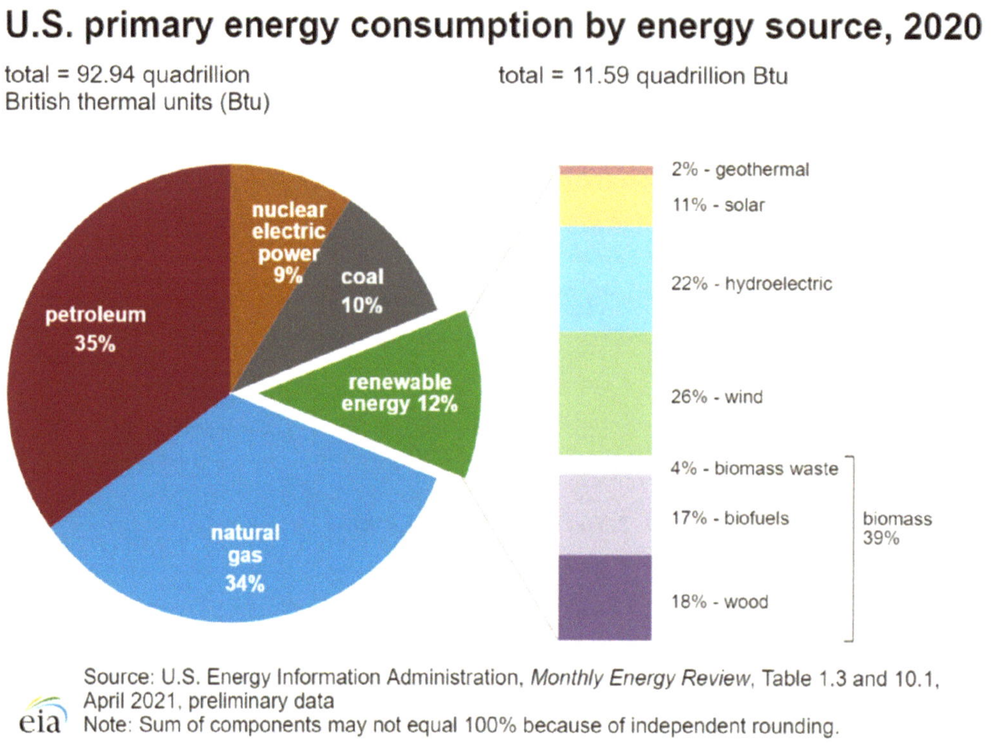


**Figure 1.12** USA primary energy consumption by energy source, 2020.[13]

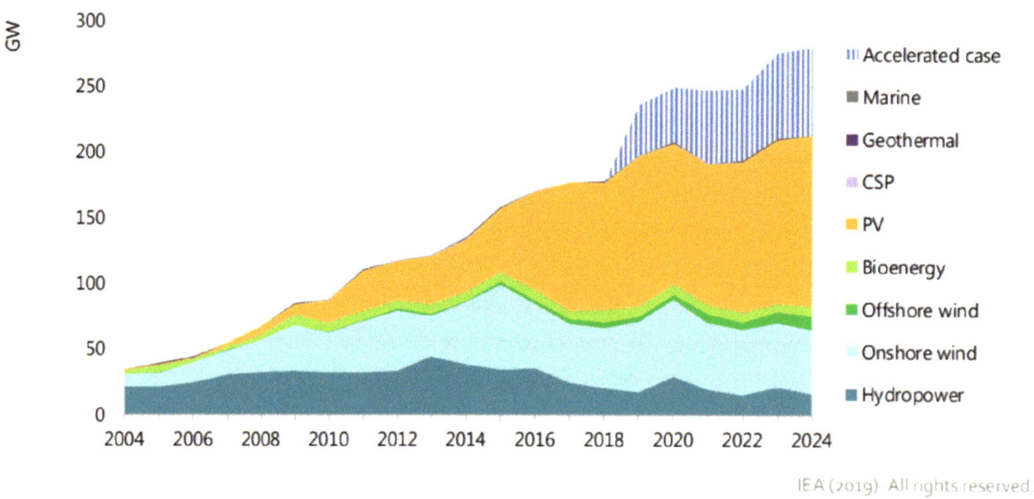


**Figure 1.13** Annual renewable capacity additions by technology, IEA Renewables 2019 Report.[14]

electricity was 3%, solar thermal heat was 1.5%, and geothermal was 0.6%.

Different types of fuels have different potential Btu outputs as well as different measurements for purchasing them. This can make it a challenge to compare output and pricing. Converting everything to similar units, such as Btu, makes this much easier. Table 1.1 has a comparison of Btu in different energy sources, liquids shown in gallon measurements. For example, 1 gallon of propane has 91,452 Btu compared to 138,500 Btu in 1 gallon of heating oil. This means that the propane has only about 66% of the Btu of heating oil. These conversion factors can be used to show payback periods for renewable energy systems compared to what people are paying for their current fossil fuel usage.

Figure 1.15 shows that the total global renewable power capacity in 2017 was 2,195 gigawatts. This has more than doubled since 2007. Hydropower currently produces over half of the renewable power in the world, although you'll notice that it didn't increase much over ten years. The sectors showing the most growth are wind and solar **photovoltaics (PV)**, which are the next highest producers of global energy.

Many charts will leave out the hydropower to more accurately compare other renewable energy sources. Figure 1.16 is an example of this. This chart is showing world totals of 1,081GW and individual countries

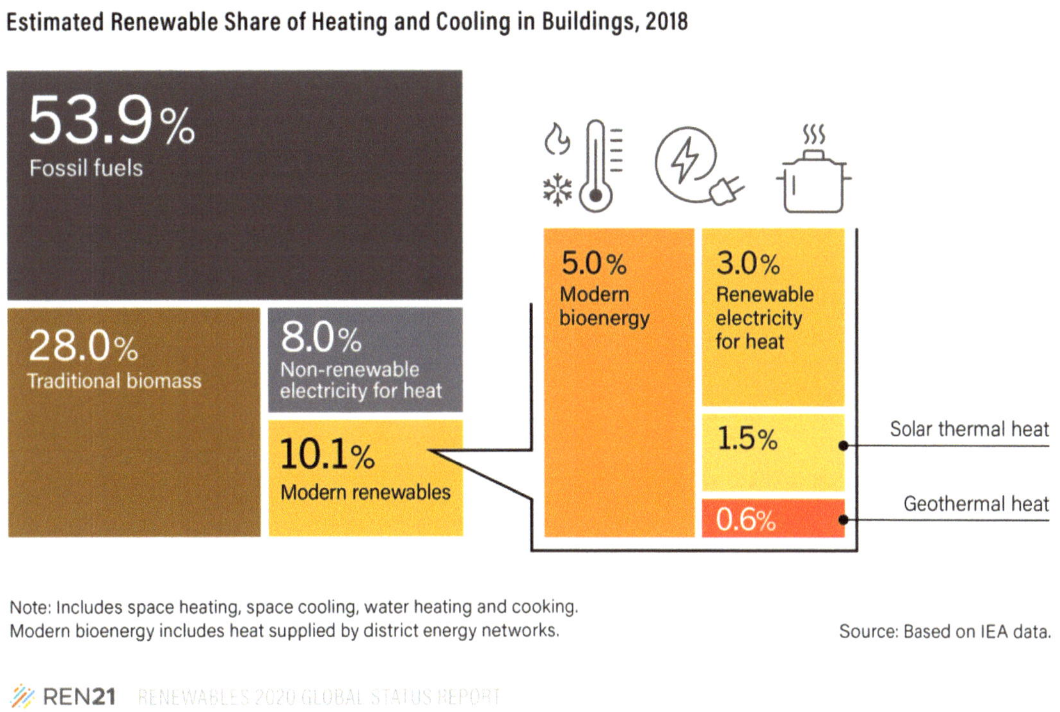


Figure 1.14 Building heating and cooling: renewable energy share of energy used globally in 2018.[15]

Table 1.1 Sample of Btu conversion factors using similar physical units, links to additional fuels and conversion factors given at the EIA website.[16]

| Energy source/fuel | Physical units and Btu[1] |
|---|---|
| Electricity | 1 kilowatt hour = 3,412 Btu |
| Natural gas | 1 cubic foot = 1,037 Btu |
| | 1 therm = 100,000 Btu |
| Motor gasoline | 1 gallon = 120,286 Btu[2] |
| Diesel fuel | 1 gallon = 137,381 Btu[3] |
| Heating oil | 1 gallon = 138,500 Btu[4] |
| Propane | 1 gallon = 91,452 Btu |
| Wood | 1 cord = 20,000,000 Btu[5] |

[1] Btu factors are for end-use consumption in 2020 from Monthly Energy Review, April 2021, excluding wood; preliminary data.
[2] Finished motor gasoline sold at retail in the United States, including fuel ethanol content.
[3] Distillate fuel with 15 parts per million (ppm) sulfur or less sulfur content.
[4] Distillate fuel with 15 ppm to 500 ppm sulfur content.
[5] This conversion is an estimate. A cord of wood is a volume unit and does not take wood density or moisture content into account. Wood heat content varies significantly with moisture content.

leading in global renewable power generation capacities: China (334GW), United States (161GW), and Germany (106GW). Compare the difference between the renewable energy generation capacities of China (334GW) and India (61GW). Both countries have a great deal of industrial manufacturing and have the world's two largest population centers.

Growth and maintenance of renewable energy systems accounts for about 11 million jobs worldwide. Figure 1.17 graphically shows these jobs broken down by type of renewable energy with each figure representing 50,000 jobs. These jobs are local and pay well. Almost five million of these jobs are in solar energy, which includes solar photovoltaics, **concentrated solar thermal power (CSP)**, and solar heating and cooling.

## Hydropower

Hydropower is created from water and generates electricity due to a volume of water flowing and dropping (the height is called **head**) through a turbine. The turbine spins a **generator**, which produces the electricity. The electricity is transformed and then distributed over power lines. Utility grade systems use dams on rivers to create large head and **flow** pressures. Figure 1.18 shows a generic section of how a utility hydropower plant works.

Hydropower plants can produce enough power for entire cities, depending on the rated power of the generator and the power demands of the city. Utility scale plants are typically rated with at least 1 **megawatt (MW)** of total generation capacity. Microhydro systems are smaller, relying on diverting a portion of the water rather than using a dam, and can power a single home or smaller community. Figure 1.19 shows a typical layout of a microhydro system. You can see that the **penstock** is a separate pipe that diverts some of the water to the turbine that is in the powerhouse. The water drops through the turbine and then returns to the river or stream. All hydropower must be located near a constantly flowing body of water.

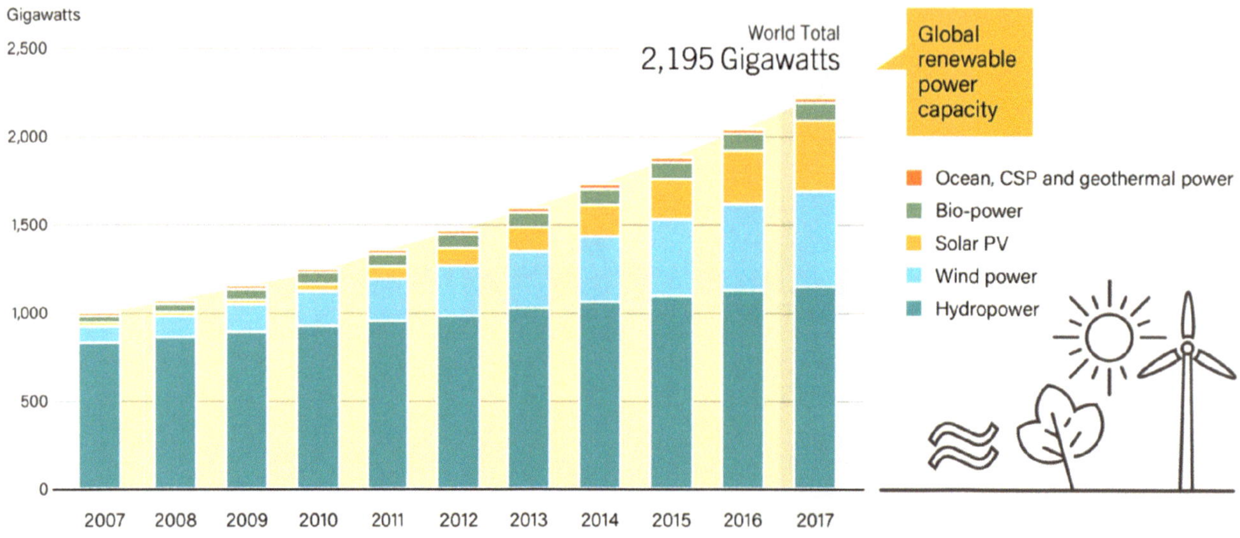


REN21 RENEWABLES 2018 GLOBAL STATUS REPORT

Figure 1.15 Growth in global renewable power from 2007–2017.[17]

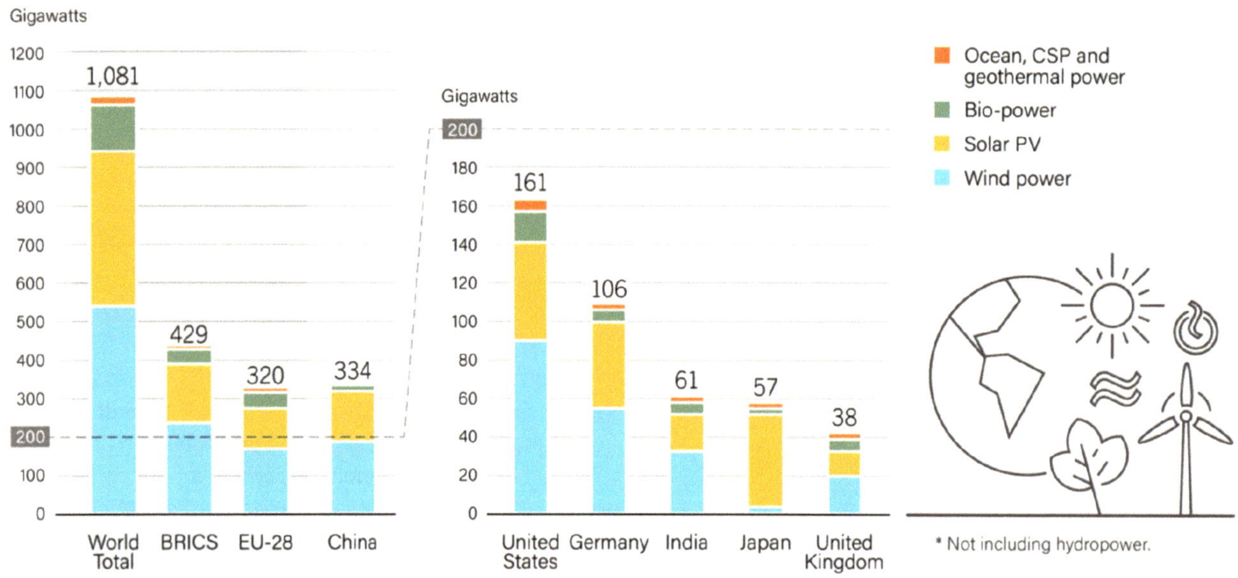


REN21 RENEWABLES 2018 GLOBAL STATUS REPORT

Figure 1.16 Renewable power capacities in the world and top 6 producing countries globally in 2017.[18]

Hydropower is the highest producing of renewable power generation as it has very little downtime. The water is flowing all the time, all year around. In the United States, the majority of utility grade power plants are located in the western states of Washington, California, and Oregon. According to U.S. **Energy Information Administration (EIA)**, Washington, Idaho, Oregon, and Vermont produce at least half of their in-state utility generation from hydropower. Many of these plants have been in service since the early 1900s, highlighting the longevity of this type of power generation. Capacity at existing plants has been improved by adding or repowering the turbine generators. In some places dams that were installed just for flood control have added turbines. **Pumped storage hydropower (PSH)** uses the same technology and has more installed capacity than all other forms of energy storage in the United States.

Oceans can also be used as energy resources through capturing the movement of **tides** and waves. Although in theory there is a lot of potential for this type of energy, it has lagged behind the development and installation of other types of renewables. Typically, the various generators use the movement of the water to

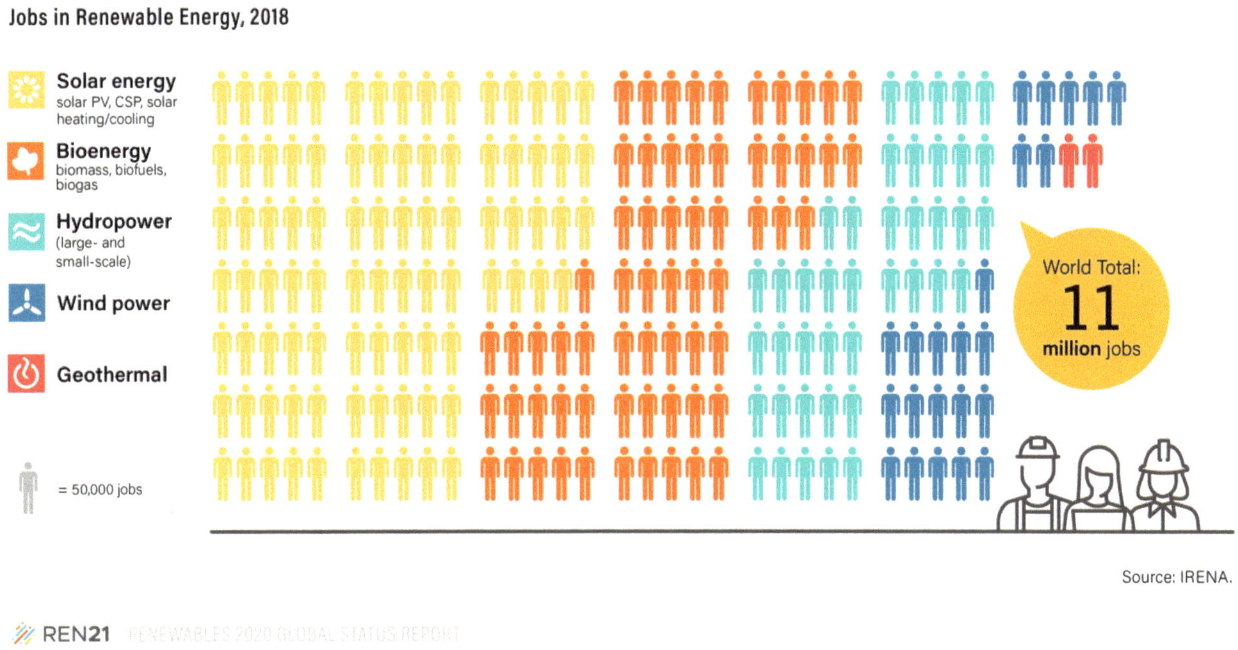


Figure 1.17 Jobs in different renewable energy technologies, each figure representing 50,000 jobs.[19]

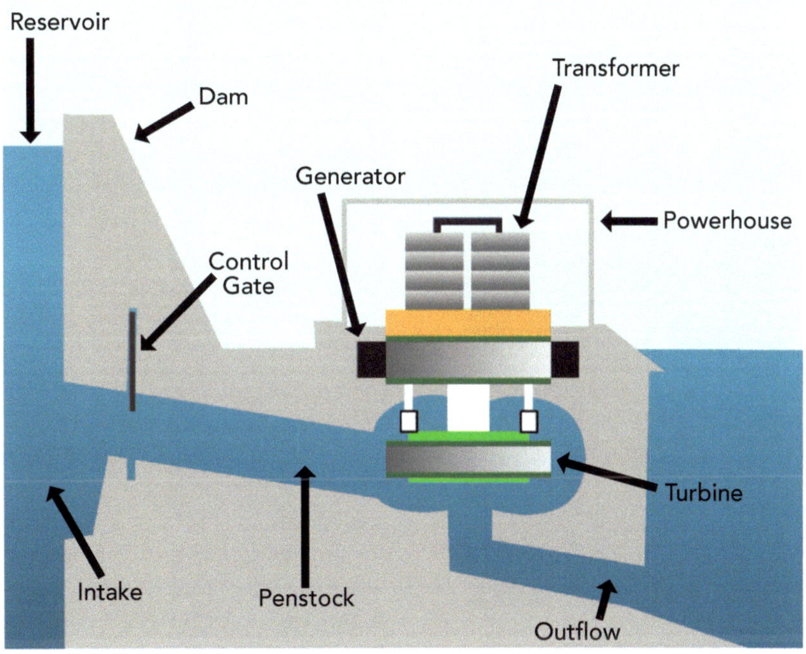


Figure 1.18 Generic section of a utility scale hydropower plant showing how water is held back by the dam and then released through the penstock to run through the turbine and then into the riverbed below.

turn turbines either because of pressure differences or gravity. The devices can be located onshore, nearshore, or offshore. On a global scale there is so little production that this type is often either left off charts or combined with other small producers such as CSP.

## Wind

Wind power converts the kinetic energy of the wind into electricity by spinning blades to turn the turbine and generator. This power was historically used to pump water and grind grains. Wind turbines can be large, utility scale generators that have a swept area of the blades over 200′ in diameter, can be almost 400′ tall, and are rated over 1 megawatt. These large utility turbines can power large communities. Smaller turbines can be used to power just one household. Figure 1.20 shows a small pole mounted wind turbine with three blades and a tail. Wind turbines need to be located where there are good wind conditions and

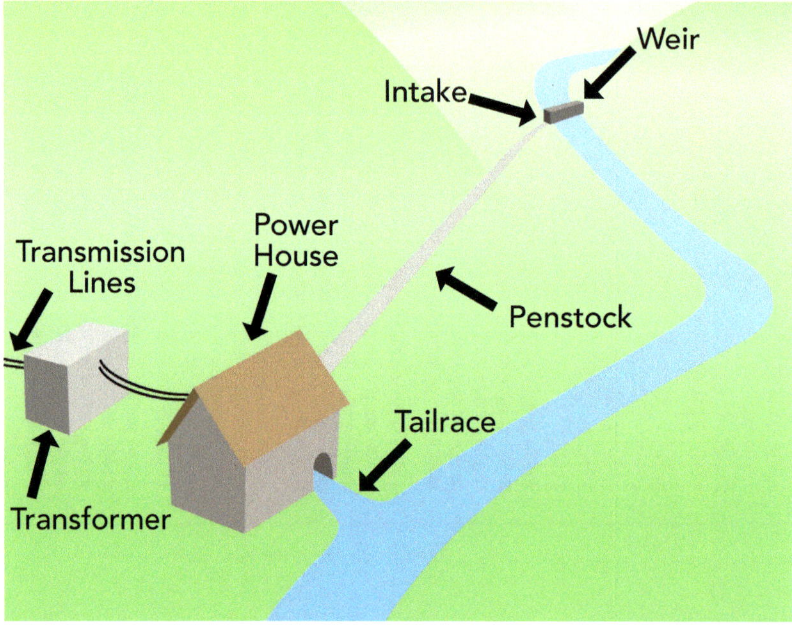


Figure 1.19 Microhydro diverts a portion of the water from a river or stream and can power small communities.

be located above or away from **turbulence**, typically away from buildings and other obstructions. Production depends on the quality and strength of the wind. There are many locations where there isn't enough wind to warrant installation of a turbine. Depending on the type of turbine, efficiencies for energy production range from as low as 5% to as high as 40%.

Figure 1.20 Small Wind Turbine.

## Biomass Power

Biomass is renewable organic material that comes from plants and animals. This material is composed of carbon atoms in the current carbon cycle, so are considered renewable because they do not add to global warming potential. Traditional biomass uses such as open fires are quite inefficient, although they are still used by many hundreds of thousands of people as primary cooking and heating fuels worldwide. Modern biomass equipment is much more efficient, burning at much higher temperatures, and has less likelihood of producing unwanted particulate matter. Biomass products include wood and wastes from wood processing, various crops (such as corn, soybeans, and algae) and their waste materials, various municipal solid waste (such as paper, food, and yard waste), as well as animal manure and human sewage. These materials can either be burned or fermented to create power, gas, or **biofuels**. Humans have burned biomass since the discovery of fire and its use remains common in semirural and rural locations throughout the globe today. It can be used directly for heating buildings and industrial processes or for generating electricity with steam turbines.

Figure 1.21 is a crazy looking chart showing the different ways various biofuels are turned into heat, electricity, and fuel. As you can see, there are many options. Remember that the more energy is put into the conversion route, the less energy there will be for the final usage because of losses along the way. Also note that the solid lines are commercial pathways that currently exist while the dotted lines represent developing bioenergy routes. Currently, most of these biomass types are being used as solid or liquid fuels. The photosynthetic microorganisms have not yet reached a solid commercial production stage but have been successful in research and small-scale settings.

## Solar PV

This technology takes advantage of the trillions of watts of solar energy that hit the earth every day. **Photovoltaic (PV) modules** convert the sun's energy into electricity. A grouping of PV modules is called an array. Figure 1.22 shows a pole mounted PV array.

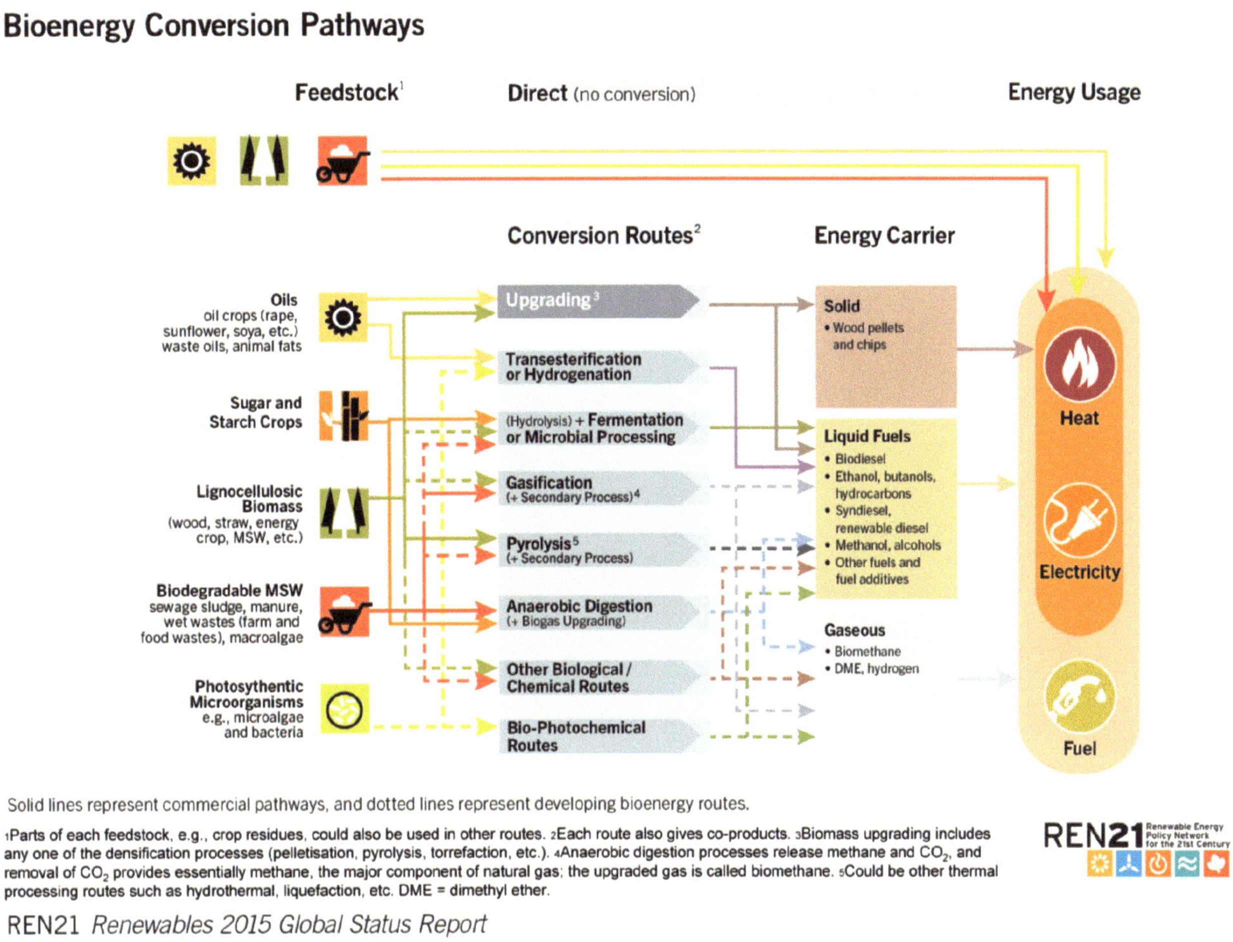


Figure 1.21 Conversion pathways for different types of biomass to final energy type.[20]

Figure 1.22 PV Array with 4 pole mounted panels. Each panel is made up of 4 PV modules bolted together.

Arrays can range in size from a single module powering a pump or battery powered light to large fields producing at utility scale. The amount of production will depend on how much sun a particular location has. This is dependent on weather patterns as well as obstacles at the site, such as buildings and trees. Most populous areas of the world are well suited for PV and system costs have continued to drop, making them competitive with the fossil fuel industry. The biggest limitation is that PV only produces electricity while the sun is shining on it. Energy storage (or other power resource) is needed if electricity is needed at night. Depending on the type of PV, it can convert between 5–20% of the sun's energy to electricity. Figure 1.23 shows total annual PV capacity

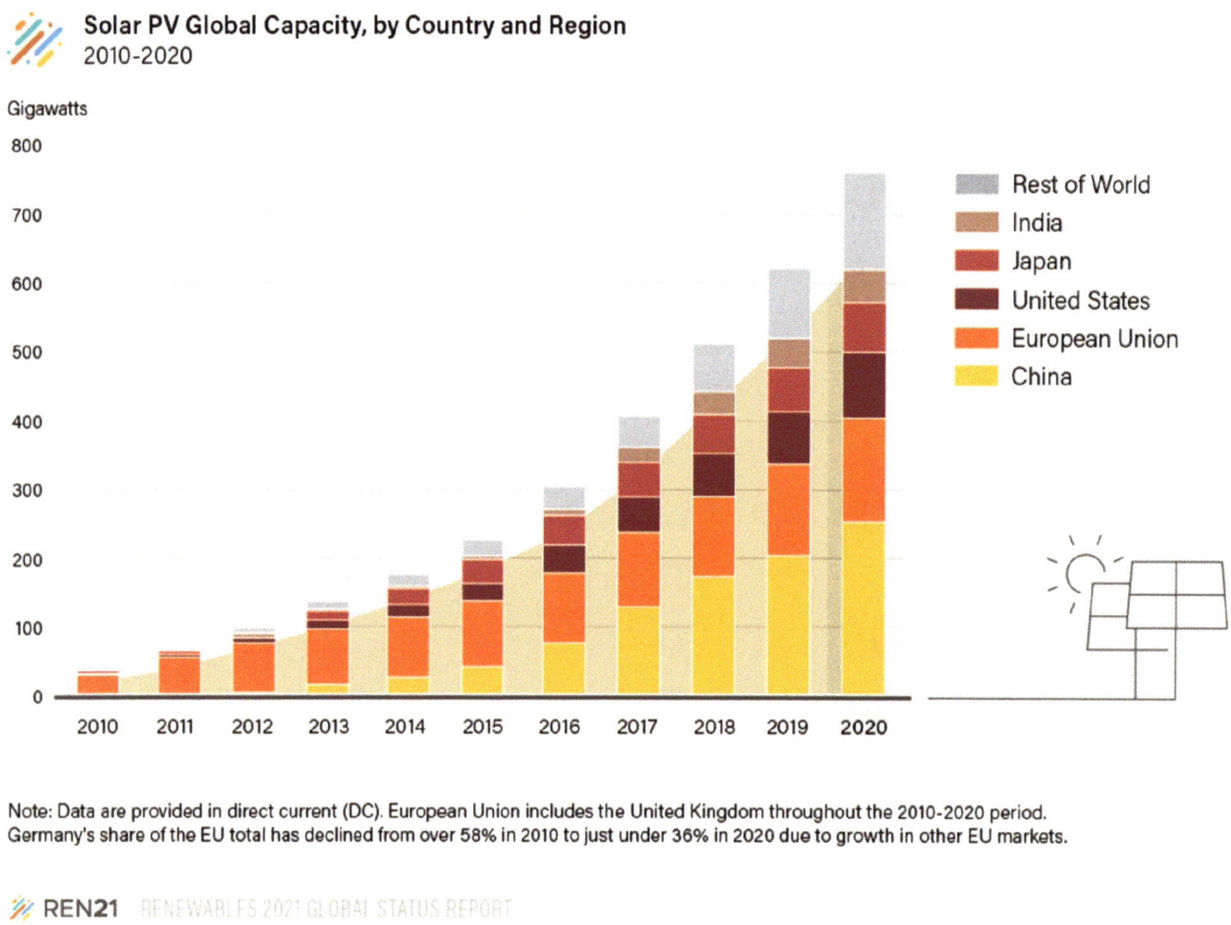


**Figure 1.23** Global solar PV capacity broken down by country and region.[21]

globally, with percentages for top producing countries. You can see that China has the most capacity. PV has had a big spike in growth over the last 10 years and it is forecasted to continue its exponential growth.

## Concentrated Solar Power

Concentrated solar power (CSP) is a utility scale technology where the sun's rays are used to create large amounts of heat. Sunlight is reflected with mirrors that focus it onto a small area to create heat. The mirrored lenses are usually either in a point focus or a linear focus configuration. The heat can be stored or used. It can create power by running the heat through a generator, usually a steam turbine. Figure 1.24 shows the point focus type where large mirrors in circular rows focus the sun's rays on the top of a tower. The tower collects the heat which is run through the system to make power which can then be distributed over utility lines.

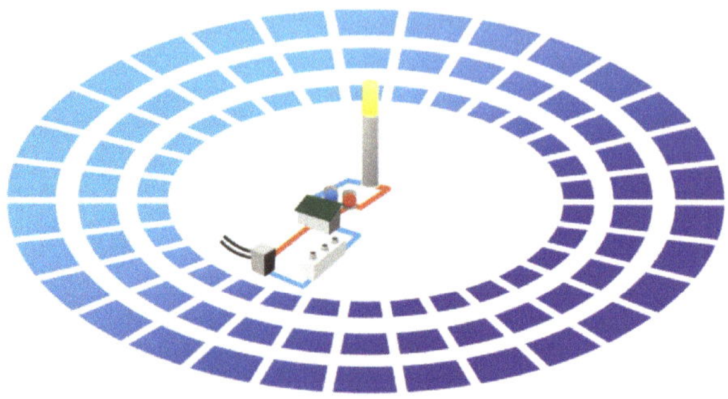

**Figure 1.24** Point power tower system for Concentrating Solar Power (CSP).

Figure 1.25 shows the linear configuration where there are lines of curved mirrors that focus the sun's heat to piping at the center of each mirror line. The piping carries the heat to the generating system to make electricity which is then distributed over the utility power lines.

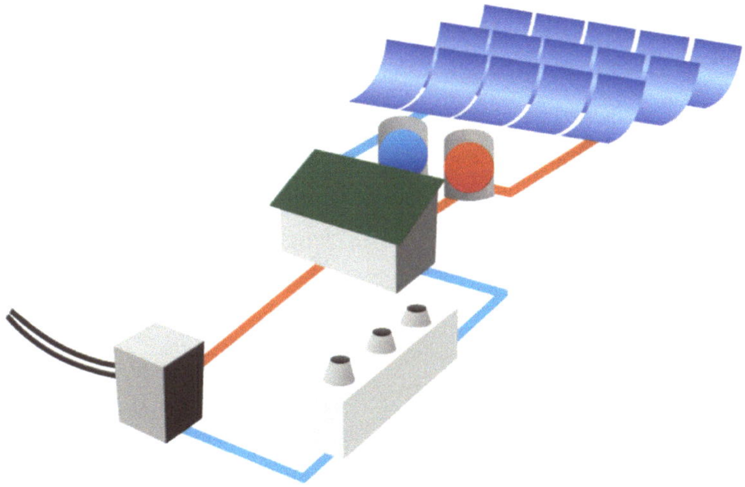

Figure 1.25 Linear parabolic trough system for Concentrating Solar Power (CSP).

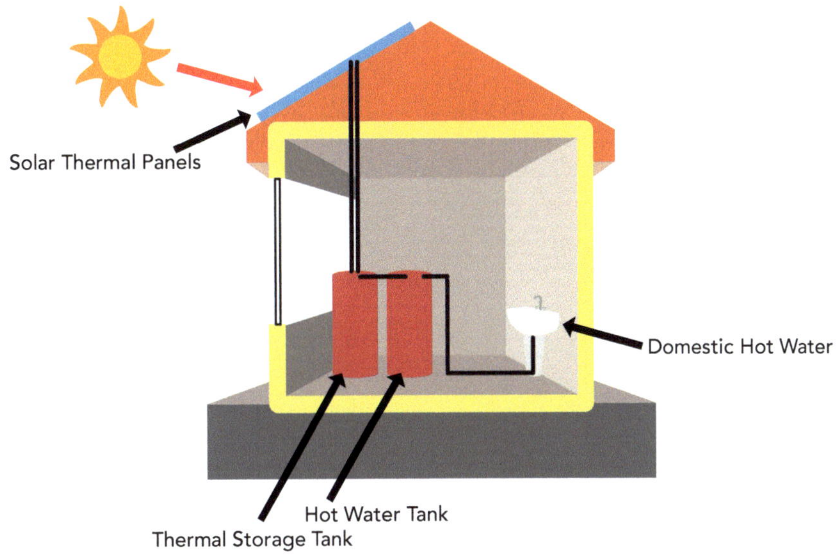


Figure 1.26 Solar thermal panels absorb heat from the sun to create domestic hot water which is stored inside in a tank.

The trough system typically uses mineral oil, while typically point systems use molten salt as the heat transfer medium. The heat can be stored in large containers for as long as 15 hours, making it possible to produce electricity at night or at peak capacity times when there is a spike in demand, such as when the sun is setting and PV production ceases. Installed production is between 15 and 30% effective at converting the sun's energy to electricity.

CSP remains one of the smallest types of renewables when looking at installed capacity as it has had technical difficulties and high initial investment costs. Spain and the United States lead the world in installed CSP, with some new installations around the globe every year. Important siting issues include the need for reliably sunny days year-round, the large amount of space required for the facility, and the need to deal with large amounts of waste heat. These plants are typically located away from population centers, so there are transmission and distribution losses similar to those of traditional fossil fuel and nuclear power plants. Newer facilities appear to have overcome technical problems and are generating reliably.

## Solar Thermal Hot Water

Solar water heating systems use a collection plate or set of tubes to gather the heat from the sun and then transfer that heat into water that can be used for domestic hot water. In the United States the collector is typically located on the roof of the building while the hot water storage is in a tank that is feeding the preheated water to a conventional water heater. This configuration is shown in figure 1.26.

Solar water heaters in many parts of the world have an integrated storage tank that directly heats the domestic hot water, as shown in figure 1.27. The storage tanks are insulated so the water remains hot.

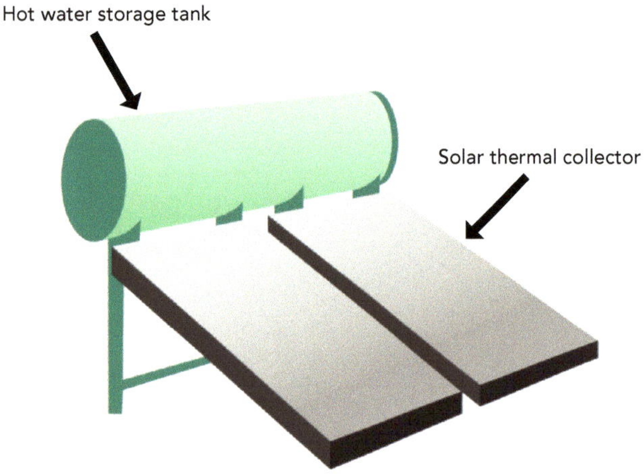


Figure 1.27 Solar thermal with integrated storage.

Solar thermal is very effective at heating water, typically collecting 73% of the sun's heat energy. Currently the fastest growth market for solar thermal is in China for the integrated storage type. Many of these units are not approved for use in other countries.

## United Nations Sustainable Development Goals

The United Nations has identified 17 sustainable development goals. Improving building performance, as well as providing renewable energy worldwide, relates to a number of the goals. Because people spend a lot of time indoors, for USA citizens it's 90%, indoor air quality is very important. Remember that we are using the most energy to condition the air inside the building (most often heating the space) and we want to be sure it is healthy for the occupants.

This book directly addresses Goal 7: Affordable and Clean Energy, "ensuring that everyone has access to affordable, reliable, sustainable and modern energy."[22] 13% of our global population lacks access to electricity. There are three billion people worldwide who need clean and safe cooking fuels. Fossil fuels have been a major energy source for power since the industrial revolution. The amount of GHG released from our activity so far has had measurable harmful impacts on people throughout the world.

In our cities many of the oldest buildings have the highest number of issues for poor indoor air quality, high energy consumption, poor quality of space, and lack of biophilia. These conditions can be crowded and unsanitary. Because these spaces can be rented cheaply, typically poorer people live there. This creates an energy burden and health issues for a segment of society who can least afford it.

Designing efficient buildings that conserve and use renewable energy not only directly improves the quality of life for the people living there but it also improves the quality of life for everyone. Reducing GHG emissions is our number one goal for slowing global climate change.

## Resources


1. U.S. Department of Energy. (2015, September). Chapter 5: Increasing efficiency of building systems and technologies. In Quadrennial technology review: An assessment of energy technologies and research opportunities. www.energy.gov/sites/prod/files/2017/03/f34/qtr-2015-chapter5.pdf
2. U.S. Energy Information Administration. (2019, October 21). Global energy consumption driven by more electricity in residential, commercial buildings. *Today in Energy*. www.eia.gov/todayinenergy/detail.php?id=41753
3. The World Bank Group. (2021, June 7). *Report: Universal access to sustainable energy will remain elusive without addressing inequalities* [Press release]. www.worldbank.org/en/news/press-release/2021/06/07/report-universal-access-to-sustainable-energy-will-remain-elusive-without-addressing-inequalities
4. Climate TRACE (https://climatetrace.org/).
5. U.S. Environmental Protection Agency. (n.d.). *Global greenhouse gas emissions data*. Retrieved June 18, 2021, from www.epa.gov/ghgemissions/sources-greenhouse-gas-emissions
6. $CO_2$-eq is used to compare GHG on the basis of their global warming potential, equal to the amount of $CO_2$ that would cause the same amount of warming. The US EPA has a GHG equivalencies calculator on their website, which also explains how the equivalencies are calculated. U.S. Environmental Protection Agency. (2021, March). *Greenhouse gas equivalencies calculator*. www.epa.gov/energy/greenhouse-gas-equivalencies-calculator
7. U.S. Environmental Protection Agency. (n.d.). *Sources of greenhouse gas emissions*. Retrieved June 18, 2021, from www.epa.gov/ghgemissions/global-greenhouse-gas-emissions-data
8. U.S. Energy Information Administration. (2021, June 23). *Use of energy explained: Energy use in homes*. www.eia.gov/energyexplained/use-of-energy/homes.php
9. U.S. Energy Information Administration. (2018, September 28). *Use of energy explained: Energy use in commercial buildings*.

www.eia.gov/energyexplained/use-of-energy/commercial-buildings.php

**10** U.S. Census Bureau. (n.d.). *U.S. and world population clock.* U.S. Department of Commerce. www.census.gov/popclock/

**11** Gaynor, T. (2020, November 30). *Climate change is the defining crisis of our time and it particularly impacts the displaced.* United Nations High Commissioner for Refugees. www.unhcr.org/en-us/news/latest/2020/11/5fbf73384/climate-change-defining-crisis-time-particularly-impacts-displaced.html

**12** Office of Energy Analysis. (2019, September 24). *International energy outlook 2019 with projections to 2050.* U.S. Energy Information Administration. www.eia.gov/outlooks/archive/ieo19/pdf/ieo2019.pdf

**13** U.S. Energy Information Administration. (2021, May 14). *U.S. energy facts explained.* www.eia.gov/energyexplained/us-energy-facts/

**14** International Energy Agency. (2019, October). *Renewables 2019: Market analysis and forecast from 2019 to 2024.* www.iea.org/reports/renewables-2019

**15** REN21. (2020). *Renewables 2020 Global Status Report.* REN21 Secretariat. www.ren21.net/wp-content/uploads/2019/05/gsr_2020_full_report_en.pdf

**16** U.S. Energy Information Administration. (2021, May 13). *British thermal units (Btu).* www.eia.gov/energyexplained/units-and-calculators/british-thermal-units.php

**17** REN21. (2018). *Renewables 2018 Global Status Report.* REN21 Secretariat. www.ren21.net/wp-content/uploads/2019/08/Full-Report-2018.pdf

**18** REN21. (2018). *Renewables 2018 Global Status Report.* REN21 Secretariat. www.ren21.net/wp-content/uploads/2019/08/Full-Report-2018.pdf

**19** REN21. (2020). *Renewables 2020 Global Status Report.* REN21 Secretariat. www.ren21.net/wp-content/uploads/2019/05/gsr_2020_full_report_en.pdf

**20** REN21. (2015). *Renewables 2015 Global Status Report.* REN21 Secretariat. www.ren21.net/wp-content/uploads/2019/05/GSR2015_Full-Report_English.pdf

**21** REN21. (2020). *Renewables 2020 Global Status Report.* REN21 Secretariat. www.ren21.net/wp-content/uploads/2019/05/gsr_2020_full_report_en.pdf

**22** United Nations. (n.d.). *Take action* for the Sustainable Development Goals. www.un.org/sustainabledevelopment/sustainable-development-goals/

## Study Questions

1. Find several examples of each of the renewable energy systems and answer the following:
   a. What part of the system is clearly visible?
   b. How big are the components you see and where are they located?
   c. How similar or different are the components for the same type of system in the examples you found?
2. Go to the IEA website and look at the section with information relating to buildings (www.iea.org/topics/buildings). Look through the items posted and select an item that interests you and answer the following:
   a. What is the article about?
   b. What kind of information is included in the article?
   c. What two items stand out the most to you – why do you find them interesting?
3. Visit the REN21 website (www.ren21.net/) and answer the following:
   a. What types of information are available on this website?
   b. After you've explored the website, what two items stood out the most to you – why do you find them interesting?

CHAPTER 2

# Basic Concepts of Renewable Systems

**Key Terms**

Alternating current (AC): voltage (current) is switching polarity (direction) at regular intervals over time – graphs as a sine wave having equal positive and negative values.

Direct current (DC): voltage (current) which flows in one direction only – graphs as a horizontal line over time.

Grid-tied RE: the building is connected to the grid and supplies surplus power to the grid.

Inverter: a piece of electrical equipment that changes DC energy to AC energy.

Off-grid RE: the building is not connected to a grid and must provide all of its own power.

Service panel: an electrical box that has power service input that is then distributed through circuits to all the building loads.

Smart grid: a utility grid that has advanced metering infrastructure which allows for fairly real-time monitoring of customer power use and two-way communication between a building's smart meter and a centralized monitoring system.

## Concepts for Success

"On the grid" means that there is a utility-run distribution power network (power lines) that connects remote generation plants to buildings. In order to bill customers, buildings have one or more meters to measure electrical usage. As competing power generation sources and individual building renewable energy generation are added to grids, it becomes more challenging to manage an even distribution and power profile across the grid. Two-way communications technology, such as smart meters, in conjunction with centralized grid monitoring helps to create a **smart grid**. Smart grids are more resilient and are being developed worldwide.

Most buildings are located in urban areas where they are connected to the grid. Designing and retrofitting buildings to be energy efficient and include renewable energy production is important for both grid stability as well as reducing greenhouse gas (GHG) emissions.

Buildings that are located where there is not a grid must supply all their own energy; this is called "off-grid". In order to reduce the size of the renewable energy system needed to power the building, off-grid buildings need to be energy efficient, designed for the climate, and be powered by electricity. Off-grid buildings that rely on fossil fuels to maintain building functions are adding to GHG climate problems, so it is important to be able to power the building with renewable resources.

There are basic components that make up renewable energy systems, based on the needs of the client and the community. Renewable energy is typically **direct current (DC)** while building loads are typically **alternating current (AC)**.

Figure 2.1 shows the symbols that will be used throughout the chapter to represent the various system components. An **inverter** is used to change between current types – DC to AC. If power is needed when none is being generated, then energy storage, such as a battery bank, must be included in the system. Batteries always have **charge controllers** to maintain appropriate charge levels. Everything in a building is powered because it has wiring that runs from a circuit on the main **service panel** to the device or a plug for the device (i.e., outlets, lights, mechanical equipment). The service panel may be labeled with different adjectives such as main, standard, essential, or DC, all of which are

DOI: 10.1201/9781003297819-3

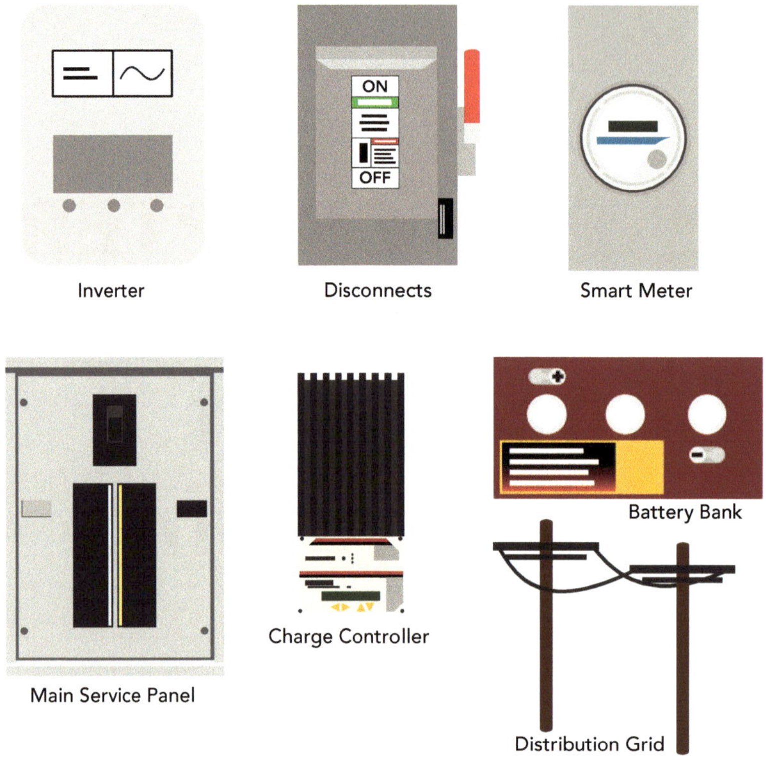


**Figure 2.1** Symbols used in this chapter to show the various components of renewable energy systems.

an indicator of what type of equipment the panel may be serving in a particular application. Grids are made of power lines and the smart meter is able to measure both the energy used and supplied to the grid. Because electrical systems for buildings have a high amount of power, disconnects are used between the main system elements so that an electrician can manually turn off the power and safely work on or replace equipment.

## On the Grid

A utility grid distributes electrical power from a generation plant to buildings. Buildings have one or more meters so consumers can be billed for the amount of electricity used. Power plants are often located away from urban centers because they are generally large, noisy and may have pollution concerns dependent on the energy source being used to generate power. The maintenance of the grid, which is made up of transformers and high voltage power lines, is done by a utility company who has the rights to serve a particular geographical area. In the United States, most states allow competition for the generation of power. This means that customers are able to make their own contract for who is generating their power but are unable to choose their distribution company. All power, no matter who is generating it, gets distributed over the grid in order to reach the consumer's buildings.

Smart grids have advanced metering infrastructure which allows for fairly real-time monitoring of customer power use and two-way communication between a building's smart meter and a centralized monitoring system. This allows the grid operator to match production to demand, creating more reliable power. Other aspects included in a smart grid are the ability to be flexible to handle the flow of energy in two directions. This is important because buildings generating electricity with renewable power typically feed their excess power onto the grid for distribution.

Smart grids have the ability to better integrate the variable production of wind and photovoltaics (PV) with power plant generation, as well as manage the feed of power onto the grid from the competing suppliers in an area. This requires that the electricity being produced from multiple sources can be matched and maintained to the grid voltage and frequency.

Smart grids are being developed all over the world with the largest markets being Asia, Europe and North America. Currently 24 states and the District of Columbia in the USA have established goals for GHG emission targets.[1] As of the end of 2020, the countries of Suriname and Bhutan have achieved net zero emissions while Sweden, UK, France, Denmark, New Zealand, and Hungary have laws to reach net zero by 2050. Proposed legislation for 2050 targets exists in the European Union (EU), Canada, South Korea, Spain, Chile, and Fiji. 80% of the world's countries have stated commitment to net zero carbon by 2050.[2]

Historically, in areas with established grids, people didn't really think about the amount of energy their building used. Energy was inexpensive and there was little incentive to create better performing buildings.

Today, there can be so much demand that it can be very difficult for the power companies to provide enough energy. In some countries people may have power to their buildings for only some hours of every day, or particular days of the week. In the United States most utilities have incentives in place to get customers to use less power.

If you have a building with very large loads, it can be impossible to design a renewable system big enough to supply all the demand. The larger the renewable system is, the more expensive it will be. Therefore, it is important to reduce load demands in every building so it is more likely that you can have a reasonably sized system that can provide enough power for the building.

## Conceptual Diagrams

The following diagrams are conceptual and do not show all the wiring, **grounding**, monitoring, and safety equipment that are part of installed systems. These items will be covered in subsequent chapters and some details, such as sizing wire and fuses, are beyond the scope of this book. There are many resources that cover electrical installations in greater depth. The diagrams of systems in this chapter show PV as the generating renewable energy resource but it could be any renewable resource. See the chapter on a particular resource for information on system components and siting. Additionally, for clarity, only one PV module is shown, rather than a whole array, and one battery represents the multiple batteries that make up a battery bank. There are other types of energy storage devices besides batteries but they are not shown in these diagrams for clarity in comparing basic system organization. More information about energy storage is covered in chapter 18 on Batteries and Energy Storage.

The purpose of these diagrams is to orient you to the typical features of different types of renewable energy system layouts and explain how they work. Note that most of the components are outside the building; only batteries and the main power supply panel are inside the building.

## Grid-Tied Renewable Energy System

The simplest renewable energy system is one that produces electricity that can be used in the building or put on the grid. The grid is used to balance supply for the building since the renewable source may not be producing all the time, may not be producing enough energy to supply the total building's demand, or may be producing more energy than needed to run the building.

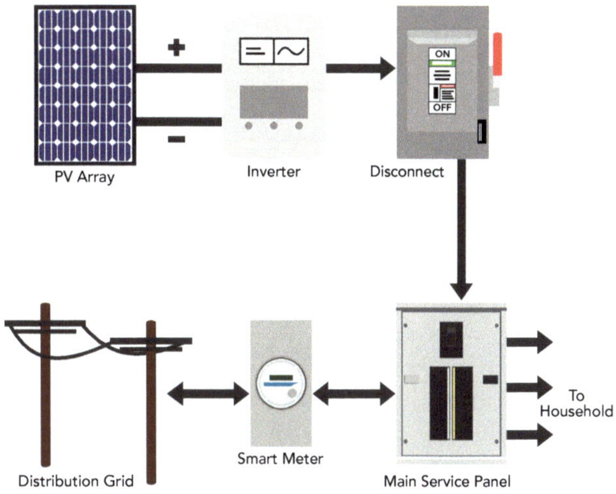


**Figure 2.2** Grid-tie renewable energy system has power production feeding to the main service panel and is connected to the grid by a smart meter.

Figure 2.2 shows a PV array connected to an inverter. The array is producing DC energy while most countries use AC for building loads. For more information on types of current and buildings see chapters 4 (Basic Concepts about Electricity) and 22 (Resiliency and Power). Electrical equipment intended to run on AC power cannot run on DC power and vice versa. The inverter changes the DC power into AC power and is always communicating with the grid to match the power profile being distributed by the grid. The inverter is wired to the main supply panel of the building. This circuit board provides power to outlets, lights, and mechanical equipment through the electrical wiring from each circuit to the various locations in the building. When power is being generated from the renewable source, that power is directly supplied to the electrical loads of the building. The main service (AC) supply panel is wired to a smart utility meter, which is wired to the grid, so the meter can measure both how much energy the consumer is adding to the grid as well as how much energy the consumer is using from the grid. If the grid goes down (no power being distributed) for any reason, the inverter will stop providing power from the renewable source to the supply panel. This is a safety feature intended to prevent anyone who would be fixing a power line from getting electrocuted because of live energy being added to the grid from a building that is producing energy. This means that when the grid is down, your building has no power, even though you have a renewable energy resource that would otherwise be making power.

## Grid-Tied Renewable Energy System with Battery Backup

Batteries can be used to back up your grid-tied system so you have some power all the time, no matter if

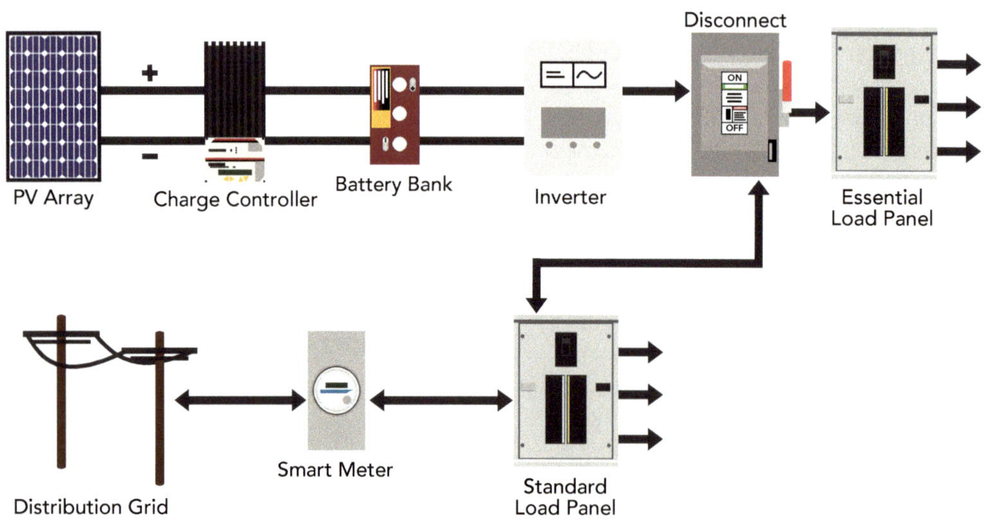


**Figure 2.3** Grid-tie system with battery backup.

the grid is powered or not. Unless your building will regularly be without power from the grid, the battery bank doesn't need to be very large. Whenever you have batteries in a system, you need a charge controller to monitor and protect the batteries. The charge controller can either be a separate piece of equipment or an additional component in the inverter.

Looking at figure 2.3, you'll see that the charge controller and battery are wired between the PV array and the inverter because batteries store DC power. The inverter will change the DC power being created by the PV array or the battery to AC power, matching the power profile of the grid. Note that the inverter powers two separate AC load panels. One is the standard panel which is then connected to the smart meter and grid. The other is the **essential load** panel. As the name implies, wiring from the essential load panel powers items in your building that you always want to be powered. Essential loads will include some lighting, plug loads for important equipment (such as refrigerators and medical devices), and power to mechanical equipment. The remainder of the building's wiring will be from the standard load panel and will not be powered when the grid goes down.

In addition to providing backup power in the event of power failure, this system can work with a smart grid to act as storage. In this case, the grid can draw power from the battery to assist in providing power to other buildings on the grid.

Another use for this type of layout, with or without an essential loads panel depending on what the building owner needs, is to be able to control the amount of energy the building draws from the grid during peak hours. Utilities may have tiered pricing based on the amount of energy used. The inverter can be programmed to provide extra power to the building instead of drawing power from the grid, keeping the building's use below a jump in electrical pricing.

## Off the Grid

Some areas are too remote to be serviced by a grid, are located where no grid exists, or have a small DC load. When off-grid, all equipment power needs for the building's operations must be met **24/7**. Remember that most renewable energy produces DC electrical current. Because an off-grid system is autonomous, it has to be able to provide all the power required to run the building at all times. Care needs to be taken to account for all of the building loads so the renewable system can be sized correctly.

If an entire building needs to be powered off the grid, it is very important to reduce energy consumption and have energy efficient electrical items. With the intention that the building should be able to run as much as possible on renewable energy, rather than fossil fuels, the building should be designed to be all electric. In rural applications it could be reasonable to use biomass for heating if it is available in the area and the client is willing/able to service the equipment (for example fill and clean a pellet stove).

The same technologies can be used in energy efficient buildings both on and off the grid. This is possible because heating and cooling loads, as well as lighting and appliance loads, have been minimized. Designing the building to be appropriate for its climate (see chapter 6, Design for the Climate) and use **heat pumps** for conditioning spaces (see chapter 11, Heat Pumps) is critical for success.

Many appliances are available that run on DC power. These have been typically made for boats and recreational vehicles but can be used in buildings. These appliances tend to be much more energy efficient than their AC counterparts, as well as much more expensive. They are usually sold by specialty dealers and are not found in regular appliance stores.

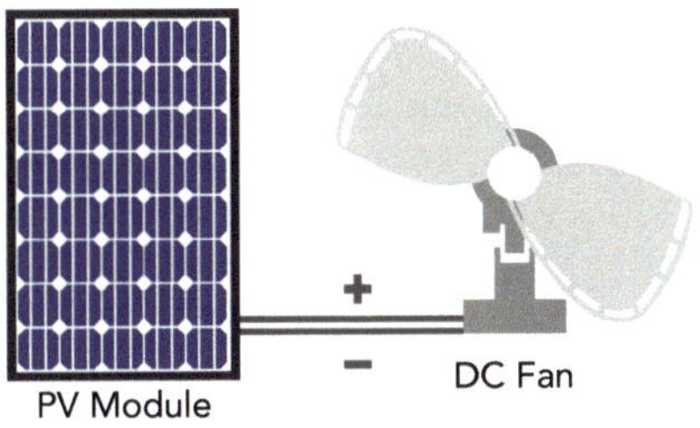


**Figure 2.4** DC power is hooked up directly to the DC equipment.

The very simplest form is using the renewable energy directly to run an item. This is common for pumping water. The electrical item will only run when the renewable resource is making energy. Figure 2.4 shows a PV module running a DC fan. The PV module would need to be large enough to produce the power needed to run the fan. Since the PV is producing DC energy, then the equipment must also be DC (note that there is no inverter). Note that power switches will be located at the equipment. For example, the fan could be turned on and off using the switch on the fan as well as being adjusted to the desired fan speed.

Of course, the problem with direct power is that you only have power when the source is producing. Resources such as wind and PV are variable: PV only makes power in daylight hours, and the wind isn't always blowing.

The next option is to add batteries to the system. Figure 2.5 shows that the charge controller and battery are located between the PV array and the DC light. The battery will be charged when power is being generated. The battery will then discharge to run the equipment when no power is being generated. The PV array needs to be sized so that while power is generated it can both run the equipment and charge the battery. Since there is no inverter, the equipment has to run on DC power. Typical types of applications are for road signs, electrified fences, lighting, and pumps. The battery can often store enough power to be used for several days without being recharged. Multiple batteries may be required depending on how much electrical demand there is vs. how long you need the equipment to run without recharging.

Off-grid systems for buildings typically have batteries as shown in figure 2.6. The battery bank has to be large enough that it can power the building's loads when there is no energy being produced. The charge controller and battery bank and will be wired between

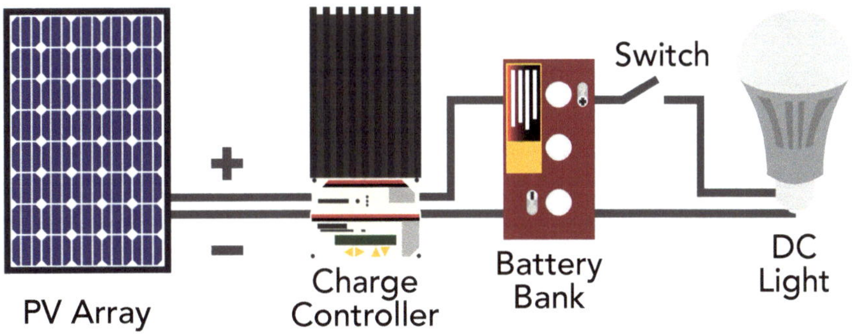


**Figure 2.5** DC system with battery storage.

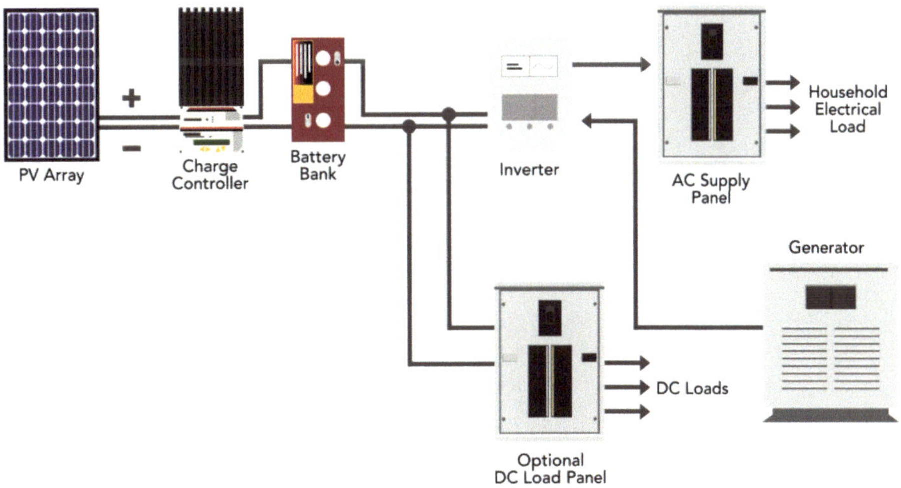


**Figure 2.6** Off-grid system for powering a building.

the renewable resource and the inverter, the same configuration as in a grid-tie with battery backup. If the building has DC loads, then a DC load panel will be wired to the battery bank and be connected to the DC loads. Notice that this optional DC service panel is running directly from DC power from the batteries. The inverter will change the DC power from the PV array or battery bank to AC power for the main AC supply panel. All AC loads are connected to the circuits on the AC supply panel. A backup generator is used when the battery bank's energy has been depleted and is not being recharged by the PV array. The generator will turn on and power the building and recharge the batteries, turning off if the renewable resource begins production again. Note that the generator is connected to the inverter because it is producing AC power.

Off-grid systems can have more than one renewable resource producing power. These **hybrid generation** systems are often a combination of wind and PV because they are complementary in their seasonal production. PV tends to produce more energy in the summer, when there are more hours of daylight, while wind tends to have better winter production. A hybrid system produces a more even amount of energy all year so the backup generator won't need to run very often.

## Resources


1 Center for Climate and Energy Solutions. (2021, March). *U.S. state greenhouse gas emissions targets*. www.c2es.org/document/greenhouse-gas-emissions-targets/

2 International Energy Agency. (2021, July). *Net zero by 2050: A roadmap for the global energy sector.* https://iea.blob.core.windows.net/assets/beceb956-0dcf-4d73-89fe-1310e3046d68/NetZeroby2050-ARoadmapfortheGlobalEnergySector_CORR.pdf


## Study Questions

1. Draw and label the components of a grid-tied system in a simple diagram.
2. Draw and label the components of a grid-tied system with battery backup in a simple diagram.
3. Draw and label the components of an off-grid system in a simple diagram.

CHAPTER 3

# Locating Renewable Resources

**Key Terms**

Degree days: heating (HDD) and cooling degree days (CDD) are a measurement of how many hours in a day (24 hours) are colder than a reference temperature (so need heat to be comfortable) or are hotter than a reference temperature (so need cooling to be comfortable).

Solar insolation: the amount of radiant energy available at a particular location due to latitude and weather.

Wind rose: a circular graph aligned with the cardinal directions that has bars that indicate the velocity, amount, and direction of wind in a particular location.

## Concepts for Success

The internet has many great repositories of weather data. This chapter is by no means an exhaustive list of sites and apps, however, these are resources that I have found to be quite useful.

The focus in this chapter is to locate information and understand what information is available that will be helpful to you in designing energy efficient buildings and knowing what renewable resources might be available at your location. See chapter 6 (Design for the Climate) for specific application of climate data for your location to your project.

## General Climate and Renewable Resource Databases

All software simulators require the use of data sets. Some programs already have weather files included or have a way to access weather station data. If you are in need of climate data for building simulations, Climate.OneBuilding.Org[1] is free and provides access to download climate location zip files worldwide. Each file contains EPW (EnergyPlus Weather Format), CLM (ESP-r weather format), WEA (Daysim weather format), DDY (**ASHRAE** Design Conditions), RAIN (hourly precipitation in m/hr) (not available for all locations), and STAT (expanded EnergyPlus weather statistics). The site clearly identifies the climate and weather data sources and what years the data sets are based on.

For USA locations, cli-MATE is a tool offered by Midwestern Regional Climate Center (MRCC)[2] which has information on **degree days**, precipitation, temperature, **wind rose**, and growing season data. All of these can be made into charts and graphs with custom dates. You can create a plot of the data on the site or download the tabular data. Once charts are created, they are interactive on the website or can be downloaded in png, jpeg, pdf, and svg formats. The site is free but requires registration to log in. The weather stations used are mostly located at airports. Figure 3.1 is an example of the output for a thermograph, which shows the average daily temperature for a location as well as the historical high and low temperatures for that day.[3]

While daily temperatures can give you a sense of the climate, to really understand when you'll need heating and cooling, you need to see data for heating degree days and cooling degree days. Figure 3.2 is an example of MRCC heating degree days output for a location, shown from 2008–2020.

Perhaps the most useful information is the wind rose data, an example of which is shown in figure 3.3.
A wind rose is a circular diagram that represents wind conditions in radiating bars. The circle represents different cardinal directions and the length of the bar, as

DOI: 10.1201/9781003297819-4

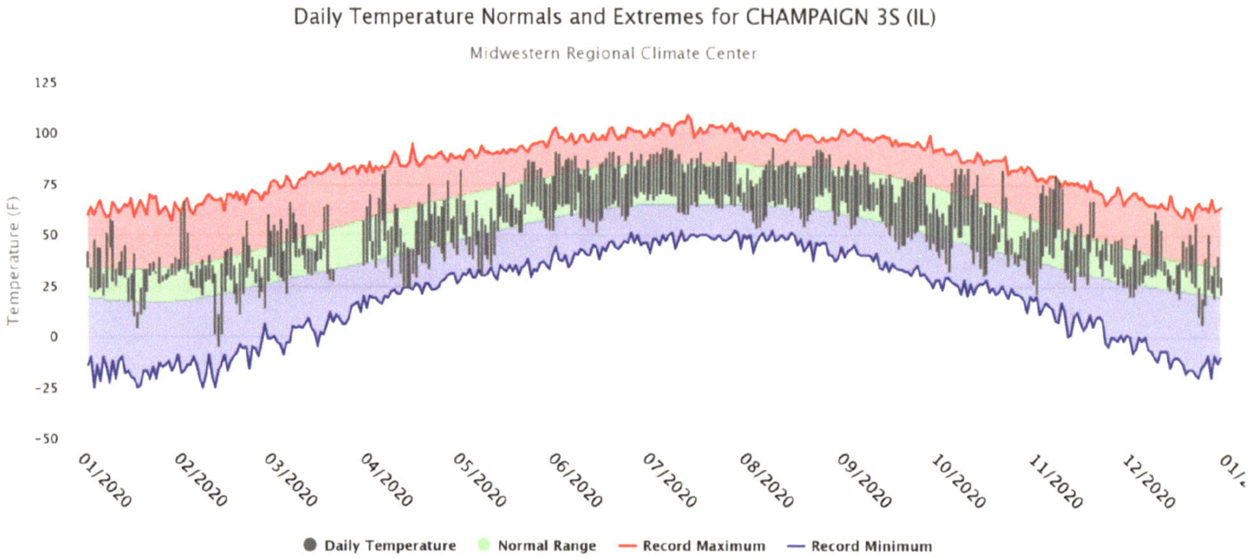


**Figure 3.1** Thermograph for Champaign, IL, USA, 2020, recorded daily temperature.[3]

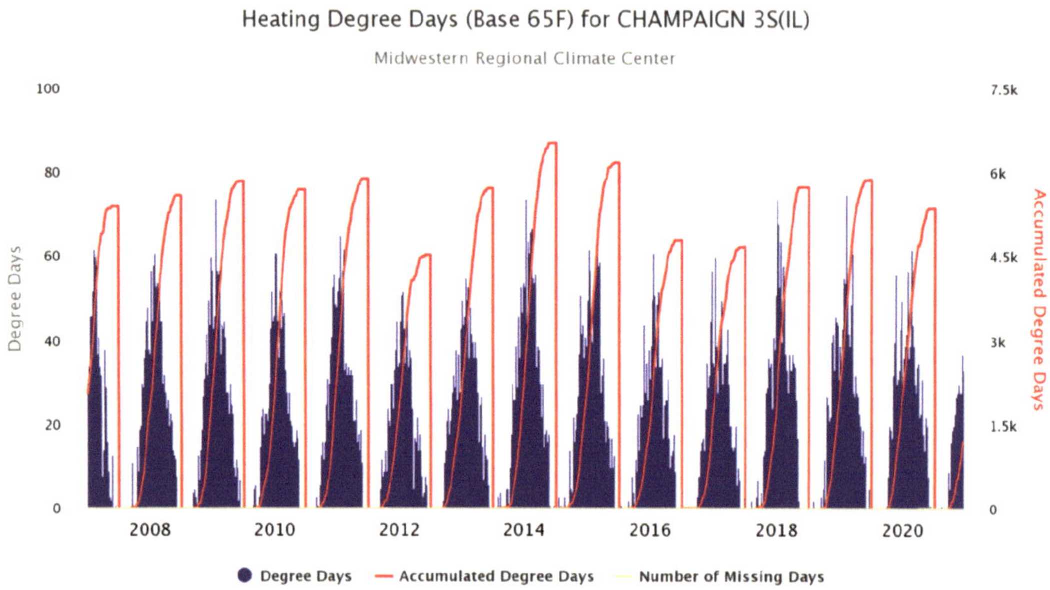


**Figure 3.2** Heating Degree Days in °F for Champaign, IL, USA, January 2008–December 2020.[3]

well as the hatching along the bar, indicate the amount of wind and its speed. This is set for an average from 2009–2020.

While graphs of heating and cooling degree days can be helpful, it can be much easier to understand the information in a list format. Degree Days Weather Data for Energy Saving, BizEE[4] has both free and professional apps that provide heating and cooling degree days around the world. It is customizable for Celsius or Fahrenheit, setting the desired base temperature and specifying output for daily, weekly, monthly, average, or custom formats. They also have great articles about heating and cooling degree days. Examples of output and its application are shown in chapter 6, Design for the Climate.

A great tool for creating wind roses for worldwide locations is the "Iowa Environmental Mesonet" hosted by the Department of Agronomy at Iowa State University.[5] Examples of output and its application are shown in chapter 6, Design for the Climate.

The National Renewable Energy Laboratory (NREL) has a website dedicated to geospatial data science[6] with data and tools accessing information for biomass, geothermal, hydrogen, marine and hydrokinetic, solar, and wind energy, as well as a crosscutting combination tool. The data is mostly focused on the United States but it does have some international locations which vary depending on the particular renewable resource. An example of the solar radiation information for the Americas is shown in figure 3.4.[7] The more intense the color is, the more **solar insolation** is available at that location.

Geospatial and earth observation data for Europe, including energy resources, are located at data.europa.eu[8] and have

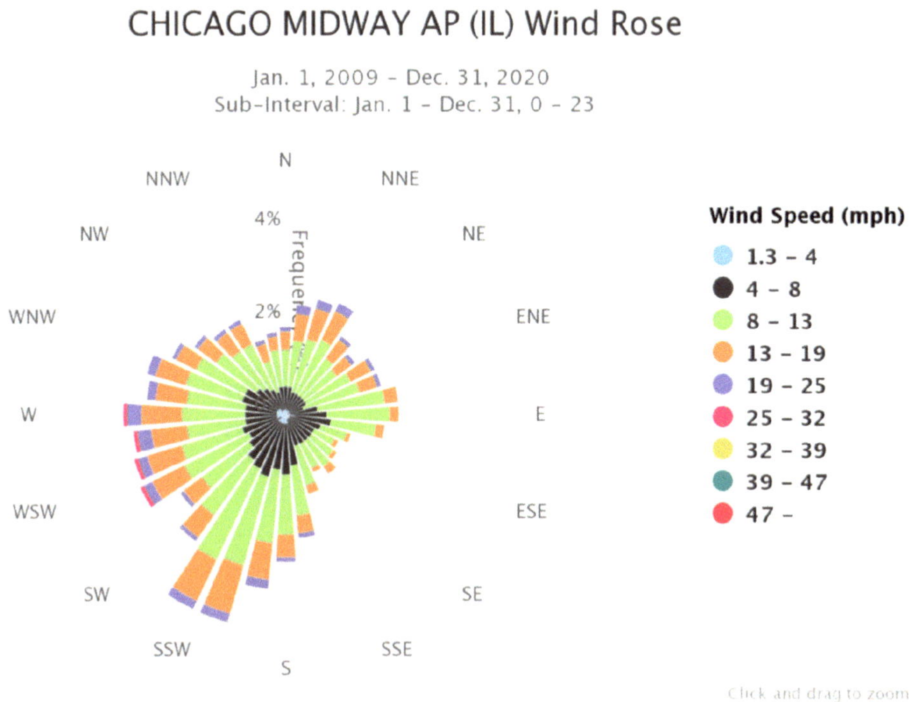


**Figure 3.3** Wind rose for Chicago, IL, USA, January 2009–December 2020, in mph with 32 compass points, Beaufort wind speed divisions.[3]

shared datasets from different European countries. It provides a lot of economic data, however, the energy section includes wave and wind information, as well as modeled geothermal resources. The files are in a variety of formats and identify the year(s) for the data.

## System Design Applications

RETScreen Expert, RETScreen Clean Energy Management Software, is renewable energy software that can be downloaded for free.[9] The free version has access to all the tools and databases that are included in the professional mode, though you can't save, print or export files for your project. Educational institutions can purchase one annual subscription and get an unlimited number of seats for the institution and current students as long as the software is used for teaching and research (non-commercial) use only. RETScreen (see figure 3.5[10]) was developed by the Canadian government with partners around the world and is available in 36 different languages. It has lots of good videos and help to learn how to use it. It has databases for products, costs, benchmarks, and climate/weather.

RETScreen Expert has a series of worksheets which allow the user to analyze a project through benchmarking, feasibility, and tracking the performance of a completed project. It has settings for all types of building occupancies (uses) from a single-family home to a power plant. It is very robust, with many options and covers all types of renewable energy technologies from solar thermal hot water to district heating and cooling. It can also be used to show greenhouse gas emission reductions between study cases.

System Advisor Model (SAM)[11] (see figure 3.6) is free modeling software which is downloaded to your computer. You do have to register it in order to use more than 15 times. It was developed and is hosted by NREL. SAM can be used to design many types of renewable energy systems including photovoltaic (PV), wind, marine energy, concentrated solar power (CSP), solar water heating, biomass, and geothermal. It has simulation options that can be used to compare scenarios and optimize input parameters. It has the ability to do in-depth financial models over specified periods of time. The financial modeling can be done on residential and commercial buildings, third-party ownership, and power purchase agreement (PPA) projects.[12]

SAM has very robust help within the program and video training is available on the website under each technology listing. The website also has a section on curriculum materials that provides links to teaching materials used in academic courses using SAM. These materials are from the institutions listed.

Weather data for simulations must be downloaded and there is an interface within SAM to access the data from the SAM website weather files. The weather files are global and in the form of a text file that has one year's data. Depending on the source of the data, it may be a single year or a historical average.

Modeling for PV and wind systems can be for residential to utility scale. Figure 3.7[13] shows a portion

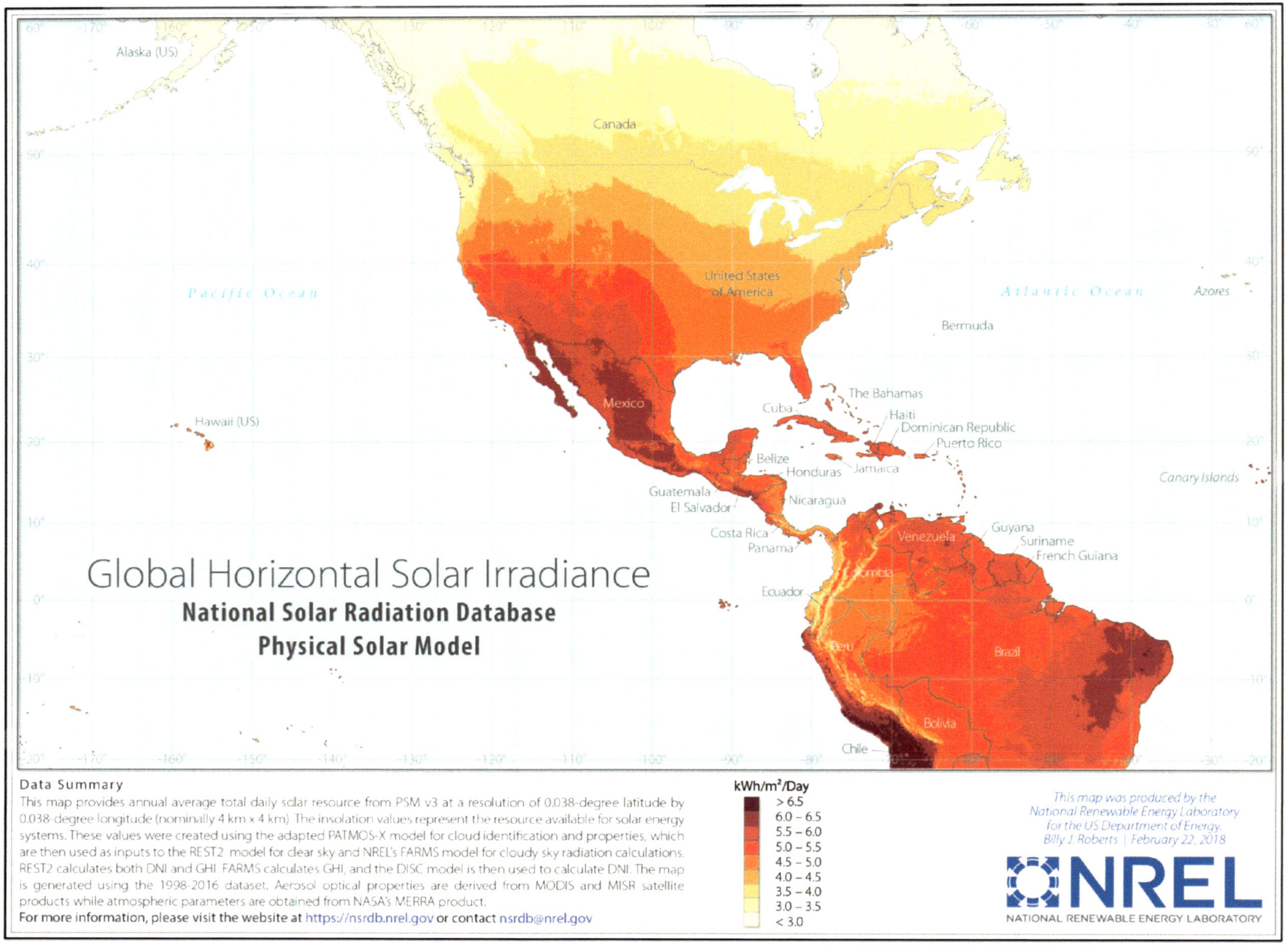


**Figure 3.4** Solar radiation, measured in kWh/m$^2$/day, for horizontal surfaces.[7]

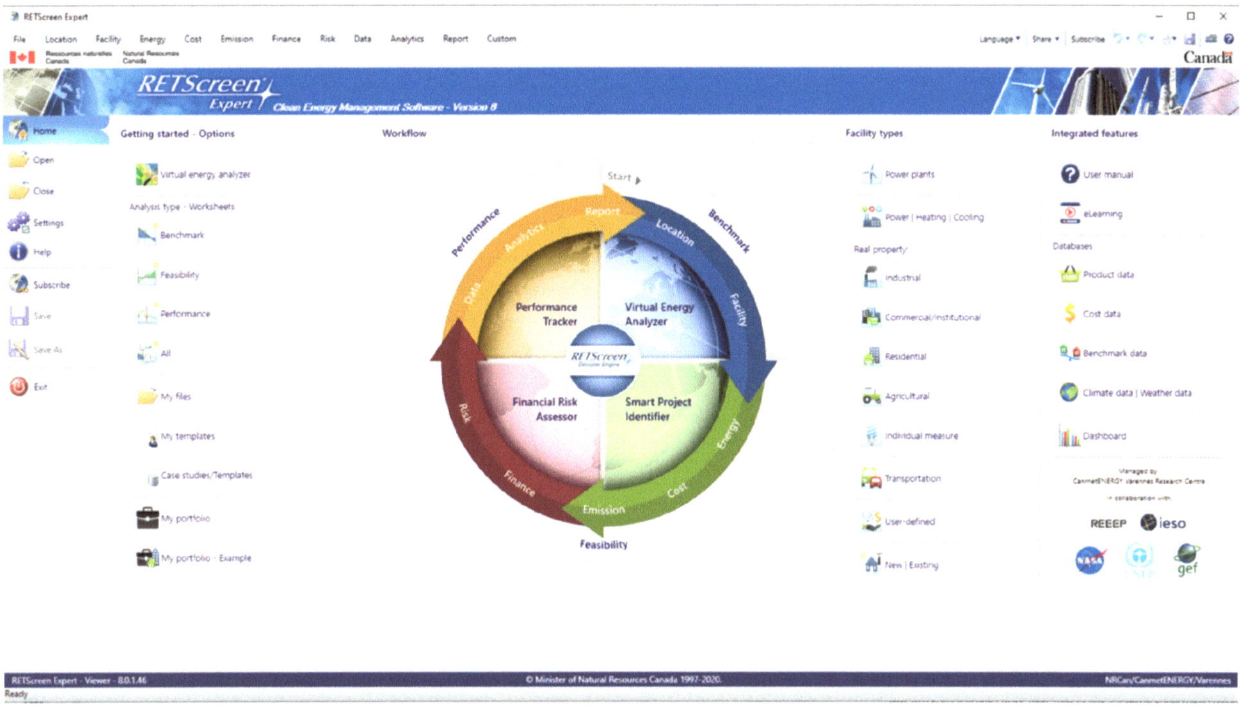


**Figure 3.5** Opening screen of RETScreen Expert.[10]

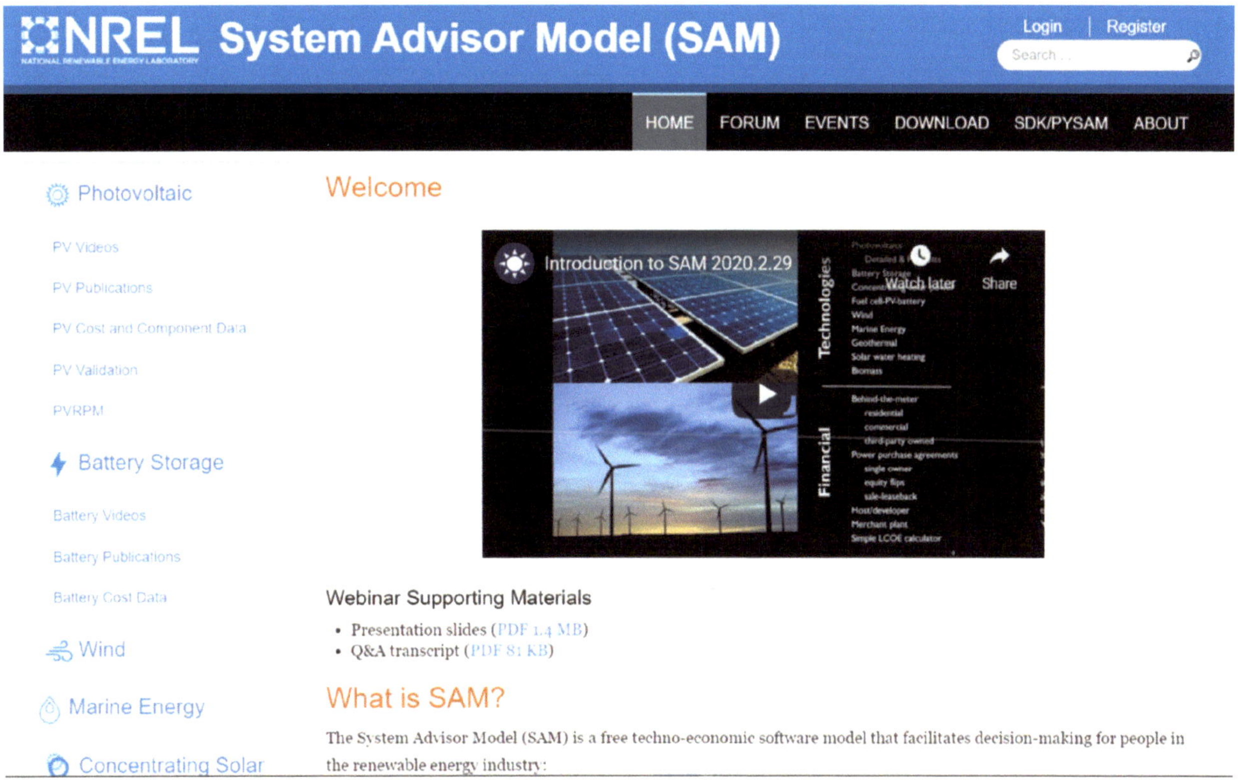


**Figure 3.6** SAM website landing page.[11]

of the PVWatts summary interface for a simulated project. Battery storage includes lead-acid, lithium-ion, and flow batteries. CSP can be modeled for parabolic trough, power tower, and linear Fresnel types of mirror applications. Marine energy can be modeled for either wave or tidal systems.

## Solar Insolation

PVWatts Calculator[14] is my favorite go-to for solar insolation. Although it is an aspect of SAM as a design tool for PV arrays which will estimate production and annual returns, PVWatts has its own website that has

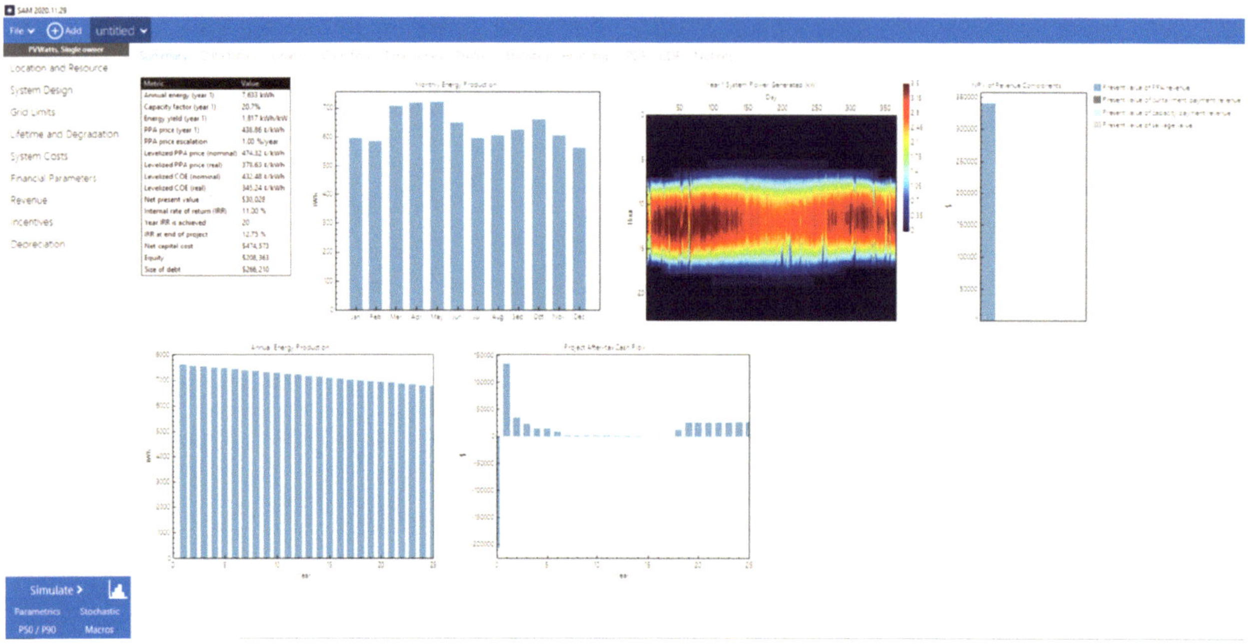


**Figure 3.7** SAM example of PVWatts, Single owner file interface simulation.[13]

excellent, easy-to-use parameters and help. Weather data is global, although as with any weather data set, there may be variations between your actual location and the nearest recording weather station.

To use PVWatts, put in the location and select the desired weather station. Because the exact amount of solar insolation available at a location depends not only on the climate but also the tilt angle and orientation of the collector, the genius of PVWatts is that it alters the solar insolation based on the parameters you enter for tilt angle and **azimuth** orientation. Most tables containing solar radiation data are for set angles and orientations, so this gives you accurate information based on your actual design. If you use PVWatts just for solar radiation numbers, then you don't need to fill out all the solar PV information, you can just leave the default settings and then ignore the other results columns.

If you are designing a solar PV system, then you'll either need to fill in the size of the system or use the "Draw Your System" tool to estimate the size of your system as shown in figure 3.8.[14] Note that the "Draw Your System" can only estimate the area of production and doesn't take into account obstacles, equipment, or access on the roof. To get accurate production results, you need to have sized your system prior to starting in PVWatts. The other system variables are the type of module (standard, premium, or thin film), how the array is mounted, system losses, and the tilt angle and azimuth of the collectors. If you update the retail electricity rate, then you'll have an accurate annual alternating current (AC) output value in the results. You can change up the tilt angles and orientation to test for best layout vs. production when working on a flat roof and looking at inter-row spacing vs. coverage. It is easy to quickly compare results.

The results are very clear, providing you with monthly and annual numbers for solar radiation, AC energy, and the value of the energy the array has produced, as shown in figure 3.9. These numbers are not guaranteed as weather patterns vary from year to year. You can print the results and you can download the data if you want to use it in a spreadsheet.

## Wind

Wind information is dependent on the hub height of the turbine. Wind can vary quite a bit over a relatively small regional area, so even if a map shows a particular type of resource, is it important to field verify the conditions over at least a one-year period. You may see resources that refer to the **IEC (International Electrotechnical Commission) Classes**. These are designated as Class 1, 2, 3, or 4 (sometimes Roman numerals are used). Class 1 winds average 10m/s (22.4mph) and have to support extreme wind gusts of 156mph (70m/s). Class 2 winds average 8.5m/s (19mph). Class 3 winds average 7.5m/s (15.7mph) and Class 4 winds average 6m/s (13.4mph). Class 1–3 winds are further broken down by turbulence, the amount the wind varies within a 10-minute period. Figure 3.10[15] shows USA states with installed wind capacity over 3.2GW, and total USA capacity, based on the wind class. The graph shows that most wind turbines are being designed and installed in Class 2 winds. Most of this capacity is utility scale.

Wind energy maps for USA are available at windexchange.energy.gov[16] with maps of the country and each state based on hub height. The maps are

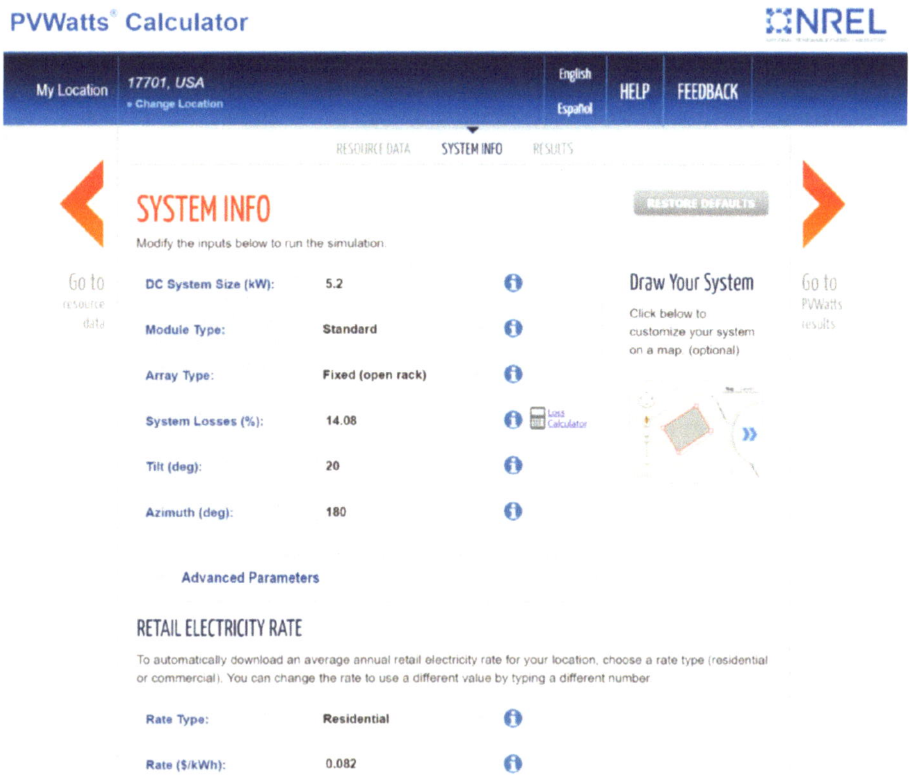


**Figure 3.8** PV settings for PVWatts.[14]

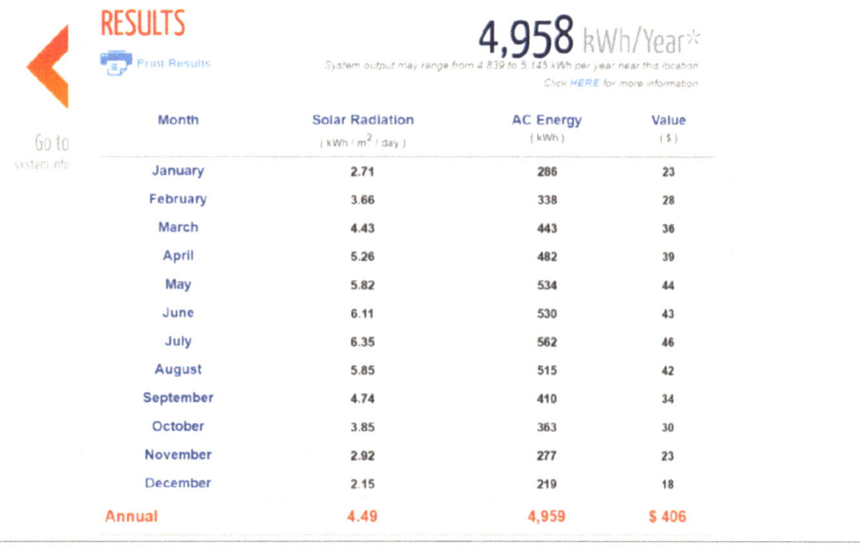


| Month | Solar Radiation (kWh / m² / day) | AC Energy (kWh) | Value ($) |
|---|---|---|---|
| January | 2.71 | 286 | 23 |
| February | 3.66 | 338 | 28 |
| March | 4.43 | 443 | 36 |
| April | 5.26 | 482 | 39 |
| May | 5.82 | 534 | 44 |
| June | 6.11 | 530 | 43 |
| July | 6.35 | 562 | 46 |
| August | 5.85 | 515 | 42 |
| September | 4.74 | 410 | 34 |
| October | 3.85 | 363 | 30 |
| November | 2.92 | 277 | 23 |
| December | 2.15 | 219 | 18 |
| Annual | 4.49 | 4,959 | $ 406 |

**Figure 3.9** PVWatts results.[14]

available for 30-meter (98′) residential (see figure 3.11[17]), 50-meter (164′) community, 80-meter (262′) land-based, 90-meter (295′) offshore, 100-meter (328′) wind speed, 110-meter (361′) potential and 140-meter (459′) potential. Not all maps are available for every state.

The Global Wind Atlas (GWA)[18] provides worldwide wind maps that are intended to be used during the exploratory and preliminary phases of wind assessment, prior to taking measurements on site. This provides a tool for governments and investors to understand wind resource potential. Governments are thus saved the need and costs for early-stage national wind mapping. It is a free, interactive map that provides global onshore coverage and offshore up to 200km (124 miles) from the shoreline. It has hub heights at 10m (32.8′), 50m (164′), 100m (328′), 150m (492′), and 200m (656′) above ground/sea level. There are video tutorials on the website. Users can create maps on the fly or download

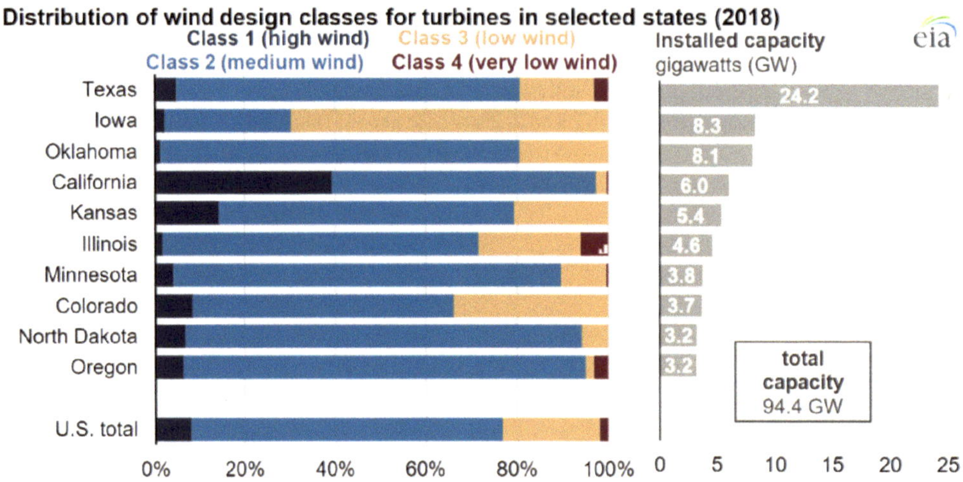


**Figure 3.10** Distribution of wind capacity generation in the United States based on IEC wind classes.[15]

high-resolution maps and data for use in GIS tools. Figure 3.12[19] shows the results of mean wind speed on land and offshore worldwide at 100m.

## Marine and Hydrokinetic Resource Data

As noted in the general section above, NREL has a section of marine and hydrokinetic (MHK Atlas)[20] resource data for the United States. These resources include interactive maps, as well as downloadable maps and data. It also includes an assessment of nonpowered dams, dams that exist but aren't making power, and tidal streams.

## Distributed Energy – Microgrid Design Tool

REopt Lite[21] is a web tool that assists designers in evaluating the economic viability of distributed PV, wind, battery storage, **combined heat and power (CHP)**, and thermal energy storage. It can also help to determine the best size for a system and estimate how long the system can sustain the critical load during a grid outage. The site has tutorial videos and clear instructions on what data the user needs to complete a financial or resilience study. The interface is easy to use and the results summarize strategies. Disclaimers include the importance of validating actual conditions at the site as well as making sure equipment at the size indicated is commercially available. The test case on the site for a hospital located in San Francisco, CA, USA, resulted in a recommended 3.47kW PV array with 969kW battery power and 7,043kWh battery capacity, which could save the hospital US$1,192,094 over 25 years. Figure 3.13 is an example of the type of interactive graph produced, showing the system performance over a day-long power outage.

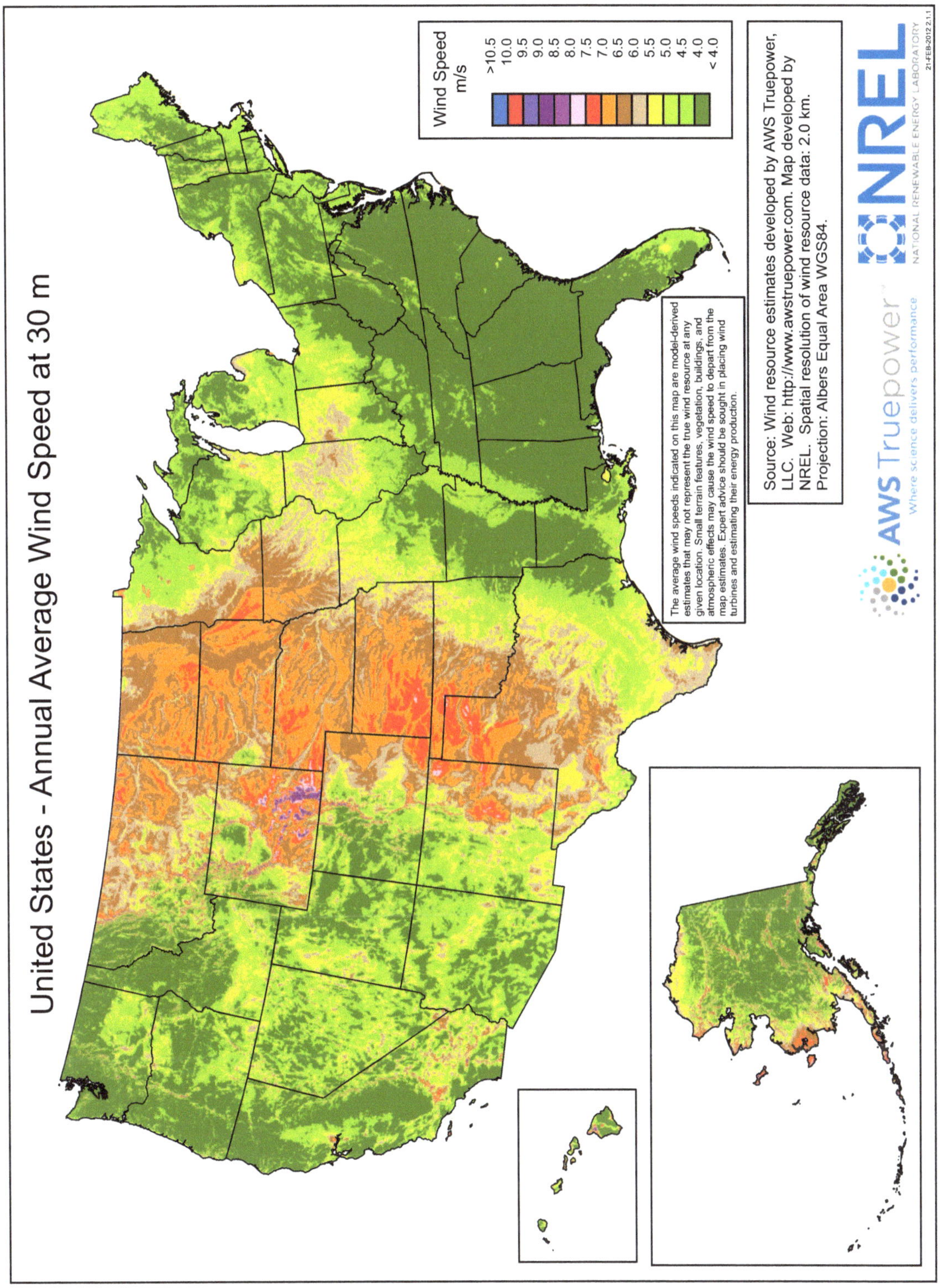


**Figure 3.11** Wind exchange example map for the USA at 30 meters hub height.[17]

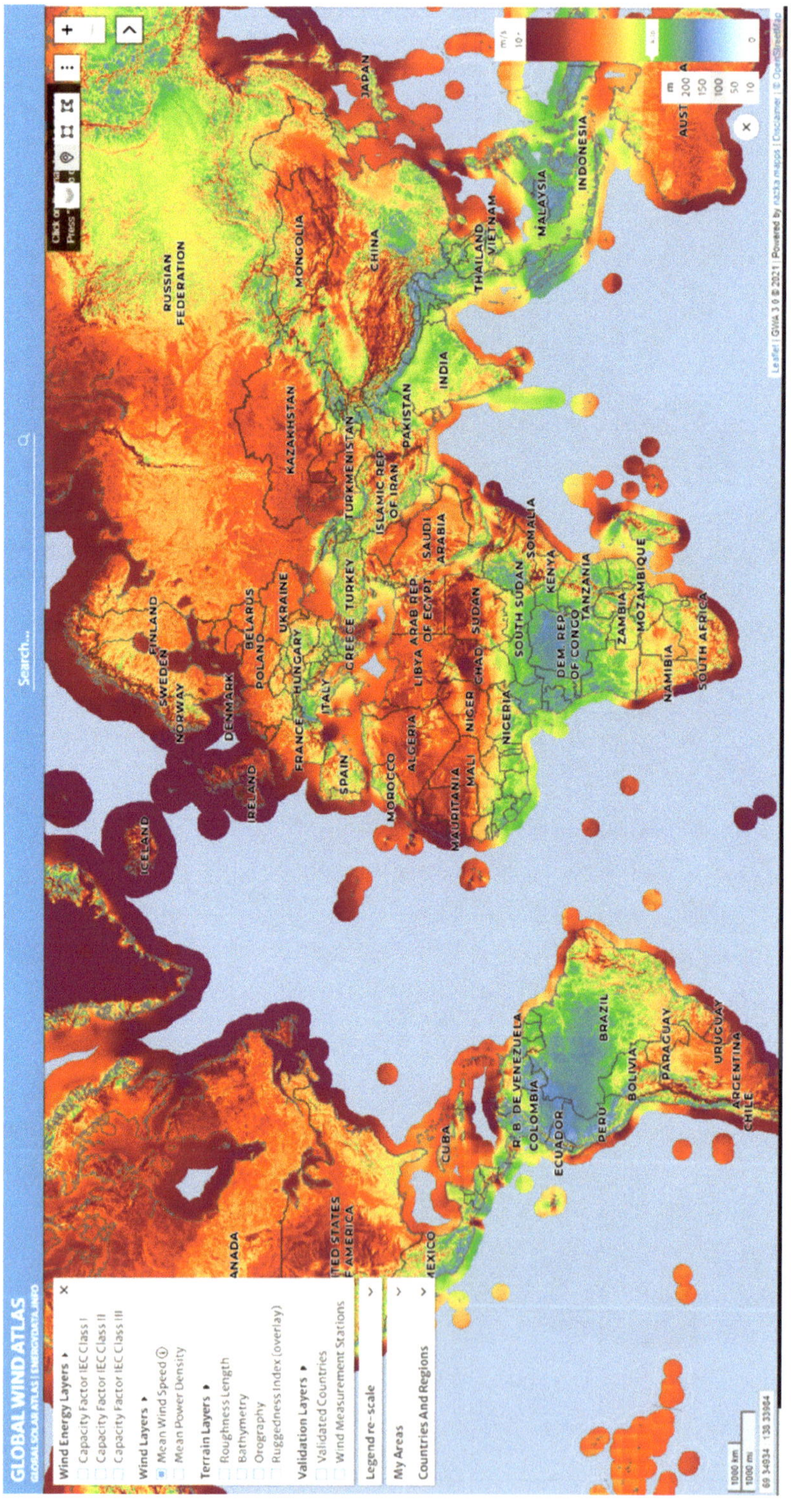


Figure 3.12 Image of GWA web interface showing mean wind speeds at 100m.[19]

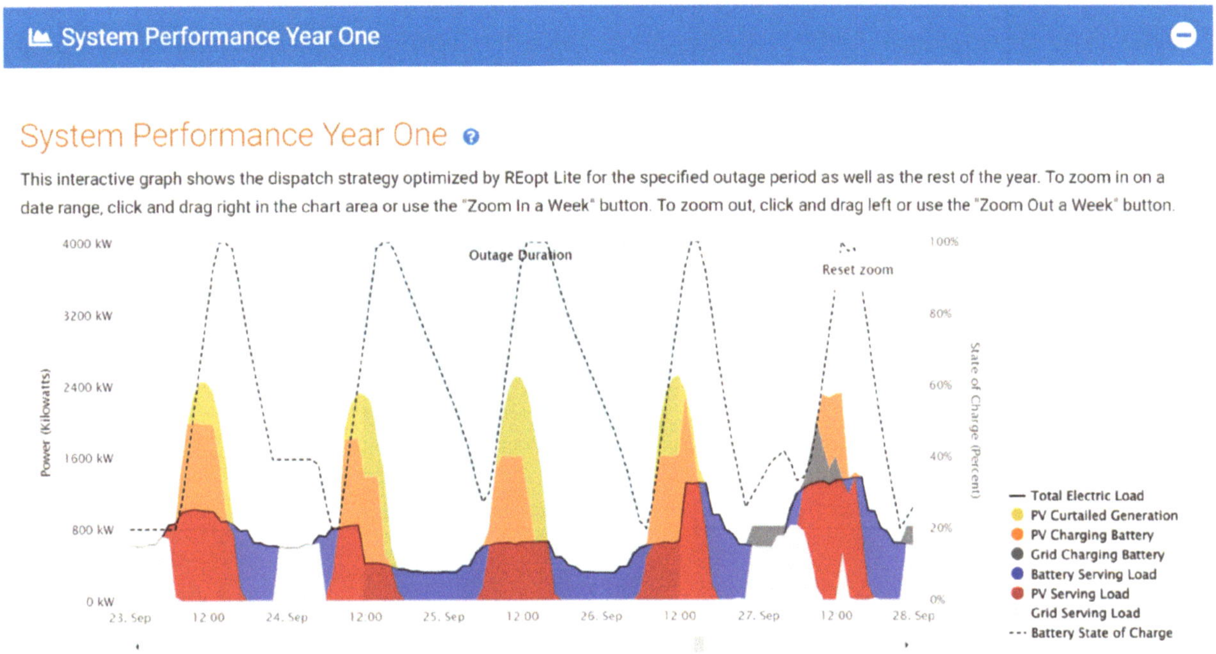


**Figure 3.13** Example of an REopt interactive graph showing projected performance of PV and battery storage system during a power outage.[22]

## Resources


1 Climate.OneBuilding (https://climate.onebuilding.org/default.html).

2 Midwestern Regional Climate Center. (n.d.). *cli-MATE: MRCC Application Tools Environment.* https://mrcc.purdue.edu/CLIMATE/

3 The data used in figures 3.1, 3.2, and 3.3 is from NOAA Cooperative Network data obtained from the Midwestern Regional Climate Center, cli-MATE (MRCC Application Tools Environment, on / CLIMATE/).

4 Degree Days (www.degreedays.net/).

5 Iowa Environmental Mesonet of Iowa State University (https://mesonet.agron.iastate.edu/sites/locate.php).

6 Geospatial Data Science Data and Tools (www.nrel.gov/gis/data-tools.html).

7 Sengupta, M., Xie, Y., Lopez, A., Habte, A., Maclaurin, G., & Shelby, J. (2018, June). The National Solar Radiation Data Base (NSRDB). *Renewable and Sustainable Energy Reviews*, 89, 51–60. https://doi.org/10.1016/j.rser.2018.03.003

8 Geospatial and Earth observation data (https://data.europa.eu/en/highlights/geospatial-and-earth-observation-data).

9 RETScreen Innovation Lab. (2021, September 21). *RETScreen* (Version 8.1) [Computer software]. Government of Canada. www.nrcan.gc.ca/maps-tools-and-publications/tools/modelling-tools/retscreen/7465

10 RETScreen Innovation Lab. (2021, September 21). *RETScreen* (Version 8.1) [Computer software]. Government of Canada. www.nrcan.gc.ca/maps-tools-and-publications/tools/modelling-tools/retscreen/7465

11 System Advisor Model (SAM) (https://sam.nrel.gov/).

12 Blair, N., Dobos, A., Freeman, J., Neises, T., Wagner, M., Ferguson, T., Gilman, P., & Janzou, S. (2014, February). *System Advisor Model, SAM 2014.1.14: General Description* (Tech. Rep. NREL/TP-6A20-61019). National Renewable Energy Laboratory. www.nrel.gov/docs/fy14osti/61019.pdf

13 National Renewable Energy Laboratory. (2020). *System advisor model (SAM)* (Version 2020.11.29) [Computer software]. https://sam.nrel.gov/

14 PVWatts® Calculator (https://pvwatts.nrel.gov/).

15 U.S. Energy Information Administration. (2019, September 27). Most wind capacity in the United States is designed for a medium wind speed environment. *Today in Energy.* www.eia.gov/todayinenergy/detail.php?id=41474

16 Wind Energy Technologies Office. (n.d.). *Wind energy maps and data.* U.S. Department of Energy. https://windexchange.energy.gov/maps-data?category=residential

17 Wind Energy Technologies Office. (n.d.). *Wind energy maps and data.* U.S. Department of Energy. https://windexchange.energy.gov/maps-data?category=residential

18 Global Wind Atlas (https://globalwindatlas.info/).

19 Map obtained from the Global Wind Atlas 3.0, a free, web-based application developed, owned, and operated by the Technical University of

Denmark (DTU). The Global Wind Atlas 3.0 is released in partnership with the World Bank Group, utilizing data provided by Vortex, using funding provided by the Energy Sector Management Assistance Program (ESMAP). For additional information: https://globalwindatlas.info/

**20** NREL has a section of marine and hydrokinetic (MHK Atlas) data available at: Marine and Hydrokinetic Resource Data, Tools, and Maps (www.nrel.gov/gis/maps-marine.html).

**21** REopt Lite™ (https://reopt.nrel.gov/tool).

**22** National Renewable Energy Laboratory. (n.d.). *REopt: Renewable energy integration & optimization.* https://reopt.nrel.gov/

## Study Questions

1. Go to the Degree Days website (www.degreedays.net/) and download average HDD and CDD for a location that you are very familiar with and one that you are unfamiliar with. Compare the monthly numbers of the two locations and notice how they are similar and different.
2. Go to the PVWatts website (https://pvwatts.nrel.gov/) and select a location. Using that location, see how much the amount of solar insolation changes when you change just one variable. Change the tilt angle. Change the orientation. How do these single changes alter the amount of solar insolation? Which variable seems to have the most impact on the available solar insolation at this location?

CHAPTER 4

# Basic Concepts about Electricity

**Key Terms**

Energy: power expended over a period of time, available in different forms, electrical power is measured in Wh or kWh.

Parallel circuit: has more than one path for the flow of electrons, adds amps while volts stay the same, positive terminals are connected to positive terminals while negative are connected to negative.

Power: how fast energy is produced or used, electricity measured in W or kW.

Series circuit: has one path for the flow of electrons, adds volts while amps stay the same, the positive terminal is connected to the negative terminal.

## Concepts for Success

Energy is available in different forms. Power is how fast energy is produced or used. Electrical equipment and systems have power ratings and are measured in **watts (W)** or **kilowatts (kW)** while energy is measured in **watt hours (Wh)** or **kilowatt hours (kWh)**. You have to memorize the symbols, meanings, and wiring rules because these items show up everywhere in power systems and manufacturer data sheets, not to mention your own power bill! Table 4.1 lists terms you need to know.

Table 4.1 Electrical terms.

| |
|---|
| Power measured in watts (W) |
| Energy measured in watt hours (Wh) |
| 1000 W = 1 kilowatt (kW) |
| 1000 Wh = 1 kilowatt hour (kWh) |
| Volts (V) are force, electrical pressure |
| Amperage or amps (A), also I, is electrical current (volume) |
| Power = Voltage × current i.e. P = V × A |
| British thermal unit (Btu) |

## Energy and Power

Power is how fast energy is used or produced and is measured in watts. The symbol for watt is W. The formula for power (P) is P = energy/t, where t is time; 1 watt does work at the rate of 1 joule in 1 second; 1000 watts = 1 kilowatt and the symbol for kilowatt is kW. If you look at a piece of electrical equipment, your refrigerator or toaster for example, it will have a plate on it stating its watt rating. You'll also see numbers for volts and amps. A **volt (V)** is a unit of force, electrical pressure, that causes electrons to flow through a wire. Current is measured in amperes (or amps) and indicates how many (volume) of) electrons are flowing through a wire. The symbols for current are either **A** or **I**.

Current is the continuous movement of electric charge and can be easily imagined as being like the flow of water through a pipe. Voltage is the force that is causing the current to flow. You can imagine it as being like water pressure forcing water to flow in a pipe. Wires work similarly to a pipe: the thicker the wire, the more energy can flow through it. A small diameter hose will only let a small amount of water through it. You need a big hose with lots of water pressure to put out a fire! As water travels through a pipe, friction is caused. Friction in electricity is called resistance and is measured in ohms (Ω). Resistance is always measured between or across two points, just as voltage is. If you are ever using Ohm's law, you'll be using the other symbol for voltage (E), which states E = IR, where I is current and R is resistance.

Energy is power multiplied by the time you used it for, so energy is measured in watt hours. The symbol is Wh; 1000 watt hours = 1 kilowatt hour and the symbol for kilowatt hour is kWh. If you have a utility bill, it will charge you for how many kilowatt hours you used in the month.

DOI: 10.1201/9781003297819-5

So, here's how this all works together. Let's say your toaster's power ratings are 850W, 120V, 7.1A. You like the highest toasting setting so your toast gets finished as quickly as possible, so it uses 850W to heat up the elements. If it takes 2 minutes to get your perfect crunch, then 850W × (2 minutes / 60 minutes per hour) = 28.33Wh of energy each time you push down the lever. If you use your toaster 40 times a month, that's 1,133.2Wh. Remember that the utility company is charging you in kWh, so converting: 1,133.2Wh / 1000kW per W = 1.13kWh used every month to get that perfect crispy toast.

Batteries are labeled with volts and amp-hours (Ah). This is to give the user an idea of how much current will be available over time. You also need to know the C-rating, which tells you the current charge and discharge rate. For example, a battery with 48Ah and a C-rating of 0.5C will provide a current of 24 amps for 2 hours.

## Series and Parallel Circuits

Electrical equipment works because it is receiving energy through a circuit. Circuits can be set up either in series or parallel. You may have had a science class where you built a circuit; if so, you are a step ahead. If not, don't stress! This will break it down for you and show you step by step how it applies to renewable energy systems.

**Series circuits**: only one path for the flow of electrons, adds volts while amps stay the same. Figure 4.1 shows a series circuit. If you trace your finger around from the positive (+) battery terminal to the switch to each of the lightbulbs and back to the negative (–) battery terminal there is only one path for electrons to follow.

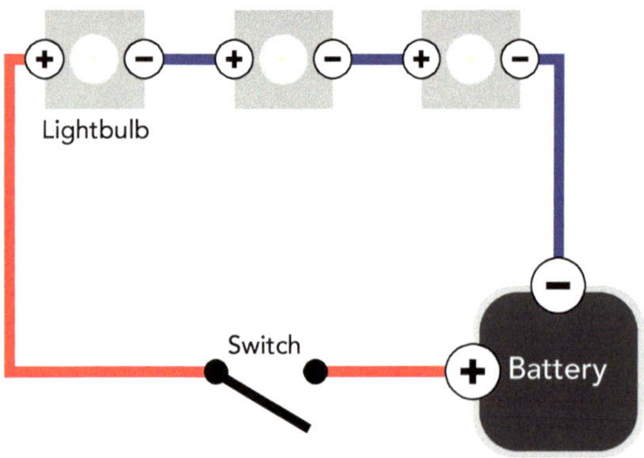


**Figure 4.1** Series circuit showing three lightbulbs (load) connected to a battery with a switch.

In this series circuit there is a battery providing the power. There is a switch to turn the circuit on and off. Note that the circuit is off when the circuit is open – this means that the electrons can't flow around the circuit, so there is no light being generated from the lights. When the switch is closed, the electrons can move around the circuit and the bulbs light up. In figure 4.1 there are three lights in a row and they are wired so that the positive pole from one light is connected to the negative pole of the next light. The lights are called loads because they are using the battery's energy to light up. If the battery is 6V, 11Ah then the voltage will be divided up between the three lights, each light using 2V (this is because of the rule that series adds volts). The Ah will remain the same. Note that if any of the lights burns out, that will open the circuit and stop the flow of electricity, so none of the lights will light up, even if the switch is on.

**Parallel circuits**: have multiple paths for electrons, adds amps while volts stay the same. Figure 4.2 shows a parallel circuit. If you trace your finger around from the positive (+) battery terminal to the switch to a lightbulb then to the negative (–) battery terminal, you'll see that you can make three separate loops around this circuit: once around the circuit through each one of the lightbulbs (loads).

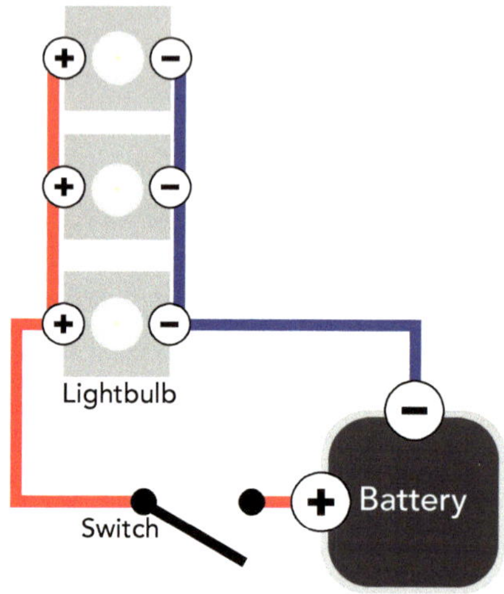


**Figure 4.2** Parallel circuit showing three lightbulbs (load) connected to a battery with a switch.

The lights are able to work independently of each other, as long as the switch is closed and the circuit is on. If one bulb burns out, the others will still light up because there are other paths for the electrons to flow through. If this is a 6V, 11Ah battery, each bulb will receive the full 6V. However, they will be splitting the current discharging from the battery (because the rule is that parallel adds amps), so they'll use up the battery's energy faster than the series circuit.

In renewable systems, circuits are often a combination of both series and parallel. Photovoltaic (PV) modules and batteries are hooked up in series to get to the desired voltage and then the groups of series circuits are connected in parallel.

## DC vs. AC Power

There are two different power profiles that describe how current flows in a circuit. They were developed in the late 1880s by Thomas Edison – direct current (DC) – and Nikola Tesla – alternating current (AC). Direct current flows in one direction and the voltage remains the same over time. Alternating current has a flow that reverses direction periodically and the voltage takes on a sine wave appearance over time. The higher the frequency, the shorter the cycle of the sine wave. Back when Edison and Tesla were battling over DC and AC, called The Current War, DC voltage couldn't be generated and transformed at high enough voltages to keep it from having high losses when sent over wires at any distance, so AC won out. There has never been a worldwide standard for power or for the voltage and frequency of AC power. That's why, when you travel to different countries, you need to use an adaptor so your electrical items will charge and work properly.

In the USA the typical power delivered to buildings is AC 120V/240V at 60Hz (hertz). Other countries use different AC voltages at 50Hz. Hertz is the unit of frequency and its unit is one cycle per second. The higher the hertz number is, the shorter the sine wave is. Most countries use AC power, although there are many items such as rechargeable batteries, LED lighting, computers, and smart phones that all use DC power. If you have a piece of electronic equipment that has a large plug adaptor (or a box near the plug) then it is changing (transforming) the AC power from the outlet into DC power.

A lot of the equipment in your building runs on AC power and there are a few things that use DC power. Renewable energy typically generates DC (the exception is hydropower) so that means that systems will need to have an inverter to change the power from DC to AC. For more on basic system layout, see chapter 2 (Basic Concepts of Renewable Systems).

## Voltage Design for On-Grid and Off-Grid Applications

Grid-tied PV systems are wired to high voltage, between 75 and 600V. The maximum limit according to the USA National Electric Code is 600V, due to the change in personal protective equipment needed to work on anything conducting more than 600V.

Off-grid systems are nominally 12V, 24V, or 48V and both the array and the battery is wired to the nominal voltage for the system. The voltage of a PV module is shown in the manufacturer's data which is printed on a label applied to the back of the module and listed on the module cut sheet.

Note that the diagrams in this chapter do not show all equipment or grounding. See chapter 2 on renewable energy systems to see overall system components. The purpose of these diagrams is to assist you in understanding how to apply series and parallel wiring to a PV system. For practice wiring, see Appendix D.

Figure 4.3 shows a high voltage array connected to a grid-tied inverter. Each of the 6 PV modules is rated at 395W and 38.5V. The modules are connected in series (positive to negative) to each other, so the volts add. The total DC voltage being fed to the inverter is 38.5V × 6 modules = 231V. The system size is 395W × 6 modules = 2370W or 2.37kW.

Figure 4.4 shows a nominal 12V, off-grid system. It has 6×100W, 12V modules. Notice that they are all wired in parallel (positive to positive and negative to negative) because they are already at the 12V needed for the system. The modules are connected in parallel to the controller; note that the connecting wires come from the opposite ends of the array to balance the wire resistance across the system. The system size is 100W × 6 modules = 600W. There are 4 batteries which are 6V, 235Ah. Because they are 6V, two must be wired in series to get to the 12V needed for the system. Notice that you end up with two sets of batteries wired in series. These two sets are then wired in parallel to each other and to the controller. The series battery voltage is 2 × 6V = 12V (series adds volts) and the amp hours to the controller (parallel adds amps) are 2 × 235Ah = 470Ah.

Figure 4.5 shows a nominal 24V, off-grid system. It has 12×100W, 12V modules. In order to get to a 24V system, two modules must be wired in series: 2 × 12V = 24V. Note that this system has to have an even number of modules so they will all be wired in groups of two in series. All the remaining wiring will be in parallel. When wiring is not connected, it is shown with an arc over the other wire – you can see this where the two vertical positive terminals second from the right are wired together vertically. The rating of this system is 100W × 12 modules = 1200W or 1.2kW. On the battery side, there are 4×12V, 215Ah batteries. To get to 24V, two have to be wired in series. There are two groups of two batteries in series. Then the groups of batteries are wired together in parallel. At the controller, the voltage is 2 batteries × 12V = 24V (which stays the

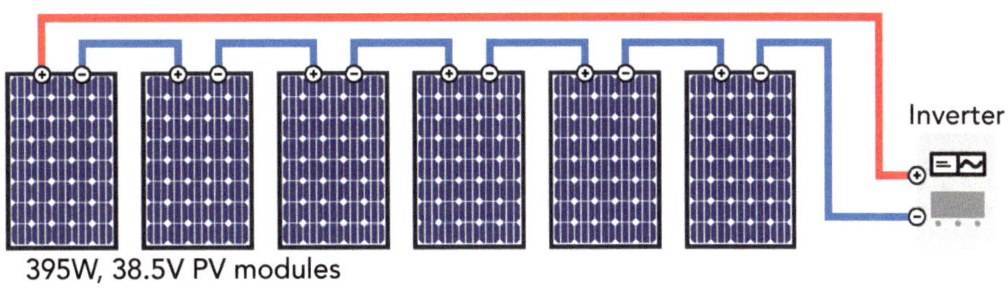


**Figure 4.3** High voltage array with 6 PV modules wired in series and connected to an inverter.

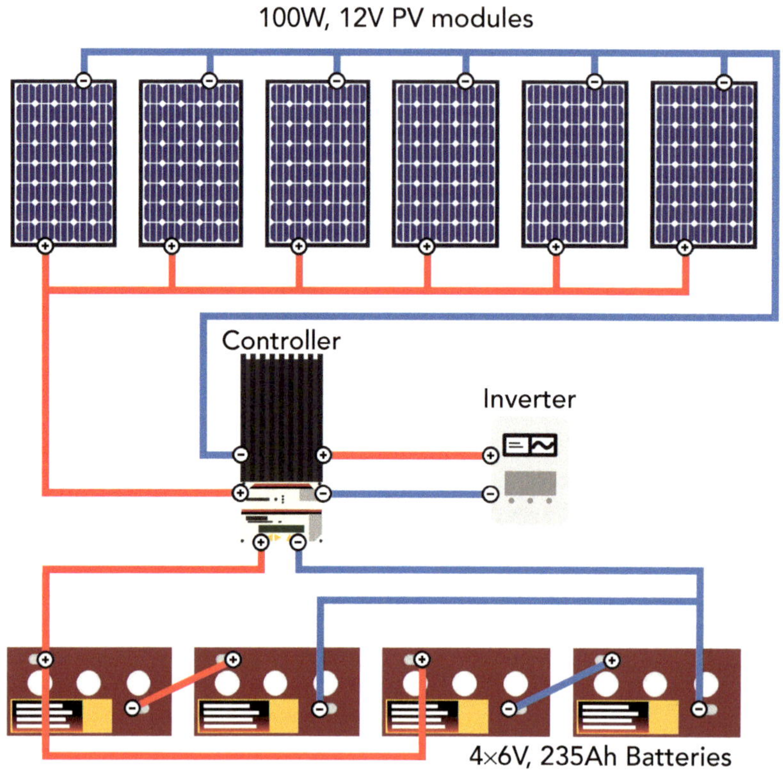


**Figure 4.4** Nominal 12V array with 6 PV modules and 4 batteries wired to a controller and connected to an inverter.

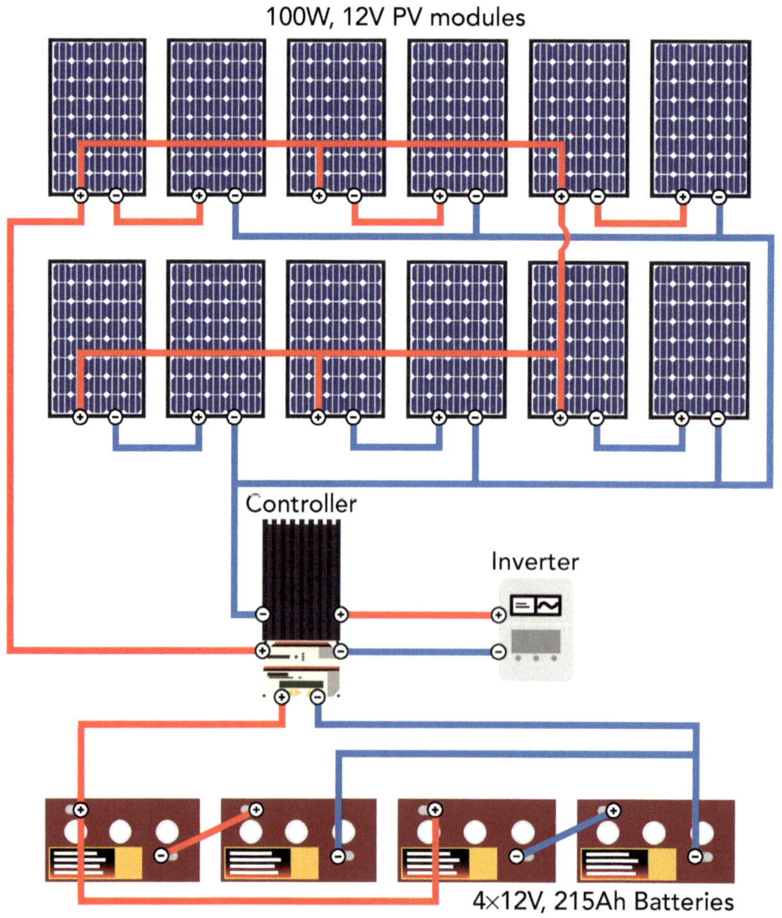


**Figure 4.5** Nominal 24V array with 12 PV modules and 4 batteries wired to a controller and connected to an inverter.

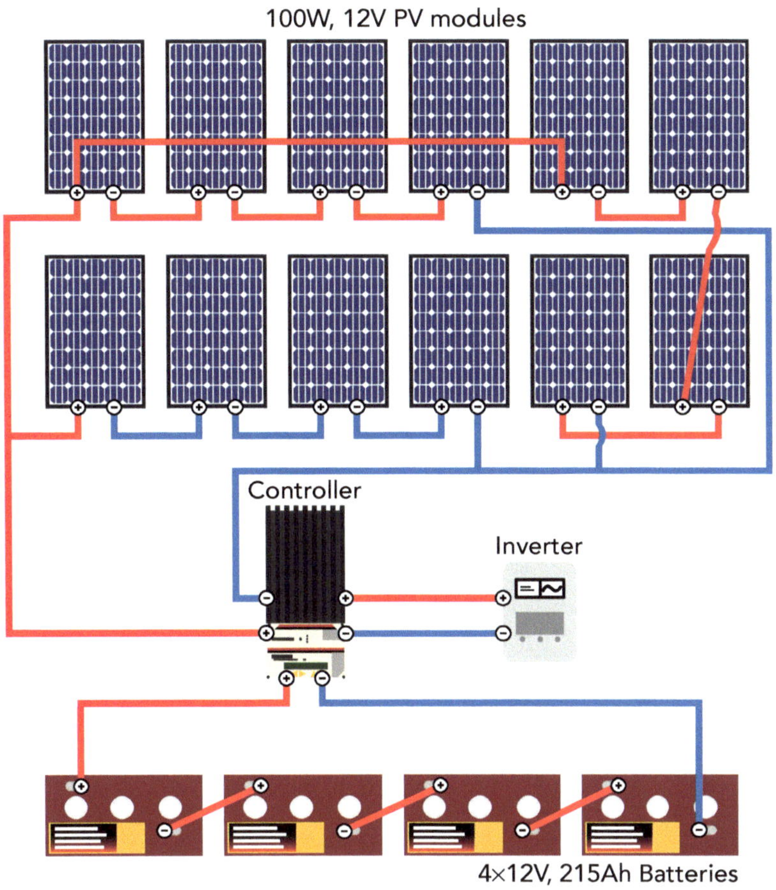


**Figure 4.6** Nominal 48V array with 12 PV modules and 4 batteries wired to a controller and connected to an inverter.

same for the parallel wiring), and 2 groups of batteries × 215Ah = 330Ah (because amps add in parallel).

Figure 4.6 shows a nominal 48V, off-grid system. It has 12×100W modules. In order to get to a 48V system, 4 modules must be wired in series: 4 × 12V = 48V. The number of modules used in this array would always have to be in groups of four because they are 12V, so if more were desired, there would have to be 16 modules in total. No matter how these 12 modules are wired, they will always be a 1200W (1.2kW) system. On the battery side there are 4×12V, 215Ah batteries. All of them have to be wired in series to get to 48V, so the controller would be processing 48V, 215Ah from the batteries (in series volts add and amps stay the same).

## Safety

Renewable energy systems can be dangerous. They are outside in all kinds of weather and sun. They produce and possibly store a large amount of energy. It isn't usually the equipment itself that causes a problem but, instead, fires and the like are caused by wiring issues or improper battery maintenance or storage. If equipment hasn't been grounded and wired together with appropriately sized and insulated wire and connectors that can withstand exterior conditions, there can be issues with arcing and grounding.

The simple diagrams throughout this book do not show all the fuses, lightning arrestors, grounding wires, and many other safety features or common elements such as junction boxes that are necessary in these systems. These items are mentioned in this book. However, for specifics on types and sizing of wiring and connectors and such, you will need to look to other resources that cover these types of specific electrical topics in detail.

It is important to have engineering documents that properly specify all of the required elements of any renewable energy system. Many companies that have been in the business of renewable energy for years have very detailed information on their websites and have customer service people who are able to answer questions.

It is vital that all the parts of the renewable energy system are located and connected properly. It is equally important to have electricians who are appropriately trained and certified installing them (at least for the electrical parts). Many countries have codes with requirements for minimum safety standards addressing renewable systems and electrical concerns. It is important to make sure that your system meets safety requirements. If you happen to be working somewhere without uniform building standards, it is a good idea to follow established standards from other countries in order to maintain safe conditions for everyone. These standards help ensure that there are not radical system failures which can result in loss of life by either building users or firefighters, or catastrophic damage to the building.

## Study Questions

1. If you have a piece of equipment that has a 550W rating and runs 7 hours a day, how much energy does it use in a day?
2. What is the V, A, and P of three PV modules wired in parallel, each rated at 320W, 32.7V, 9.79A?
3. What is the V and Ah of 4 batteries wired in series, each rated at 12V and 215Ah?
4. What are the nominal voltages used for off-grid systems?

## Study Answers

1. 550W × 7 hours per day = 3,850Wh or 3.85kWh
2. Parallel adds amps, volts stay the same: 32.7V; 3 modules × 9.79A = 29.37A; 3 modules × 320W = 960W.
3. Series adds volts, amps stay the same: 4 batteries × 12V = 48V; 215Ah.
4. 12V, 24V, 48V

CHAPTER 5

# Solar Insolation

**Key Terms**

Altitude: the vertical height of the sun in the sky measured above the horizon.

Azimuth: the angled location of the sun based on the horizon/cardinal directions.

Equinox: when the plane of the earth's equator passes through the geometric center of the sun (the center of the sun's annual path through the sky) where there are equal hours of sunlight and darkness, generally around March 21 and September 21.

Latitude: imaginary lines that run east–west around the earth, where 0° is at the equator and 90° is at the poles.

Magnetic declination: the variation, measured in a horizontal plane, between magnetic north and true north.

Solstice: when the sun reaches maximum and minimum declination (the extreme edges of the sun's annual path through the sky). Summer solstice occurs at the maximum declination (the greatest number of daylit hours), around June 21 in the northern hemisphere or December 21 in the southern hemisphere. Winter solstice occurs at the minimum declination (the least number of daylit hours), around December 21 in the northern hemisphere or June 21 in the southern hemisphere.

## Concepts for Success

The items that will impact your system's output are potential solar insolation, shading specific to the site, orientation and tilt angle of the collector, and the efficiency of the collector.

The earth's magnetic poles are tilted at an angle and the earth's orbit is elliptical, and not quite centered, around the sun. This causes the change of seasons on our planet. The sun's path across the sky at any location is predictable. The height of the sun measured from the horizon is its **altitude**, while the location of the sun related to a flat circular plane is its azimuth.

There are two main types of graphs that depict the path of the sun. One has a circle representing the horizon with an overlay of the sun's path and the other is an elevation drawing of the sun's path. Both of these graphs are helpful for determining the exact location of the sun at specific times during the year.

We use **latitude** (imaginary lines circumventing the globe east–west) and longitude (imaginary lines circumventing the globe north–south) to locate places on the earth. The equator is the 0° latitude line with the northern hemisphere to the north and the southern hemisphere to the south. The sun appears to be in the southern sky in the northern hemisphere and the northern sky in the southern hemisphere.

True north is always the same but magnetic north changes dramatically across the planet. Correcting for the difference is called **magnetic declination**. This is important because you want your collectors to face south in the northern hemisphere and north in the southern hemisphere. You want to be sure that when you are physically at your site you are actually facing the right direction.

Solar insolation (also called radiation) is the amount of energy available from the sun and is measured in $kWh/m^2$. Solar insolation varies by location because of weather patterns and the number of available daylit hours.

The tilt angle and the orientation of your collector has a big impact on the available solar insolation.

Shading also impacts the amount of solar insolation that will hit your collector. It is best to have an open solar window for at least 3 hours on either side of solar noon (when the sun is directly overhead).

DOI: 10.1201/9781003297819-6

## Earth–Sun Relationship

The earth rotates around the sun in a predictable orbit. As shown in figure 5.1, our planet's magnetic poles are at a slight angle which causes either the northern or southern hemisphere of the earth to be closer to the sun's rays at any given point in the orbit. That extra exposure creates the seasons because half the globe warms up while the other half cools off a bit. Hence, when it is summer in the northern hemisphere, it is winter in the southern hemisphere and vice versa. Note that the earth's orbit is an ellipse which isn't quite centered on the sun, and the earth's north–south axis is tilted at 23.4°. The warming and cooling of seasons lag behind the changing of the sun's location, so seasonal temperatures are typically not symmetrical at the **equinoxes**, even though the sun is in the same position.

In modern times, humans have partitioned the globe with bisecting lines called latitude and longitude. Latitude lines run east–west and longitude lines run north–south. The equator is at 0° latitude while 90° is at the poles and it is recorded as either north or south of the equator. In northern latitudes the sun appears to be located more in the southern sky, while it is just the opposite in southern latitudes where the sun appears to be in the northern sky. This is important because, for **passive design** strategies, the building needs to be oriented with a major face to the sun.

We tend to talk about the sun rising and setting, although really it is the earth turning on its axis, one full turn each day and one full orbit each year. If we track the sun's path across the sky from any particular latitude, over the course of a year the sun makes a predictable journey. The outer boundaries of the path are the summer and winter **solstices**: the days with the longest and shortest number of hours of sunlight –shown at the outer bounding lines for June and December in figure 5.2. When the sun gets to a solstice, it "changes direction" and appears to travel back across the sky. Exactly in the middle between the solstices are the spring and autumn equinoxes, the curved line labeled September 21, when there are exactly the same number of sunlit and dark hours in the day. Armed with this information you are ready to impress all the druids and priests from ancient religions!

The height of the sun in the sky is called altitude (easy to remember because we know altitude is a height measurement) and the sun's location along the horizon is called azimuth. This is shown in figure 5.3. Solar noon is when the sun is directly overhead, at the highest altitude for that day. Azimuth can be measured as degrees to the east and west of south

**Figure 5.1** Earth–Sun relationship.

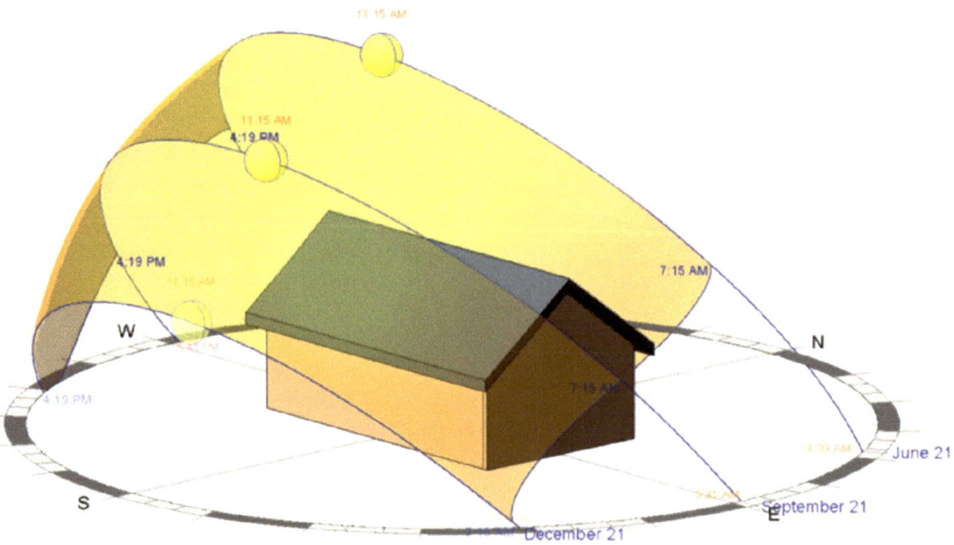


**Figure 5.2** Annual path of the sun through the sky, 41°N Latitude.

or north, or on a scale of 360° starting from north as 0°. Azimuth can be applied to your collector as well. If your collector is fixed, it should be within 15° of south (northern hemisphere) or north (southern hemisphere) for optimal collection. The sun moves across the sky at 15° per hour, so facing your collector toward solar noon maximizes your ability to catch the solar radiation.

If you stood in a single location for a year and recorded the sun's path across the sky in an elevation drawing it would look like figure 5.4. Each of the months are shown in blue, while the times of day are in red. These are plotted against a grid of the azimuth (horizontal axis) and altitude (vertical axis).

You'll notice that the sun is always highest in the sky at solar noon. In this diagram, north is 0° and 360°, so south is 180°. The location of the sun at solar noon is at 180° and creates a vertical line. This chart is for the northern hemisphere, so December 21 has the shortest curve (least number of hours of sun). On the longest day, the June 21 solstice, the sun rises at 55° in the



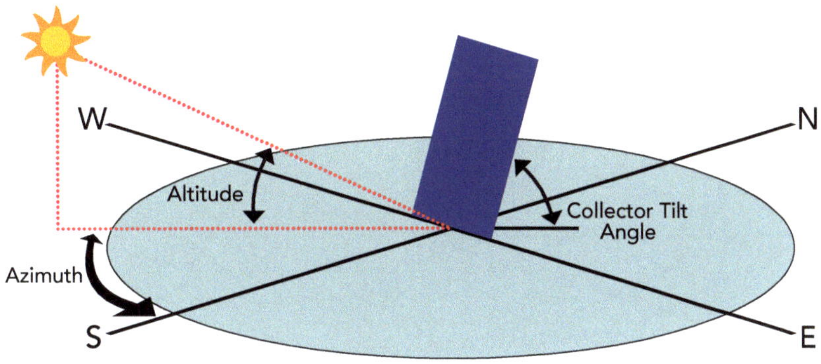


**Figure 5.3** Solar terminology.

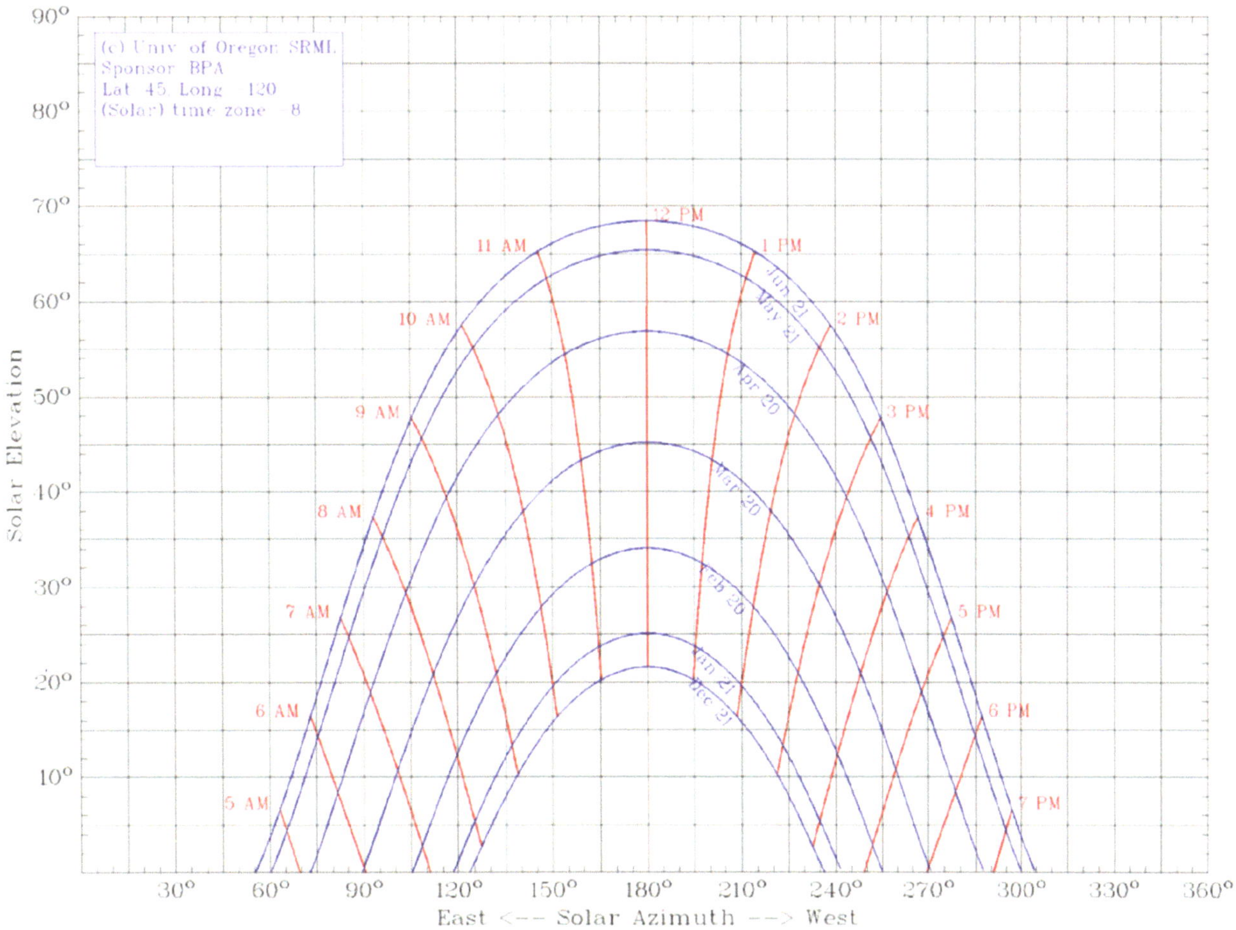


**Figure 5.4** Annual sun path chart at 45° N latitude. Azimuth is the horizontal axis with north on the two edges and south (180°) in the middle. East is to the left of south. Altitude is the vertical axis with the horizon at the bottom and 90° at the top.[1]

north-east and sets at 315° in the north-west. The sun is low in the sky in December and it is rising in the south-east and setting in the south-west.

## Latitude and Sun's Path

Your project's latitude is the determining factor for where the sun will be located in the sky. How do you understand the information you may see in a solar path diagram or on a 3D computer display?

Your location is always at the center and you must have the correct location, which includes latitude and orientation for your building (or system). If using a solar path diagram, such as shown in figure 5.5, pick a latitude close to yours. The ground plane is presumed to be flat and the horizon is the bounding circle. Figure 5.5

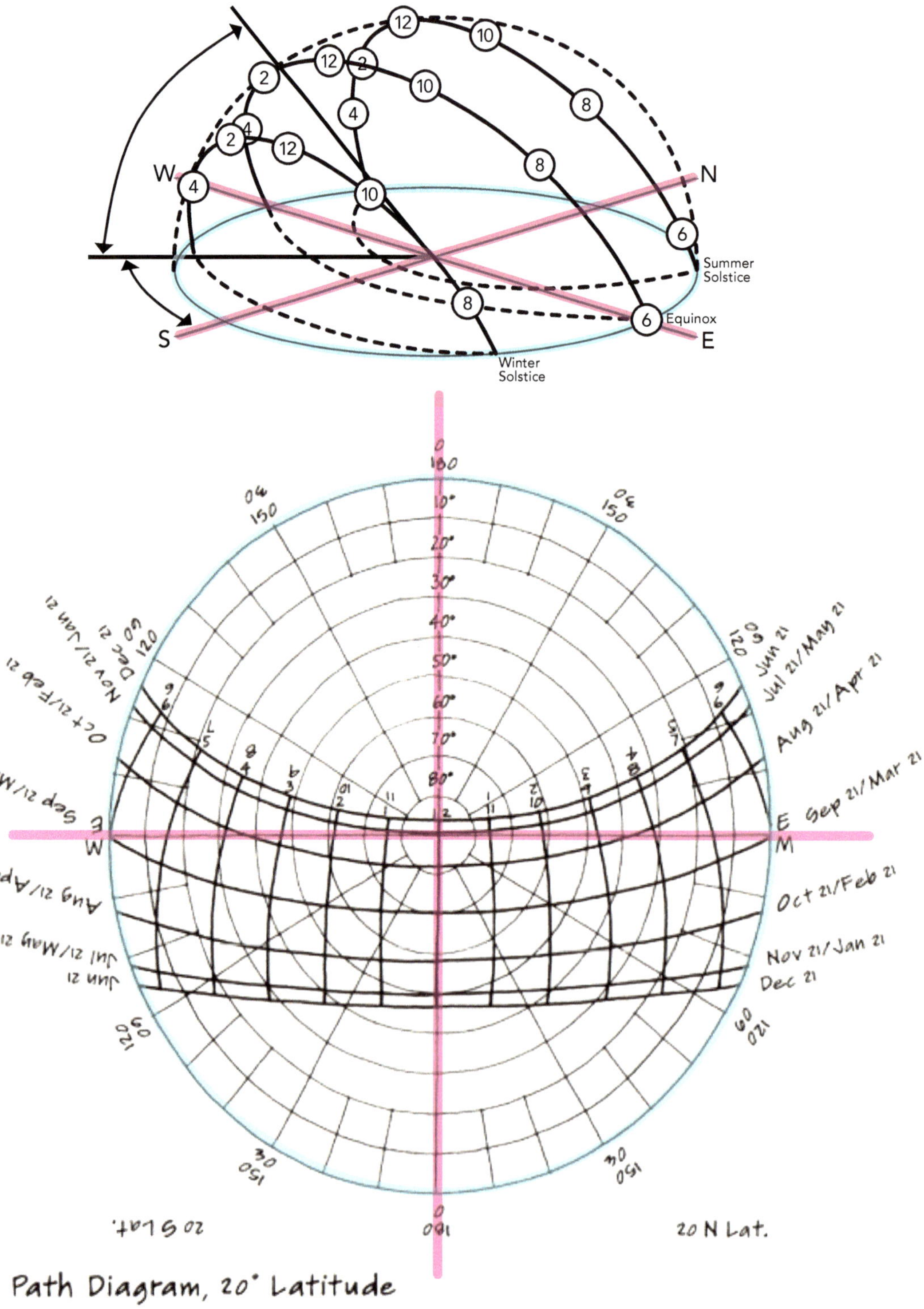


**Figure 5.5** Sun path diagram with horizon lines in blue and cardinal directions highlighted pink. 3D representation on top and flattened information on the bottom.

shows a 3D representation, similar to figure 5.2, on the top and the same information as a flattened diagram on the bottom. The cardinal directions are highlighted pink while azimuth degrees, representing the horizon, are highlighted blue.

The sun's path is shown in arcing lines as you can see in figure 5.6. The solstices are the bounding edges, highlighted pink and yellow, while the equinoxes are in the middle, highlighted blue. Remember that the sun's path sweeps back and forth over the months of the year, so each one of the arcing lines, except for the solstices, represent two different months.

The sun's location at any time of day can be identified by the azimuth and altitude, as seen in figure 5.7. Remember that the center of the diagram that aligns with north–south is solar noon, where the sun is highest in the sky. Clock time and solar noon rarely match. This doesn't matter if you are just collecting energy



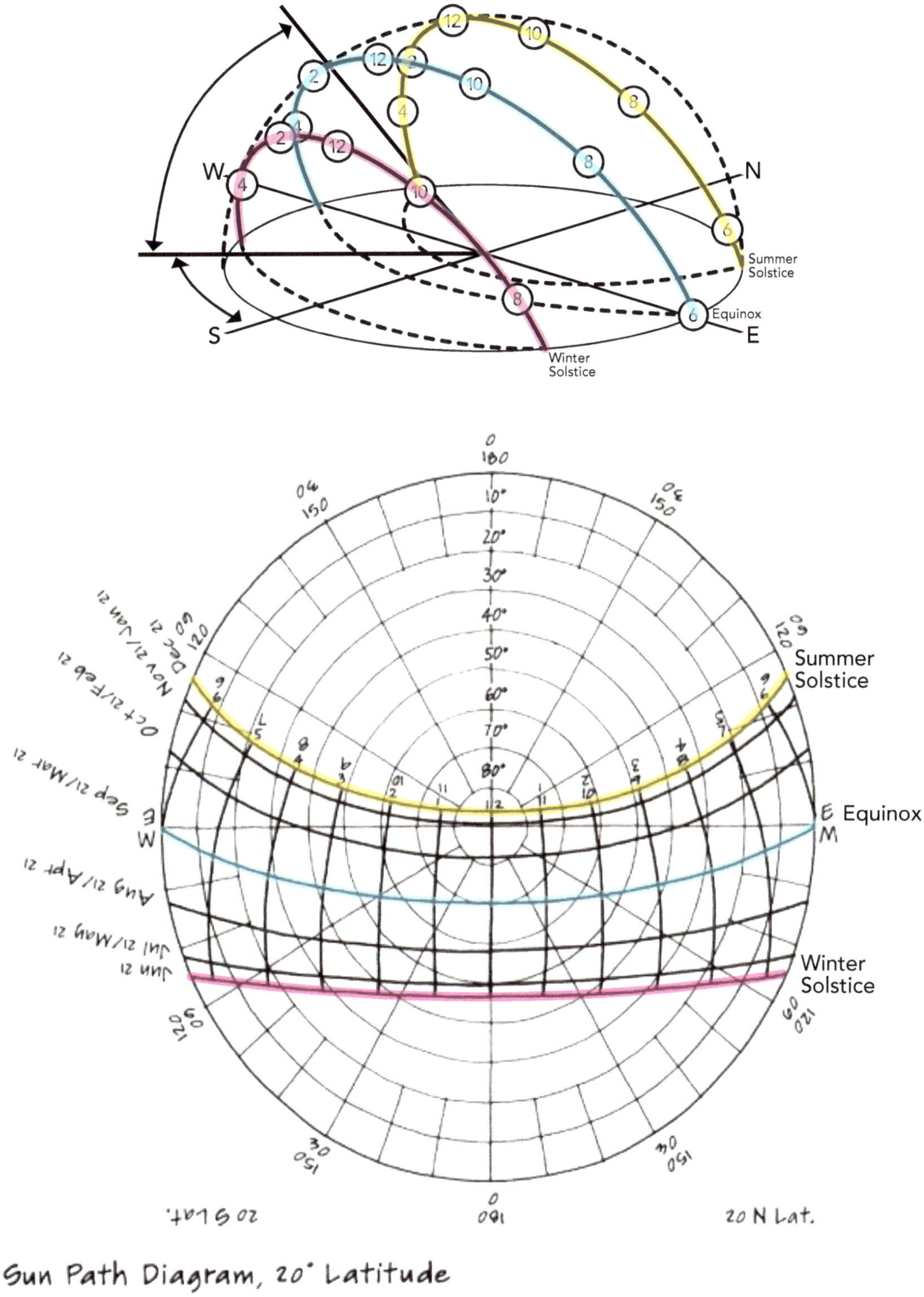


**Figure 5.6** Solstices and equinox path lines highlighted in 3D and 2D.[2]

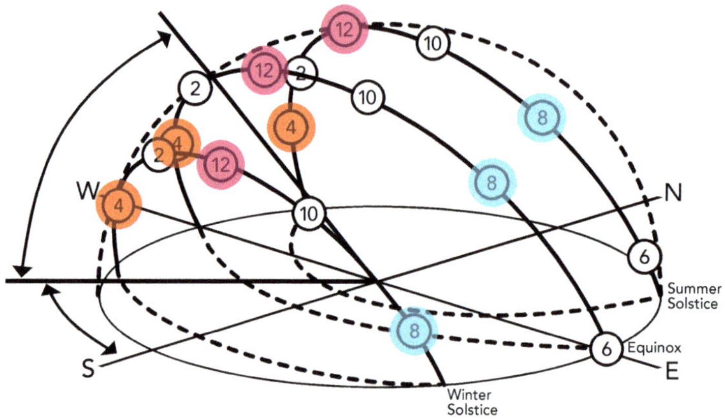


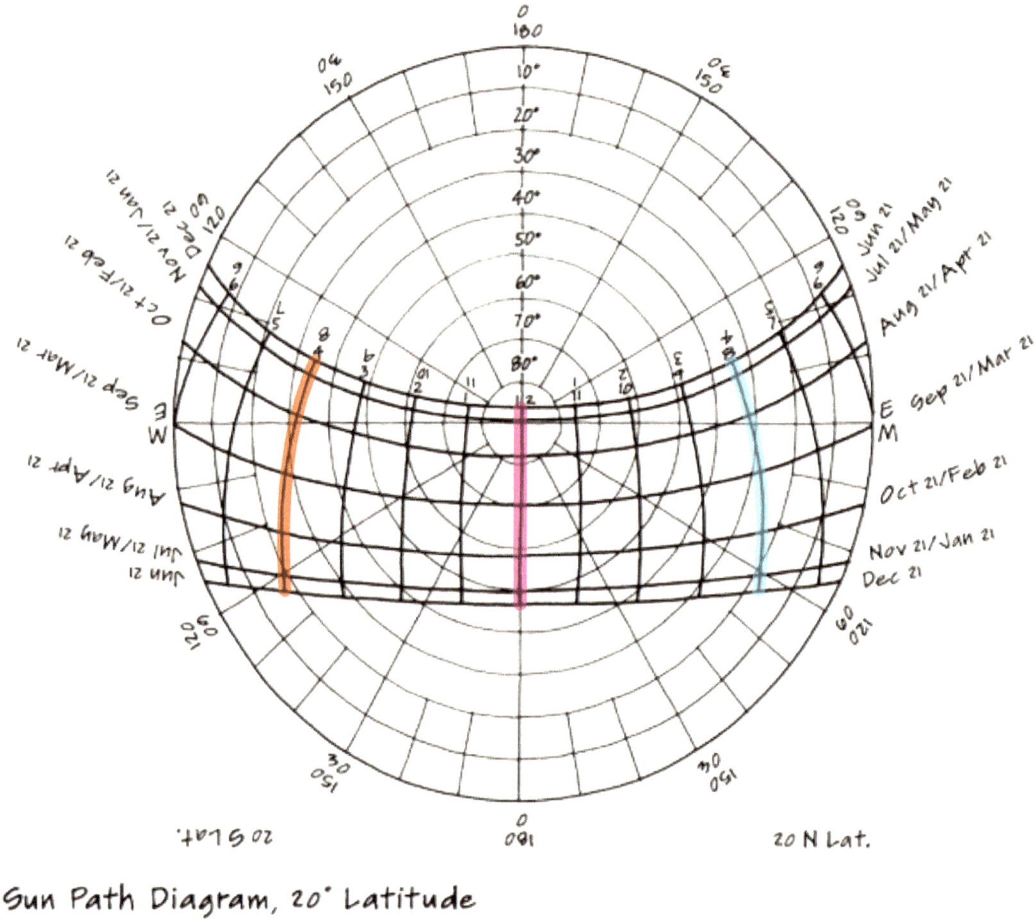


**Figure 5.7** Time of day is shown in somewhat vertical lines. The same time of day is highlighted in the same color: 8am blue, noon magenta, 4pm orange.[2]

or shading your building. It does matter if you want to have sun's rays hit a particular part of your design at a particular clock time.

These diagrams show a single vertical gently curved line for each time of day, however, that is a simplification. Because of the earth's elliptical orbit and 23.4° tilt, the sun's location is an elongated figure "8" with unequal loops. This is called an analemma curve. Some diagrams and computer-generated 3D visuals show the analemma curve but you still read the information the same way.

Figure 5.8 shows the appearance of an analemma curve created in Autodesk Revit.

Determining the sun's location requires finding the intersecting points of the time of year and day and then reading the azimuth and altitude. This is shown in figure 5.9. The azimuth is found by following the diagonally radiating lines out to the horizon, highlighted magenta. The altitude is found by reading the radiating circles, highlighted blue, where the horizon is 0° and the very center of the diagram is the zenith of the sky's dome at 90°.

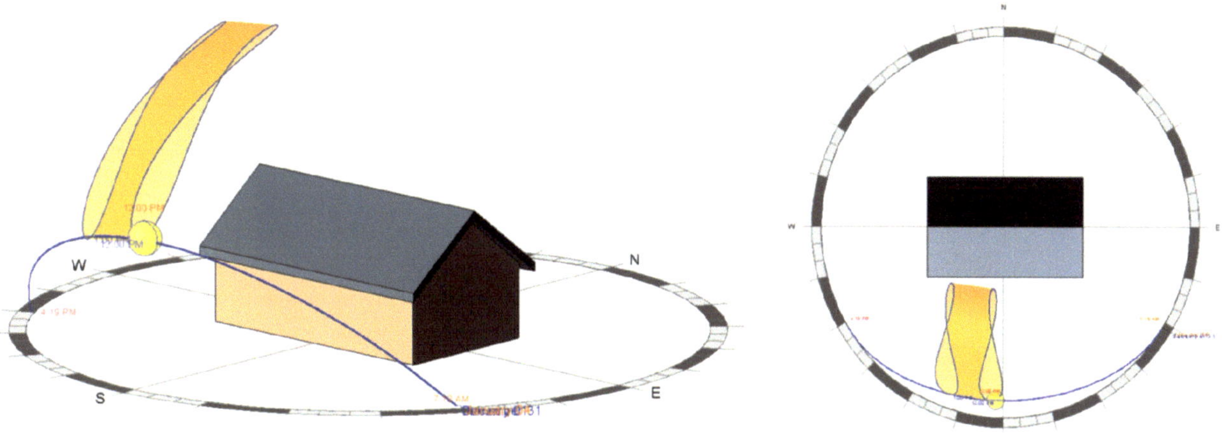


**Figure 5.8** Analemma curve for solar noon–1pm at 41° N Latitude, 3D on the left and 2D top view on the right.

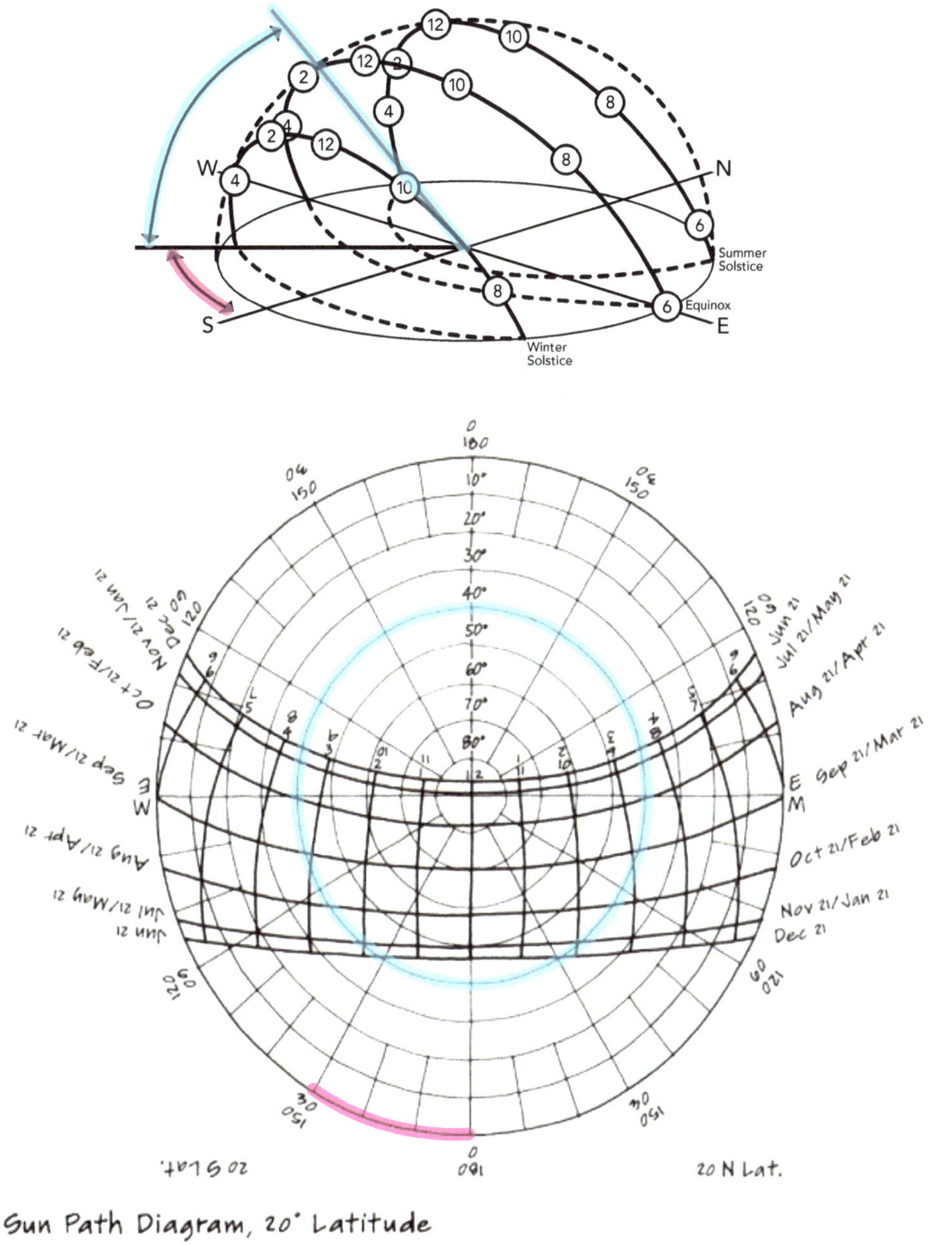


**Figure 5.9** Reading azimuth (0–30° west of south highlighted magenta) and altitude (40° highlighted blue) on a sun path diagram.[2]

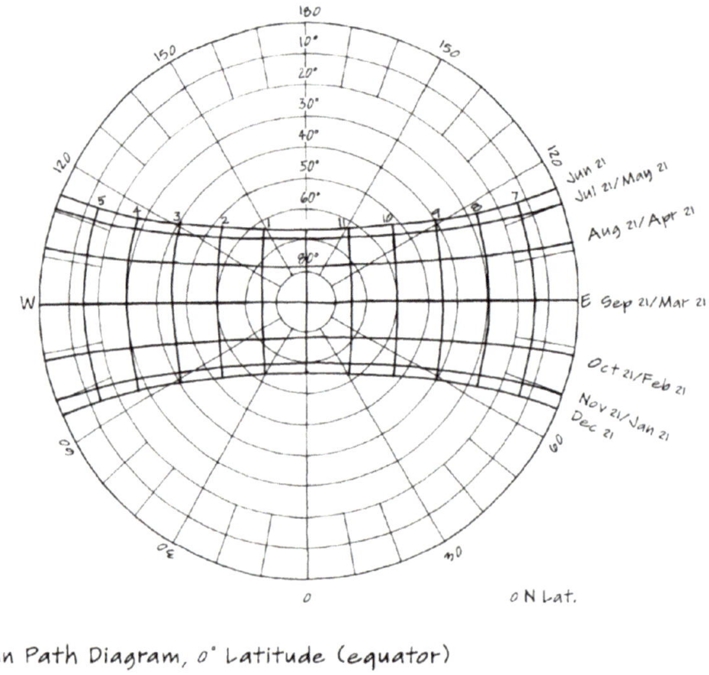


**Figure 5.10** Sun path diagram for 0° Latitude. Courtesy Mark DeKay.[3]

If you are at the equator, the sun is directly overhead at the equinox and moves north to the June solstice, then into the south for the December solstice. Figure 5.10 shows the equatorial solar path with the equinox as a straight horizontal line in the middle. A building at the equator will have an equal amount of sun on both the northern and southern sides, depending on the season.

As you move away from the equator, the sun's arced path appears to move either into the northern or southern part of the sky's dome. Compare the sun path diagrams shown in figure 5.11 for latitudes 20°, 44°, 68°, and 90°. Notice how the sun's path moves away from the zenith of the sky dome until, at 90°, the sun is only around the horizon and never overhead. At 68° notice that the sun never sets at the summer solstice, creating a full circle in the sky dome. Note that the sun never rises above the horizon in the winter, there isn't a line for the winter solstice because the sun is below the horizon.

The sun path patterns are the same for the same northern or southern latitude – the paths just flip whether the sun's path arcs in the north or south. Figure 5.12 shows this difference. North is up in both of these diagrams, northern hemisphere on the left and southern hemisphere on the right.

In a northern latitude, the sun is going to be located to the south of any building, so the majority of the solar gain will be on the southern façade. In a southern latitude the sun is in the northern sky, so the majority of the solar gain will be on the northern façade.

## Magnetic Declination

Magnetic declination is the variation, measured in a horizontal plane, between magnetic north and true north. Compass readings need to be corrected for this difference because it can be very large. Figure 5.13 shows a world magnetic declination map from the year 2021. The green line shows where magnetic north and true north align. The curving lines are called isogonic lines and each of the blue and red lines are indicating how much magnetic variation, moving away by degrees, is occurring compared to the location of true north. The exact location and value of the isogonic lines change over time and can be very unreliable around the poles.

To put this in context, Williamsport, PA, USA, northern hemisphere, has a declination of 11° 13′ W which changes 0° 1′ E per year (as of June, 2021). Let's say I'm recording shading conditions or building my building and I want to be within 15° of south: well, I'm already 11° off if I don't correct for magnetic declination. It is easy to check and apply declination corrections.

## Solar Insolation

The sun's radiation passes through our atmosphere to get to the ground. The angle of the sun compared to the ground changes the intensity – think about a hot summer sun at noon – it's way more intense than at 8am. Some of the sun's radiation is absorbed by our atmosphere, and some of it is diffused by air molecules, aerosols, and the like. Clouds can block some or a lot of the solar radiation.

Solar insolation is measured in kW/m$^2$. Historically collected weather data is used to estimate what the probable amount of solar insolation will be in a particular area. Of course, there is variation in weather patterns from year to year, so it is just an estimate. Computer programs and charts usually cite about a 10% variation

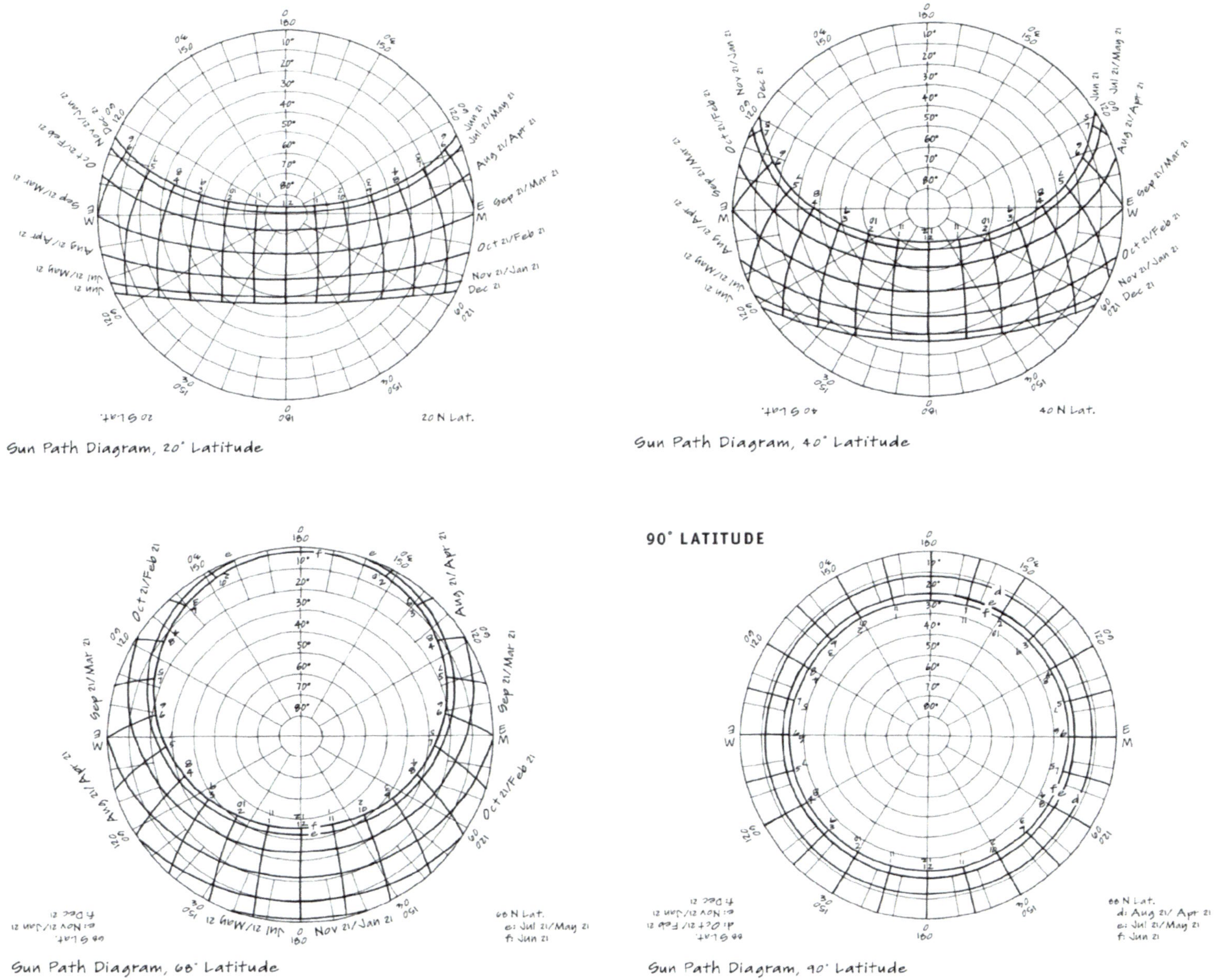


**Figure 5.11** Sun path diagrams for 20°, 40°, 68° and 90° Latitude. Courtesy Mark DeKay.[3]

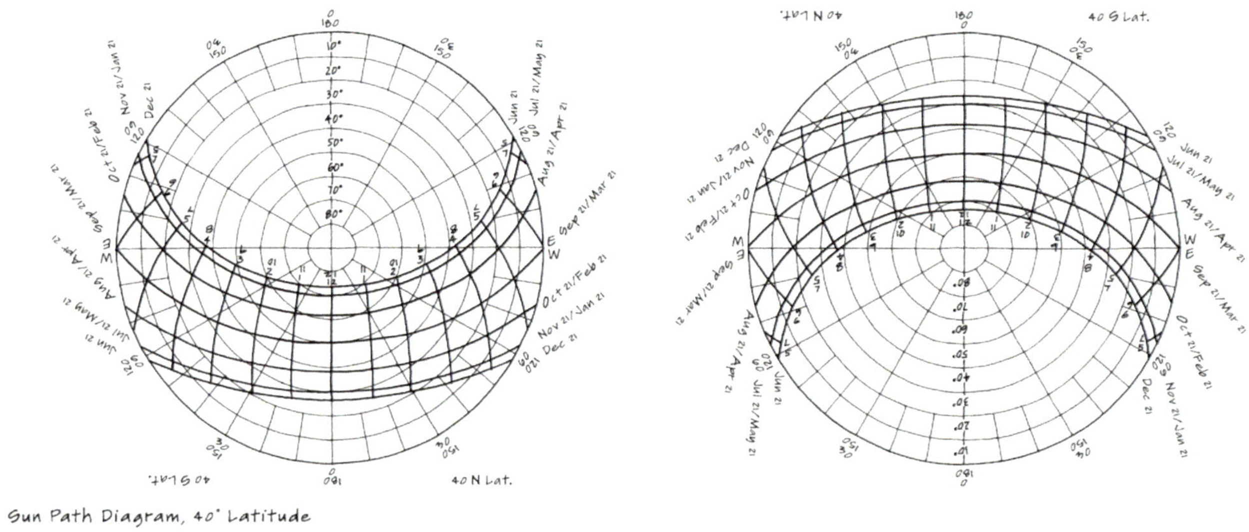


**Figure 5.12** Sun path for northern (left) and southern (right) hemispheres is inverted. Courtesy Mark DeKay.[3]

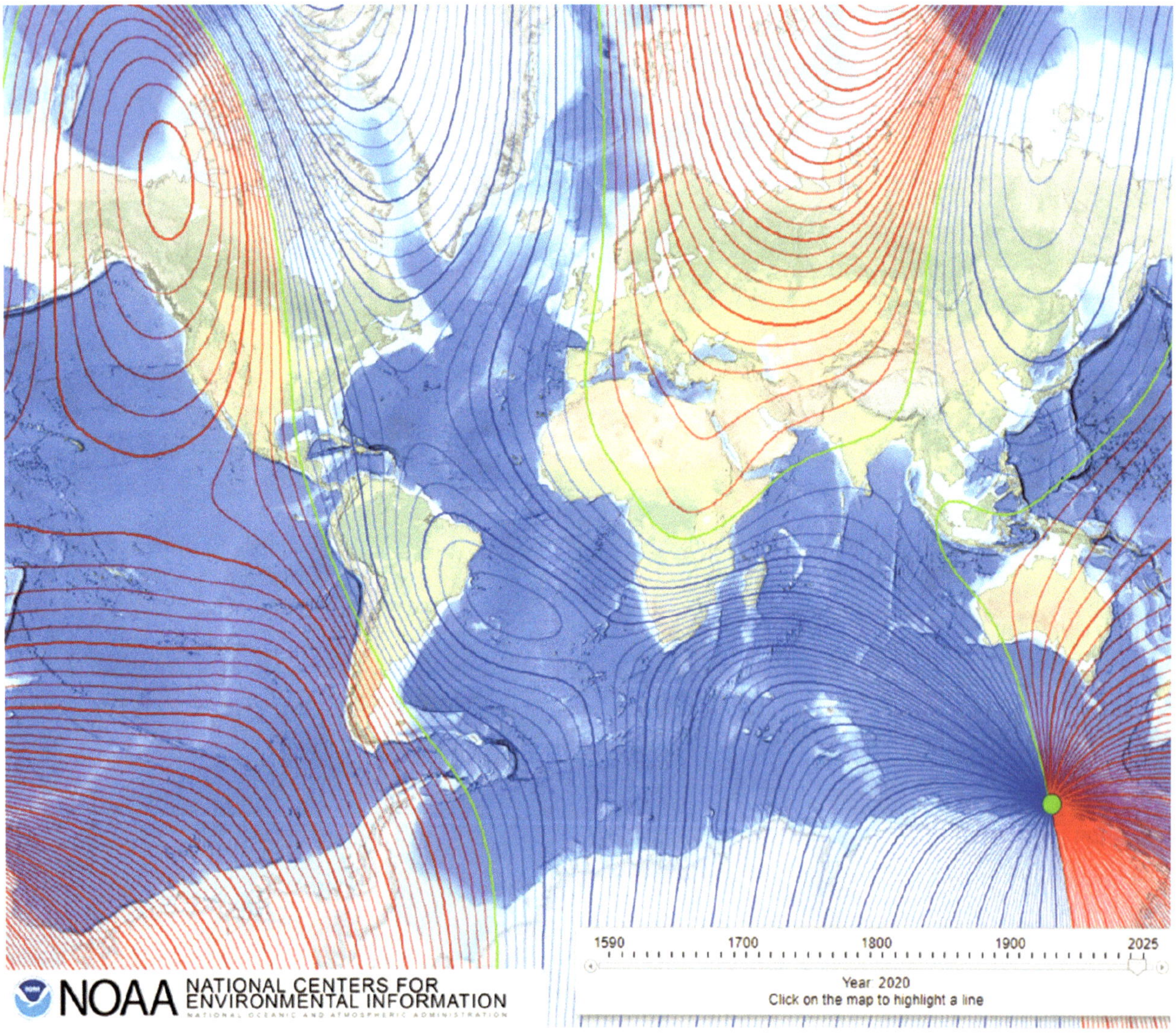


**Figure 5.13** Magnetic declination across the earth: the green line shows where magnetic and true north are the same.[4]

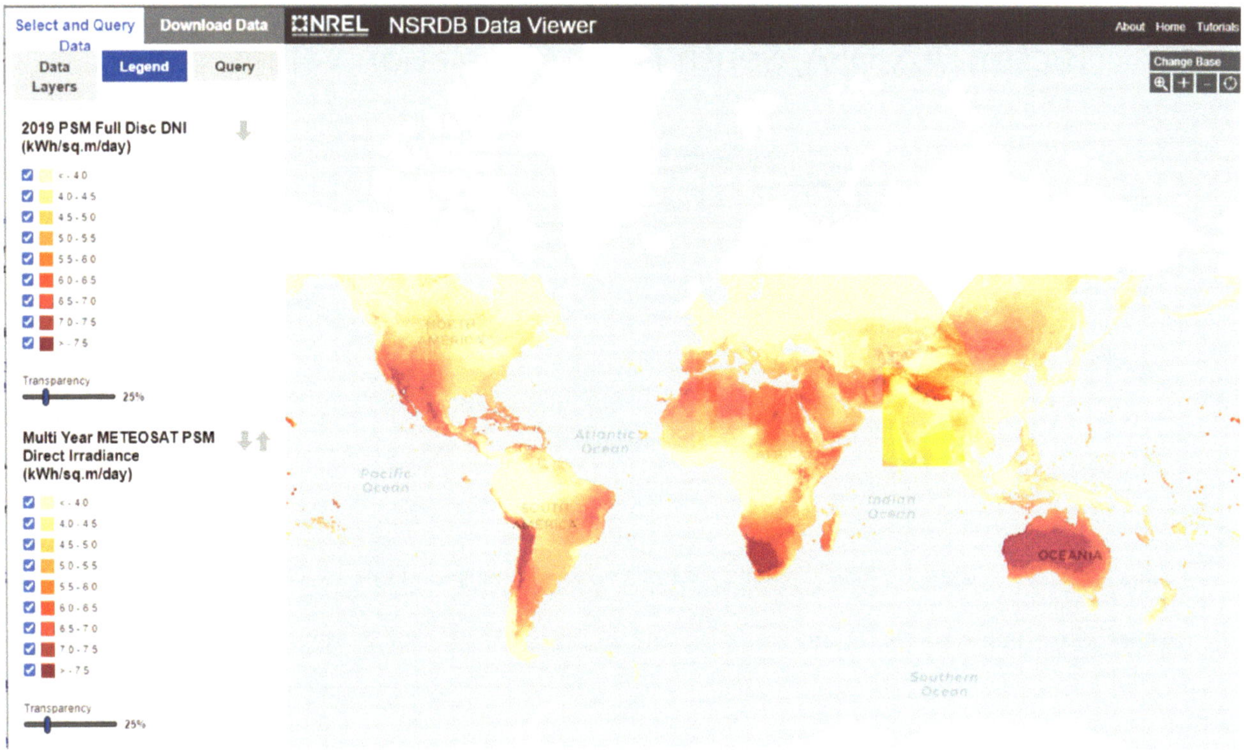


**Figure 5.14** Daily solar insolation worldwide, the darker the color, the more solar energy is available.[5]

(uncertainty). Figure 5.14 is a general map showing annual solar insolation worldwide.

Note, as you look at the map, that the places with the most amount of solar insolation (dark red) are places that are the deserts. Because solar insolation is the amount of potential energy that you can collect, it is very important in determining what the output of your system will be.

## Tilt Angle and Insolation

The amount of insolation you are going to be able to convert to heat or electricity is dependent on the orientation of the collector and its tilt angle.

For illustration, figure 5.15 shows monthly and annual insolation values for different collector tilt angles all facing south located in Tucson, Arizona, USA, at 32.21° N latitude. This is a hot, dry desert climate.

Figure 5.15 shows that tilt at latitude has the most solar insolation at 6.6kWh/m$^2$/day. In this particular climate, there isn't much difference between latitude – 15° (17.21°) and latitude + 15° (47.21°) tilt angles. Vertical is only 4.1, a lot less. Remember that this energy is for each day, so the larger the insolation number, the better production there will be.

Changing the orientation of the collector to different azimuths impacts insolation numbers (figure 5.16). This chart compares possible insolation values keeping the tilt angle at latitude, 32.21°, and orienting the collector in different directions.

Note that for Tucson, the azimuth orientations for south and south-east are very similar. The south-west orientation still has good production but other orientations have quite a bit of loss. If you use a poor tilt angle and have poor azimuth orientation, you've just lost a lot of potential insolation.

## Determining Shading at a Site

Shading happens because buildings and collectors are often surrounded by other buildings, street lights and towers, trees and geography. Shading can also happen from the building itself due to its geometry compared to the collector placement, as well as items located on the roof such as machinery, vents, and cables. Collectors also need to be mounted and spaced so they don't shade each other.

A general rule for best production is that your collector should be free of shading for three hours on either side of solar noon. Shading on the collector early and late in the day often doesn't have much impact on production. This is because the low incidence angle of the sun bounces off the atmosphere and just doesn't have as much energy. These times of day also produce long shadows that are difficult to avoid.

Management of all the equipment on a building's roof has become an art form. Collectors require a fair amount of space. They are rectangular and you have to provide space for whole units. You also have to provide access around the collectors. There is a growing trend to mount collectors well above the height of flat roofs

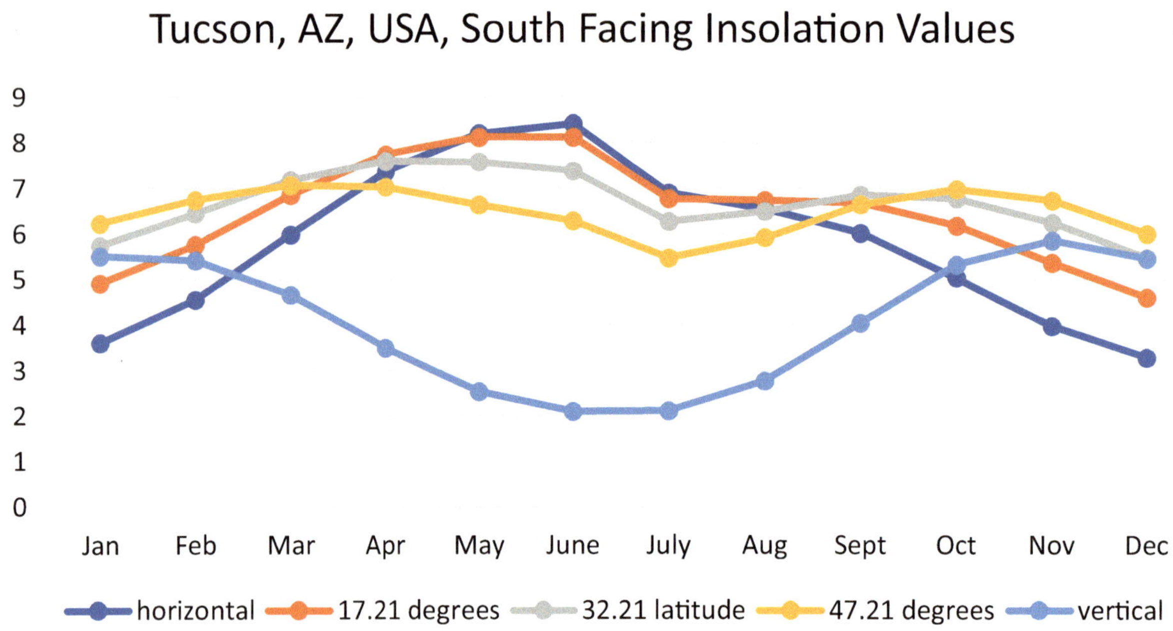


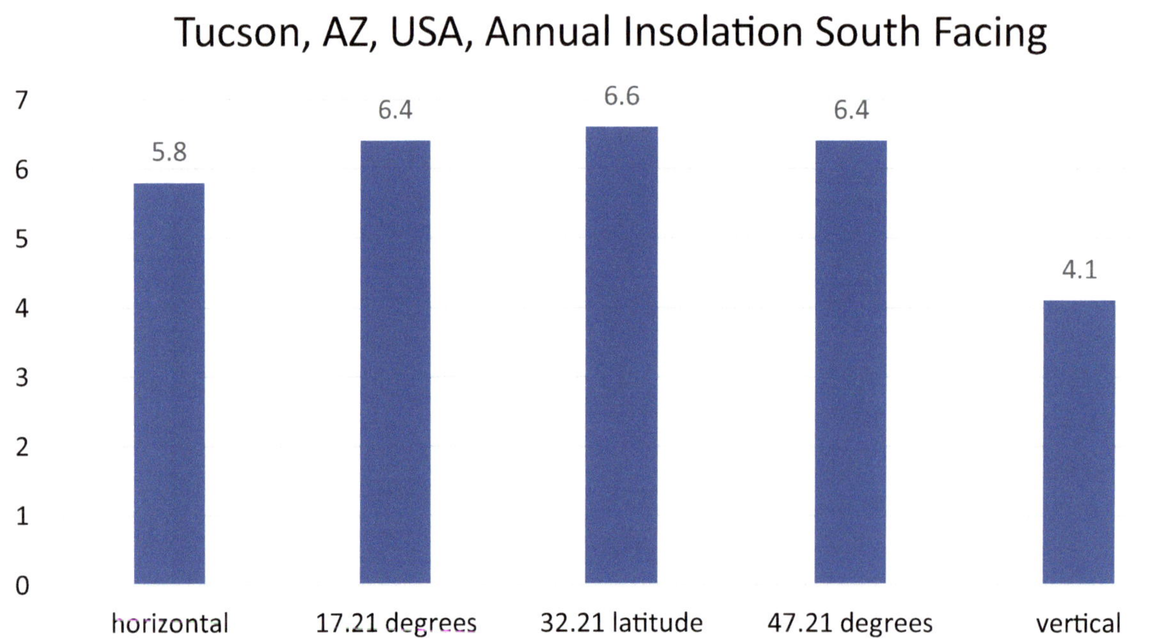


**Figure 5.15** Annual solar insolation at varying tilt angles for Tucson, AZ.[6]

so that it shades equipment and there are fewer issues with space management on the roof's surface.

## Inter-row spacing

To prevent inter-row shading, you need to know what the size of the collector is (M), the tilt angle of the collector (Θ), and the azimuth and altitude of the sun at the winter solstice at 9am. Refer to figure 5.17 for a visual reference for terms.

Using trigonometry (be sure your calculator is set to degrees, not radians), the height difference at the back of the module is found using this formula: $H = \sin\Theta \times M$.

The module inter-row spacing is based on the height difference divided by the tangent of the sun's altitude at 9am: $B = H/\tan(A)$. Rounding this number up to a whole number means added safety for keeping those shadows off your modules.

This doesn't account for the fact that the azimuth of the sun isn't directly south or north, it is farther to the east and west at 9am and 3pm. This means that your rows can actually be a bit closer together. Use a sun path

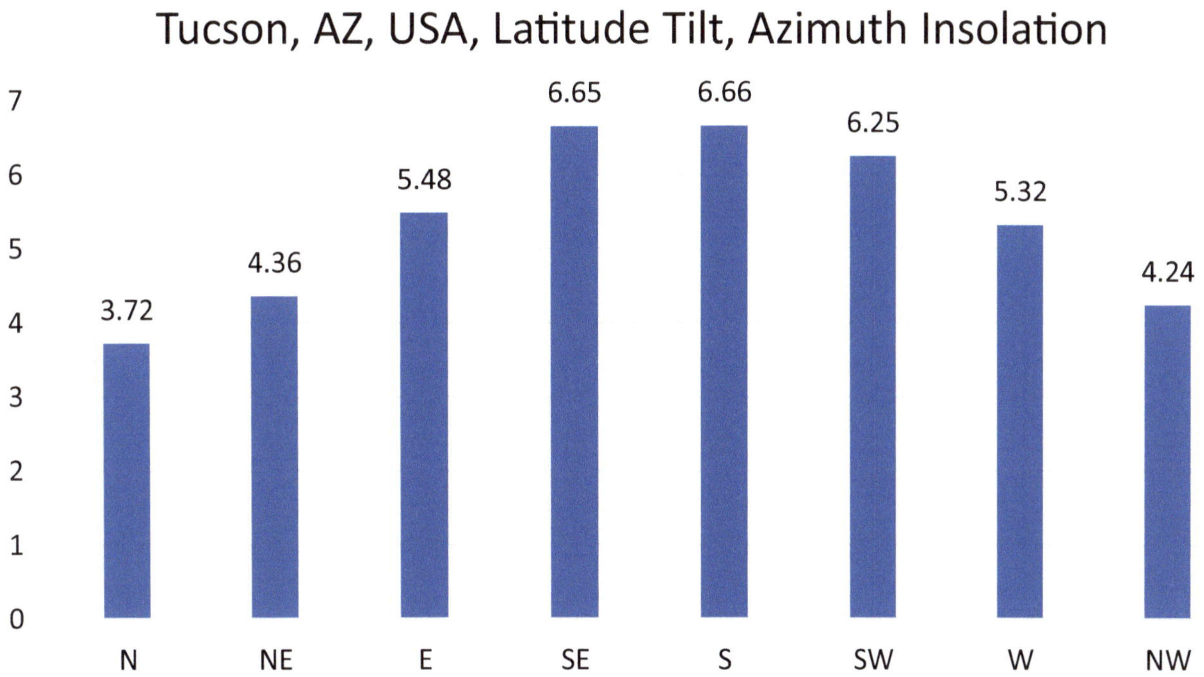


**Figure 5.16** Annual solar insolation tilted at latitude at different orientations for Tucson, AZ.[6]

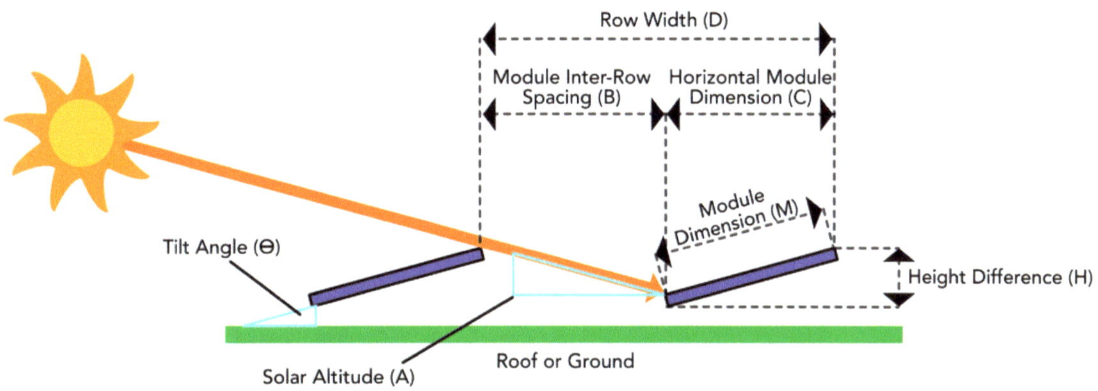


**Figure 5.17** Terms for determining inter-row spacing and row width.

diagram to determine the number of azimuth degrees between 9am and noon. The correction amount will be the cosine of these degrees multiplied by the inter-row spacing: $B_{min}$ = B × Cos (azimuth correction angle). Again, round up to the nearest whole number. The row width is: D = (M × CosΘ) + $B_{min}$.

Figure 5.18 demonstrates how to find the azimuth degrees and altitude of the sun on an elevation type sun chart while figure 5.19 shows this on a plan type. Both identify the location of 9am on December 21 on the diagram. Note that they have slightly different altitudes and azimuths because one is for 45°N and the other 48°N.

Referring to figure 5.18, at 9am on December 21 (winter solstice) the sun's altitude is 10°. The azimuth degree difference between south and 9am is about 40° (each mark across the bottom is 5°). Let's say I've picked out a SolarTech Universal EPIQ325W PV module. Its dimensions are 39.25″ (997mm) × 65.4″ (1663mm). Usually on flat roofs, modules are mounted in rows with the smallest dimension vertically, although there are many options for **racking**. Both wind loads and inter-row spacing increase with higher tilt angles, and 45° is very steep for a flat roof (usually flat roofs are racked between flat to 30° maximum). Lots of times it is better to use a slightly lower tilt angle so I can fit more modules on the roof. I need to refine my production numbers with the tilt angle, getting an idea for the row width so I can figure out how many modules will physically fit on my roof. It may be that trading a lower tilt angle for more modules works better with the roof layout and gets me to my desired production number.

Let's just see what the results are for a 30° tilt. The height difference from front to back of the panel will be: H = SinΘ × M

H = Sin (30°) × 39.25″ (997mm) = 19.625″ (498.475mm)

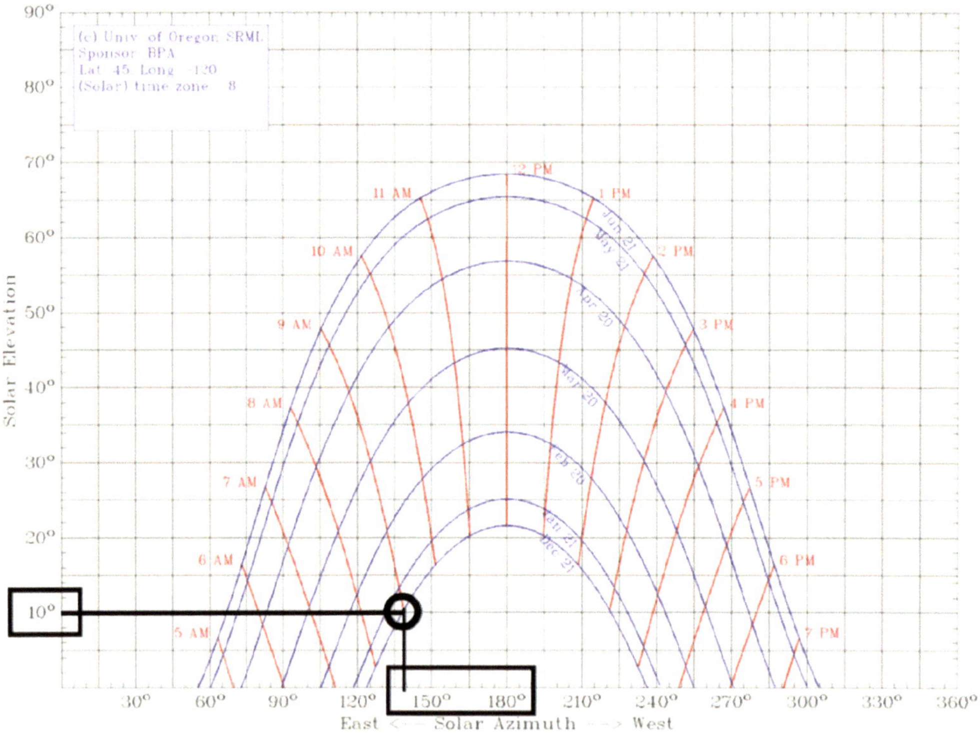


**Figure 5.18** Sun's altitude (10°) and azimuth (140°) for December 21 at 9am for 45° N.[1]

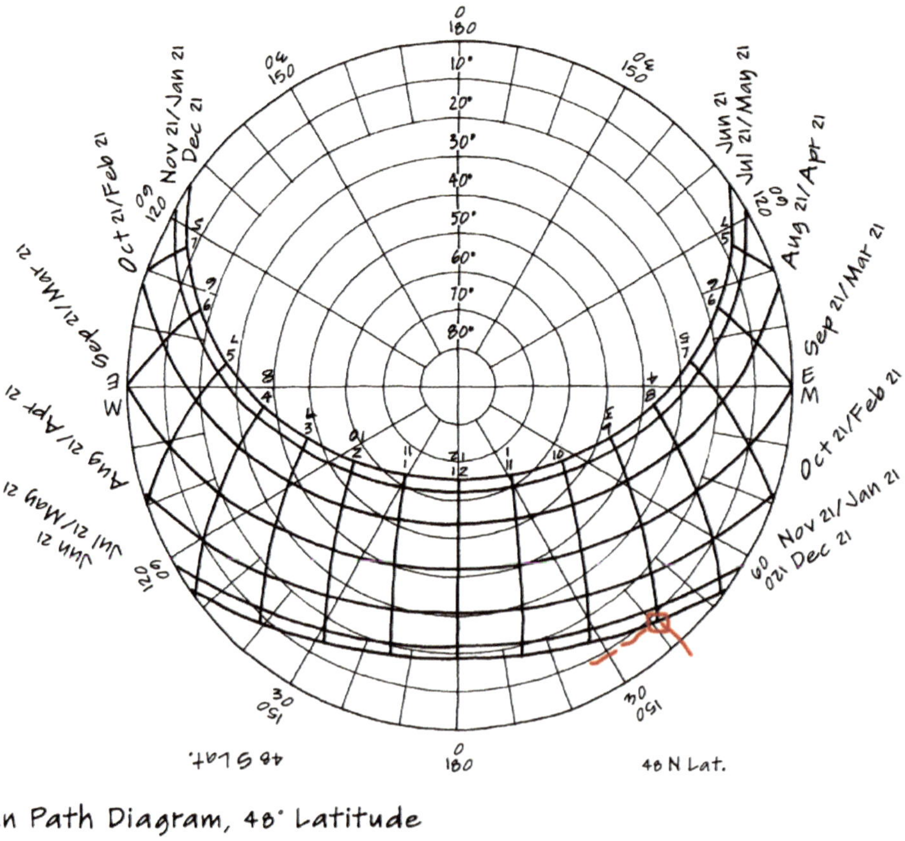


**Figure 5.19** Sun's altitude (9°) and azimuth (138°) for December 21 at 9am for 48° N.[2]

The inter-row spacing: B = H / Tan (A)

B = 19.625″ (498.475mm) /Tan (10°)
= 111.3″ (2826.9mm) or 9.28′ (2.83m)

The correction for the azimuth at
9am: $B_{min}$ = B × Cos (azimuth correction angle)

$B_{min}$ = 111.3″ (2826.9mm) × Cos (40°) = 85.26″ (2165.6mm), which could be rounded to 86″ (2184.4mm) or 7.16′ (2.18m)

The total row width: D = (M × CosΘ) + Bmin

D = (39.25″ (997mm) × Cos (30°)) + 86″ (2184.4mm) = 120″ (3048mm) or 10′ (3.048m)

The options for getting more modules to fit on the roof would be (and we could use any combination): lower the module tilt angle (10° will shade a lot less than 30°); change the shaded solar window time (i.e., say 10am instead of 9am – the most shading will be in December and January, so might not really affect overall annual production very much); and/or pick out a smaller module (which would have a lower power rating and therefore make less energy). Of course, all these options need to be tested for how they actually fit on the roof of my building vs. the production desired for the building. Remember that more modules will cost more money (and require more labor time to install) than fewer modules, so getting the best production for the investment is important.

If these collectors are solar hot water collectors, they can be a bit shaded and not hurt the production. A small amount of shading on PV can have a dramatic reduction in production.

## Computer Modeling for Shading

If you have a 3D computer model of your building, it can be fairly easy to include appropriately scaled, blocky forms representing the environment around your building (even trees can be made of simple geometry). Some programs will automatically import buildings in cities. It is always a good idea to check that the building forms are actually the correct size and in the right location. Make sure that you have the location and azimuth of the site and building set correctly. Then you can do sun studies and see when your collectors will be shaded.

## Professional Tools for Shading Analysis

Professional tools that are used for shading analysis include the Solar Pathfinder, Solmetric's SunEye, and Meteotest's Meteonorm Horicatcher. These tools can be taken to a site and then used to capture the existing shading conditions at particular locations. If the building hasn't been built yet, readings can be adjusted to the correct collector height.

The Solar Pathfinder is the least expensive and uses a latitude specific sun path diagram under a clear dome to reflect the environment. It reflects the environment in the dome and can be used at any time of day under any lighting condition. Anything reflected in the dome that covers a part of the sun's path will shade the collector at that time of year and day. It is easy to either calculate shading by hand or use software.

The SunEye is a hand-held electronic device with a digital calibrated fisheye lens camera. It has a rechargeable battery and downloads information to your computer for analysis. You can take up to 100 images and record notes with them. You need to input the site's location prior to taking images. It has a live survey mode where you can see the annual sun paths imposed over the image as you move around. The images can be used to determine shading at any particular location and height using the software.

The Meteonorm Horicatcher has a digital camera and a camera mounting device with a horizon mirror that attaches to a tripod. Images are downloaded into analysis software loaded on your computer. Photographs are taken at the site. The software has many weather parameters and a hefty database of weather stations around the globe. The analysis will provide solar insolation taking the shading from the images into account.

## Using a Solar Pathfinder

The Solar Pathfinder is easy to set up and understand the results. It uses a sun path diagram that relates to a set of latitudes as shown in figure 5.20. You can compare these examples with the sun path diagrams shown earlier in the chapter and see that they represent the arc of the sun in the sky dome. East is to the left, west to the right and solar noon directly in the middle. The solar time is broken down into half hour increments, each with a number representing the intensity of the insolation at that time of day. The largest numbers are around solar noon.

The Solar Pathfinder, shown in figure 5.21, can easily be moved around to discover the best sunny location. The results can be modified for height if the roof isn't built yet or you plan a pole mounted collector. The more area that is shaded, the worse that location is for having good solar gain on a collector.

Figure 5.22 shows the appearance of reflections in the dome. Note that clouds are visible but are not traced as shading elements. Only permanent obstacles are considered shading: if there are no leaves on the trees then you'll need to trace around where the full tree would create shade. You can compare the reflection of the shading in the dome on the left to the line drawn in red on the right.

Figure 5.23 explains the information shown on the Solar Pathfinder diagram.

Solar Pathfinder has software that you can upload the dome image into and it will figure out the potential at

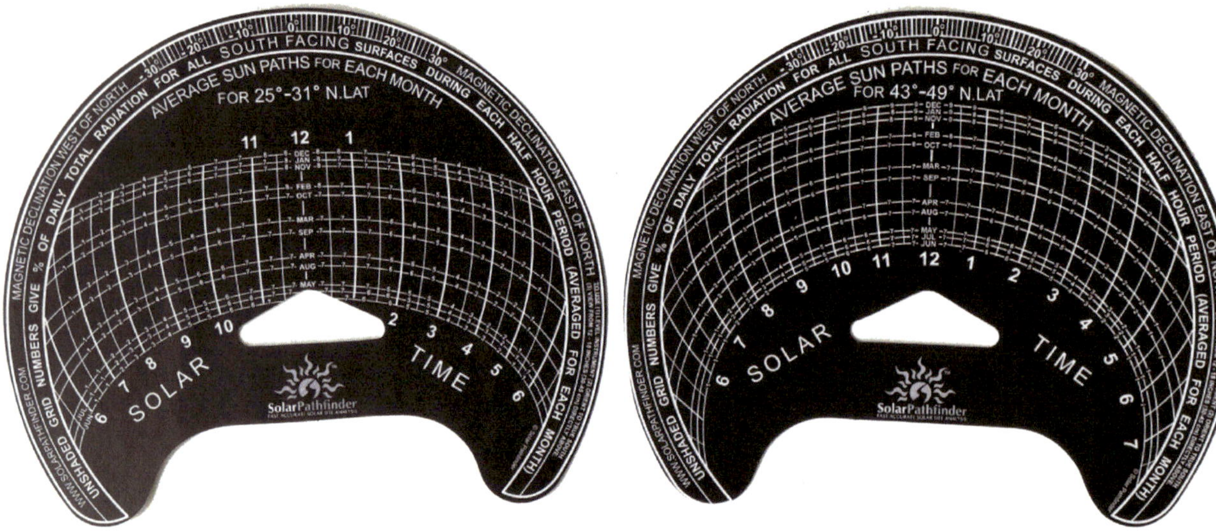


**Figure 5.20** Solar Pathfinder sun path diagrams for different latitudes showing arcs for time and month.[7]

**Figure 5.21** Image of a Solar Pathfinder with base unit stand and dome.

the location. You can also determine it by hand. Although figure 5.24 is for 37–43°N latitude, since the Solar Pathfinder is set up facing the sun, the short winter days will always be at the top, while long summer days are at the bottom. If you are designing an off-grid building, your collectors would need to provide enough power every day, all year, so you'd need to check how much collection you have in the winter. At this location in figure 5.24 there is quite a lot of shading in the winter mornings but the afternoon gets sun. This could make daily net zero difficult to achieve in the winter months.

When determining sun/shade, you can either count up the numbers in the shaded or unshaded (sunny) areas of the diagram. You can use partial numbers, for instance if it looks like half of a half hour increment is shaded, count it as half the value.

The numbers along a monthly arc generally add to 100, so 100 represents 100% of the solar insolation available in the month. If you count up the shaded numbers from

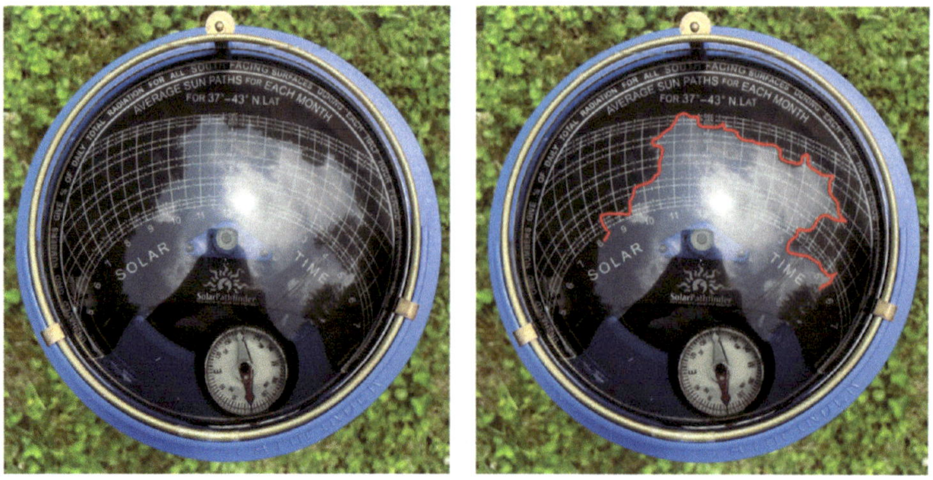

**Figure 5.22** Solar Pathfinder dome with reflections (left) and with shading outline marked (right). Note that shading photo must be taken directly above the bubble level.

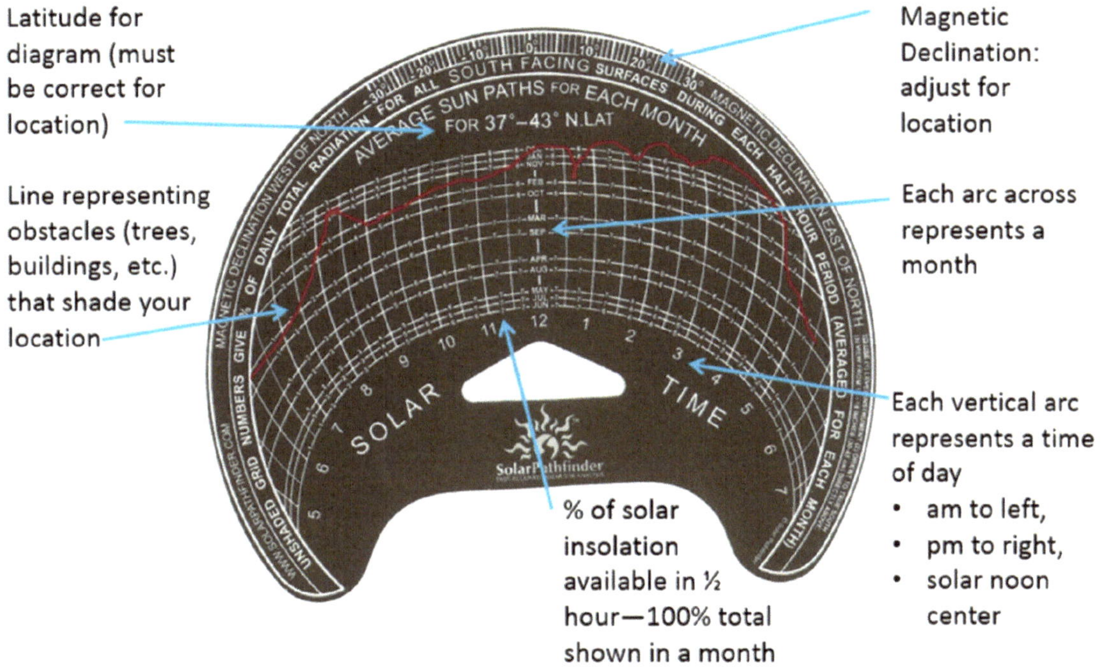


**Figure 5.23** Information on a Solar Pathfinder diagram includes the sun's path, time, and intensity of the sun's rays. It also has adjustment for magnetic declination and indicates what latitudes the diagram should be used for.[7]

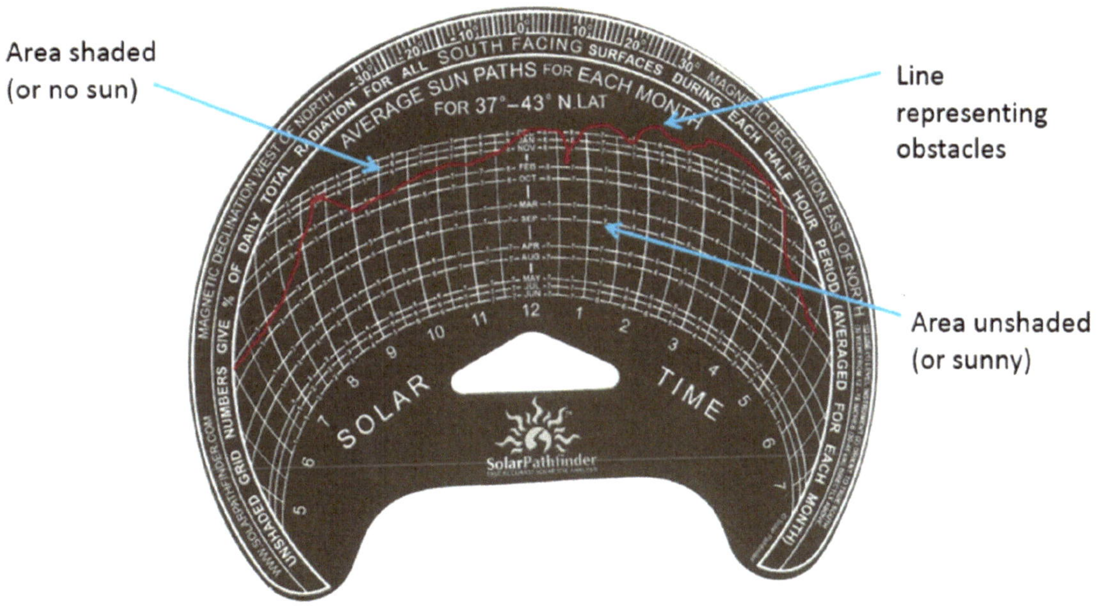


**Figure 5.24** The Solar Pathfinder diagram shows what is shaded and what is in the sun due to obstacles at the site. The shaded times of year will be on the upper side of the line while the sunny times are on the lower side of the line.[7]

the diagram, then the sunny hours equal 100 minus total shaded hours. If you counted the sunny hours, you are set – that's how many there are.

In figure 5.24, December only has sun in the afternoon. Since there are fewer sunny numbers than shaded ones, I counted them. I counted 34 hours in the sun. That means that this site receives 34% of the solar insolation available in December.

Many projects are aiming for annual net zero energy production. That means that instead of counting along a month, all the hours on the diagram will be counted. Because each month has 100 hours, a year has 1200 hours. For the year, I counted the shaded areas because there were a lot less of them. I counted 187 shaded hours. To determine the percentage of hours that are shaded, divide the shaded areas by 1200.

187 / 1200 = 16% shaded. Annually there will be 100% − 16% = 84% actual sunny hours. That means that over the year this location for my collector will receive 84% of the local solar insolation. If I counted the sunny hours, the way to get the percentage is to divide the sunny hours by 1200. If I were to use a program like PVWatts to determine solar insolation values, I'll need to add 16% to the shading variable because the site isn't getting 100% sun.

## Resources


1 University of Oregon Solar Radiation Monitoring Laboratory. (2019, July 11). *Software tools.* http://solardat.uoregon.edu/SoftwareTools.html

2 Base sun path diagram courtesy Mark DeKay.

3 Brown, G. Z. (2012). *Sun: Architectural design strategies.* Chichester: John Wiley and Sons Ltd.

4 National Centers for Environmental Information. (n.d.). *Magnetic declination.* National Oceanic and Atmospheric Administration. https://ngdc.noaa.gov/geomag/declination.shtml

5 National Renewable Energy Laboratory. (n.d.). *NSRDB data viewer.* US Department of Energy (DOE)/NREL/ALLIANCE. https://maps.nrel.gov/nsrdb-viewer

6 These charts use insolation values produced from PVWatts® Calculator (https://pvwatts.nrel.gov/).

7 Solar Pathfinder paths are used with permission from Solar Pathfinder (www.solarpathfinder.com/).


## Study Questions

1. Referring to figure 5.4, what is the azimuth and altitude of the sun for the following dates and times?
   a. December 21 at 4pm
   b. March 20 at noon
   c. June 21 at 11am
2. Referring to figure 5.11, identify the altitude of the sun at solar noon on the equinox for 20°, 40°, and 68° latitude.
3. Go online to the University of Oregon Solar Radiation Monitoring Laboratory sun path chart program (http://solardat.uoregon.edu/SunChartProgram.php) and create a sun path chart for a specific latitude (be sure to enter the verification code in step 7).

## Study Answers

1. Figure 5.4 is for a northern latitude so the winter solstice is in December.
   a. Each azimuth mark is 6°, with vertical lines every 18°. 4pm on December 21 appears to be about 8° less than 240°, so the azimuth is 232°. Each altitude mark is 2 degrees, with horizontal lines every 5 degrees. 4pm on December 21 appears to be located at about 3°.
   b. March 20 at noon: azimuth is 180°, altitude is about 45°.
   c. June 21 at 11am: azimuth is 144°, altitude is 65°.
2. Figure 5.11 shows the altitude of the sun at solar noon on the equinox (September 21/March 21): 20° latitude = 70° altitude, 40° latitude = 50° altitude, and 68° latitude = 22° altitude.

CHAPTER 6

# Design for the Climate

**Key Terms**

Passive design: strategies that reduce energy usage of building (mechanical, lighting) without using any energy.
Psychrometric chart: a chart that shows the physical and thermal properties of moisture in the air.
Relative humidity: the amount of water vapor in the air expressed as a percentage compared to saturation (100%) at the same temperature.
Ventilation: fresh air introduced into the building.

## Concepts for Success

A building needs to be designed to fit the climate. Types of information that define a climate are precipitation, temperature, humidity, and the amount of sun and wind. These are easily searchable through many websites that use historical data to create a variety of useful charts.

Buildings are supposed to create comfortable conditions on the interior, no matter what the weather is outside. The **psychrometric chart** is used to identify temperature and **relative humidity** of air and is one of the tools used to size mechanical systems. Another tool is knowing the number of **heating degree days (HDD)** and **cooling degree days (CDD)** at a particular location. High numbers of HDD and/or CDD indicate a potential for needing a larger mechanical system, which will use energy to condition the air. Knowing the HDD and CDD, as well as other climate information, points the designer to what passive strategies could be appropriate for the building.

Passive design strategies are about orienting the building toward the sun, providing windows that can allow solar heat gain through them, having some thermal mass, and making sure the windows are shaded when heat gain would be undesirable. Passive strategies combined with fans can be effective for both heating and cooling. Natural **ventilation** can provide fresh air and comfortable temperatures when climatic conditions allow. Natural ventilation can be achieved through stack effect or cross ventilation. Stack effect works because air is forced through the building due to unequal pressure between the bottom of the building and the top of the stack. Cross ventilation works because windows on different façades open and the wind blows through. Natural ventilation works best when the windows are perpendicular to prevailing winds.

Earth tubes use the ground as a **heat sink** to preheat or cool outside air before it is introduced to the mechanical system. It can work well anytime there is a large temperature difference between the air temperature and the deep ground temperature.

Shading windows is vitally important anytime there are CDD. Heat gain through the windows when the building is already overheated makes the mechanical system work harder and wastes energy. Shading for early and late in the day needs to be vertical to block the sun, while midday shading is achieved with horizontal devices. These can be combined to create effective shading on any window in any climate.

## Defining Climate

The most important issue for designing an energy efficient building is understanding the climate at the location where it will be built. That means you need to know what natural conditions your building is going to have to thrive in. Topping the list are precipitation and temperatures throughout the year. Precipitation includes rainfall and snow. It is important to know if there are extremely dry or wet parts of the year, as well as how long a typical depth of snow stays on the ground. Temperatures can vary widely both daily and yearly.

There are many sources for good historical records for precipitation and temperatures: be sure you are looking at something that is combining years of recent weather

DOI: 10.1201/9781003297819-7

data so that you have an accurate picture of what the location is like. Just looking at what the weather is today isn't helpful enough!

## Psychrometric Chart and Comfort

The whole purpose of constructing buildings is to separate us from undesirable conditions outside. To achieve comfort, we have to heat, cool, dehumidify, and humidify the outside air to create desirable temperatures and humidity levels inside the building. The psychrometric chart can help us to see what it will take to get us to comfort zone inside the building. The psychrometric chart can be a bit daunting but it is a great tool for understanding the relationship between temperature and moisture.

Figure 6.1 is a simplified psychrometric chart. The horizontal line across the bottom is the dry bulb temperature. This is what you think of as air temperature: it is what your thermometer reads. It is called dry bulb because it doesn't tell you anything about the humidity in the air. The temperature of 20°F (–6.7°C) is on the left and the chart ends at 130°F (54.4°C). The dry bulb temperature is read vertically up the chart. Hot air can hold a lot more water molecules than cold air. The lines on the chart for humidity are curving lines. The horizontal line that is shared with the dry bulb temperature is 0% relative humidity (RH). The top curved line is 100% RH. The air can't hold more than 100% humidity. If the air continues to cool down, the "extra" humidity falls out of the air: we call this dew. That means that the 100% RH line is also labeled the saturation or dewpoint line. Relative humidity is measured with a hygrometer. The simplest hygrometer is a sling psychrometer which has both a dry bulb and a wet bulb (a wet wick on the tip) thermometer. It is then spun (slung) around the handle which causes evaporative cooling on the wet bulb. The temperature on the wet bulb starts with the numbers along the 100% RH line and then is read diagonally downwards and to the right. Comparing the two temperatures gives the relative humidity.

Let's plot a point and find the RH on figure 6.2. Imagine it is a warm, sunny summer day. If the dry bulb temperature is 80°F (26.7°C) and the wet bulb temperature is 70°F (21.1°C), then following the lines for these temperatures, you find where they cross. That is where you find that there is currently 60% RH. This likely feels comfortable to you, at least in the shade because, even though it is a bit warm, sweat is able to evaporate from your skin to cool you.

Figure 6.3 is a psychrometric chart with a "box" from about 68°F (20°C) to 72°F (22.2°C) and 40% to 60% humidity. These are regular thermostat targets for buildings in the USA. The bigger you can make this box, the less energy it will take to condition the air. There are many strategies for this but they have to be appropriate for the climate and culture.

Let's see what has to happen to get that 80°F (26.7°C) 60% RH air into our target comfort zone (figure 6.4). If I just cool the air, I move horizontally across and the dotted line shows that the RH would go way

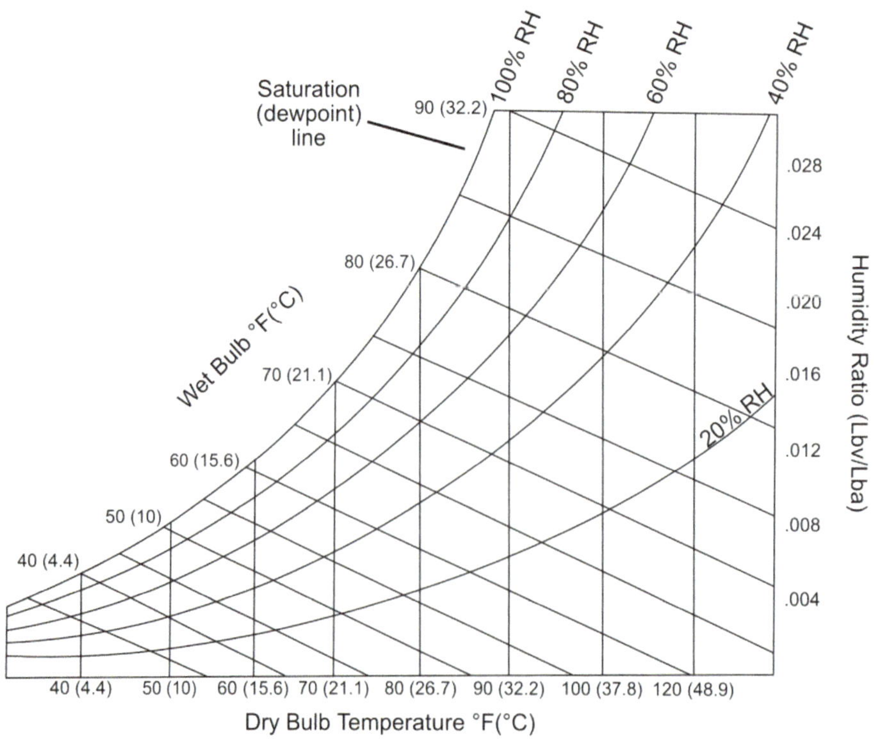


**Figure 6.1** Simplified psychrometric chart elements.

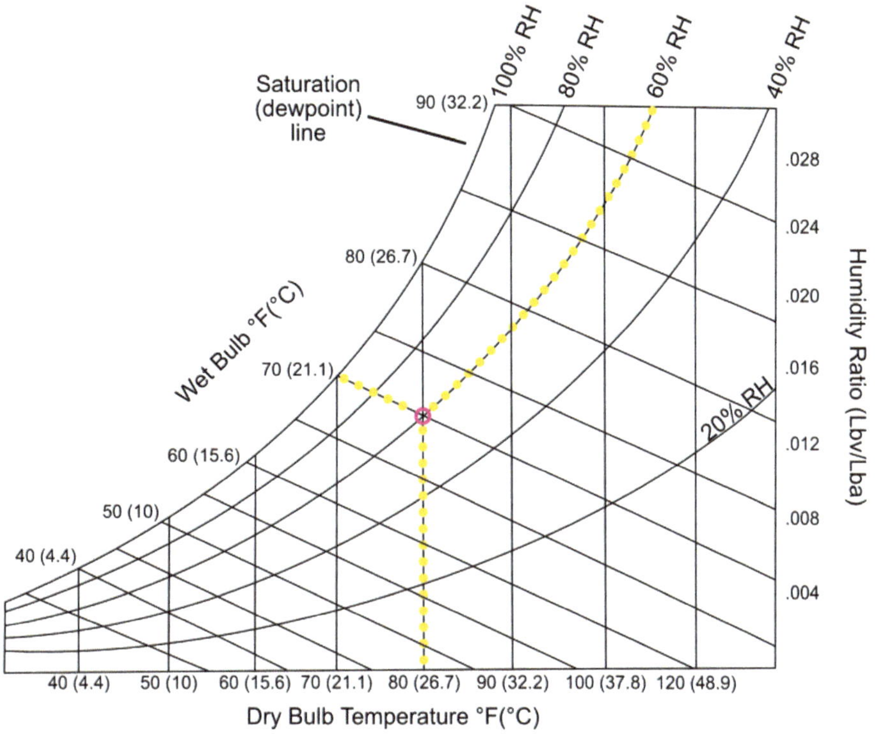


**Figure 6.2** Psychrometric chart point location.

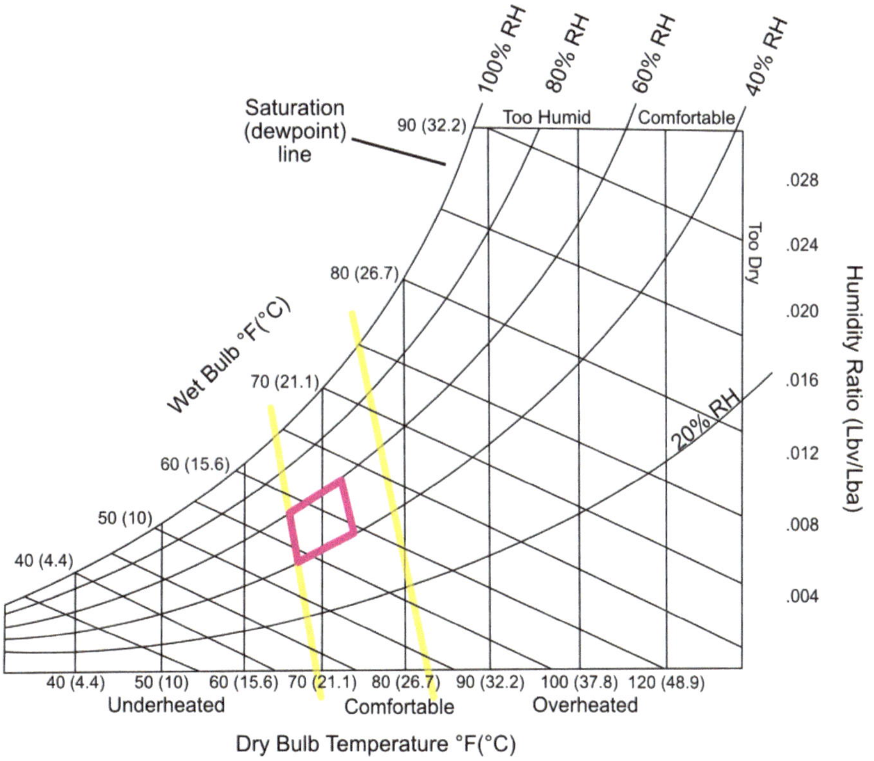


**Figure 6.3** Psychrometric comfort zone.

above 60%. This means the mechanical equipment has to dehumidify as well as cool the air to achieve comfortable indoor temperature and humidity.

To complicate things a bit more, the people and equipment in the space are also continuously adding heat, carbon dioxide, water vapor, and other contaminants to the inside air. This means that usually the new air being added to the space must be either a bit warmer or colder or drier or moister so that, when it mixes with the old air in the space, the result is comfortable (and healthy). Also, different people have different perceptions about what is comfortable, so there can be issues with trying to control individual comfort with a general thermostat.

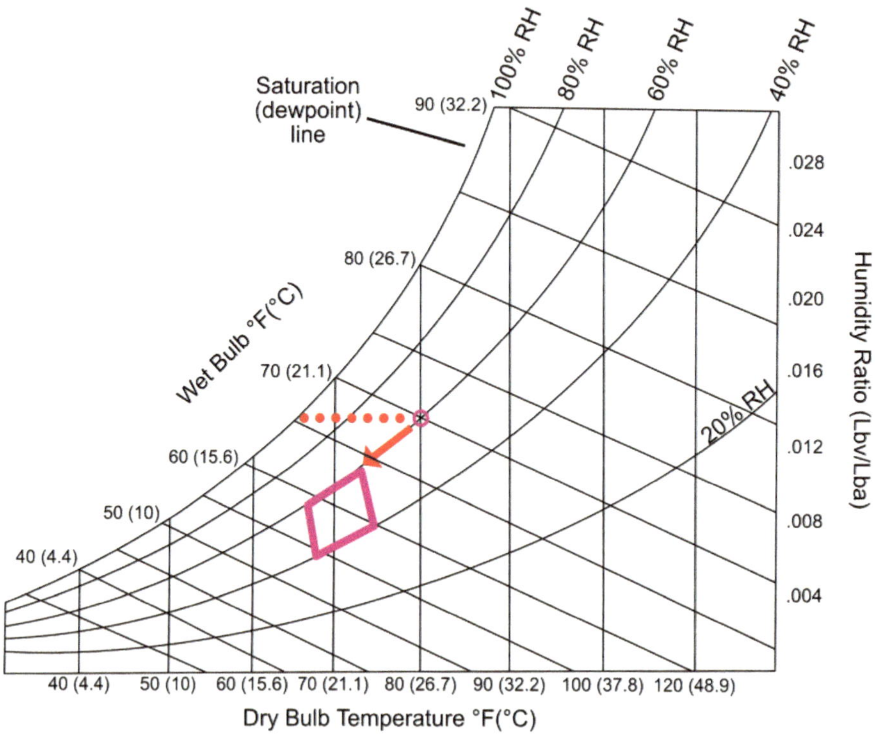


**Figure 6.4** Psychrometric chart conditioning air.

## Heating and Cooling Degree Days

A climate is all about what the outside air temperature and humidity is like. Across the world there are various ways that similar climates are grouped together. A simple example is cold, temperate, hot and arid, and tropical. Biologists would group climates by biomes which have similar types of plants and animals such as grassland, forest, and desert. For the mechanical systems of a building, climates are defined by heating degree days and cooling degree days. Because both HDD and CDD are based on daily temperature averages compared to 65°F (18°C), the average mean temperature used in the USA, there can be a lot more "days" in an HDD or CDD number than there are days in the year. The more HDD and/or CDD you have in a climate, the more uncomfortable the exterior conditions are all the time. If there are very few HDD and/or CDD, that indicates you are in a great tropical climate!

In the USA, HDD are measured as the number of degrees that a day's average temperature is below 65°F (18°C). If the high on a winter day is 34°F (1°C) and the low is 15°F (–9.4°C) then the daily average is 24.5°F (–4°C). Use the mean temperature and subtract the daily average: 65°F – 24.5°F = 40.5 HDD.

CDD are days with average outdoor temperatures above 65°F (18°C). First find the average temperature, say of a hot summer day with a high of 95°F (35°C) and a low of 72°F (22°C). The daily average is 83.5°F (28.6°C). Use the daily average and subtract the mean temperature: 83.5F (28.6C) – 65°F (18°C) = 18.5 CDD.

Generally, there are resources that list historical HDD and CDD for many locations, or at least average daily temperatures. One such resource is the National Oceanic and Atmospheric Administration (NOAA)[1] website which has historical weather data from around the world. BizEE Degree Days[2] is a free website which is easy to navigate and allows you to change the mean temperature. You need to create and download the heating and cooling degree days separately.

It is very helpful to know what the number of HDD and CDD are because they give an indication of how much energy may be required to make buildings comfortable. Let's examine table 6.1 which compares HDD and CDD of some cities around the world.

In Capri, annually there are practically the same number of HDD as CDD, with the highest HDD in January and February and highest CDD in July and August. Overall, most of the degree day numbers are relatively small numbers. This indicates a fairly mild climate where solar gain in the winter would go a long way towards making a building comfortable. You'd have to be careful not to overheat though, because some months have both heating and cooling degree days. Shading would be helpful in the summer: you'd have to look at overnight temperatures and humidity levels to determine the suitability of overnight ventilation to cool down the building. You might also be able to use earth tubes to cool down air before bringing it into the building.

Williamsport annually has many more HDD than CDD, and a lot fewer CDD than Capri. The HDD are

**Table 6.1** Comparative degree days for cities around the world.[2]

| City | Capri, Italy | | Williamsport, PA USA | | New Delhi, India | | Harbin, China | | Vitoria, Brazil | | |
|---|---|---|---|---|---|---|---|---|---|---|---|
| Latitude | 40.5° N | | 41.2° N | | 28.6° N | | 45.8° N | | 20.3° S | | |
| | HDD 65 | CDD 65 | HDD 65 | CDD 65 | HDD 65 | CDD 65 | HDD 65 | CDD 65 | HDD 65 | CDD 65 | CDD 72 |
| January | 500.1 | 1.2 | 1107.1 | 0 | 277.9 | 24.6 | 2001.1 | 0 | 0 | 512.2 | 295.3 |
| February | 404.2 | 2.8 | 869.9 | 0.7 | 125.1 | 91.7 | 1557.4 | 0 | 0 | 454.1 | 256.6 |
| March | 352.6 | 14.8 | 776.2 | 2.7 | 26.9 | 304.2 | 1088.3 | 0.5 | 0 | 468.8 | 253.8 |
| April | 206.3 | 47.2 | 462.7 | 18.6 | 0.2 | 677.4 | 555.6 | 9.9 | 0.1 | 375.2 | 175.2 |
| May | 85.5 | 141.6 | 203.8 | 85.8 | 0 | 850.2 | 217.5 | 88.6 | 2.5 | 290.8 | 109 |
| June | 11.6 | 349.7 | 60.3 | 192.8 | 0 | 850.4 | 52.3 | 165 | 2.5 | 254.5 | 88.9 |
| July | 0.3 | 530.1 | 11.8 | 327.6 | 0 | 704.8 | 5.3 | 337.3 | 11.6 | 193.2 | 60 |
| August | 0.1 | 517.8 | 20.6 | 265.9 | 0 | 655.1 | 27.8 | 215.5 | 6.9 | 203.8 | 53.2 |
| September | 4.8 | 339.7 | 100.1 | 138 | 0 | 618.3 | 192.6 | 52.1 | 0.8 | 278.8 | 99.6 |
| October | 40.5 | 172.7 | 341.1 | 33.5 | 1.4 | 475.6 | 675.6 | 5.1 | 0 | 352.2 | 147.7 |
| November | 179.1 | 33.7 | 709.2 | 0.9 | 56.3 | 168.1 | 1296.9 | 0 | 0.1 | 348.7 | 148 |
| December | 393.5 | 2 | 974.1 | 0 | 244.7 | 45.1 | 1870.5 | 0 | 0 | 435.9 | 220.9 |
| **Total** | **2178.6** | **2153.3** | **5636.9** | **1066.5** | **732.5** | **5465.5** | **9540.9** | **874** | **24.5** | **4168.2** | **1908.2** |

Downloaded from www.degreedays.net/ 6.30.2021

moderately large in the winter months from November through February. It has a problem with the shoulder seasons. That means that the sun is in the same location in March as in September but there are radically different numbers of HDD and CDD in those months. In March, the solar gain would be very desirable but, in September, I'll be overheating my building if there is solar gain. This indicates that I'll want to very carefully address how I'm shading windows and consider something that can be seasonally adjusted.

Annually, New Delhi has a very large number of CDD and not many HDD. This climate will require a careful look at how to keep the building cool; it will be really easy to have an overheated building. The HDD numbers are so low that there may not be a need to provide heating if the building is well insulated and can maintain comfortable temperatures through massing. A building in this climate probably needs to exclude solar gain throughout the year.

Harbin annually is a very cold location. It has huge numbers of HDD from November through March. It has very few CDD. Probably a building in this climate wouldn't need any cooling, the focus would be on how to maximize solar gain and insulate the building to maintain the heat once it is warmed up.

Vitoria may be a bit warm annually with a fair number of CDD 65 and virtually no HDD. The CDD 65 are very similar every month. This is an example where changing the CDD mean from 65°F (18°C) to 72°F (22°C) practically eliminates all the CDD. You can see this in the last column labeled CDD 72°F. It seems likely from the very low numbers in the CDD 72 column, depending on the actual temperature and humidity, fans might just keep building occupants comfortable.

## Passive and Passive–Active Design Strategies

Passive strategies are building design elements that do not require power to work, i.e., they don't use pumps or fans. Active strategies require power to work. The two most common passive design strategies are direct solar gain and natural ventilation. The amount of sun that is available during winter months depends on the latitude and the weather. Solar insolation is measured in $kWh/m^2/day$ and its value is dependent on the angle of the receiving surface compared to the sun's location in the sky.

### Thermal Mass

Solar heat gain can either be direct, indirect, or isolated. The heat gained is often stored in a mass of material, called a thermal mass, which then will radiate stored heat into the space over time. Thermal mass can be provided by a variety of items such as water, concrete, or **phase change materials**. For solid materials, darker colors on the surface absorb more heat, while lighter colors will reflect heat. The mass needs to have thickness to store the heat. For example, a concrete floor slab should be 4–8" (100–200mm) thick. Any time you are using solar gain for heating, you need to pay attention to the HDD vs. the CDD and provide appropriate shading. It is quite easy to overheat quickly in shoulder seasons.

Direct gain (figure 6.5) means that the sun's rays are shining through your window and your space is heating up because of it. Direct gain will also heat up thermal mass. Direct gain thermal masses are typically thinner in comparison to indirect masses. These thinner types of thermal masses that are located in floors, walls, and

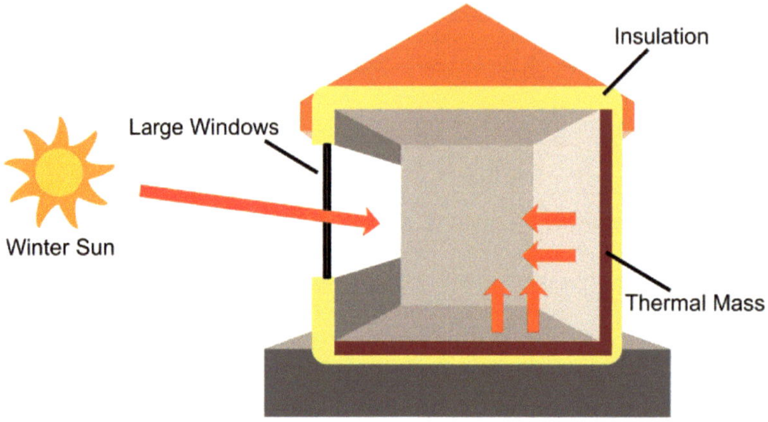


Figure 6.5 Direct gain thermal mass.

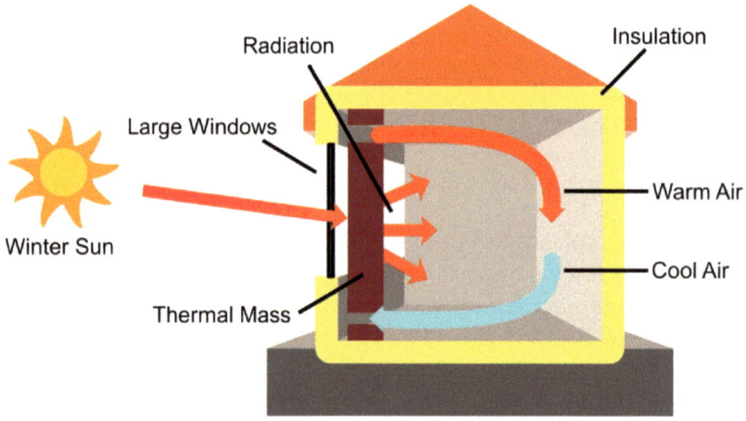


Figure 6.6 Indirect gain thermal mass.

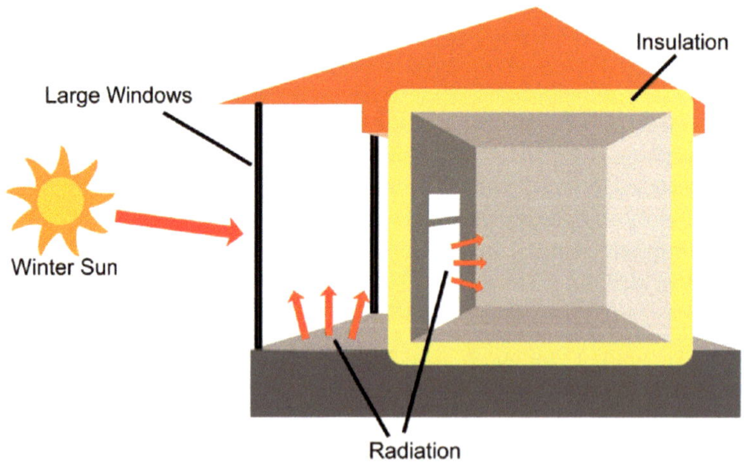


Figure 6.7 Isolated gain thermal mass.

ceilings defining the building's thermal envelope are typically insulated from the exterior conditions to keep the gain on the interior of the building.

Indirect gain (figure 6.6) means that the sun's rays are heating up the thermal mass and an air space rather than a room. The air space is located in front of the thermal mass which has vents at the top and bottom. The sun directly heats the air space and the thermal mass. Due to convection, the heated air rises and exits out the top vent, simultaneously pulling in cooler air through the bottom vent. Fans may be incorporated in order to more evenly distribute the heated air over a larger area. Typically, the vents are closed at night, relying on the radiant heat from the thermal mass to keep the space warm overnight.

Isolated gain (figure 6.7) uses a separate sunspace or greenhouse to collect the solar gain. The thermal mass is often the wall between the sunspace and the building but it can also be other masses located in the sunspace. As in the indirect gain system, the thermal mass can be vented at top and bottom. In many climates having sloped or horizontal glazing results in too much heat gain and can be very challenging to shade when it is undesirable to have heat gain, i.e., on already comfortable or hot days.

For all of these passive systems, the heat coming into a space can be collected into the mechanical system and distributed throughout the building. Coupled with a **heat recovery ventilator (HRV)** or **energy recovery ventilator (ERV)** these passive gain and thermal mass systems can very easily provide most of the heating energy required, as long as there is enough solar insolation to heat the thermal mass in the winter months.

Thermal mass can be used to mitigate temperature swings across days and months. This is helpful for resiliency (continued comfort even in a power outage) and reducing energy needs. Historical examples would include the great stone cathedrals in Europe that have very thick masonry walls. Albi cathedral, the largest brick cathedral, has walls 8.2′ (2.5m) thick. There are many other examples of thermal mass building from around the world made with materials such as adobe, rammed earth in tires, and other types of alternative materials (alternative compared to modern steel and stick frame construction methods). When the exterior building envelope is a mass wall that is thick enough, typically 3′ (914mm) or more, heat transfer through the walls happens so slowly that the summer heat is just getting inside when winter comes along, and it has just cooled off in time for summer. While there are residential and small commercial buildings that are built this way, it can be difficult to build thick thermal mass buildings. This is because of the need to maximize floor area in buildings as well as limitations of height and weight imposed by the building material being used as the thermal mass.

Thermal mass can also be used for cooling if night temperatures are low enough and air can be moved around it enough to cool it down. Often this strategy is called night ventilation and requires fans to ensure that enough air is exchanged that the thermal mass cools off enough that it will act as a heat sink (absorbing heat from the space) during the day. In cooling mode, a thermal mass shouldn't be heated by the sun.



## Natural Ventilation

Natural ventilation can be as simple as opening windows. If people open the windows, it is a passive system. If motors open the windows, it is an active system. Natural ventilation relies on either stack effect or cross ventilation to be effective. Stack effect (figure 6.8, left) means that there are windows located at ground level and windows or openings located much higher in the building. Air moves from the lower to higher openings and is assisted because of pressure differences and convection. Cross ventilation (figure 6.8, right) means that the open windows are located at the

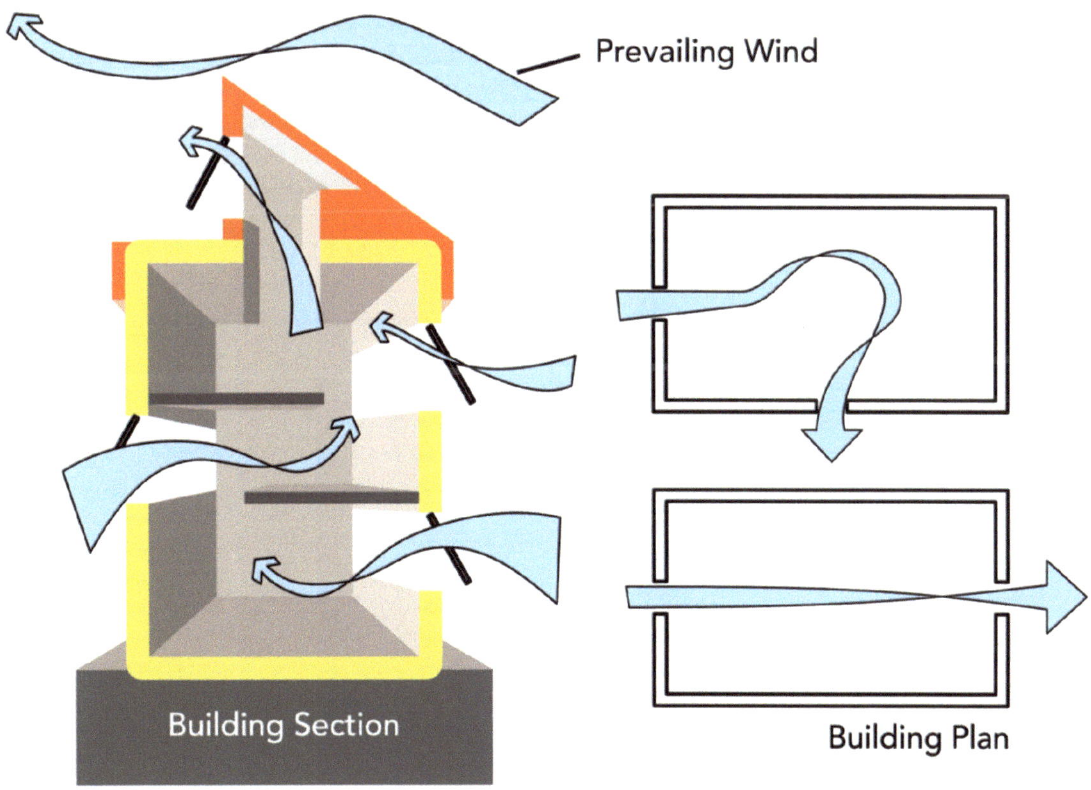


**Figure 6.8** Stack effect and cross ventilation.

same building level and are on different walls. Cross ventilation doesn't happen if the windows are on the same wall. Cross ventilation relies on air movement (wind) to move the air through the building. Windows that open all the way are more effective at capturing the prevailing wind, which should be perpendicular to the window. Operable window orientation may be tricky because winds may or may not be seasonally consistent. Fortunately, any wind blowing on your building will cause pressure differentials that can help to pull air through, and you can design features to capture the wind as well (this could be as simple as a casement window).

To find out where the wind is coming from, it is important to look at a wind rose. A wind rose is a circular diagram that represents wind conditions in radiating bars. A great tool for creating wind roses of worldwide locations is the Iowa Environmental Mesonet.[3] Just select the location and weather station and then create the wind roses: you can generate annual and monthly ones based on historical averages or set custom dates. Figure 6.9 shows yearly and monthly wind roses for Capri, Italy.

Wind rose diagrams always show the direction the winds come from by proportion. The longer the radiating bar, the more wind blows from that direction. The wind speed is color coded so you can get a sense of how strong the winds are from a particular direction. Looking at figure 6.9 (annual rose on the left), Capri's winds are calm 26.3% of the time during the year with a lot of gentle winds at 2–5mph (0.89–2.23m/s). Mostly, the wind in Capri is from the west-south-west. The whole set of 12 months is shown on the right, starting with January in the upper left and ending with December in the lower right.

Knowing the prevailing winds over a cooling season could be important. For this particular location, the winds don't change direction much but some locations have very different wind patterns over the course of the year. Capri, like Vitoria, is a location where CDD 65 is artificially creating more CDD than would be actually experienced, especially if the design were able to leverage natural ventilation wind patterns. Care would have to be taken to pay attention to humidity levels as they can have humid summer days. Perhaps the windows could be open overnight and early morning, then closed as the day heats up. These are questions that would be answered by looking more closely at the temperature and humidity conditions and doing computer modeling. Because the summer wind is mostly from the west-south-west, the best cross ventilation would require operable windows or vents on the west-south-west and east-north-east.

If you are passively cooling your whole building using either stack effect or cross ventilation, the air needs to pass through all the main areas of your building to be effective.

## Earth Tube Tempered Air

The earth can be used as a heat sink to temper intake air for mechanical systems. This is an active system because it uses fans to move the air. This can reduce energy costs by tempering the incoming air. For instance, cold outside air will be preheated because it absorbed heat from the ground or warm outside air will cool off. The greater the temperature difference between the outside air and the ground temperature, the more effective the heat exchange is. You'll need to know what the deep earth temperature is. Deep earth temperature is the depth where the ground temperature is stable, not fluctuating over seasons. The deep earth temperature changes depending on the climate. Generally, because of consistent temperatures near the equator, the earth is warmer nearer the equator. Because of the tilted angle of the earth as it rotates, the poles are colder – the earth temperatures are cold and remain frozen as permafrost year-round. Earth tubes work by pulling outside air through tubes that are buried in the ground (figure 6.10). The intake tube extends above grade high enough to keep it from flooding and is capped to keep wildlife out (figure 6.11). Small systems may be just one small diameter tube, usually not smaller than 4" (102mm), while large systems may have quite large diameter tubes, as much as 3' (914mm) which are laid out in parallel lengths (figure 6.10). Air should be filtered and UV treated to kill off microorganisms and mold prior to being introduced into the mechanical system.

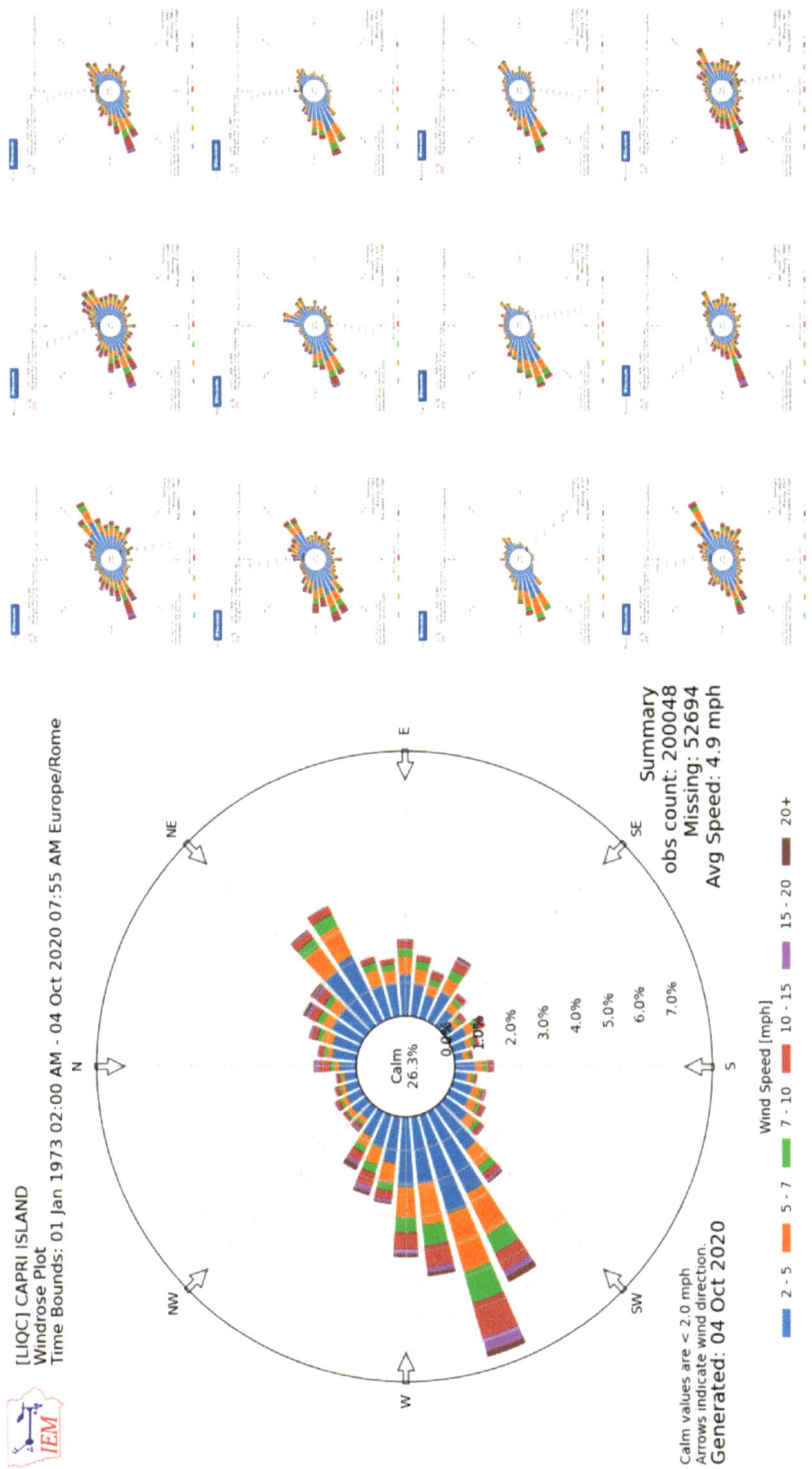


**Figure 6.9** Capri annual and monthly wind roses.[4]

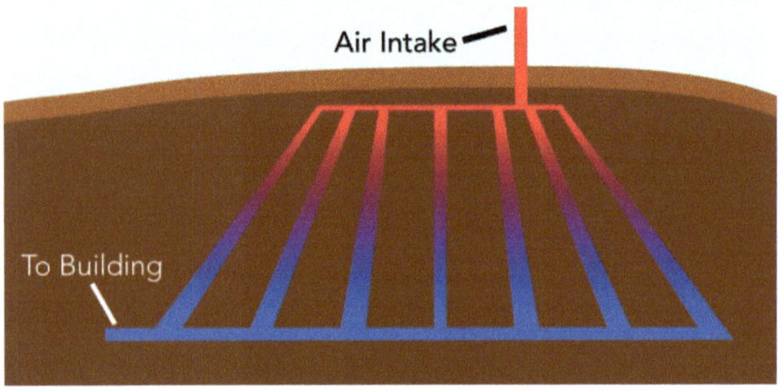


**Figure 6.10** Earth tube harp-type piping layout.

**Figure 6.11** Air intake for earth tube system.

## Understanding Shading

Solar gain is desirable when the building has HDD. However, the windows need to be shaded when adding heat is undesirable, that is, when it is comfortable or there are CDD. Shading needs to be addressed at an early point in conceptual design as an integral part of controlling the building's energy use. Shading can be on the interior of the building but interior shading has already let the heat gain into the building. The best solution is to use exterior shading which is integrated into the building's design, so the building shades itself. Shading devices can become mounting points for photovoltaics. There can be advantages to having adjustable shading devices, though they will cost more than fixed shading. Adjustable shading can be passive, where a person has to adjust it, or active, where sensors signal motors to move it. There are pros and cons to each. If relying on people to adjust it, the shading has to be reachable and easy to change, people need to understand why and when they need to adjust it, and then they have to follow through and adjust the shade. This is okay for a house but is not necessarily going to work for a commercial building, especially a larger one. Automated shading adjustment can have issues with sensors, servomotors, and computer programs interfacing with each other.

Recall that the sun's position changes throughout the day, both the altitude and azimuth, as well as changing across the seasons. In the morning (east) and afternoon (west) the sun is lower in the sky and the sun's rays strike the building more horizontally. That means the best shading element for the east and west façades is oriented vertically (figure 6.12). The length of the device depends on the actual angle of the sun at the latitude of the building, the time of day/year you want to keep the sun out, and how far apart the shades are spaced. It may be desirable to have adjustable shading depending on the climate. For example, there may be a few HDD which occur mostly in the morning on fall days. Heat gain might be desirable on one day but the next is warmer and the heat gain would be undesirable. With adjustable shading you could optimize the heat gain for just the mornings that you need it.

In the northern hemisphere the sun is in the southern sky, so the main heat gain occurs on the southern façade. In the southern hemisphere, because the sun is in the northern sky, the main heat gain occurs on the northern façade. In equatorial locations, both the northern and southern façades receive direct sunlight depending on the solstice, southern in December and

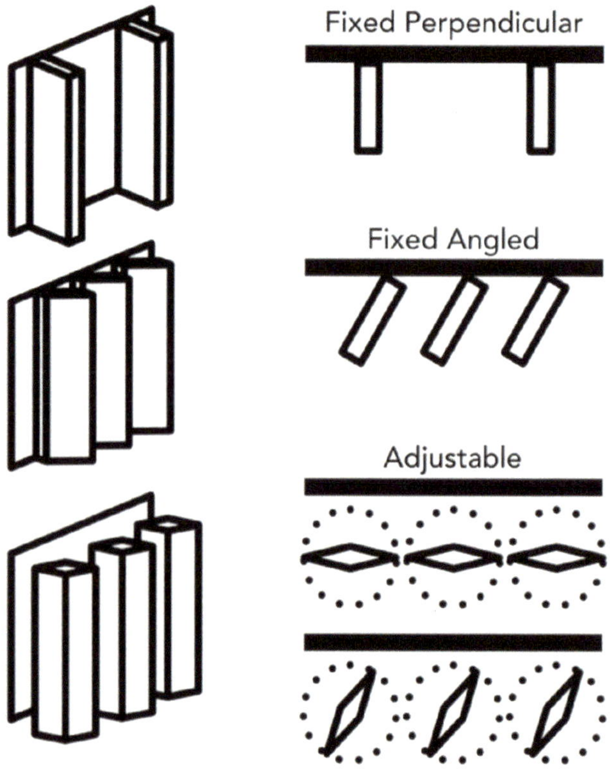


**Figure 6.12** Vertical shading elements for east/west sun angles.

northern in June. When the sun is in either the northern or southern sky, it tends to be at a higher altitude than when it is in the east or west. The higher altitude means that horizontal shading devices will do the best job at keeping out the sun (figure 6.13). The exact length of the horizontal element depends on the location of the sun at your latitude, how tall the window is, and whether the projection is horizontal or angled. Angled shades require less depth, as do horizontal shades with an added vertical element. If the shading device projection is too much, then additional elements can be added over the same window. The length of these shelves will be shorter, the length will depend on how far apart they are spaced. Multiple horizontal shading devices can be fixed or adjustable and can be directly attached to the building or on a framework in front of the façade.

Depending on your climate, latitude, and the orientation of your building, you may need both horizontal and vertical shading devices to appropriately control solar heat gain. These can be individually located at each window (figure 6.14) or can become an "egg-crate" over a larger window (figure 6.15). The length of the shade elements is determined by the latitude, climate, and spacing of the horizontal and vertical elements.



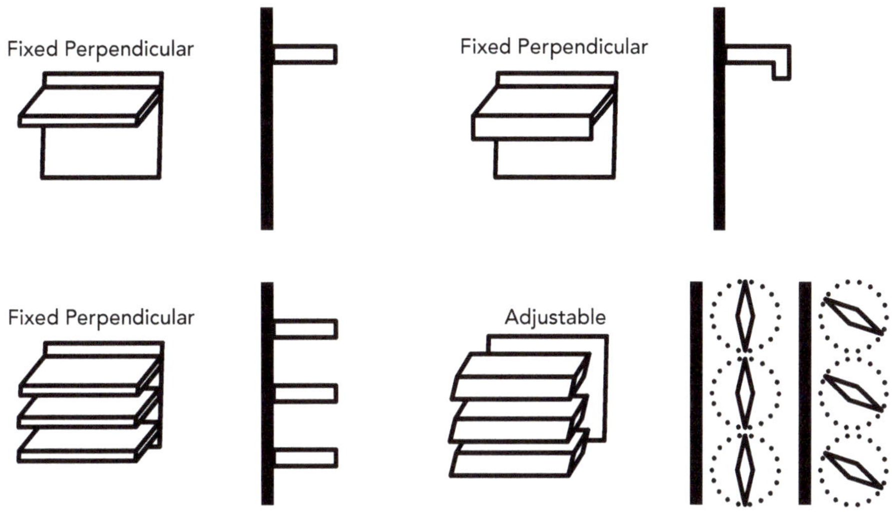


**Figure 6.13** Horizontal shading elements.

**Figure 6.14** Vertical and horizontal shading at NREL in Golden, Colorado, USA.

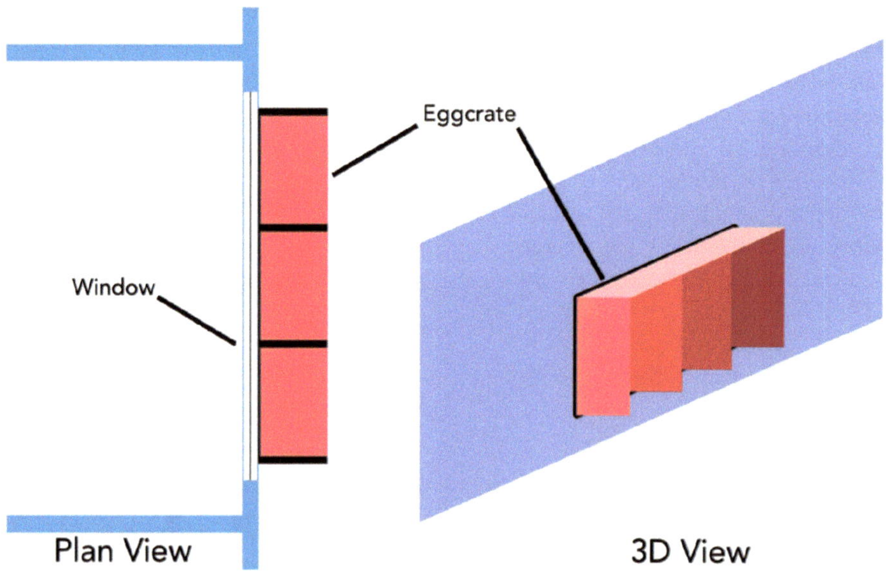


**Figure 6.15** Combined horizontal and vertical shading "egg-crate".

## Resources

1 National Centers for Environmental Information (www.ncei.noaa.gov/).
2 Degree Days (www.degreedays.net/).
3 Iowa Environmental Mesonet of Iowa State University (https://mesonet.agron.iastate.edu/sites/locate.php).
4 Capri, Italy data generated from Iowa Environmental Mesonet of Iowa State University (June 3, 2021).

## Study Questions

1. Select a location and look up climate information about it. Record the precipitation, temperatures, HDD, and CDD. Interpret the information answering these questions:
   a. How comfortable are the outside conditions in this climate?
   b. Will your building need heating or cooling?
   c. Will the amount and type of precipitation have an important impact on your design?
2. Go to the Iowa Environmental Mesonet of Iowa State University (https://mesonet.agron.iastate.edu/sites/locate.php) and create a wind rose for a particular location. Look at and describe the patterns at different times of the year. If you used the same location as your climate information, how would the wind pattern impact comfort throughout the year?

CHAPTER 7

# Building Performance and Energy Efficiency

**Key Terms**

Infiltration: air that passes through all the holes and discontinuities in the building envelope, i.e., air leakage.

Thermography: use of thermal imaging, which visually displays infrared energy emitted, transmitted, or reflected by an object.

## Concepts for Success

The building designer needs to provide an envelope that will insulate, be air sealed, control water vapor, shed water, and support the building. This is true for the roof, walls, and floor of the building. All of these envelope components need to be as continuous as possible to be effective.

The thermal envelope insulates the building. It is measured in R-value or U-value. The continuity and performance level of the thermal envelope has a direct impact on how much energy the building uses to maintain comfort.

The air barrier keeps conditioned air inside and unconditioned air outside. Leakage through holes in the building envelope is called **infiltration**. Many older buildings rely on infiltration for fresh air. There are a number of problems that occur by having unregulated air exchange. There is no way to control the volume of air exchange, the quality of the air coming in, nor control of water vapor. Infiltration can lead to uncomfortable and unhealthy conditions for building occupants.

The thermal envelope and air barrier must be next to each other to be effective. Defining the location of the building's boundary is important and can be tricky. Any section cut through the building should be able to trace a continuous line around it to locate the boundary. The simpler the boundary is, the easier it will be to construct the details to keep the insulation and air barrier continuous. Any building element that touches both the outside and inside of the building, without being insulated, creates a thermal bridge. Thermal bridges waste energy and can create uncomfortable conditions inside the building.

Thermal imaging and blower door testing are used in existing buildings to help identify envelope problems. Better insulation and tighter envelopes make a building more comfortable, healthy, and reduce energy use.

Buildings use energy to be comfortable and occupants use energy to perform tasks. Your building will use different amounts of energy depending on the time of day and year. This demand is generally higher during the day and less overnight. Renewable energy is generated depending on the type of resource. Resources are not always available, for instance, the sun has limited daylight hours in any location and there might be clouds obstructing the sun on a particular day. Achieving net zero energy use can be measured either annually or daily and requires the designer to understand how to appropriately leverage the renewable energy resource for the particular project.

Energy usage patterns for different building types can be compared through energy rating systems such as the **Home Energy Rating System (HERS)** index and Energy Use Intensity (EUI). HERS Index is used for houses while EUI is used for commercial buildings. Both systems compare energy usage to building area, so it is possible to directly compare the energy efficiency of buildings.

DOI: 10.1201/9781003297819-8

## Buildings as Systems

Buildings are complex systems that balance occupant needs, building infrastructure, and climate. Occupants expect to be comfortable and healthy and achieve their needed tasks. The building has to stand up, keep the weather out, condition air and distribute it, provide power, and create great spaces where the occupants can thrive. The climate provides daylight, varying temperatures and humidity, precipitation, and a biome full of life.

The building envelope is the outside of the building, all the way around. Like an eggshell, on the inside it contains an environment different from that on the outside. Unlike an eggshell, buildings have lots of holes. Some are intentional, such as windows and doors, and others are not, such as where different materials meet. The building envelope must shed water, insulate the conditioned space, keep air in or out, look good, and support the building.

Besides windows and doors there are many items that pierce through the building envelope. Power cables, water and sewer lines, vents from equipment and plumbing, and ductwork for mechanical systems, not to mention attachments that hold lights, finish materials, and other exterior building features, all create holes and thermal bridges. Places where materials meet or abut have to be sealed in order to prevent air and possibly water leaks.

### Thermal Barrier

Insulation creates a thermal barrier. Insulation slows down the transfer of heat energy between inside and outside. Nature is trying to balance conditions inside your building with the conditions outside: you might recall your physics teacher telling you that nature abhors a vacuum. The more dramatic the temperature difference between the outside conditions and the interior conditioned air, the more insulation is needed to slow down the transfer of heat by conduction. The effectiveness of insulation is measured in R-value and sometimes identified as **R SI-value (RSI)** for metric units. The larger the R-value number, the better the insulation. R-value can either be a unit measure for a material, such as per inch (mm) of thickness, or for an assembly, such as a roof or door. R is measured in (ft$^2$ °F h)/Btu and RSI is in (m$^2$ K)/W. Some computer programs and building elements use U-value instead of R. U-values are the inverse of R-values. Because U-value is the inverse of R-value, the smaller the number, the better insulator it is. Windows are almost always listed by their U-value. U-values in the United States are Btu/(ft$^2$ °F h) while SI units are W/(m$^2$ K). See table 7.1 for R- and U-value conversions.

It is easy to visualize R-value (figure 7.1). Which coat do you pick for biking on a cool spring day vs. skiing on a cold winter day? Your ski coat is warmer because it

Table 7.1 R- and U-value conversions, US units and SI units, RSI = R-value in SI units, R = R-value in US units.

| R-value and U-value Conversions | |
|---|---|
| R to RSI | R ÷ 5.678 |
| RSI to R | RSI × 5.678 |
| UUS to USI | UUS × 5.678 |
| USI to UUS | USI ÷ 5.678 |
| R to U | 1/R |
| U to R | 1/U |

has more insulation, it is probably visually puffier and weighs more than your lightweight jacket. Note that when you are wearing the winter coat, on its interior the temperature matches your body and at its exterior the temperature matches the outside air. That temperature change is constant through the depth of the coat. If you were headed out for a long adventure into very cold weather, you might also have ski pants and boots, maybe even a special hat, so you can completely cover up and stay warm. It is the same with your building: you want the insulation to be continuous because, if there is a gap in it, it is going to be cold where the gap is. You wouldn't buy a winter coat that had a big area of stuffing missing. The problem with buildings is that it is difficult to see missing insulation.

Figure 7.1 A thermal barrier insulates "inside" from "outside" similarly to clothing keeping you warm.

### Air Barrier

The component keeping the air in or out of the building is called the air barrier. The air barrier seals up the building envelope and reduces the amount of natural air exchange. Differences between inside and outside temperatures, wind, and the operation of fans and mechanical systems inside the building all create pressure across the building envelope. Any holes in the building will exchange air (leak) if they aren't sealed up. Although you might think that there aren't many holes in a building envelope, there can actually be many. In the past, for residential homes in the USA, fresh air was provided by air exchange through all these holes. Air is coming in one set of holes (infiltration) and exiting through another set of holes (exfiltration). You might say, "oh, that makes sense, it is important to have fresh air." While it is true that fresh air is important, the problem with air leakage is that you have no control over how much or where air exchange is occurring, nor can you control the quality of the air coming in since it is randomly coming in from everywhere.

Let's visualize that winter coat again. Imagine that there is stitching on the coat that goes through all the layers, including the insulation. Where the needle pierced through it has left a hole. A small hole but it is still a hole. And there's a lot of holes because the stitching is all around the coat. It is important to note that the stitching didn't change the insulative value of the coat. You probably won't notice the air exchange through the small needle holes but imagine that they are large enough to put your finger through, like the torn clothing shown in figure 7.2. Where there aren't holes, the insulation is still going to be working to slow down your body's heat transfer to the cold snowy day. Unfortunately, you are not going to be very comfortable because there is going to be air exchange happening through the holes – cold air! You wouldn't purchase a winter coat that had a bunch of big holes in it. The problem with buildings is that you can't easily see the holes.

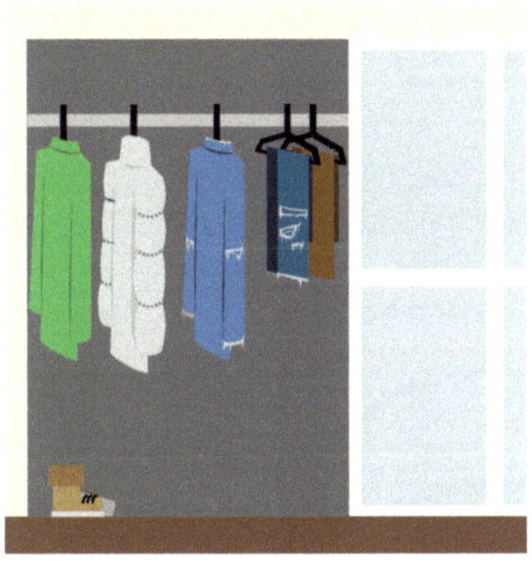

**Figure 7.2** An air barrier prevents free transfer of "inside" and "outside" air; coats with holes aren't warm.

Air leakage causes direct heat loss as well as drafts. The other big problem is that air has water vapor, measured as relative humidity (RH). The moisture is going to travel with the air because it is very small and suspended with all the air molecules. If you read about the psychrometric chart in the Design for the Climate chapter (6), you'll remember that hot air holds a lot more moisture than cold air. Our lungs and materials in our buildings are happy when the interior air is 40–60% RH. If it is a cold day out, as interior warm air moves through the wall, it will pass through the dew point: water vapor will fall out of the air and into the wall assembly (figure 7.3, left). The exact location of the dew point inside the wall varies because temperature and humidity are always changing. In hot, humid conditions where the interior is air conditioned, the incoming hot, humid air can shed water vapor into the wall assembly (figure 7.3, right). The water vapor will condense on the first cold surface it encounters and that surface will act as a dehumidifier. Dependent on the volume of air moving through the wall assembly, the interior of the wall cavity can become wet and grow mold. Mold can cause mild to severe health issues for people and pets as well as cause damage to building materials. The mold is growing where the air leakage is occurring, so it will be distributed with the air as it leaks.

If we go back to wearing the winter coat with all the holes in it, it is actually possible that, as the wearer breaks a sweat skiing down the hill, little bits of frost could develop around all the little holes. In a building, this might look like frost on the sharp ends of the nails that are holding on the roofing product and protruding into the attic space. A web search will show you many pictures of frosted nails as well as sheathing! The reason for this is that warm, humid air from the house is leaking into the attic and the first cold surface is the nail and the sheathing. An attic is a place you can see the sheathing and frost – the same thing is happening in the walls where you can't see it.



## Continuity and Relationship of Thermal Envelope and Air Barrier

The thermal envelope and air barrier must be located next to each other and be continuous around the entire building. In theory this sounds straightforward. In practice it takes a lot of attention to detail to achieve a continuous envelope.

Let's start with the thermal envelope. Buildings have roofs with and without attics, walls, floors, and foundations of various types. All these elements are made of different materials and have different exterior finishes. It is very important to identify where the thermal envelope is located. Energy efficient buildings have all mechanical ductwork, mechanical systems, and the water heater inside the thermal envelope.

Foundations can be slab-on-grade, crawlspace, or a basement. Basements (figure 7.4A) are deep and have headroom for people to walk around in. Full basements are common in areas where the ground has a freeze–thaw cycle so that the bottom of the foundation is below the frost line. Basements can be either in or out of the thermal envelope. If you are putting mechanical equipment down there, or using it as extra living space, the basement has to be inside the thermal envelope. Slab-on-grade (figure 7.4B) typically has just a thicker edge on the slab rather than a separate foundation and is often used in areas with no frostline. The slab becomes the defining edge of the thermal envelope. Crawl spaces allow for access under the building but are short spaces. Crawl spaces are typically outside the thermal envelope and shouldn't have any type of mechanical equipment or ductwork located in them. The location of the thermal envelope for crawl spaces is the ground floor framing. Below-grade crawl spaces (figure 7.4C) are mostly located below grade and are often used in climates that don't have a deep frost line. They reduce the height difference from grade to the

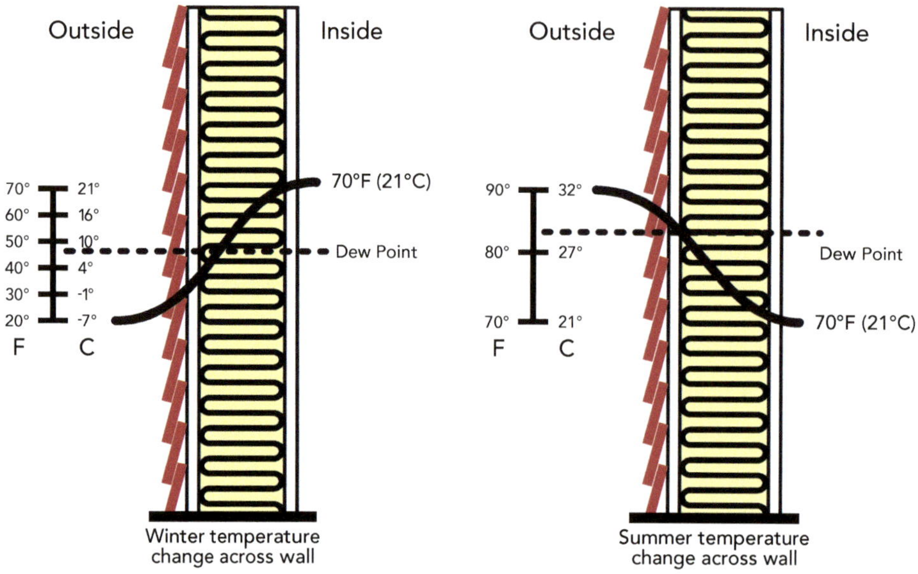


**Figure 7.3** Wall assemblies change temperature: inside is conditioned while outside isn't. The left shows temperature and dew point in cold weather, while the right shows hot conditions.

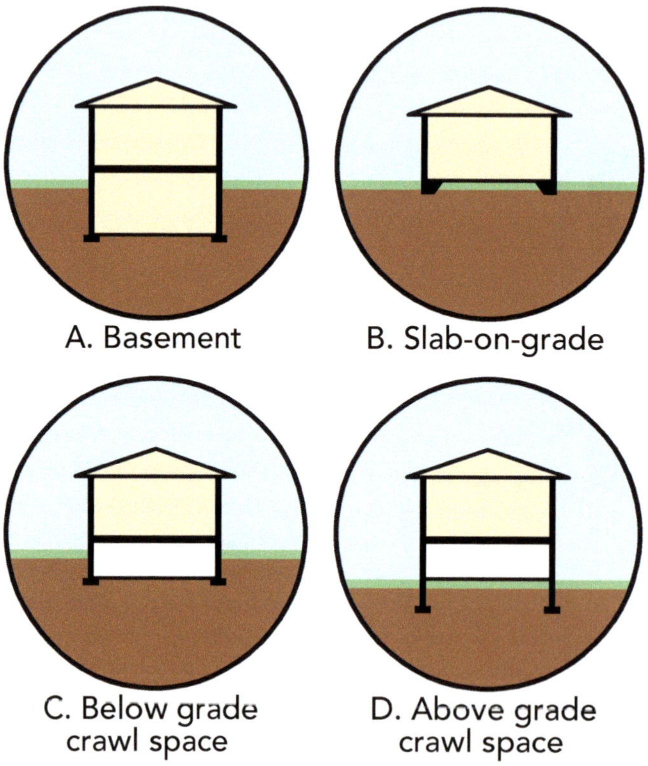


**Figure 7.4** Foundation types: basement (A), slab-on-grade (B), below-grade crawl space (C), above-grade crawl space (D).

ground floor. Above-grade crawl spaces are often used in climates with no frost line (figure 7.4D). They increase the height difference from grade to the ground floor, making accessibility without stairs to the ground floor an issue.

Slab-on-grade and crawl space foundations may be used in colder areas as long as they are frost protected. Buildings may have a combination of foundation types depending on the terrain, climate, soil type, and building design.

The walls define the vertical part of the thermal envelope – there may be indoor/outdoor rooms that will not be insulated. The next decision is locating the thermal envelope at the roof. Attic spaces, half walls, cathedral ceilings, dormers, and other design elements may be part of your roof. If any mechanical equipment or ductwork will be in an attic space, then the attic should be inside the thermal envelope.

No matter what your building looks like, you should be able to trace a continuous line all the way around a building section, cut in any direction, that defines the thermal envelope and air barrier.

Figure 7.5 shows a building cross section where the basement is inside the thermal envelope and the attic is outside the thermal envelope.

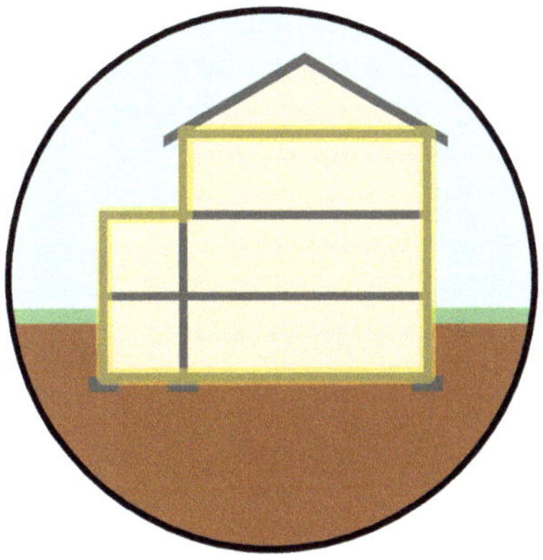

**Figure 7.5** Continuous envelope defined with basement in, attic out.

Figure 7.6 is the building with both the attic and the basement inside the thermal envelope.

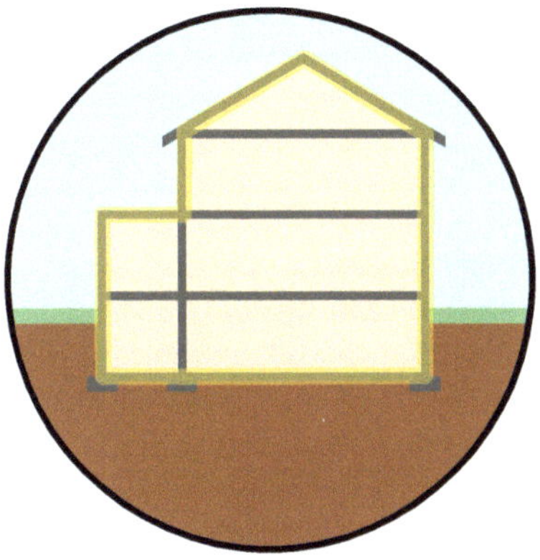

**Figure 7.6** Continuous envelope defined with basement in, attic in.

Figure 7.7 is the building with a below-grade crawl space that has the thermal envelope at the ground floor line and the attic is inside.

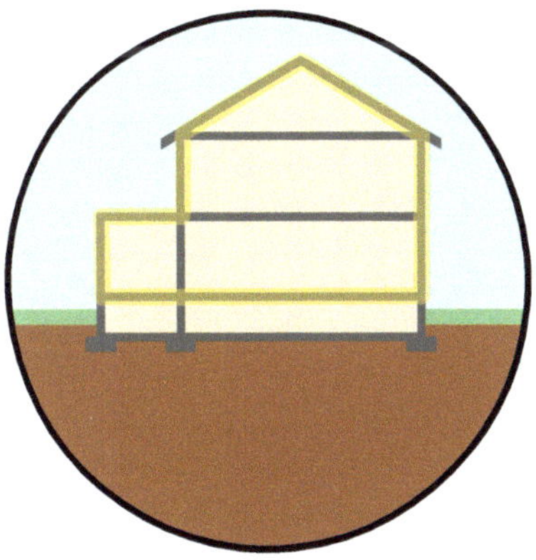

**Figure 7.7** Continuous envelope defined with crawl out, attic in.

Figure 7.8 is a cross section with a bonus room under the pitched roof. Often there is not enough headroom at the edges of the bonus space, so shorter walls (knee walls) are put in. In this diagram the thermal boundary follows the habitable space of the bonus room. Notice that the shape of the thermal envelope is a bit complicated. Remember that the thermal envelope is supposed to be continuous. Everywhere that there is a change in the orientation of the thermal boundary, or a connection between different materials, there is likely a lot of framing. The designer will have to figure out how to insulate all that framing, which can be a challenge. Remember that the air barrier is going to follow this same line. The more complicated it is, the more difficult it will be to achieve continuity of the thermal envelope and air barrier.



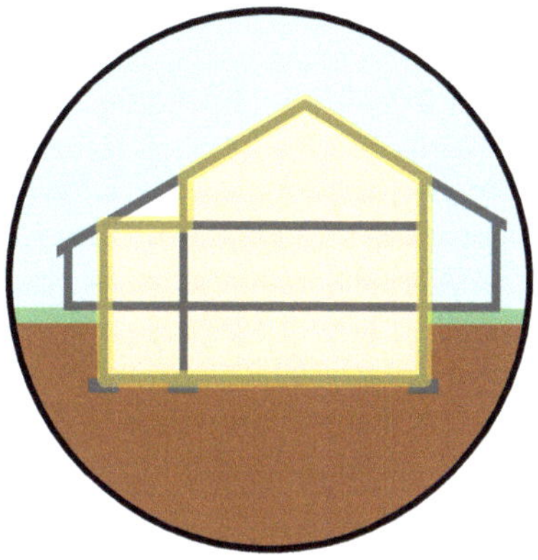

**Figure 7.8** Continuous envelope following habitable space at attic knee wall, basement in.

To simplify this building, the thermal envelope could follow the roof line to the exterior walls as shown in figure 7.9. Including the smaller attic spaces inside the thermal envelope could provide advantages for locating ductwork or plumbing. A big advantage to this simplified envelope is that various building elements have different R-values. Ceilings have the highest R-values. In this project, it would be much easier to continuously insulate the roof than figure out how to partially insulate the roof and floor as outlined in figure 7.8.

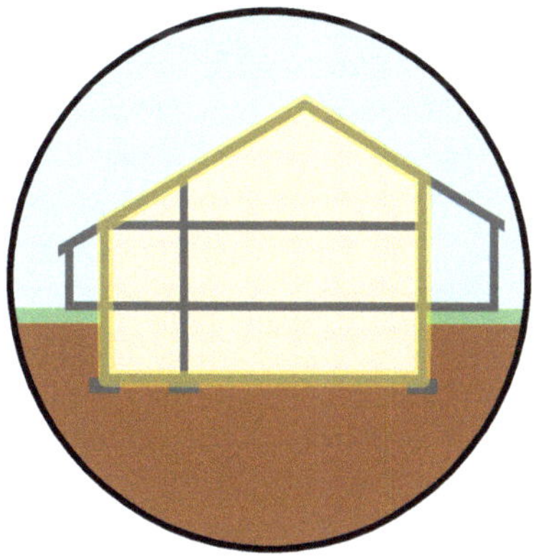

**Figure 7.9** Continuous envelope including attic storage simplifies buildability, basement in.

**Figure 7.10** Interior of computer room: photo (left) and thermal image (right).

## Designing and Testing for Continuity of the Thermal Envelope and Air Barrier

All buildings should have a defined thermal envelope and air barrier. Identifying the location is the first step. The second step is to determine how to make them as continuous as possible through every connection and attachment, joint, and material change. Any building element that touches both the outside and inside of the thermal envelope is called a thermal bridge. There are computer programs that can test the heat transference through assemblies with thermal bridges, such as THERM. This can help a designer to identify how much loss will happen through fasteners or areas with less insulation because of complicated framing and material joints.

In existing buildings, the continuity of the thermal envelope can be examined through **thermography**. Thermal imagers today come in a range of options. They are available as fairly expensive hand-held units that will store not only images but also voice notes. Smaller units are less expensive and have fewer options. There are also attachments for mobile phones that are very reasonably priced. Thermal images record heat differences. If the thermal envelope is continuous and well designed and built, then the exterior of the building should be the same temperature as the exterior air temperature and the interior surfaces would all match the interior air temperature. These images would all be one color.

Figure 7.10 is a pair of photos that shows the interior of a computer classroom taken on a cold winter day. On the left is a regular photo of the space. You can see the window in the exterior wall with shades half drawn, computers on tables, and chairs. On the right is the same space in a thermal image. There is a thermal gradient scale on the right side of the image ranging from white/yellow warm at the top, hottest reading 72°F (22.2°C), to blue/black cold at the bottom, coldest reading 54°F (12.2°C). The majority of the interior surfaces and furniture are around 69°F (20.5°C). The window frames are about 57°F (13.9°C). The shade is blocking the view of the upper part of the window frame, although you can still see colder lines. Most interesting are the vertical lines that can be clearly seen at equal intervals along the wall: the metal stud framing. It is not well enough insulated on the exterior to keep it from acting as a thermal bridge, although it is better insulated than the window frame. You can also see that near the floor in the room the walls are cold between the studs. This could indicate air infiltration or missing insulation. Remember that cold air transferring through this area would cause direct heat loss as well as generally cooling down the area around the leak through conduction. It is quite likely that sitting near the wall and the windows in this room would be uncomfortable due to all this heat loss.

Figure 7.11 shows this same area on the exterior of the building, the photo on the left and thermal image on the right. The computer room window from figure 7.10 is the one to the farthest right. This room is at a building intersection where the abutting wall is taller and extends out from the computer room wall. The gradient on the figure 7.11 thermal image ranges from a high of 48°F (8.9°C) to 30°F (−1.1°C). Note that the tree leaves are the same as the ambient air temperature so, if the exterior of the building were well insulated, it would all be the same color as the leaves. Instead, it measures 39.3°F (4.1°C) and the window frame and floor framing are clearly visible.

Thermal imaging is a very effective tool to help identify problems of missing insulation, air leakage, and possible moisture issues in existing buildings.

Blower door testing (figure 7.12) is used to determine the amount and location of air infiltration in existing buildings. The blower fan is set up in a door of a building and then blows air out. The target is to reach a pressure differential of 50Pa (pascal) which is equivalent to 20mph (8.9m/s) wind on all faces of the building. The pressure causes all the leaks in the building to leak. Readouts on the manometer indicate what the amount of leakage is. The person doing the test can find the location of air leakages by checking around the building. Leakage results are recorded in ACH50, meaning air changes per hour at 50Pa. These

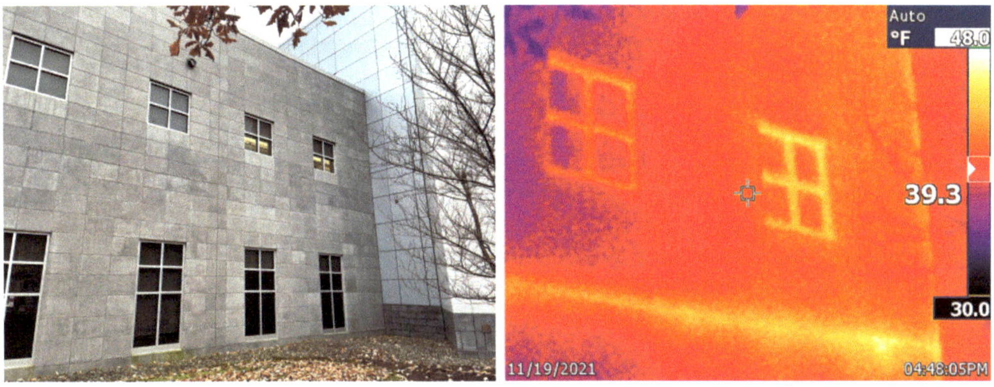


**Figure 7.11** Exterior of building at same computer room: photo (left) and thermal image (right).

**Figure 7.12** Architecture students setting up blower doors at Penn College's Clean Energy Center.



can be converted to natural air changes, which is how much the building leaks under normal conditions. The more air changes there are an hour, the leakier the building. Identifying the location of the largest areas of infiltration allows the owner to fix them and achieve better building comfort.

When a building is very tight, with little air leakage, fresh air is provided by ventilation through the mechanical system. If an existing building doesn't have a mechanical ventilation system, either it has to be kept at a leakage level where the occupants are getting enough air changes through infiltration or it needs to have ventilation added. It is more efficient to condition air with ventilation. Ventilation provides control over the quality of the incoming air. Eliminating infiltration creates a healthier and more comfortable environment for occupants.

## Energy Usage in Buildings vs. Renewable Energy Generation

Depending on the use of the building, the energy needs vary over time. They vary day to day and seasonally. The shaded part of the graph in figure 7.13 indicates when a household might be using energy. Time is across the bottom and the amount of power consumed is on the vertical axis. It is zero at the bottom, using more as you read upward. This graph could represent the electrical energy used in a household where the residents are sleeping at night, getting up in the morning, perhaps some people stay home and others head off for the day, and then everyone is home for dinner in the early evening. The spikes on the graph are from the start-up draw of compressors on a refrigerator or freezer.

Imagine that this house has a cooling system. In the summer months, the overall power use would be higher, especially from noon on. This would increase the energy use across the whole day as seen in figure 7.14.

Similarly, if this house were in a cold climate, there would be increased energy demand in the heating season. The exact appearance of this graph would depend on how energy efficient the home is and how hot or cold the climate is.

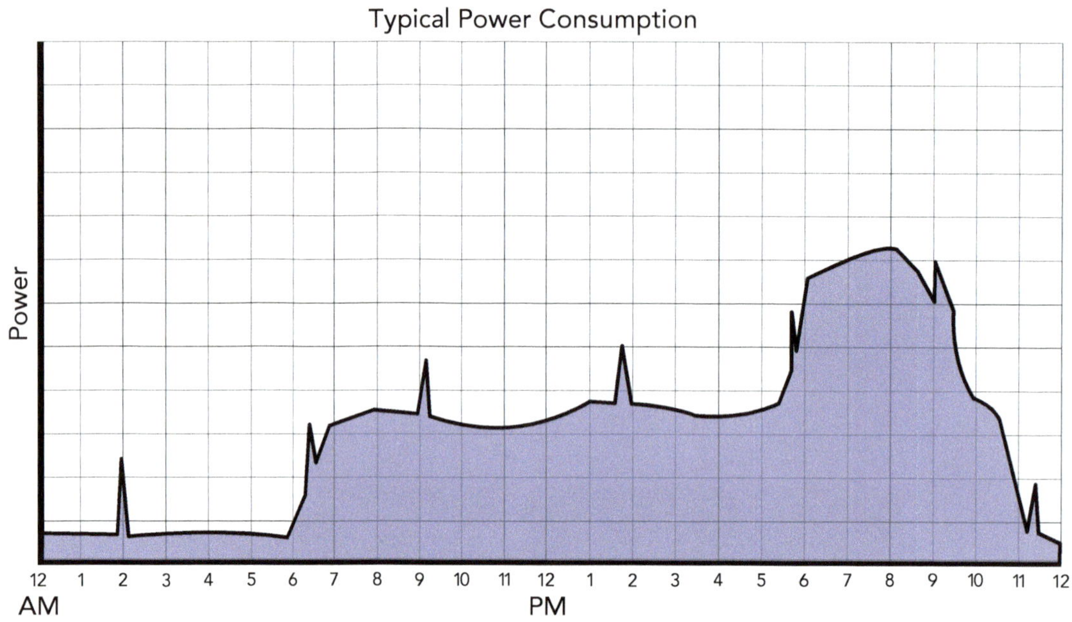


**Figure 7.13** Hypothetical residential electrical energy use graph over 24 hours.

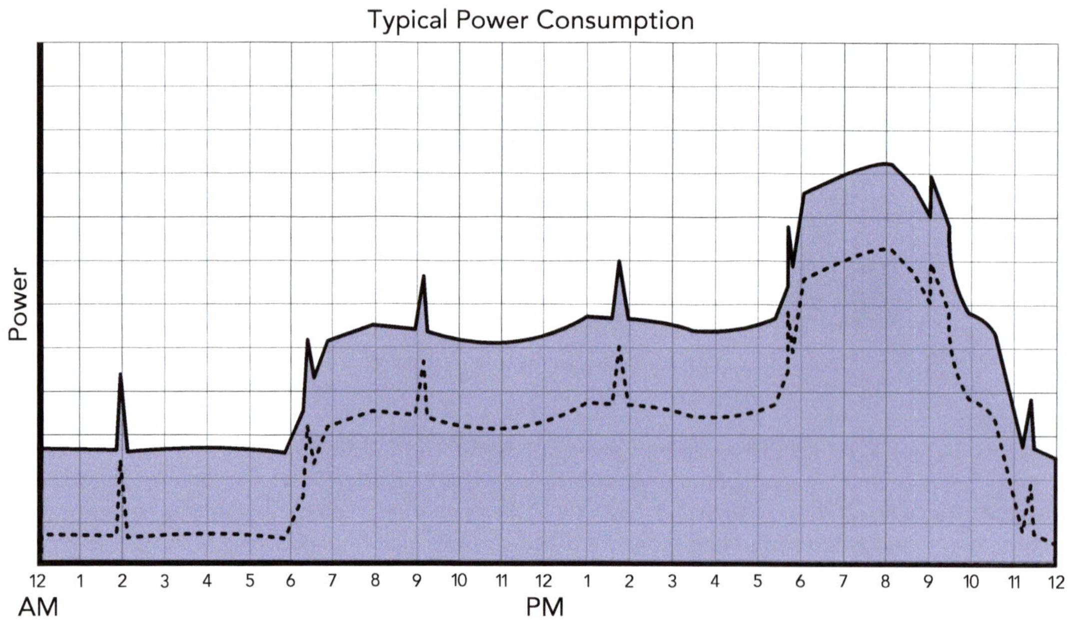


**Figure 7.14** Increased residential demand during cooling season in a hot climate.

Power demand is going to be based on what the building is used for, the type of equipment in the building, and when people are using the building. A hospital will use a large amount of energy all the time but more during the day when surgeries are happening and the kitchen is cooking food. An office building most likely has very little overnight power consumption but has a big increase in the morning that stays high until the evening. The office building is not likely to have the same occupancy on weekends and holidays so, on those days, the building would not use much power, perhaps similar to overnight use. Schools may have similar consumption patterns to an office building but they may be mostly closed over the summer months.

Building performance computer modeling programs have many choices for occupancy patterns in order to predict energy usage. Typically, commercial buildings have energy saver modes for when the building is unoccupied and then ramp up heating and cooling when the people arrive. People are going to turn on lights, computers, and equipment, all of which increase demand.

Figure 7.15 shows average hourly electricity load during a typical day in various regions of the USA. This is based on data collected since 2015. Note that January has a morning and evening peak, July has increased afternoon to early evening demand with a peak around

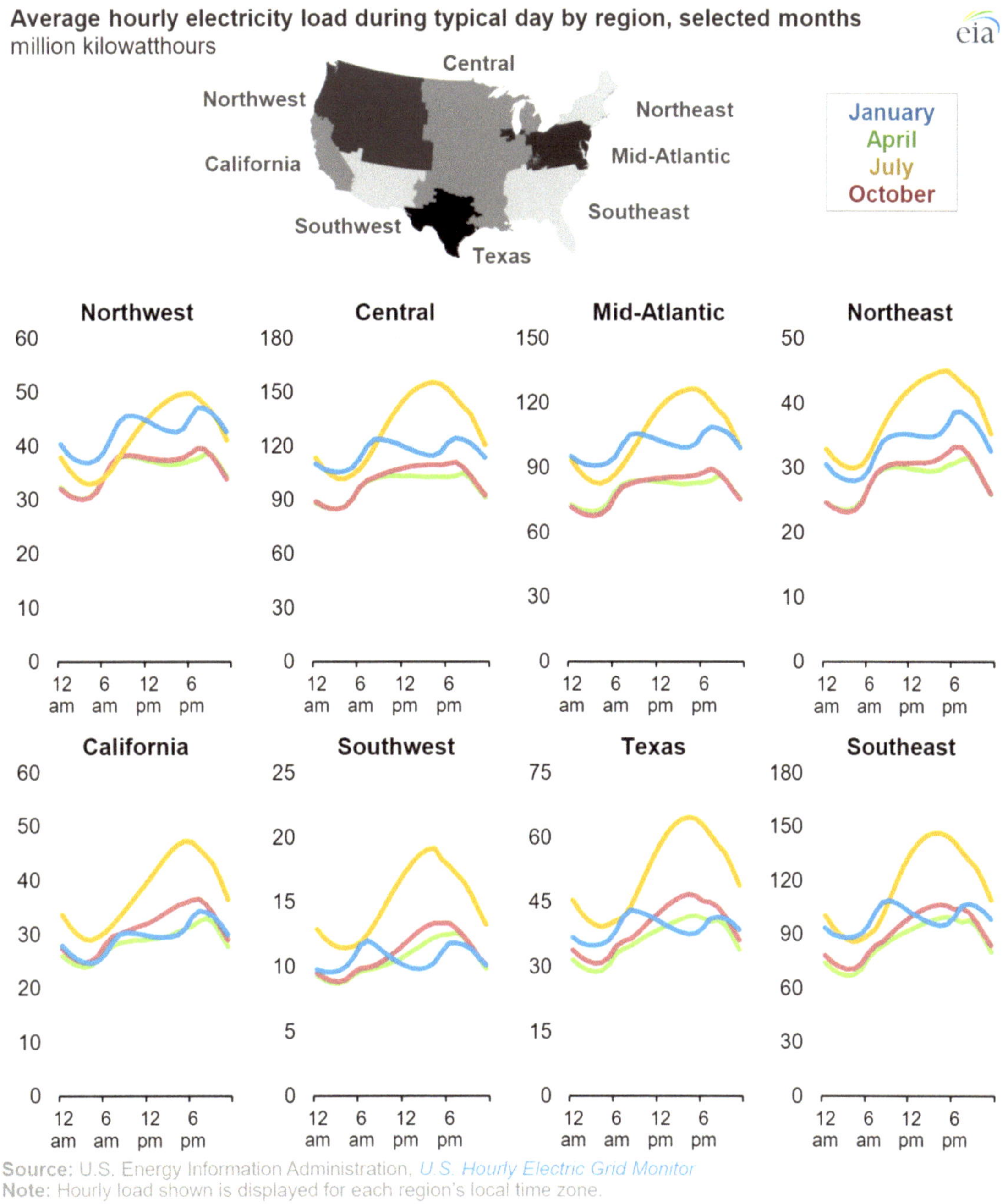


**Figure 7.15** Average hourly electricity load during typical day by region in the USA.[1]

5 or 6pm. Shoulder seasons, April and October, have fairly even demand with increase to 6pm, depending on the regional climate. Overall, the highest demands are in the hot summer months. This graph shows the actual usage of all buildings connected to the grid. Utilities consider weekday hours from 7am to 11pm as on-peak hours, while everything else, including holidays, are off-peak hours; they may charge higher prices for energy use during on-peak hours. These graphs have time across the bottom and million of kWh used on the vertical axis. Note that the vertical axis units of the various regions are different, with Central and Southeast using the most energy and Southwest using the least energy.

When designing renewable energy systems for a building, it is important to determine what portion of the energy you want to be producing on site. If you want to produce 100% of the energy annually, then it may not be an issue if you have variation in generation seasonally, because you can make more energy than your building needs when you have the resource.

## PV Power Production

For instance, figure 7.16 shows the summer production for a PV system (pink) compared to household power use (blue). The most amount of energy is produced midday, peaking at noon (solar time when the sun is directly

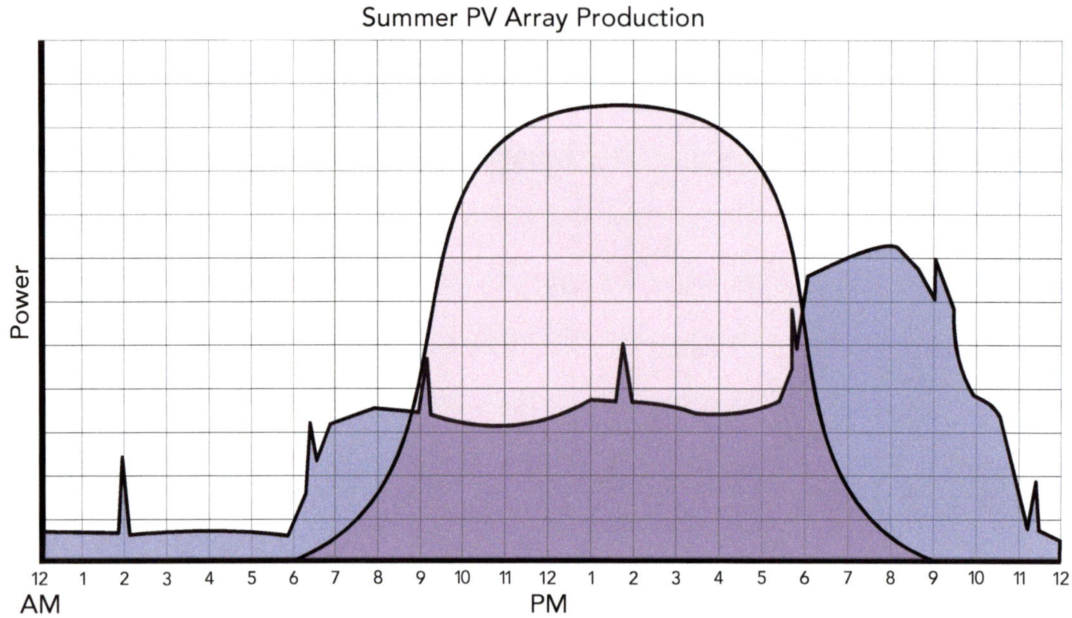


**Figure 7.16** Summer PV production exceeds household demand.

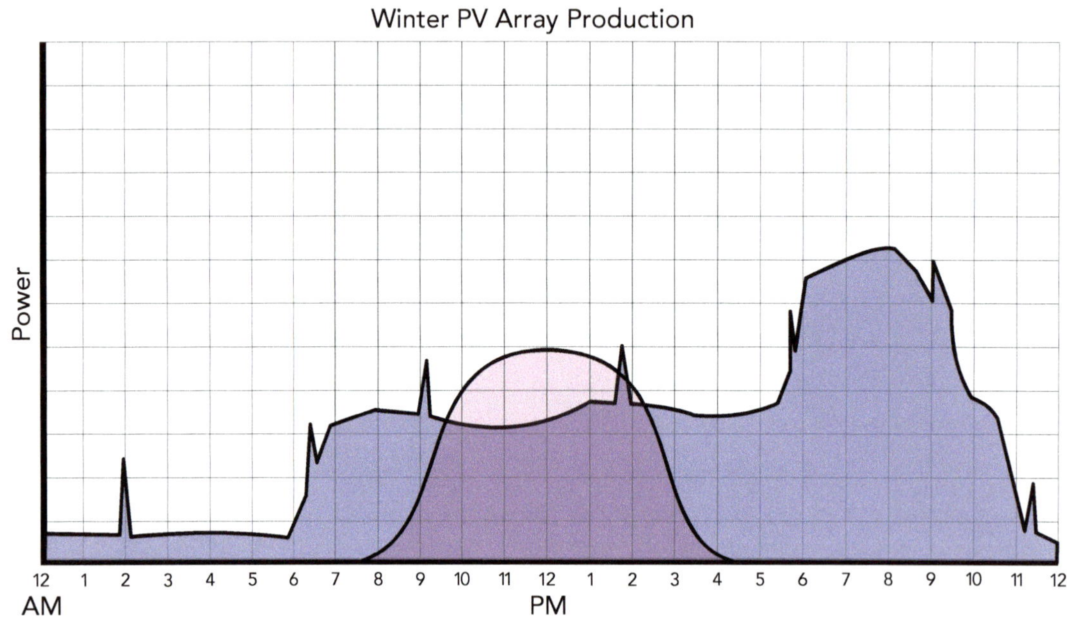


**Figure 7.17** Winter PV production just meets midday power demand.

overhead). The array is producing energy in the area under the rounded curve and it is making a lot more energy than the house is using. The extra energy can be put onto the grid to help support daytime demand. Although the array is overproducing compared to the house loads during the day, it is not producing during the time of the peak demand in the early evening or overnight.

Winter production will probably be less, perhaps like figure 7.17. Because there are far fewer hours of winter sun (depending on your latitude) the array is only producing a little more than the house is using in the middle of the day. Mostly the energy demands of the home are being met by using energy from the grid.

If the average energy used over the year equals the average energy produced over the year, this is an annual net zero house.

Because the house has such a widely varying demand over the day, and it has a fairly high power demand, it could be difficult to make this house net zero on a daily basis. A daily net zero would require batteries to store energy to power the house when the array is not producing. That would be very easy for the summer production diagram. However, the winter diagram has

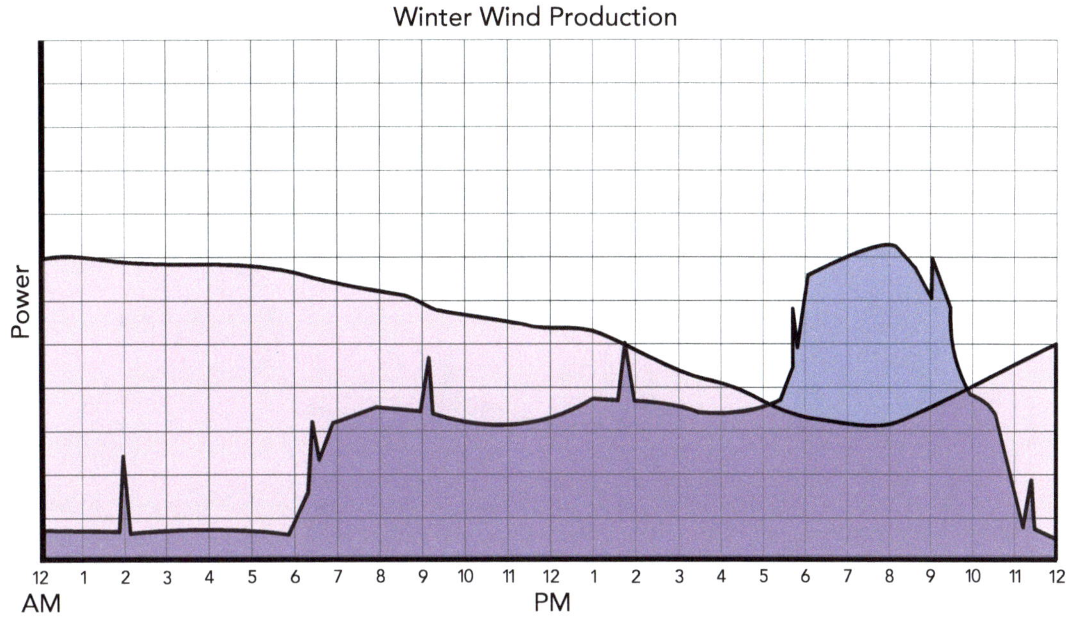


**Figure 7.18** Winter wind production profile compared to household loads.

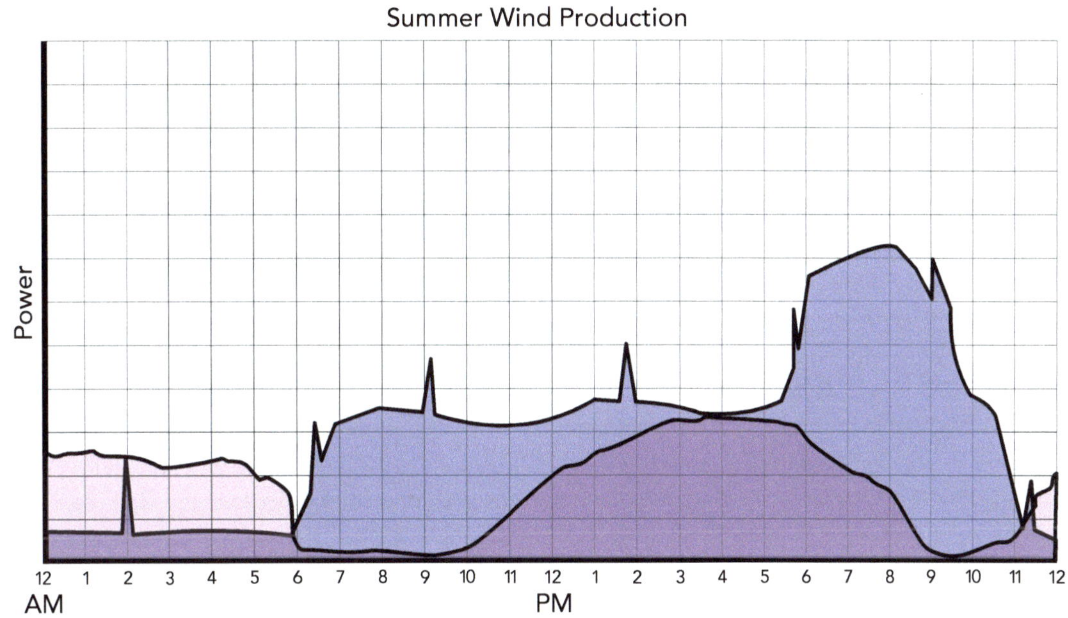


**Figure 7.19** Summer wind production profile compared to household loads.

very small production compared to consumption, the small amount that is extra midday isn't enough to run the house for the rest of the day.

## Wind Power Production

Wind varies during the day and the seasons. Typically, there is more wind in the winter than the summer. In the winter, the wind might have a profile production like figure 7.18. You'll notice that it is producing more power (pink) than is needed (blue) for the home.

A summer wind might produce power as in figure 7.19. Notice that, even at lower overall production, there is a bit of overproduction overnight. If the power demands of the house could be evened out, this system could provide most of the power the house needs if the system included batteries.

Actual production will depend on the wind resource and the size of the wind turbine. If the site has a steady wind resource all year, it is possible to design a daily net zero building.[2]

Hybrid systems including both PV and wind are quite effective since wind produces more energy in the winter when there is little PV resource, while in the summer the PV can produce a lot of energy when there

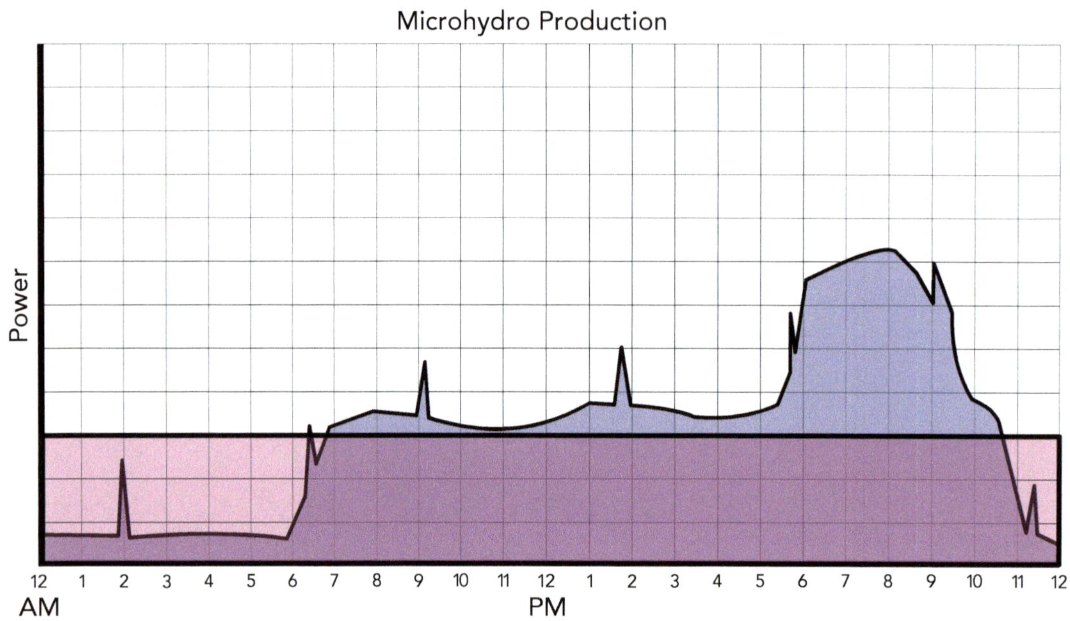


**Figure 7.20** Microhydro production profile compared to loads.

is little wind. This balances out well for meeting daily net zero.

## Microhydro Power Production

Microhydro produces the same amount of power every day (figure 7.20), all year long. If you have a hydro resource, it is a great opportunity to offset some, or all, of your power needs. As long as the resource has the ability to match your daily power needs, it is easy to design a daily net zero building.

## Comparing Energy Efficiency

Energy use in buildings in the USA is compared using the Home Energy Rating System (HERS) Index for houses and Energy Use Intensity (EUI) for commercial buildings. These standards were developed so that it is easy to compare the annual energy use of your building to other buildings.

The HERS Index is based on a scale where 0 is a net zero energy building and 100 is a home built to the 2009 **International Energy Conservation Code (IECC)**. Older homes in the USA typically use 40% or more energy than the 2009 code minimum. Those homes have a rating of 140 or more. Prescriptive standards in the 2015 IECC now require a HERS rating of 65. A home built to the 2015 code uses 35% less energy than a 2009 code-built house.

Our buildings use energy because we have lights, computers, refrigerators, and many other electrical appliances and equipment. Currently, very efficient homes can be built at a HERS rating of around 20. A building with this low energy demand can have a small energy generation system that will take the rating to zero or even net positive. Net positive means your building is generating more energy than it uses over the year.

EUI is expressed as energy per square foot per year and is calculated from the total energy used annually by a building divided by the gross floor area of the building. US units are kBtu/ft$^2$ and SI units are GJ/m$^2$. The lower the EUI, the more energy efficient the building is. Energy use in buildings is not only a function of the quality of the envelope and systems but also the use of the building. A hospital is going to use more energy than a retail store. Because of this, EUI is generally compared between similar building uses. Also, because there is a lot of energy lost in transmission, EUI is measured in terms of both site and source energy. For a current list of Median EUIs in the USA, see the Energy Star website.[3]

## Certifications

Current codes in the USA require blower door testing and/or computer modeling to show that the building and its ducted mechanical system are performing at least to code minimum standards. Blower door tests are regulated under ASTM E779 standard and duct leakage tests under ASTM E1554-07. The **Building Performance Institute (BPI)** has training and certification programs.[4] They have a Building Science Principles certificate as well as Certified Professionals in a variety of areas including AC and Heat Pump, Building Analyst, Energy Auditor, Multifamily Building Analyst, and Quality Control Inspector. Certification is earned through training, experience, and passing a rigorous exam.

## Resources


1 U.S. Energy Information Administration. (2020, February 21). Hourly electricity consumption varies throughout the day and across seasons. *Today in Energy*. www.eia.gov/todayinenergy/detail.php?id=42915

2 With California's power incentives and windy regions, the new Bergy 15 has been successful at powering in off-grid areas at pricing less than PV, see this article: Gipe, P. (2021, March 5). Bergey 15 Installed in Tehachapi—New Era for Bergey. www.linkedin.com/pulse/bergey-15-installed-tehachapinew-era-paul-gipe/?published=t

3 ENERGY STAR. (n.d.). *What is energy use intensity (EUI)?* U.S. Environmental Protection Agency & U.S. Department of Energy. www.energystar.gov/buildings/benchmark/understand_metrics/what_eui

4 Building Performance Institute (www.bpi.org/).


## Study Questions

1. Go online and search "building section drawing". From the image search results, choose four different building sections. Draw a continuous line (sometimes called a "pen test") around the whole section that indicates where the thermal and air boundaries are (you can just make it up based on the information you have). Remember that this "line" is where you separate inside from outside. As you draw your line, notice the implication of locating the line there and answer these questions:
   a. Does the line go around or through elements, such as the foundation/footing?
   b. Is the attic/basement/ancillary space in or out?
   c. How would you make transitions between building elements (floor/wall/ceiling) that have different R- or U-values and construction – will continuity be easier or harder to achieve for this building? (Remember that easier = simple.)
2. Go online and search "thermal image of building" and select four different images. Answer the following questions:
   a. What do you see in the image? Is the building one uniform color? Remember: if the building has a good continuous envelope, it should all be the same color.
   b. How would you describe the changes of the colors in the image? Are they regular or irregular changes?
   c. What does the person who posted the image say about what they found?

CHAPTER 8

# Energy Modeling

**Key Terms**

Energy efficiency measures (EEM): any practice that reduces energy usage – this can be a system, an approach, or a particular piece of equipment or building material – the combination of which will result in the best synergies.

Green building certifications: rating systems that are used to indicate the performance and sustainability achieved by a building project, usually with ranked levels based on the outcomes and number of issues addressed.

## Concepts for Success

Buildings take resources to construct. Very energy efficient buildings have faster payback times and continuously save the owner money, year after year. Only human resources cost more for a company than the operation their building. Energy modeling is used to estimate how much energy a potential design might use. When energy modeling is done in an integrated design process, the team will set performance benchmarks, brainstorm and test possible solutions, and compare results to develop the building's design.

Building code compliance, documentation for incentives, and third-party green rating systems all have requirements for building performance and modeling. This is a growing field and working with building modelers is here to stay. There are many modeling software apps available and they are continuously being improved.

## What is Energy Modeling?

**Building energy modeling (BEM)** is the process of using a computer model of a building to analyze and investigate the building's performance. Energy modeling is done at various points in the design process. In an integrated design process, BEM can be a very effective tool to evaluate the impact of possible design decisions. These can start as very basic, box-like building forms which are tested for massing and orientation; then glazing percentages and shading parameters are added. The team can brainstorm possible **energy efficiency measures (EEM)** which are then tested through multiple modeling runs to study the resulting impact of each EEM. Teams then select combinations of EEM that meet the benchmarked project goals. The **AIA** (American Institute of Architects) has a free guide on this topic titled *Integrating Energy Modeling in the Design Process*[1] which is informative and helpful.

Energy modeling is required under the 2015 IECC (International Energy Conservation Code), which is the model code in the USA. According to the IECC website, energy efficiency in regards to code requirements are a path to net zero carbon.[2] The 2021 IECC has a greater increase in the stringency of energy requirements than previous editions in order to better align with the net zero buildings by 2049 target.

All third-party **green building certifications** require some amount of energy modeling. Typically, the point values received in energy efficiency depend on the modeled performance: the better the performance, the more points are earned.

Because building energy use is dependent of a multitude of factors, performance simulation can be done as a whole building or as single aspects. The simulations may be done by architects or engineers on the team. Whole building simulation requires simple box models that replicate the massing and shape of the building, the location (for climate weather files) and orientation (solar heat gain and loss), and other systems information depending on where the team is in the design and model testing process. Typical types of analysis include: massing and building orientation; lighting, daylight, and glare; solar access and window

DOI: 10.1201/9781003297819-9

shading; natural ventilation and night flush; envelope and façade components with individual and overall insulative values; **heating, ventilating, and air conditioning (HVAC)** options; and carbon emission impacts of material choices.

When using modeling as a design development tool, the first models are quite simple. As decisions for siting, massing and orientation are made, models become more finely tuned and further details are added. Modeling is always driven by determining how to best meet desired design outcomes by exploring a variety of options. The better the selection of EEMs, the more options there are for finding a combination of EEMs that provide a clear advantage for overall building performance.

The IECC has two compliance paths. Mandatory items must be met whether you follow the prescriptive or performance path. The prescriptive path has a list of requirements that must be met by every building design. If the requirement is not met, then there are lists for trade-offs to make up for the missing requirement. The prescriptive path limits aspects of design and loses synergy between building systems because they are seen as separate and not related. All trade-offs must be modeled. The **Department of Energy (DOE)** has free software that can be used for modeling: REScheck[3] for residential buildings and COMcheck[4] for commercial buildings. The performance path provides design freedom and innovative solutions because it requires complete system energy simulations.

All BEM uses a comparison between at least two iterations of the same building (figure 8.1). The first is the reference building, which is based on meeting minimum code performance requirements or specific

**Figure 8.1** Twin to reference comparison buildings.

parameters designated by a rating system. The others are a geometric twin which incorporates all the information about building options (orientation, insulation, etc.). Results will compare the reference building and the twin building. Energy modeling software usually has an easy means of making as many twins as you want and changing various parameters. You can then compare all the twins to each other as well as the reference case to determine which set of EEMs has the best outcomes. Reports are generated based on the twin model setup. It is very important to accurately input the information about the twin building in order to get accurate results. Modeling results will not be correct if variables are not completed to the accuracy desired and needed for decision making or documentation.

## Modeling vs. Detailing

It is easy to input variables into models that are not possible to build, either because materials can't achieve the performance values or the construction of materials attachments and/or supports don't match the variables. The saying "They just don't build like they used to" is true – and it's a good thing! If you've been in the design and construction field for a while, you may be taking ways of detailing for granted. There are many great resources on high performance detailing. The Building America Solution Center[5] has building components and guides in addition to program checklists for Energy Star, Water Sense, and others. It also has computer aided design (CAD) files showing fully annotated building details that can be searched by climate zones. There are case studies, images, presentations, videos, and optimized climate solutions.

There are modeling programs such as THERM,[6] a free program which was developed at Lawrence Berkeley National Laboratory, that model two-dimensional conduction heat transfer through assemblies. This type of modeling can show temperature patterns within an assembly, such as a wall or roof, that can highlight if there will be issues with condensation or thermal bridging. Figure 8.2 shows a window detail study for The Alfred Zimberg School, Public School 2 in New York City, NY, USA. The existing wall is a 100-year-old masonry building with single pane windows. Using THERM, Energy Shrink[7] was able to test different combinations of insulation and high performance windows with thermally broken mounting details along with adding air and vapor barriers. The final detail is shown with the major components in the new detail on the left, the infrared in the middle shows color distribution of heat transfer, while the isotherm on the right shows lines with the temperature values.

## Energy Modeling Programs

The BEM field is constantly updating and changing because of improvements needed to create better models that more accurately represent how the built building performs. Many modeling programs can use 3D files exported from building information modeling (BIM) programs, such as Revit, for the basis of the model. Others require inputs that are more like spreadsheets to collect information on the various building components.

The modeling program you choose will depend on what outcomes you are looking for and what kind of computing expertise you have. A good resource about software has been published by the USA Office of Energy Efficiency and Renewable Energy.[8] Currently they list and provide information about: EnergyPlus, Spawn-of-EnergyPlus (Spawn), OpenStudio, and Building Energy Modeling Project Portfolio. These are all programs developed with the support of DOE.

To obtain a HERS (Home Energy Rating System) index rating, you must use REM/Rate[9] or EnergyGauge.[10] EnergyGauge also has commercial building software which can be used for **Leadership in Energy and Environmental Design (LEED)** documentation and IECC compliance. IECC also lists EnergyPlus, Sefaira,



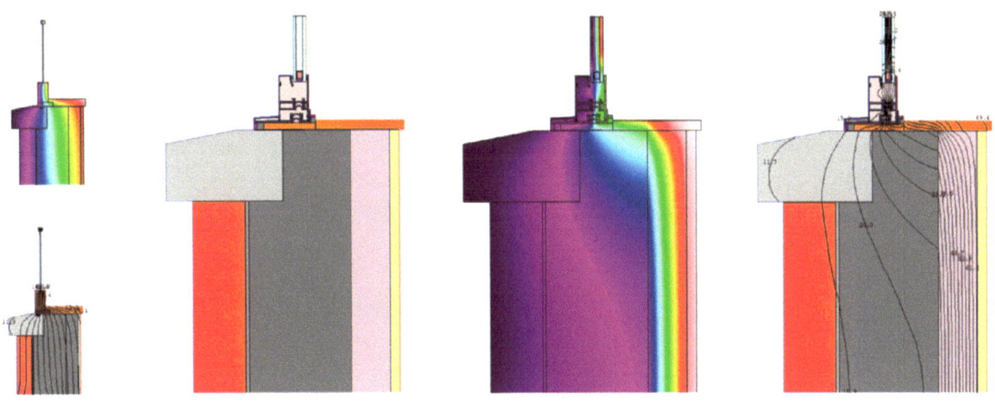


**Figure 8.2** THERM sample screen shots of heat transfer results. Project images courtesy of Energy Shrink.

eQuest and OpenStudio as approved software for commercial project modeling documentation. There are many other modeling programs that can provide documentation for meeting code requirements, receiving incentives, and earning third-party green building certification.

## Accounting for Energy and Carbon

Different kinds of buildings inherently use vastly different amounts of energy. For instance, a hospital with beds will run 24/7 compared to an office building that mostly runs 9am–5pm Monday–Friday. Typically, buildings are compared to similar building types and their **energy use intensity (EUI)**. For a deeper discussion, see chapter 7 on Building performance and Energy Efficiency.

Besides EUI, another factor that comes into account is site vs. source energy. Site energy is exact amount of energy that is used to power your building. We tend to be very wasteful in the way we distribute energy. The exact amount that is lost in transmission depends on factors such as how far energy is being distributed, the climate, and the age of the power infrastructure. Source energy is how much energy was generated in order to provide the site energy. Government agencies and third-party certification programs provide numbers for use in modeling.

In the end, what we are really concerned about are carbon atoms. We need to drastically reduce the number of carbon atoms in our atmosphere and stop adding more. A net-carbon building is one that equalizes out the embodied carbon and carbon over its life-cycle. As we make more energy efficient buildings, the carbon that is part of the construction materials has a greater environmental impact. There are new software programs that are being introduced to model impacts. Some of them include Pharos (hazardous chemicals in building products),[11] Tally® (carbon life-cycle assessment)(paid),[12] and Embodied Carbon in Construction Calculator (EC3) (free).[13]

## Resources


1 American Institute of Architects. (2012). *Architect's guide to integrating energy modeling in the design process*. www.aia.org/resources/8056-architects-guide-to-integrating-energy-modeli

2 Energy Efficient Codes Coalition. (n.d.). *About the International Energy Conservation Code (IECC)*. https://energyefficientcodes.org/iecc/

3 Building Technologies Office. (2021, November 30). *REScheck* (Version 4.7.2.1) [Computer software]. U.S. Department of Energy. www.energycodes.gov/rescheck

4 Building Technologies Office. (2021, May 25). *COMcheck* (Version 4.1.5.3) [Computer software]. U.S. Department of Energy. www.energycodes.gov/comcheck

5 Building America Solution Center (https://basc.pnnl.gov/).

6 Windows and Envelope Group of the Building Technology and Urban Systems Division. (2019). *THERM* (Version 7) [Computer software]. Lawrence Berkeley National Laboratory. https://windows.lbl.gov/software/therm

7 THERM, Energy Shrink (www.energy-shrink.com)

8 BEM is all here, with some software recommendations: Building Energy Modeling U.S. Department of Energy. (www.energy.gov/eere/buildings/building-energy-modeling).

9 NORESCO. (n.d.). *REM/Rate™* (Version 16.3.2) [Computer software]. www.remrate.com/

10 Florida Solar Energy Center. (n.d.). *EnergyGauge USA* (Version 7.0.3) [Computer software]. University of Central Florida. www.energygauge.com/

11 Pharos (https://pharosproject.net/).

12 Building Transparency. (2022). *Tally®* (Version 2021.11.01.01) [Computer software]. https://choosetally.com/

13 Embodied Carbon in Construction Calculator (EC3) (www.buildingtransparency.org/).


## Study Questions

1. Choose two different green building rating systems and compare their requirements.
   a. What is the basic structure of the rating system?
   b. What kinds of things earn points in the rating system?
   c. What kind of energy modeling is required in the rating system?
2. Open the AIA's *Architect's guide to integrating energy modeling in the design process* located at www.aia.org/resources/8056-architects-guide-to-integrating-energy-modeli and answer the following questions about Section 4 – Performance Analysis and Modeling:
   a. What is an energy model?
   b. What levels of modeling and their inputs are listed?
   c. Why is monitoring and reporting of a project's energy usage important?
   d. What kinds of questions should the architect ask about the modeling tool being used?
   e. What are important areas to investigate during energy modeling and how do these areas change in the different stages of a project?

CHAPTER 9

# Indoor Air Quality

**Key Terms**

Carcinogen: a substance or radioactive material that can cause the formation of cancer in living tissues.

Volatile organic compound (VOC): gasses that are emitted into the air at room temperature from an organic (carbon) compound – many are human-made chemicals used for a wide variety of processes and emitted by a wide variety of products – and are often found in much larger concentrations inside buildings.

## Concepts for Success

Indoor air quality is affected not only by the construction materials used, but also by the activities and products that we bring with us into the building. Although the building's mechanical system is important for providing good air quality, toxins and **volatile organic compounds (VOCs)** brought inside can make it difficult to achieve good indoor air quality.

Designers must pay attention to materials specified and the appropriate design of mechanical systems. Having a tight enough building envelope to keep out poor air quality, radon, and pests is important. Users need to pay attention to how their everyday activities bring in contaminants as well as doing appropriate housekeeping.

## Causes of Indoor Air Pollution

There are a number of causes of indoor air pollution. In broad terms, these can be simplified into groupings about the building's design, what conditions are outside of the building, and how people use their building.

The building's design includes what materials are used and attention to mechanical systems. Materials can be an issue because manufacturers use chemicals and ingredients that are known to be a serious health risk to humans and our ecosystem. Mechanical systems can be lacking, installed incorrectly, and improperly maintained. Worst case scenarios can be that contaminants are not exhausted or that contaminants are distributed throughout the building.

### Materials

Manufacturers have not been transparent about what they are using in their products. This includes not only building products, but also all the things we put inside our buildings and use to maintain the building. Older buildings may have lead paint, asbestos in insulation and finishes, and mercury in various products such as appliances, thermostats, electronics, batteries, medical equipment, and antiques[1]. Green rating systems have pushed for transparency and award points for selecting healthy materials. The most complete guide is the **Living Building Challenge's (LBC)** Red List and Watch List List CAS Registry Number (CASARN, a unique number assigned to a chemical substance) which is updated annually. Market pressures have helped to create better labeling and declarations from manufacturers but it remains a time-consuming endeavor to check out products. LBC encourages specification of materials that have an active Declare label of "LBC Red List Free", "LBC Red List Approved", or "LBC Compliant".[2] This labeling is updated annually by the manufacturer with LBC.

Many products are still available that contain VOCs, formaldehyde, and the like. These can off-gas, releasing fumes that can make people ill. This issue was first identified as sick building syndrome in the late 1980s as buildings were made more energy efficient and more of these fumes stayed inside. Although better air changes can help to mitigate issues from off-gassing materials, the healthiest option is to eliminate the materials made with dangerous items all together. Choosing safe materials also makes it easier to deal with them at the end of their useful life in the building.

DOI: 10.1201/9781003297819-10

## Mechanical Systems

The **World Health Organization (WHO)** cites 3.8 million premature deaths annually due to household exposure to smoke from dirty cookstoves and fuels. There are 3 billion people who are poor and live in low- and middle-income countries. These open fires have inadequate, or no, ventilation which results in high levels of pollutants. These include fine particulate matter, carbon monoxide, and nitrogen oxides. This affects women and young children the most. Deaths are caused due to pneumonia, ischemic heart disease, **chronic obstructive pulmonary disease (COPD)**, stroke, and lung cancer[3].

All combustion equipment should be directly vented outside, well away from the fresh air intake. The building's mechanical system should be able to provide enough supply air that when all fan exhausts and the combustion equipment is running, they are all able to exhaust. Backdrafting occurs when the building depressurizes and pulls supply air through the combustion exhaust pipe (figure 9.1). Combustion air that should be exiting the building is instead being pulled inside and distributed throughout the building.

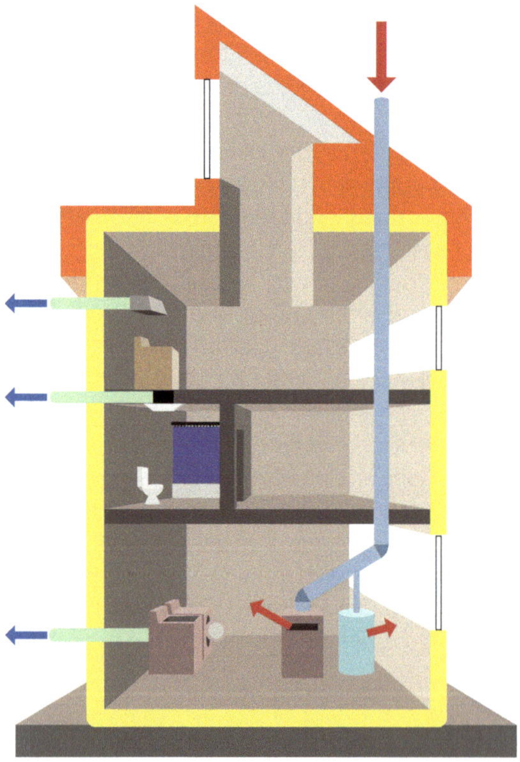

Figure 9.1 Backdrafting of combustion equipment due to inadequate supply air.

HVAC (heating, ventilating, and air conditioning) equipment needs to be properly installed and maintained. **Commissioning** of HVAC systems should be done to make sure that they are working correctly. Commissioning means that equipment is checked to ensure it was properly installed and is operating as expected. Maintenance includes regularly changing filters, cleaning, and inspecting equipment regularly. Recommissioning of systems can be done to check that equipment is still working properly. All coils, equipment, and ductwork need to be kept dry to prevent mold and bacterial growth.

Very tight buildings often eliminate all combustion equipment. Typically, these tight buildings will have balanced ventilation systems. This means that all exiting and entering air is equalized so that the building is never depressurized.

**Carbon monoxide (CO)** detectors are inexpensive and are easy to include in a project. In existing homes, CO detectors can plug in to a typical electrical outlet. In commercial buildings, both CO and $CO_2$ monitors can be included in the overall building monitoring. $CO_2$ monitors can be used to automatically increase the mix of fresh air; this can be helpful for keeping healthful conditions in meeting rooms and assembly areas that have varying numbers of people.

## Conditions Outside

### Filters

WHO reports 4.2 million deaths every year due to outdoor air pollution. In fact, 91% of the world's population live in areas where the air quality is worse than maximum WHO guidelines. If the building is relying on infiltration for fresh air exchanges, then it's probably not getting fresh air. Tightening the building and adding ventilation that has a MERV (minimum efficiency reporting value) filter rating as defined by ASHRAE (American Society of Heating, Refrigerating and Air-Conditioning Engineers) standards on intake air can improve air quality dramatically. MERV ratings are numbered 1 to 16 where lower MERV ratings remove fewer contaminants than larger number ratings. Filters rated MERV 11 and above remove bacteria, pollen, dust mite debris, mold spores, pet dander, smog, smoke, and cough/sneeze. MERV 13 and above add removal of bacteria and virus carriers. HEPA (high-efficiency particulate air) filters rate higher than MERV 16 and can remove viruses, carbon dust, radioactive material, and 99.97% of small dust particles. Forced air systems typically have at least one filter and can have several layers so that the lower filter catches the majority of pollutants and is changed most frequently. These systems have at least a MERV 6 filter, which removed dust, lint, and pollen. Filters need to be replaced regularly and disposed of properly. Commercial building HVAC systems always have ventilation systems with intake air and filters, although the building may still have a fair amount of infiltration.

### Radon

Radon is a radioactive gas that is naturally occurring in soil and water all over the world. According to WHO, no studies have yet linked radon in drinking water with increased risks of stomach cancer. However, radon is released from the water into indoor air. WHO recommends the use of granular activated carbon filters or aeration by the water distribution system to reduce radon concentration.

The larger problem is radon in the rocks and soil. As uranium decays, it becomes radium and then radon, see figure 9.2. Radon gas percolates up through the soil and is released into the air. Buildings can pull in the radon gas through the foundation and basement or slab-on-grade, creating high concentrations in the indoor air. Radon cannot be seen, tasted, felt, or smelled. The reason radon is so dangerous is because the particles are a small size that fit in the lung's alveoli, the tiny air sacs that exchange oxygen and carbon dioxide while breathing. The radon continues to decay and damages DNA, which makes it difficult for tissue to heal and it can become cancerous. There is no known safe level of radon exposure and people who smoke are at higher risk than non-smokers. It is estimated that radon causes between 3–14% of all lung cancers, depending on the country's radon level and how popular smoking is.

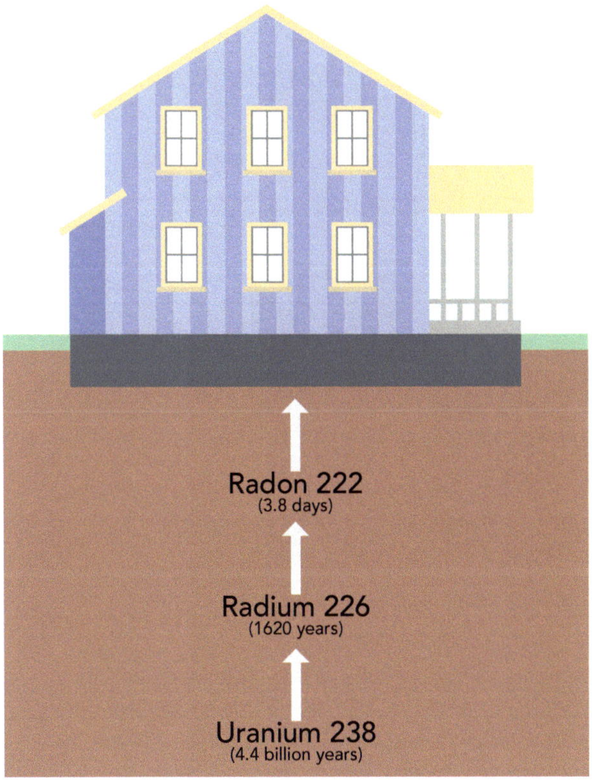


Figure 9.2 Uranium decays into radon gas that can enter a building.

Many building codes around the world require new buildings to have radon mitigation measures in place. Jurisdictions may have alternate detailing due to soil types or increased radon levels. In new construction, passive control measures depressurize soil under a building by adding 4" of gravel under the basement or slab-on-grade (figure 9.3). This layer allows the soil gasses to move freely below the slab so they can be collected in a pipe and vented to the exterior. Heavy plastic (6 mil. polyethylene) (6 mil = 0.006", 152μm) or a vapor retarder is installed on top of the gravel to seal it and keep the concrete pour from filling in any of the gravel. A 3–4" (76–101 mm) solid PVC Schedule 40 pipe is installed vertically from below the slab through the roof. The pipe can also extend outside through a basement wall but the top of the vent pipe should be up high and away from any fresh air intake and operable windows. All openings and cracks in the joints in the concrete slab and between the slab and foundation need to be sealed with polyurethane caulk. In areas with high concentrations of radon, it may be necessary to add a vent fan to the pipe to actively pull soil gasses out from under the slab. Additional information is available at the EPA (Environmental Protection Agency) website.[4]

Installing a radon sump pump pit in a basement can be effective for mitigation in an existing building (figure 9.4). It needs to be ventilated to remove soil gasses and needs a sealed, removable lid so it is easy to maintain the sump pump. Cracks and joints in the concrete floor and at the foundation should be sealed.

Other ways to help reduce radon in existing buildings are: sealing the upper levels from the basement; sealing floors and walls between either the soil and interior space or the basement and living areas; and tightening the building overall and adding ventilation with MERV 16 or better filters. WHO cites at least 50% reduction in indoor radon levels from the use of passive ventilation systems.

Radon testing can be done in existing buildings using specialty test kits. The US EPA recommends under 4pCi/L (picocuries per liter of air) in indoor air. In some jurisdictions, radon testing is required when selling a building. Typically, the seller is responsible for paying for the testing and radon mitigation if needed.

## Stuff We Bring Inside and Biology

There are many issues that are created in our buildings because of the way we live and work in them. We breathe out carbon dioxide, shed skin and hair, and generally transport a host of bacteria and viruses around with us. We bring in furniture, plants, pets, food, and all sorts of products and equipment. All of these can have an impact on the quality of the interior air.

In green rating systems, commercial buildings require attention to how to keep from tracking in stuff on our shoes. Some solutions can be grates, grilles, walk off mats, or other systems that allows for shoes to shed contaminants at entry ways and not track them throughout the building. These have to be regularly

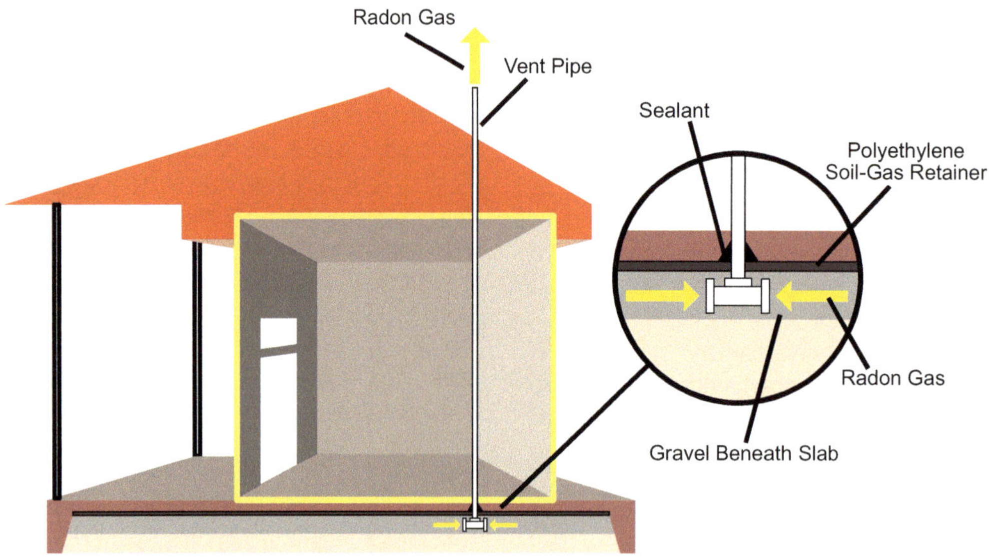


Figure 9.3 Under slab venting for radon mitigation.

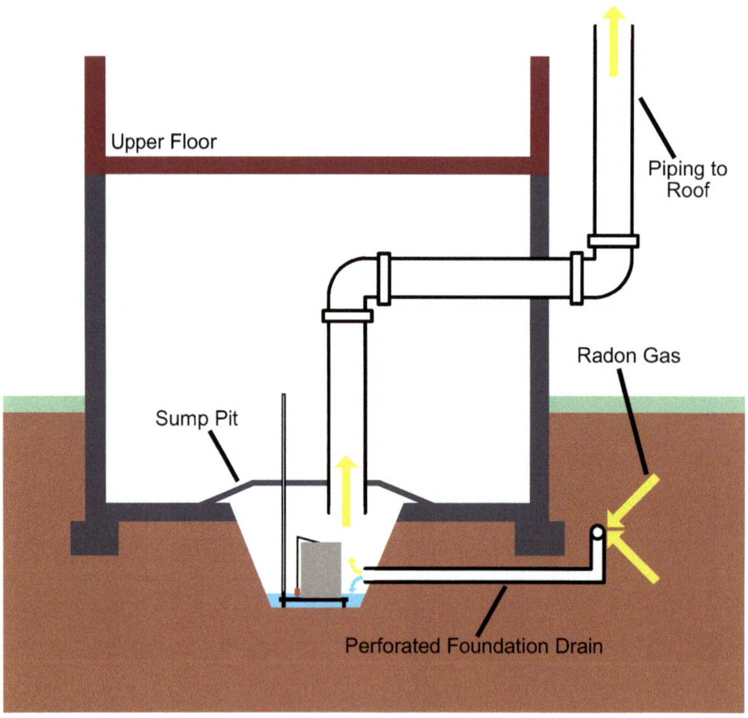


Figure 9.4 Sump pump pit used for radon mitigation in an existing building.

cleaned, at least once a week. Similarly, homes should have walk off mats or even an area to take off and leave shoes to help maintain cleanliness and prevent shoe-borne particulates from getting mixed into the interior air.

## Relative Humidity

It is important to maintain proper relative humidity (RH). To reduce mold growth and keep lung function happy, indoor humidity should not be too high or too low.

Air will be too dry below 30% RH and can cause respiratory issues due to the air drying out the mucous membranes in nasal passages, which can lead to coughs and sore throats. Skin can also dry out and become itchy or cracked. Besides health effects, dry conditions can also affect materials in your building. Wood can shrink and warp, ruining furniture and causing cracking of moldings and door frames. You may also experience static cling and static electricity due to low RH. Dry conditions can exist because of climate or season. Humidification may be needed to keep air at a good RH level. However, vapor moisture should be mixed into all of the air. In cold climates windows should not be acting as dehumidifiers because they are cold and water vapor condenses on the glass. There also needs to be consistent air and vapor control layers to keep moisture from getting into walls and causing issues.

Air is too moist over 60–65% RH and can encourage the growth of mold, fungus, and dust mites. It can also attract unwanted pests. All of these items can lead to increased allergies and asthma attacks. High humidity can feel uncomfortable because of limiting skin evaporation to feel cooler. In addition, high humidity can damage and warp wood products due to swelling. To keep from adding vapor moisture to the air, sources of water vapor should be directly exhausted outside. This means that exhaust vents need to be installed and used over stoves, run during hot showering, and clothes dryers properly vented. Any other processes that produce steam or water vapor should be vented. In humid climates or seasons, it may be necessary to dehumidify the air to keep it in the comfort range.

## Smoking and Tobacco Product Use

WHO identifies tobacco use as a worldwide epidemic that kills 8.2 million people every year.[5] Of these, 1.2 million deaths are non-smokers exposed to second-hand smoke. In the USA, 1 in every 5 deaths is related to tobacco use, making it the leading preventable cause of death. Tobacco in any form is harmful and kills about half of its users.[6] While smoking cigarettes is the most common form of tobacco use, other tobacco products include waterpipe, smokeless products, cigars, cigarillos, pipes, bidis, kreteks, rolled, heated products, and e-cigarettes. There is no safe level of tobacco exposure. Often, those least able to afford it smoke, which increases poverty issues. Life expectancy for smokers is at least 10 years shorter, with increased risks for dying of lung cancer and heart disease. Smokers are also at 40–50% increased risk for death or severe complications from COVID-19.[7]

All green rating systems have requirements for eliminating smoking inside the building. Many municipalities, employers, and organizations do not allow smoking indoors in commercial settings.

Studies show that at least half of people who understand the health risks of tobacco use want to quit but lack resources to help them. WHO has a global campaign for helping people quit tobacco products called "Commit to Quit" and its Quitting Toolkit[8] is available for free and includes many helpful items including text messaging.

## Pests and Pesticides

Pests can range from unwanted insects to rodents. Pests are attracted by the availability of food and water. Pests can carry numerous types of diseases and leave wastes behind that can be damaging or harmful. Pests can damage buildings and create unsafe conditions, such as mice or rats chewing through electrical wiring sheathing which can cause fires.

Pesticides are used inside and out to kill pests. All pesticides are toxic, classified as semi-VOCs, and are very dangerous if not used according to the product labeling. According to a study published in December 2020, it is estimated that there are likely 325 million cases of unintentional, acute pesticide poisoning which resulted in about 11,000 fatalities. Globally, about 44% of farmers are accidentally poisoned each year.[9]

There are both short-term and long-term health risks from exposure to pesticides, all of which are dependent on the type of chemical used in the pesticide and severity of exposure. Short-term effects could include headaches, dizziness, muscle twitching, weakness, tingling sensations, and nausea. Long-term risks include birth defects, learning disabilities, behavioral changes, organ damage, cancer, and asthma.[10]

Tighter buildings are harder for pests to access and there is not a risk that pesticides used outside are drawn inside through infiltration. Removing areas of safe havens, food, and water will discourage pests from coming in and breeding. Consistently doing good housekeeping discourages pests. Use traps, vacuums, gels, and baits instead of pesticides. Pay attention and monitor for pests, eliminate food, and water sources.



## Cleaning Products

Many of the cleaning products that are available in the marketplace have toxic ingredients. It is very important to properly clean surfaces to keep healthful conditions in buildings. Improper cleaning can lead to mold and fungus growth, spread of bacteria and virus spores, and the dust and dirt can cause allergies and asthma.

Reading the ingredient list on the product label is important. Green building rating systems have sections that address the care and maintenance of the building without using products that add toxins to the interior air.

## Pets

Our furry, swimmy, scaly, and feathered friends brighten our days but add to housekeeping. Pet dander, fur, feathers, and the like can aggravate allergies and asthma. Food and water left out can attract pests. Pet wastes must be controlled and disposed of in an appropriate fashion to prevent damage and spread of disease. Fish tanks add humidity to the air. Scaly and shelled pets may feed on insects and mice, which need to be properly stored so they don't become pests. Pets that go outside can track in dirt and pests. Pets can have very pungent odors. Paying attention to proper care of the animal as well as cleaning up with non-toxic materials is important for everyone's health. Any products used for flea and tick control should be evaluated for safety and used carefully following the product's labeling.

### Carpets

Although carpet can be great in a room, providing a soft surface that can also help to reduce noise, it is also able to trap a lot of contaminants that can cause respiratory problems.[11] Contaminants will be anything that is shed from living beings in the building, large and small: skin and hair, pet dander and fur, particulates, mold spores, and leavings from any pests, as well as dust and dirt. Pretty much anything that is airborne will settle into carpets, including toxic materials, heavy metals, and microorganisms. More of these are contained in carpets than on hard floors. Because children and pets are breathing low to the floor, they are more likely to be affected by contaminants that are stirred up from walking or crawling across the carpet. Studies show that carpets in young children's rooms can add risk for development of asthma.

According to the American Lung Association, carpets should be vacuumed at least three times a week using a vacuum that has a HEPA filter. However, it can be difficult to remove all contaminants, so carpets should also be dry steam cleaned annually.[12] Care should be taken to read the labeling of any cleaning products used on carpets. Carpets should be kept dry so they don't develop mildew.

High traffic and wet areas such as entryways, kitchens, and bathrooms should never have carpet. Instead, they should have durable and easily cleanable materials which can dry out quickly so as not to grow mold. You can also use easily cleanable mats and rugs that can be taken outdoors for cleaning so that contaminants do not get airborne inside during cleaning.

Carpets and backings also are a source of VOC emissions, so check labeling. Due to consumer awareness of VOCs, there are products available with no or low VOCs.

### Personal Care Products

Personal care products include everything you use for personal hygiene, skin care (including sunscreen), fragrances, and make-up. Many of these products fall into a cosmetics category that has few regulations or oversight on their content, besides color additives. There can be concerns about skin absorption, breathing in sprayed product, and ingestion. Product that gets airborne can combine with particles and water vapor in the air. Run the bathroom exhaust vent to help move air out of circulation. Read product labels and be aware of what the health impacts of ingredients are.

### Hobbies and Maintenance Products

Depending on your hobby, it may require the use of hazardous paints, chemicals, and/or solvents. There may be VOC products or other known **carcinogens**. Read labels carefully and follow directions. If the product says to use in a well-ventilated area, you shouldn't be using it in a room that doesn't have an exhaust vent. Many buildings have equipment that needs maintenance or uses combustion products for power. All chemicals and fuels need to be stored in a location where they cannot contaminate indoor air.

### Plants

While it is wonderful to have biophilia inside, it is very important to check that the kind of plants you are including are not a cause of allergies. Problems with plants can include airborne pollens, mold growth, and insect infestation. Researching the kind of plant that will do well for your household is important as some plants are toxic to pets.[13] Some plants with large leaves can accumulate dust. Generally, the worst plants for dermatitis are chrysanthemums, common ferns, lilies, and geraniums. Others that can cause respiratory issues are ornamental plants, hoya compacta (wax plant succulent vine), and holiday pines.

### Machines and Furniture

Printers and photocopiers can off-gas from the print materials during and after printing. Depending on the printing process, there could be different types of chemicals and VOCs that are getting airborne. Other contributors to the off-gassing mix are melamine furniture, foam cushions, and pressed wood products (medium density fiberboard). If you have these items, look at how the offending vapors can be vented to the outside or consider putting these items in an area where you can provide an air cleaner. Paying attention to the material content of furniture and avoiding those with VOCs will improve indoor air quality.

## Resources


**1** U.S. Environmental Protection Agency. (2021, July 14). *Mercury in consumer products*. www.epa.gov/mercury/mercury-consumer-products

**2** International Living Future Institute. (2021). *About the LBC Red List*. https://living-future.org/lbc/red-list/

**3** World Health Organization. (2021, September 22). *Household air pollution and health* [Fact sheet]. www.who.int/news-room/fact-sheets/detail/household-air-pollution-and-health

**4** U.S. Environmental Protection Agency. (n.d.). *Radon-resistant construction basics and techniques*. www.epa.gov/radon/radon-resistant-construction-basics-and-techniques

**5** World Health Organization. (2021, July 26). *Tobacco* [Fact sheet]. www.who.int/news-room/fact-sheets/detail/tobacco

**6** Centers for Disease Control and Prevention. (2020, April 28). *Tobacco-related mortality* [Fact sheet]. www.cdc.gov/tobacco/data_statistics/fact_sheets/health_effects/tobacco_related_mortality/index.htm

**7** World Health Organization. (2021, May 28). *WHO supports people quitting tobacco to reduce their risk of severe COVID-19* [News release]. www.who.int/news/item/28-05-2021-who-supports-people-quitting-tobacco-to-reduce-their-risk-of-severe-covid-19

**8** World Health Organization. (2021). *Quitting toolkit*. www.who.int/campaigns/world-no-tobacco-day/2021/quitting-toolkit

**9** Boedeker, W., Watts, M., Clausing, P. & Marquez, E. (2020). The global distribution of acute unintentional pesticide poisoning: Estimations based on a systematic review. *BMC Public Health*, 20(1875). https://doi.org/10.1186/s12889-020-09939-0

**10** Canadian Centre for Occupational Health and Safety. (2017, December 1). *Pesticides – health effects*. www.ccohs.ca/oshanswers/chemicals/pesticides/health_effects.html

**11** Becher, R., Øvrevik, J., Schwarze, P. E., Nilsen, S., Hongslo, J. K., & Bakke, J. V. (2018). Do carpets impair indoor air quality and cause adverse health outcomes: A review. *International Journal of Environmental Research and Public Health*, 15(184). https://doi.org/10.3390/ijerph15020184

**12** American Lung Association. (2020, February 12). *Carpets*. www.lung.org/clean-air/at-home/indoor-air-pollutants/carpets

**13** Poslusny, C. (2019, June 6). Best plants for indoor allergies: Can they help or cause more problems? *Molekule*. https://molekule.science/best-plants-for-indoor-allergies-can-they-help-or-cause-more-problems/

## Study Questions

1. Go online and search for a world air pollution map. What do you observe about air quality around the world?
2. Go online and search for radon maps/information about radon in a particular location/region. Is radon an issue in the location?
3. Using the headings listed under the "Stuff we bring in and biology" section, do an indoor air quality check of your living space. What problems did you discover? What can you do to fix the issue?

PART II

# Renewable Energy: Systems and Design

CHAPTER 10

# Biomass

**Key Terms**

Biofuel: biomass that is turned into a liquid fuel such as ethanol and biodiesel.

Biomass: organic matter from plants and animals that are used as fuel to generate electricity, power vehicles, heat buildings, and cook.

Combined heat and power (CHP): concurrent generation of both heat and power (from burning a single resource) which is located at or near the site where the heat and power is used. This is also called cogeneration. It captures the heat that would normally be lost in the production of electricity and is generally used for industrial and large commercial applications.

## Concepts for Success

Biomass is created from organic plant and animal material. It can be directly burned or processed to create biofuels. It is considered a renewable resource (and carbon neutral) because it is part of the current carbon cycle, as opposed to fossil fuels which, when burned, add sequestered carbon to the atmosphere.

Ethanol is fermented from plant starches and sugars and then mixed with petroleum to power vehicles. Biodiesel is made from vegetable oils and animal fats and is compatible with diesel engines. Biodiesel can either be rated R or B depending on how it is made. Renewable diesel (R) is a chemical match to petroleum diesel, B types are not.

Modern biomass burners have a wide variety of sizes and options. They can create heat, electricity, or combined heat and power (CHP). You need to plan on enough space to contain the system as well as to store the biomass.

## How Biomass Fits into Renewable Energy Resources

Biomass consists of organics such as wood, alcohol fuels, crops, and methane gas. They often include waste products from agriculture, animals, and industry. Biomass is typically burned for cooking, heating, and producing electricity. Because biomass carbon atoms are part of the current carbon cycle, they are not adding carbon to global warming. Biomass is considered to be carbon neutral. It could be argued that due to the massive overload of carbon in the atmosphere, we shouldn't be burning anything. Wood sequesters carbon for as long as it doesn't decompose, so using it for buildings is a great way to help the environment. However, there are a lot of woody products that are not suitable for building materials or are left over from manufacturing processes. Sawdust and chips can be used for layering in landfills but this uses only a small amount of the waste produced. Many cellulose agricultural by-products, like straw, are regularly burned in the field, causing air pollution and wasting a resource. Methane is a by-product of decaying organics, especially wastes, and is better captured and burned than released into the atmosphere where it adds to our GHG emissions.

Many places in the world use traditional biomass for cooking and heating. These types of open fires are often burned indoors without proper ventilation and are a cause of premature deaths of women and children. Traditional biomass is fairly polluting because the flames do not burn at high temperatures.

Modern biomass is much more energy efficient and clean-burning. Because it burns at higher temperatures, it causes less pollution and waste and large burners have scrubbers to reduce emissions.

DOI: 10.1201/9781003297819-12

## Biofuels

Biomass that has been manufactured into a liquid is called a biofuel. To convert biomass, it has to be deconstructed into usable chemical components. This is done either at high temperature (using extreme heat and pressure) or low temperature (using enzymes or chemicals). After deconstruction, the chemical components must be upgraded by adding microorganisms, sugars, bio-oils, or other chemicals needed to clean and stabilize final product.

### Ethanol

Ethanol began major production to replace whale oil in lamps in the mid-1800s. In the USA it has had bumps in production: first when taxed to raise funds during the Civil War (repealed 1906) and then banned during Prohibition (1919–1933) as it is a grain alcohol. Ford's Model T automobile ran on an ethanol-gas mixture.

Ethanol is a clear, colorless alcohol and is made from fermenting plant starches and sugars. The main crops used for ethanol are corn grain (USA), sugar cane (Brazil), and sugar beets (worldwide). Although ethanol can be made from cellulose in plant fibers (crop residues and wood), it currently takes more steps and is therefore more costly than using starches and sugars.

Ethanol has compatibility limits with engine valves and other components. Most gasoline today has ethanol in it. In the USA and Europe 10% ethanol (E10) is commonly mixed in with petroleum gasoline. E85 should only be used in flex fuel vehicles and has 51% to 83–85% ethanol with the balance petroleum. Because denatured ethanol (98% ethanol) has about 30% less energy than gasoline per gallon, an E10 mix results in a 3.7% drop in gas mileage.

The manufacturing of ethanol, especially from corn, has been energy intensive, and in the mid-1970s it was a negative energy balance (took more energy than the resulting ethanol had in it). More recently, due to better efficiencies in manufacturing, in the USA production from corn has become a positive energy balance: for every 1Btu required to create ethanol from corn, 2.3Btu are created. Agricultural corn yields have increased by 39%, requiring fewer acres of land in production for the same amount of fuel.[1]

Figure 10.1 shows the comparative life-cycle carbon emissions of petroleum gasoline (highest value at 98,000 gCO2e/MMBtu) and ethanol in 2005 (59,766 gCO2e/MMBtu), as well as projections for ethanol based on business-as-usual (54,588 gCO2e/MMBtu) and improvements due to high efficiency hybrid cycle (27,852 gCO2e/MMBtu). The units MMBtu are $10^{12}$Btu.

### Biodiesel

Biodiesel is made from vegetable oils, recycled restaurant grease, and animal fats. In the USA the most common blend is 20% biodiesel (B20) and 80% petroleum diesel. B100 (100% biodiesel) may

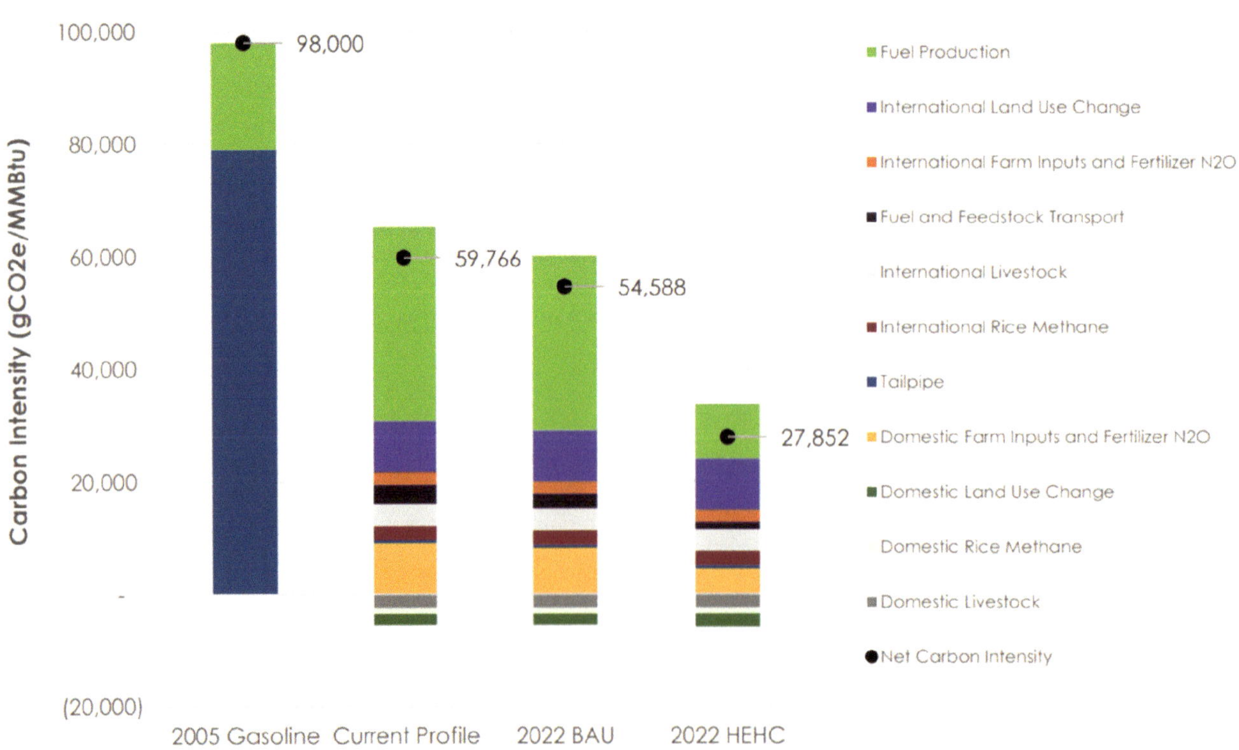


Figure 10.1 Full Life-Cycle Corn Ethanol GHG Results for the Current Conditions, 2022 Business-As-Usual (BAU), and 2022 High Efficiency Hybrid Cycle (HEHC) Scenarios.[2]

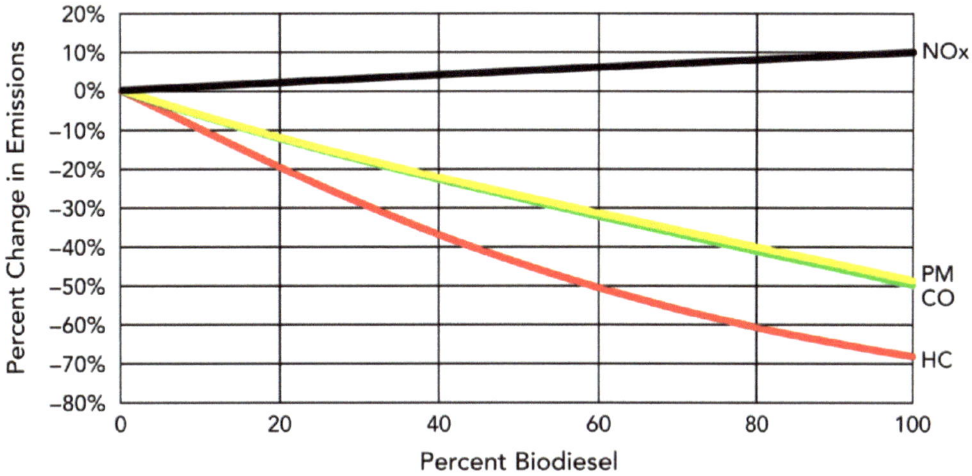


Figure 10.2 Average emissions impact of biodiesel for heavy-duty highway engines.[3]

be compatible with diesel engines built since 1994 (check vehicle manufacturer recommendations). B100 has a solvent effect and can require frequent filter replacement when converting from petroleum diesel. Biodiesel biodegrades easily and is non-toxic. In fact, it will biodegrade if handled or stored improperly – which includes being exposed to heat, oxygen, sunlight, or stored in nonferrous metals. Like regular diesel, biodiesel can gel in cold weather, so additives are required to prevent winter gelling.

B100 has about 8% less energy per gallon than petroleum biodiesel, while B20 has 1–2% less.

Emissions from burning biodiesel are all reduced, getting better with higher percentages of biodiesel in the fuel, except for nitrogen oxides (NOx) which increase (figure 10.2). Hydrocarbons (HC) have the greatest reduction at −20% for E20 and −68% for E100. Particulate matter (PM) and carbon monoxide (CO) are almost tied at reductions of about −13% for E20 and almost −50% for E100. Nitrogen oxides increase from about 4% for E20 to 10% for E100. Unfortunately, this family of gasses are highly reactive and harmful to humans because they contribute to the formation of fine particles and ozone smog, acid rain and oxygen depletion in waterways, and haze air pollution.

The USA Department of Energy has an interactive biodiesel cost calculator (figure 10.3) that will estimate your annual fuel cost savings (USD) and emissions (pounds of $CO_2$) between using diesel vs. biodiesel.

Table 10.1 is a chart showing that, in 2016, the world consumed 9.3 billion gallons of biodiesel: USA at 22%, followed by Brazil and France at 10%, Indonesia at 9%, and Germany at 7% of the total.

## Renewable Diesel

Renewable diesel, also called green diesel, is made to match ASTM D975 chemical properties for petroleum diesel. Because it is the same, it can be used in the same distribution systems as petroleum diesel. It is created from cellulose using high-temperature deconstruction. R100 (100% renewable diesel) can be used in any diesel engine. It also is produced in blends of R20 (20% blend with 80% petroleum diesel) and R5 (5% blend). In the USA in 2019, renewable diesel use was about 900 million gallons.[6]



## Modern Biomass

Modern biomass is a term used for production of electricity, heat, or CHP from burning biomass in modern fuel-efficient equipment. It accounts for about 50% of the world's renewable energy production.[7] Electricity is generated by turbines powered from high-pressure steam created by a boiler from burning biomass. The biomass used is often agricultural waste, fast growing woody crops, or wood production by-products such as sawdust, pellets, and chips. Biofuels and methane produced from decomposing organic waste materials can also be used as fuel.

Biomass power plants are about 20% efficient while CHP is 80% efficient, depending on seasonal heating needs.

Biomass systems can vary from a small stove for a house to a large power plant. The size and space requirements of the system will depend on the desired output. Biomass systems require fuel storage which is usually outside and protected from the elements. There has to be a clear path to move the fuel to the combustion equipment. Combustion equipment has to have supply air and directly vented exhaust. The biomass is burned at high temperatures so there is very little ash compared to the volume burned. Ash is collected and either manually or automatically removed. Heat is either used directly or transferred to water to make steam. System controls are typically automated and can be quite sophisticated depending on the complexity of the machinery.

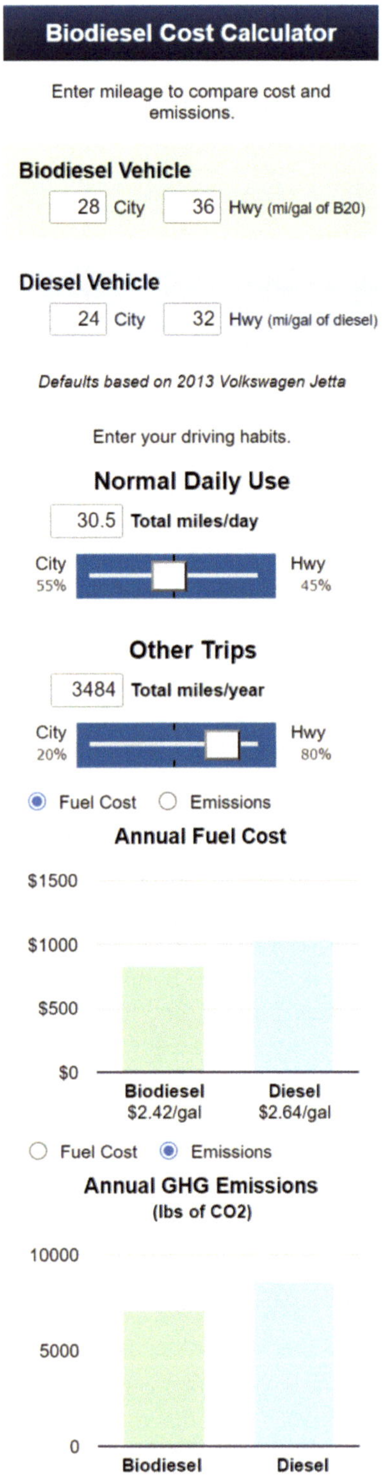


Figure 10.3 Annual cost (USD) and $CO_2$ emissions comparison between diesel and biodiesel calculator.[4]

## Heating System for Large Building

The East Lycoming School District, located in rural Pennsylvania, USA, installed a biomass steam heating system in 2010 in the Hughesville Junior/Senior high school. The school serves about 750 students. Biomass replaced oil and natural gas-powered boilers, saving about US$50,000 in fuel costs per year. The school is located on a large property and was able to grow hybrid willow (figure 10.4) on 40 acres (16.1 hectares), harvesting one-third of it on a rotating annual basis. They have since found a local source of wood chips that was less expensive than renting a harvester.

Biomass is delivered in large quantities with a tractor to a large bin just inside the roll-up door shown in figure 10.5.

Six to seven hundred tons of wood chips are moved during the heating season by augers to a conveyer belt that automatically feeds the boiler (figure 10.6).

The boiler burns at almost 2000°F (1093°C) and creates about 5 gallons (19L) of ash per 20 tons (18 metric tons) of chips burned. The front of the boiler can be seen on the left in figure 10.7 and the ash barrel with the side of the boiler behind it is visible on the right.

The space needed for this system is large and two storeys tall. There are many pipes for supply and exhaust air as well as connecting to the building's heating system (figure 10.8).

## Pellet Stove for a Small Building

Many homes in the USA burn heating oil or natural gas to heat their homes. This pellet stove is used to heat about 2500ft$^2$ in a climate with 5611 HDD annually (base 65°F). The stove (figure 10.9 left) is in a small room that has a door to the vestibule to bring in pellets which are stored outside. The stove can be set to burn hardwood or softwood pellets and is controlled by an automatic thermostat. The stove must be loaded manually and has a hopper that will hold about 1.5 bags of pellets (figure 10.9 right). Pellets are purchased in 40lb bags. Heated air is blown from the unit with a built-in fan.

This home uses 2 tons (1814kg) (100 bags) to heat annually, which replaced burning 600 gallons (2271L) of heating oil (13,404lb (6080kg) of $CO_2$ annual emissions). On the coldest day in the winter, the stove burns a bag of pellets. The ashes are emptied from a small bin at the bottom of the stove, when the interior of the stove is also cleaned, about once a week.

Table 10.1[5]

| World biodiesel consumption in 2016. | Billion Gallons | Share of World Total |
|---|---|---|
| World Total | 9.3 | |
| United States | 2.1 | 22% |
| Brazil | 1 | 10% |
| France | 0.9 | 10% |
| Indonesia | 0.8 | 9% |
| Germany | 0.7 | 7% |
| All others | 3.9 | 42% |

Figure 10.4 Field of fast-growing hybrid willow for use as chipped biomass.

Figure 10.5 Roadway to biomass delivery door.

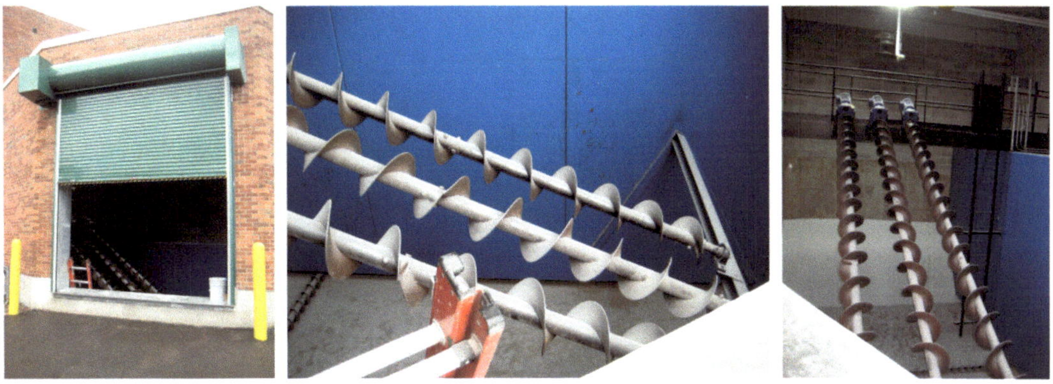

Figure 10.6 Augers inside the biomass storage bay move fuel to a conveyer belt that goes to the boiler.

Figure 10.7 The front of the boiler (left) and the ash barrel with the side of the boiler behind (right).

Figure 10.8 The large 2-story boiler room has mezzanines and stairs. Exhaust piping (left) and all the piping and equipment needed to hook up to the building's heating system is shown on the right.

Figure 10.9 Pellet stove for a single-family home.

## Resources


1 Chapin, B. (2017, February 21). *USDA report shows improving corn-ethanol energy efficiency*. U.S. Department of Agriculture. www.usda.gov/media/blog/2010/09/21/usda-report-shows-improving-corn-ethanol-energy-efficiency

2 Rosenfeld, J., Lewandrowski, J., Hendrickson, T., Jaglo, K., Moffroid,K., & Pape, D. (2018, September 5). *A life-cycle analysis of the greenhouse gas emissions from corn-based ethanol* (Contract No. AG-3142-D-17-0161). ICF for the U.S. Department of Agriculture. www.usda.gov/sites/default/files/documents/LCA_of_Corn_Ethanol_2018_Report.pdf

3 U.S. Department of Energy. (n.d.). *Biodiesel vehicle emissions*. Alternative Fuels Data Center. https://afdc.energy.gov/vehicles/diesels_emissions.html

4 U.S. Department of Energy. (n.d.). *Biodiesel vehicle emissions*. Alternative Fuels Data Center. https://afdc.energy.gov/vehicles/diesels_emissions.html

5 U.S. Energy Information Administration. (2020, August 6). *Use of biomass-based diesel fuel*. www.eia.gov/energyexplained/biofuels/use-of-biodiesel.php

6 U.S. Energy Information Administration. (2020, August 18). *Biomass-based diesel fuels*. www.eia.gov/energyexplained/biofuels/biodiesel-in-depth.php

7 International Energy Agency. (2018, October 8). *Modern bioenergy leads the growth of all renewables to 2023, according to latest IEA market forecast*. www.iea.org/news/modern-bioenergy-leads-the-growth-of-all-renewables-to-2023-according-to-latest-iea-market-forecast


## Study Questions

1. Look around your community and determine what kinds of biomass are available. Can you purchase products like wood pellets or chips? Do gas stations have biofuels? Are there industries that have organic wastes?
2. Go online and research "masonry heater" and answer the following questions:
   a. What is a masonry heater?
   b. What makes a masonry heater different from a fireplace?
   c. Why are masonry heaters so effective?
3. Explain the differences between ethanol and biodiesel:
   a. What are they made of?
   b. How are they made?
   c. What types of vehicle engines can they be used for?
4. Do some research about CHP and its applications. A good article from the Whole Building Design Guide www.wbdg.org/resources/combined-heat-and-power-chp has a clear presentation of the types of technologies, applications, and emerging issues relating to CHP. Answer the following questions:
   a. How does CHP work?
   b. What types of renewable energy are used in CHP systems?
   c. What are the pros and cons of CHP?
   d. What types of projects are good candidates for CHP?

CHAPTER 11

# Heat Pumps

**Key Terms**

Heat pump: a device that uses mechanical energy to transfer heat from one location to another.

Heat sink: a mass that absorbs or dissipates heat without much change of its own temperature – for heat pumps used in building HVAC this is typically air, water, or earth.

## Concepts for Success

Heat pumps move heat between the interior and exterior of the building. They are very efficient at moving heat because they use the refrigeration cycle to transfer the heat. Heat pumps can provide both heating and cooling, called bi-directional, or may provide only heating or cooling.

The exterior heat sink, which can either provide heat or receive heat, depending on whether the building needs heating or cooling, can be air, ground, or water. Air source heat pumps can either provide heating and cooling for the whole building, using ducts distribute conditioned air, or small ductless mini splits. Ground source heat pumps have exterior closed loops for earth and water. Water can alternatively use open loops. There are various configurations for the way the loops are installed.

Heat pumps can also be used for heating **domestic hot water (DHW)** and for clothes dryers. Heat pump water heaters use the heat from the room they are located in to heat the water. Heat pump clothes dryers do not need to be vented because water is condensed from the circulating air by the heat pump coil.

## How a Heat Pump Works

Heat pumps use the physics of compression and state change of refrigerants (referred to as the refrigeration cycle) to move heat from one place to another. It takes a small unit of electrical power, say 1kW, to move 4kW worth of heat, which is a net gain of 3kW.

Heat pumps need a compressor, two coils, and an expansion valve. One coil is indoors and the other is outdoors. The outdoor coil is located in a heat sink – a mass of either air, ground, or water – that is able to accept or provide heat.

Heat can be moved either in or out of the building. We tend to talk about heating and cooling spaces but actually it is just moving heat around. When we move heat out, it feels cooler! The amount of heat moved is regulated by the thermostat.

## Air-to-Air Heat Pumps

Figure 11.1 shows a heat pump with the refrigerant flow in heating mode. The heat sink is air (outside the building). The refrigerant is a state change material, changing from a liquid to a gas and back again. Heat energy is absorbed and released as it changes state. The basic parts are the outdoor and indoor coils, the compressor, and an expansion valve. Operating in heating mode, heat is collected from the exterior air by the exterior coil, acting as the evaporator, because the liquid is changing into gas as the refrigerant heats. The compressor compresses the refrigerant, which causes the refrigerant to become hotter. The hot, gaseous refrigerant circulates through the interior coil. A fan blows air over the coil (called a fan coil), moving the heat into the air of the room or ducts of a mechanical system. The interior coil acts as the condenser because, as the refrigerant cools off, it condenses into a liquid. The expansion valve slows the flow of the refrigerant.

In the heating and cooling mode diagrams (figures 11.1 and 11.2), the exterior heat sink is air, so it is also a fan coil. Fan coil units have coils made of copper tubing and typically the coils are surrounded by aluminum fins to aid the transfer of heat. Notice in figure 11.2 that the

DOI: 10.1201/9781003297819-13

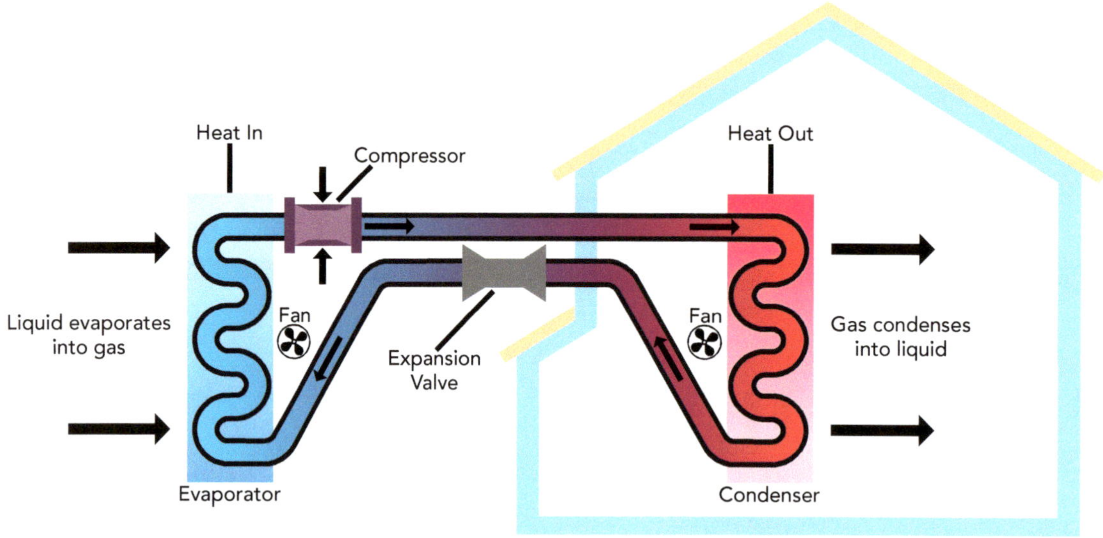


**Figure 11.1** Heating mode operation of heat pump.

cooling mode of the heat pump circulates the refrigerant in the opposite direction from the heating mode. This causes the interior coil to act as the evaporator, absorbing heat from the air, and the exterior coil acts as the condenser, releasing heat to the air.

Heat pumps can condition a whole building or one space. Whole building air-to-air (fan coils both inside and outside) heat pumps are larger, heavier, and noisier than small ones. Whole building heat pumps are typically mounted on a concrete pad and connect to an air handler on the interior (figure 11.3). The heating or cooling is then distributed through supply air ducts to the building's rooms. Building HVAC systems may have additional components for heat recovery and controlling how much air at what temperature is being distributed into zoned spaces. Remember that all ducts should be insulated and sealed as well as be located inside the building's thermal envelope. The actual size of the heat pump depends on the heating and cooling demand for the building.

Mini split heat pumps are ductless. They are great for retrofits in older homes where it would be difficult to run ducts. In less efficient buildings, a mini split can condition one or two rooms, depending on the size of the unit and spaces. Larger heat pump units can be connected to air handlers in separate rooms. In energy efficient buildings, a mini split may be able to provide all of the heating and cooling needs for the building or tenant space. As shown in figure 11.4, mini splits have a smaller air-to-air heat pump located outside, condenser piping which generally is covered with a casing, and an interior air handler (fan coil unit). The air handler is fairly small and is mounted near the ceiling of the room. A drain line to the exterior is installed to remove the condensation from the interior air handler. The mini split is controlled with a wireless remote.

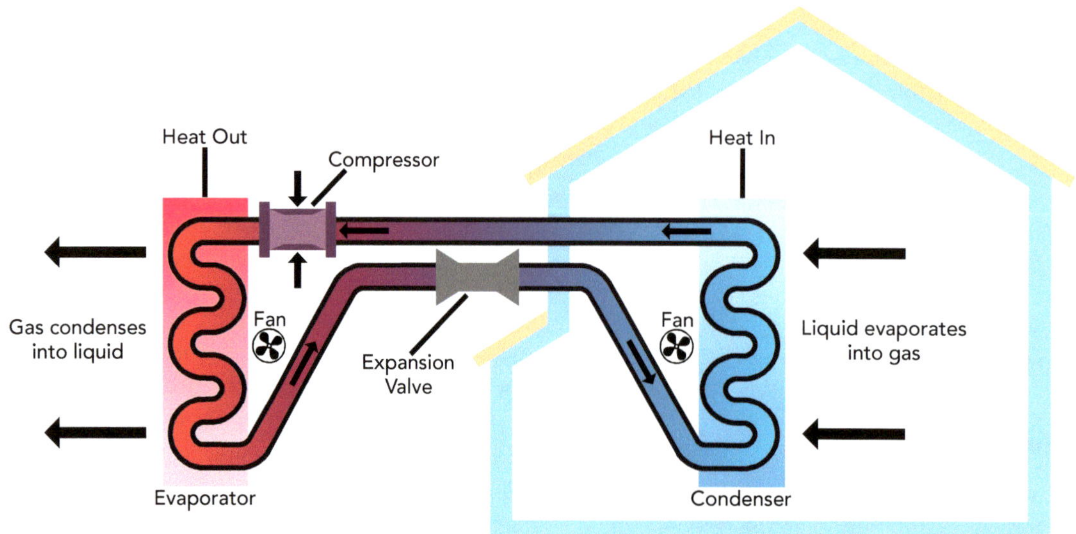


**Figure 11.2** Cooling mode operation of heat pump.

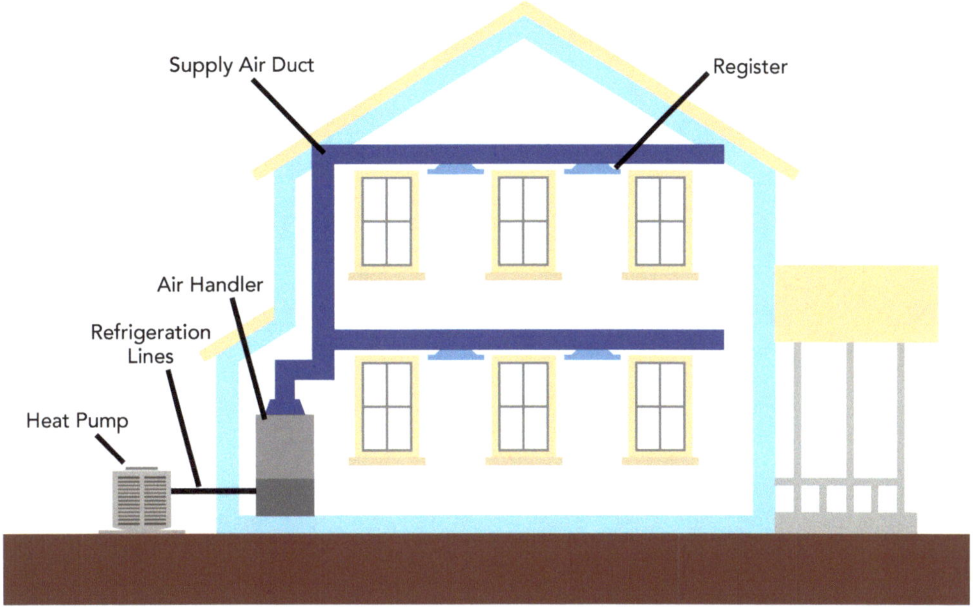


Figure 11.3 Whole building heat pump connects to an air handler and ductwork.

Single mini split exterior units are installed near the location of the interior unit. They can be mounted on the ground or supported by brackets to keep it above snow. In areas where there is a lot of snow, you may consider installing a protective covering, such as a shed roof, over the outside unit. Remember that the unit has to be exposed to the air, so it cannot be enclosed. In very cold weather there isn't much heat in the air. The building may require a backup heating source when temperatures are below 17°F (–8°C). The exact efficiency depends on the particular unit.

When multiple mini splits are used on large commercial buildings, the outside units need to be spaced a bit apart from each other on the roof to make sure they each have adequate air exposure as well as to provide maintenance access.

## Ground Source Heat Pumps



Ground source heat pumps use the earth, or a large body of water or aquifer, as a heat sink. They can be either closed or open loop; closed loop are more common. The loops are usually made of high-density polyethylene plastic tubing and piping. Closed loops circulate a non-toxic antifreeze mix while open loops circulate water. Closed loops are buried deep enough to avoid freezing and fluctuating ground temperatures. Deep earth temperatures vary by location and

Figure 11.4 Mini split heat pump has a small exterior heat pump unit and a fan coil on the interior.

**Figure 11.5** Ground source closed loops: horizontal, pond loops, and vertical.

climate: generally, the earth is warmer at the equator and colder as you move toward the poles. Geothermal systems typically have very little maintenance.

If cooling loads are quite a lot larger than heating loads, geothermal may be combined with a cooling tower.

Closed loop geothermal systems have varying loop configurations (figure 11.5). Which one you use depends on the heating and cooling load, the soil type, and the ground temperature. All loops must be contained within the property line and must meet code and zoning requirements. Jurisdictions usually have requirements that loop locations must be kept away from wells, septic systems, and water and sewer lines.

Vertical bores are holes about 4" (102mm) in diameter that are located at least 20' apart from each other. The more heating and cooling demand there is, the farther apart the holes need to be so that you don't change the temperature of the heat sink. The vertical height of the hole can vary from 100–400' (30–120m) deep. As you can see in figure 11.5, there is a horizontal loop to/from the interior heat exchanger. Each vertical bore creates a loop that attaches to the horizontal piping in the opposite location along the supply/return loop to equalize pressure. Vertical bores can be made in rocky soil or on smaller sites where there isn't enough room for a horizontal-type loop. Larger commercial buildings use vertical bores.

Horizontal loops can be effective if you can dig trenches 4–6' (1.2–1.8m) deep and 100–400' (30–120m) long. The piping can either use two pipes that are buried about 2' (610mm) vertically apart from each other or can be coiled in a loop that can be buried either flat or vertically. Coiled loops take up less horizontal area than a straight set of pipes.

Closed loops in ponds require a large body of water within 200' (61m) of the building. The water must have enough mass that it stays a fairly consistent temperature at the bottom. Depending on exactly how cold your climate is, you probably need about a half-acre of water that is 8–10' (2.4–3.05m) deep. Coils of piping between 300–500' (91–150m) long are anchored at the bottom of the pond.

Open loop systems are used with larger bodies of water, a well, or deep aquifer (figure 11.6). Open loops pull the water directly from the source, run it through the heat pump, then return it to the source. In some instances, a recharge well or surface discharge may be used. There are often a lot of permit requirements because of the possibility of water contamination. Open loop systems need to have a consistent supply of fairly clean water. Particulates, sediments, and microorganisms can clog the system, causing down time for maintenance. It can also be difficult to know the exact temperature of deep aquifers.

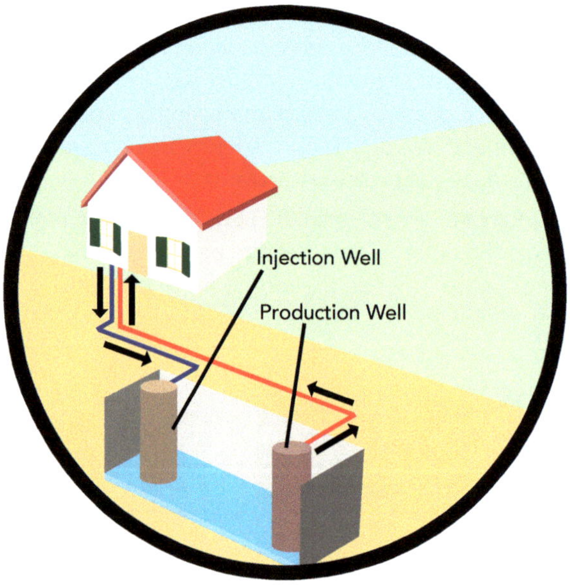


**Figure 11.6** Ground source open loop draws and returns water directly from a source.[1]

## Sizing Ground Loops

The main considerations for the amount of bore or horizontal tubing depends on the load demand, the deep earth temperature, soil type, and characteristics of the site. A general number for estimating runs for heating-dominated climates is 150–200' (46–61m) of vertical bore per ton of heating demand. The size of the piping loop depends on the flow rate needed for the heat pump, the size of the circulating pump, and the demand. Each ton of heating capacity needs a flow of 3 gallons per minute (11 liters per minute).[2]

How do you know how much the heating and cooling load is? You have to calculate the heat losses and gains throughout the year. In the USA, residential loads are calculated using **ANSI/ACCA** Manual J.[3] This takes into account the insulative values of the building envelope, climate zone, relative humidity, infiltration, shading, and performance of ducts if using a ducted system. Computer simulations can also estimate loads. It is important to have the right sized equipment because improperly sized equipment is inefficient. If undersized, it will run too much, causing stress that can lead to breakdowns, costly repairs, and reduced lifespan. It may also void the warranty. Undersized equipment may not be able to properly heat, cool, or dehumidify the building. Oversized equipment uses more energy than needed and has higher initial costs. Traditionally, mechanical engineers have oversized equipment to ensure that building occupants are comfortable. Right-sizing is important for an energy efficient building.

## Performance Labeling of Heat Pumps

Heat pumps are rated based on their efficiency for heating or cooling. The higher the rating number, the more efficient the heat pump is. That is because all of the ratings are based on output divided by input energy.

Air source heat pumps use seasonally adjusted ratings for heating and cooling. Air temperature varies over the season, so ratings are based on fluctuating temperatures. The heating is rated by the **heating season performance factor (HSPF)** while cooling rated by the **seasonal energy efficiency ratio (SEER)**. HSPF = output heating energy in Btu over a season/ input electrical energy in Wh during the same season. SEER = output cooling energy in Btu over a season/ input electrical energy in Wh during the same season. HSPF is not tested in temperatures below 17°F (18.3°C), so care must be taken to properly design backup heating systems when the air temperatures are very cold.

Ground source heat pumps use the **coefficient of performance (COP)** for heating ratings. COP = power output/power input. Cooling is measured by the **energy efficiency ratio (EER)**. EER = output cooling energy in Btu/input electrical energy in Wh. The difference between EER and SEER as well as COP and HSPF is that EER and COP are measured at a specific temperature because the ground is going to remain a constant temperature.

In the USA, the Department of Energy (DOE) has established minimum performance requirements for heat pumps. These numbers increase over time. Third-party rating systems require much higher efficiencies than the minimum requirements. Currently the minimum cooling efficiency is 14 SEER for residential air-source heat pumps, although some mini split units are available at 27.2 SEER. The minimum for heating is 8.2 HSPF. For geothermal units, heating is 3.1 COP and cooling is 16.1 EER for closed loop systems.

It is important to match your heat pump to the climate it is being installed in. Colder climates need HSPF ratings over 12 in order to work at colder air temperatures. The **Northeast Energy Efficiency Partnerships (NEEP)** has a list for cold climate heat pumps.[4]

## Heat Pump Water Heater

Domestic hot water can be heated with a heat pump. The **heat pump water heater (HPWH)** transfers heat from the interior air into the water in the storage tank (figure 11.7). When hot water is in high demand, the HPWH uses standard electric heating elements to quickly heat water. HPWH has to be located in an insulated space that remains between 40°–90°F (4.4°–32.2°C) all year. Because the heat pump is moving heat out of the space, it needs to have at least 1,000 cubic feet (28.3 cubic meters) of air space around it and be located in a room that has excess heat. If multiple HPWHs are in the same space, they can cool the space too much, resulting in poor performance. Cooled air can be exhausted to the room or outdoors. HPWHs can be used with a geothermal system by adding a desuperheater. HPWHs can also be integrated with preheated water from a solar thermal hot water system.



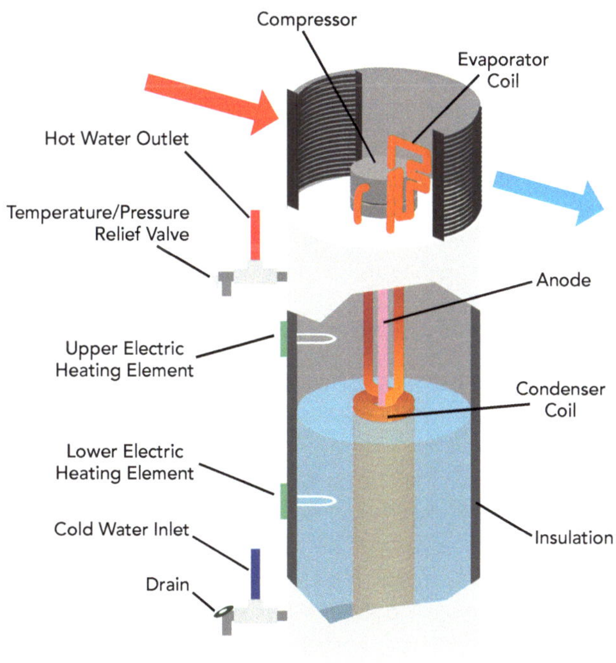


**Figure 11.7** Heat pump water heater.[5]

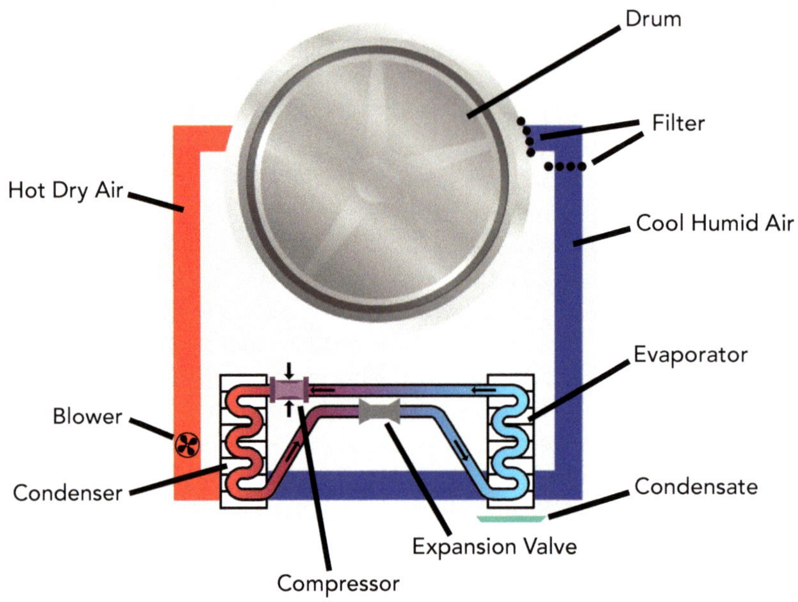


**Figure 11.8** Heat pump clothes dryer.

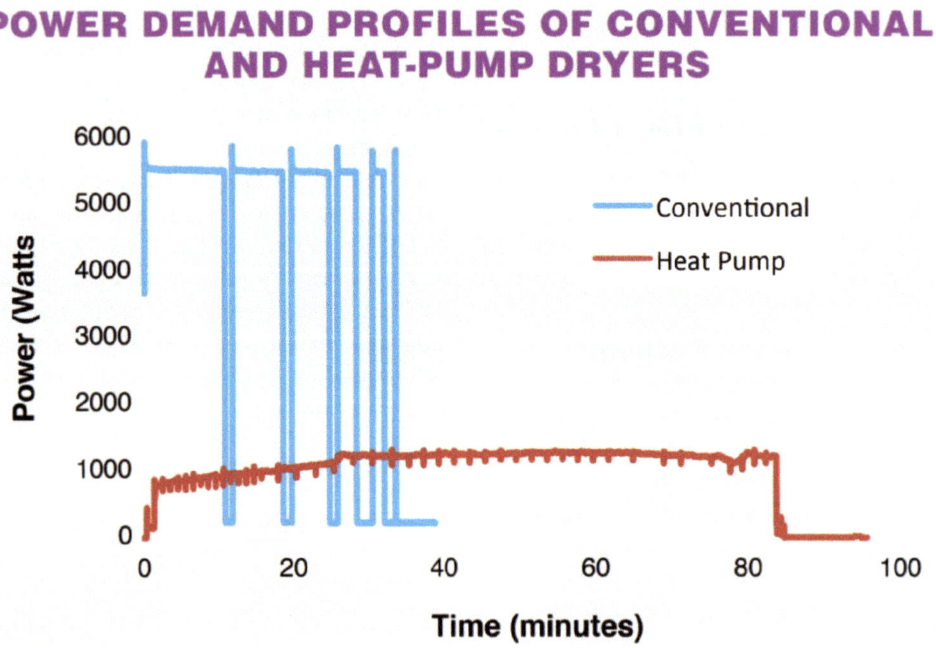


**Figure 11.9** Comparison of power demand and drying time between electric resistance (conventional) and heat pump clothes dryers.[6]

## Heat Pump Clothes Dryer

Electric clothes dryers use a lot of power by using resistance heat to make air hot enough to evaporate the water in the wet clothes. Heat pumps in clothes dryers are located at the bottom of the unit (figure 11.8). The dryer is unvented and uses the air in the room. Air is drawn into the dryer and then across the evaporator and condenser and through the clothes in the drum. Then it continues to circulate around the loop. The air is dried out by the evaporator because the cold coils cause the water vapor to condense and drip into a drained pan. The dry air then passes over the condenser coil, which is hot, so the air is warm and dry as it is reintroduced into the dryer. The coils are set up so that the refrigerant only flows one direction. The compressor and expansion valve work as they normally would in any heat pump. Heat pump clothes dryers can use 40–50% less energy than electric resistance dryers. According to EnergyStar, 6% of US residential electrical consumption is used for drying clothes.[6]

Heat pump dryers can take longer to dry clothes. However, they use much less energy at any given time to complete the drying. This helps to reduce peak load electrical demands (figure 11.9).

## Resources


1 Office of Energy Saver. (n.d.). *Geothermal heat pumps*. U.S. Department of Energy. www.energy.gov/energysaver/geothermal-heat-pumps

2 Williams, C. (2011, June 29). 4-step guide to designing geothermal systems. *HeatSpring Magazine*. https://blog.heatspring.com/how-can-i-design-geothermal-systems-a-4-step-guide-to-designing-geothermal/

3 ANSI/ACCA *Manual J* https://www.acca.org/standards/technical-manuals

4 Northeast Energy Efficiency Partnerships. (n.d.). *Heating electrification*. https://neep.org/smart-efficient-low-carbon-building-energy-solutions/air-source-heat-pumps

5 ENERGY STAR. (n.d.). *How it works — Heat pump water heaters (HPWHs)*. U.S. Environmental Protection Agency & U.S. Department of Energy. www.energystar.gov/products/water_heaters/high_efficiency_electric_storage_water_heaters/how_it_works

6 Super Efficient Dryer Initiative. (2012, October). *Potential energy savings from heat pump dryers in North America* [Fact sheet]. www.energystar.gov/ia/partners/pt_awards/SEDI_Fact_Sheet_H.pdf


## Study Questions

1. Look at a map (or climate information) that shows deep earth ground temperatures. How does deep earth ground temperature change as you move from 0–90° latitudes? How does deep earth ground temperature compare to air temperature?
2. Find images of mini split heat pump installations showing installation of exterior unit on blocks, a concrete pad, wall-mounted, and protected from snow.
3. Draw sketches of a different geothermal system loop layouts: one with 4 vertical bores, one with 4 horizontal loops, and one of an open loop using deep well water.

11

CHAPTER 12

# Solar Thermal Hot Water

**Key Terms**

Expansion tank: a small tank installed in closed loop hot water systems to protect the piping and equipment from excessive pressure.

Phase change material (PCM): stores heat energy through state changes, absorbing and releasing heat when changing between solid and liquid or gaseous state.

Scale (water chemistry): a white, chalk-like substance of minerals that is deposited on surfaces, left behind after water evaporates.

## Concepts for Success

Heating water with the sun is very effective. Water heating is often a large expense in a US household. Electric water heaters are often used and run for 3–5 hours a day to keep water hot in the tank. Solar thermal collectors heat up water using the heat from the sun. They are located outside, typically on roofs, although they can also be ground mounted (see chapter 14, Racking for Solar Thermal and PV). Hot water is stored in a tank which can be used for domestic hot water (DHW), radiant floor heating, or anything you need hot water for.

There are three types of collector that can be configured in a few different ways. The collector types are flat plate, evacuated tube, and unglazed. Unglazed collectors are uninsulated and don't reach high temperatures. These are most often used for pools but are quite inexpensive and can be used for preheating industrial process water. Flat plate and evacuated tube collectors reach high temperatures and can either be used in a thermosiphon, where water is stored in a tank located above the collector outside, or in building integrated configurations of either closed loop or drain-back.

Flat plate collectors are an insulated box that heat up and trap the heat under the glazing. Evacuated tubes are individual tubes that insert into a manifold. The tubes are glass and use the evacuated volume as an insulator. Evacuated tubes achieve higher temperatures than flat plate collectors but cost more. Evacuated tubes come in two types – direct flow, which are plumbed into the manifold, and heat pipe, which transfers heat through a bulb that has a dry fit into the manifold.

Closed loop systems have a heat transfer loop to the storage tank from the collector. In climates that have any frost or freezing potential they use a glycol solution to prevent damage to the collectors and piping. Drain-back systems have a drain-back tank that provides an extra space for the water in the heat transfer loop. When the pump stops, the water drains back out of the collector and piping, filling the drain-back tank and protecting equipment from freezing damage because no water stays in the outside portion of the loop.

The largest system components are the collectors and the storage tank. The collectors are always outside and the storage tank is always inside. They should be located no farther than 150′ apart in order to reduce heat loss through the piping. The piping needs to fit into the building in an unobtrusive manner, provide vertical chase space and horizontal runs.

Storage tanks come in two types, pressurized and non-pressurized. Pressurized tanks look a lot like water heaters. Non-pressurized tanks tend to be larger tanks from 200–5000 gallons (760–18900L) and are wide cylindrical shaped insulated containers.

Designing a solar thermal system requires that you know how much hot water is being used, what temperature it needs to be, and when the hot water is being used. Systems are designed to produce a portion, but not all, of the hot water needed. Because

DOI: 10.1201/9781003297819-14

most latitudes experience longer sunny summer days and shorter hours of sun in the winter, the best tilt angle is one that orients the collector more toward the winter sun. For the most consistent production year-round, the best tilt angle for solar thermal collectors is latitude + 15°.

Certification of solar thermal equipment and systems is available through the Solar Rating and Certification Corporation (ICC-SRCC™).

## What is Solar Thermal Hot Water?

Solar thermal collectors heat up in the sun and transfer that heat to water. The hot water can be used for domestic or industrial applications. The hot water can also be used for low-temperature heating such as in a radiant floor.

The most common collector systems in the USA have the collectors located on the roof, and a storage tank next to the water heater, see figure 12.1. There can be an in-line, on-demand water heater instead of a tank-type water heater. The storage tank has a heat exchanger coil located in the bottom half or third of the tank. Potable water is heated in the tank because the heat from the coil is absorbed by the water. The coil is part of a loop to the collectors on the roof.

The storage tank has potable cold water added in at the bottom and hot water supplied out from the top. The water system is pressurized, so no pumps are required. When hot water is needed at a sink or shower, the tap is turned on and hot water comes out. As hot water is used, it flows out of the hot water heater or through an in-line heater. The water for the heater is pulled from the top of the solar thermal storage tank, so the water is already hot. The heater will only have to run if the water in the storage tank is not at the desired temperature. Because the water being supplied to the heater is already warm, the heater doesn't have to

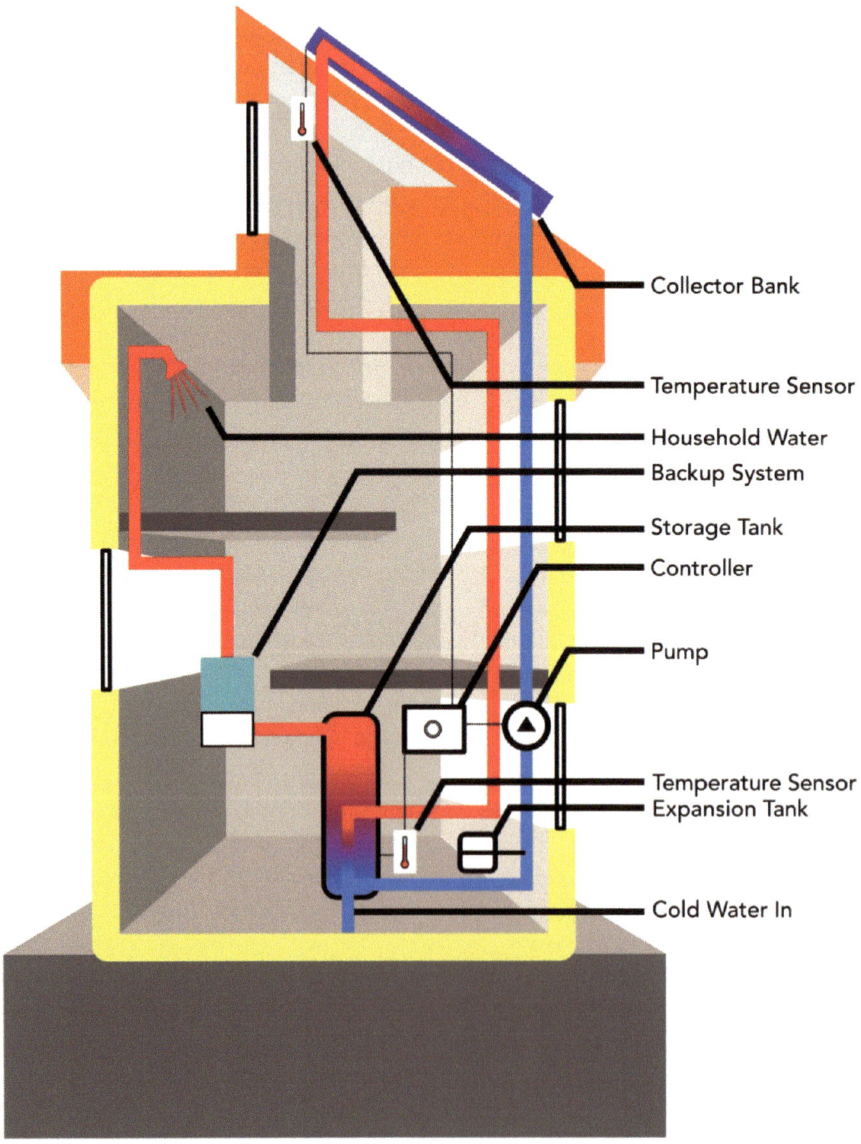


**Figure 12.1** Closed loop solar hot water system.

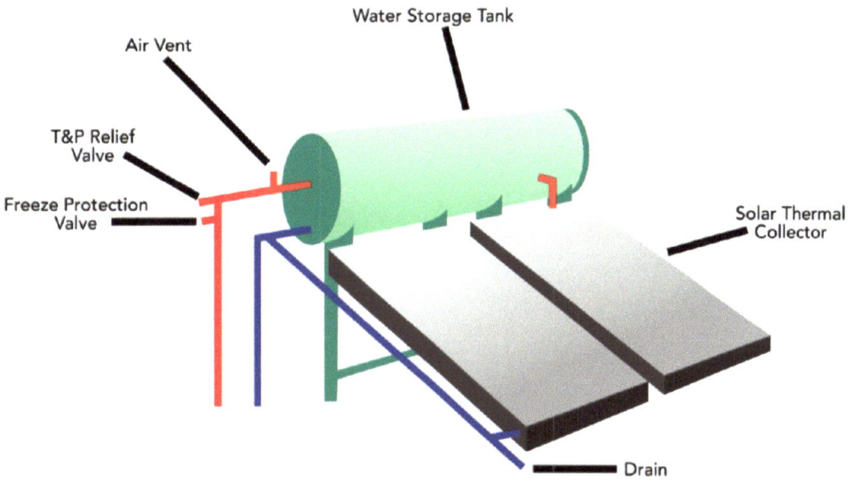


**Figure 12.2** A thermosiphon stores the hot water in a tank above the collector.

provide much additional heat, so it eliminates a lot of energy use.

The coil is located in the bottom of the solar storage tank because colder water is at the bottom of the tank. The reason for this is the water in the tank stratifies because hot liquid rises. Cold water is added in at the bottom of the tank because there is faster heat exchange between hot coils and cold water. Note that the top of the coil has the hot liquid circulating from the collectors and the bottom of the coil is colder and returning liquid to be heated again. The cold side of the system has an in-line pump, the symbol is a circle with a triangle in it where the tip of the triangle points the direction of the flow. The other item located on the cold side is an **expansion tank**. Its symbol is a somewhat oval shape with an elliptical type shape in the middle. This represents the actual appearance of expansion tanks, which have a bladder in the middle to allow for release of pressure in the line.

The controller turns the pump on and off. The controller is wired to thermometers in the storage tank and the collectors. If heat is called for in the storage tank, i.e., the water in the storage tank is below the desired hot water temperature, and the collectors are hot, then the controller will turn on the pump. The pump moves the liquid in the closed loop around so that the cold liquid is heated by the collectors and then supplied to the coil in the tank where the heat is transferred to the water in the tank. The pump continues to circulate until either the water in the tank reaches the desired temperature or the collectors are no longer hot enough to heat up the circulating liquid.

An alternative to having the storage tank inside is to have it outside at the collector. These systems are called thermosiphons, shown in figures 12.2 and 12.3. Thermosiphon solar heaters can be set up to either directly heat the water or have an internal transfer loop. Thermosiphons use natural circulation due to convection to heat the water and the tank must be located above (higher than) the collector in order to work. These systems do not require a pump. Thermosiphons work best when attached to only one or two collectors, so have limited application if large amounts of hot water are needed. It may also be difficult to place the tank on a steep roof. Simple thermosiphons are only used in frost-free climates, as water could easily freeze and harm the collector and/or piping. Thermosiphons are popular in the Mediterranean and China, as well as places with uncertain or no electrical supply. Thermosiphons can be configured with either flat plat or tube type collectors and can be pressurized or non-pressurized.

**Figure 12.3** A tube type thermosiphon.

## Pool Solar Water Heaters (Unglazed Collectors)

Pool solar water heating is done by adding a circulation loop through a set of collectors. When the pool pump is running, heat is needed, and the sun is shining, then the pool water is directly pumped through the collectors and returned to the pool. The collectors are always black to increase their solar absorption. Figure 12.4 shows some pool heater collectors in storage, hung on a wall.

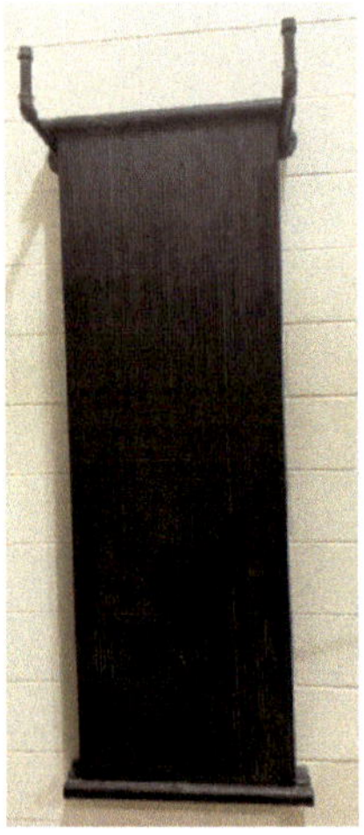

**Figure 12.4** Pool heater (unglazed) collectors hung in storage during winter season.

Figure 12.5 shows how the controller is connected to a temperature sensor on the roof at the collectors and an auto valve that can divert the water through the collectors or bypass the collectors. The pool pump needs to be set up so it is operating during the best solar collection hours of the day. The collectors are generally attached to a roof that is near the pool and receives good solar radiation.

Note that the cold side of the piping, shown in blue, will be the longest run and connects to the bottom of the collector system. The hot water piping, shown in red, exits from the top of the system and should have the shortest run back to the pool, so the heat doesn't dissipate into the air instead of reaching the pool. Water needs to be able to run diagonally across the collectors; note that all the collectors are connected top and bottom to each other. This configuration is the same as the harp flat plate connectors which will be discussed below.

Pool solar collectors are made of inexpensive materials and have a limited lifespan. They are manufactured as either as a single rigid molded and/or thermally welded plastic collector or they can have rigid tubing at the top and bottom with flexible piping connecting between. Units should be made from plastics that will withstand hot temperatures and UV on roofs.

These types of systems are sized based on the surface area of your pool. The area of the collector should be anywhere between 50–100% of the surface area. If the pool is located in a cold climate, where it is closed down for the winter, the pool collectors can be taken down and stored to extend their lifespan.

Pool collectors are inexpensive, efficient, and easy to install. They only work in warm months as there is no

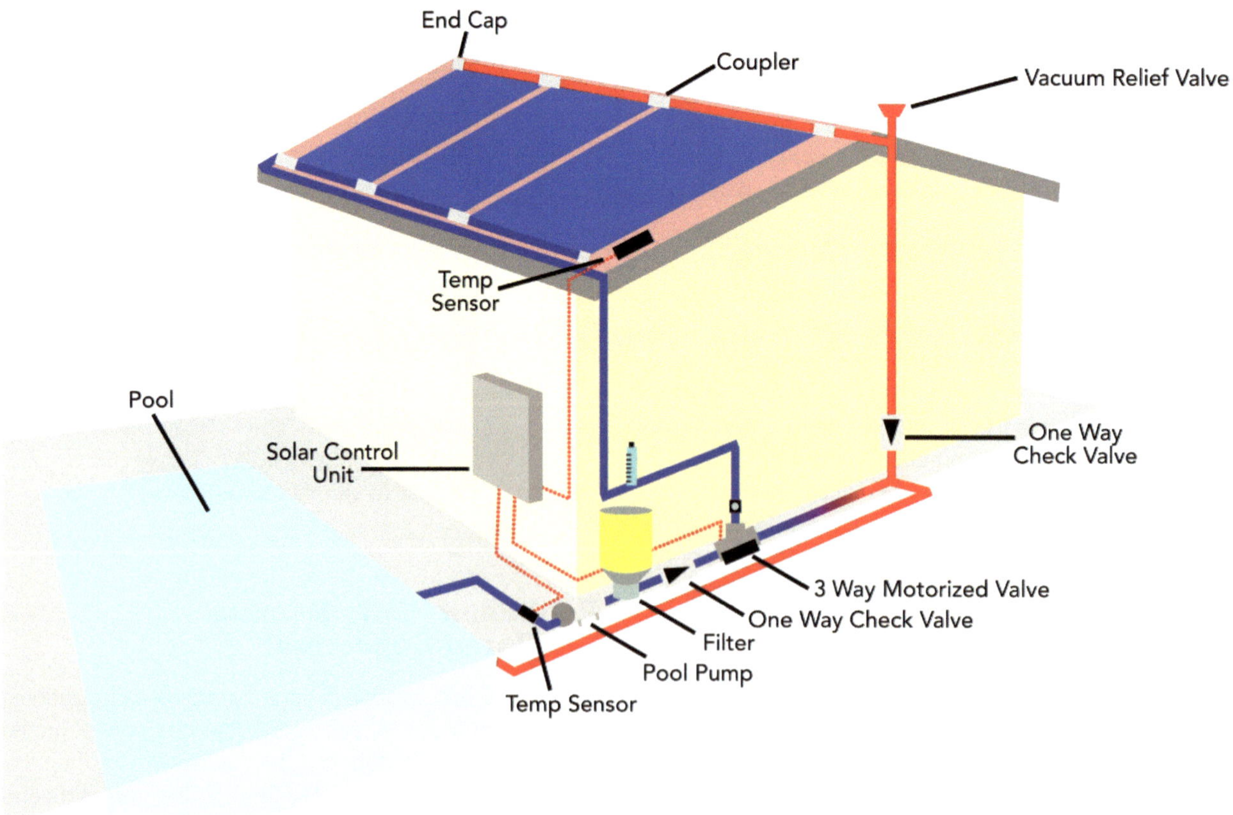


**Figure 12.5** Pool solar water heater layout that circulates water from the pool through the collectors.

insulation in the piping. There are no incentives available for heating pools, nor are there points awarded in any third-party rating system.

## Flat Plate and Evacuated Tube Collectors

For domestic or industrial hot water production, there are two types of collectors: flat plate and evacuated tube. Flat plate collectors are rectangular boxes with a single piece of glazing over the top, see figure 12.6, and generally appear dark gray. They have a similar appearance to some photovoltaic (PV) modules and skylights, although usually flat plate collectors are larger. Evacuated tube collectors have multiple tubes that connect to a manifold at the top of the collector and generally appear blue as shown in figure 12.7.

**Figure 12.6** Glazed flat plate collector with black frame.

**Figure 12.7** Evacuated tube collectors mounted vertically on a wall.

The heating principle for all solar thermal collection is to collect the solar radiation, trapping the heat inside a glassed volume. Heat absorbing plates have heat absorbing piping attached. As the sun shines on the collector, it gets hot and traps the heat inside. This heat is transferred to the circulating fluid and then to the water in the storage tank.

Flat plate collectors have an insulated box to help keep the interior hot. There is a single backplate, which is a dark gray color to increase its heat absorption, and a pattern of tubes. The whole box is covered by a sheet of glazing. Figure 12.8 shows their typical construction in a cut-away section. Flat plate collectors must be installed at an angle as they are not watertight and there are drainage holes at the bottom of the collector. They need to be installed oriented as a vertical rectangle and cannot be turned sideways. All collectors have built-in threaded inserts to receive bolts so that they can be easily mounted to rails or **standoffs**.

Piping can be on top, in the middle, or under the backplate, depending on the manufacturer. There are two different piping layouts used in flat plate collectors, see figure 12.9. The most common is a header-riser or harp-type. This configuration has slightly larger piping that runs horizontally across the top and bottom (called headers or manifolds) of the inside of the collector box and vertical piping running perpendicularly (called risers). Fluid flow happens diagonally across the unit because there is equal pressure across the collector. The cold fluid is always in the bottom pipe and it is heated up as it crosses the collector through the risers.

The second piping layout is called meander or serpentine because the piping is one continuous curved pipe. Note that the fluid enters in at the top and exits from the top on the other side. Because the fluid has to travel the entire length of the piping, this layout absorbs more heat than a harp-type layout. If two or more meander-type piping collectors are connected together, they will connect at the top.



Harp-type collectors need diagonal flow. If they are hooked up so the headers are blocked on the same side, then the liquid will only flow through the first few risers, not the entire collector. Single collectors must have cold input at the bottom on one side and hot output on the diagonally opposite side (see figure 12.10) in order to maximize heat gain.

Multiple harp-type collectors are connected side by side (figure 12.11)so that both the top and bottom collectors are connected, while the input and output are always located diagonally across the entire grouping from each other. Remember that the hot always comes off the top and has the shortest run to the storage tank. The ends that are not connected to the circulation loop are capped.

If flat plate collectors are damaged, the whole unit needs to be replaced. Flat plate collectors are fairly large, as much as 4′ × 10′ (1.2m × 3m) and weigh up to 150 pounds (68kg), which can make them a bit difficult to install on steep roofs. Over time, the interior face of the glazing can have **scale** buildup, which can interfere with solar heat gain. Flat plate collectors are

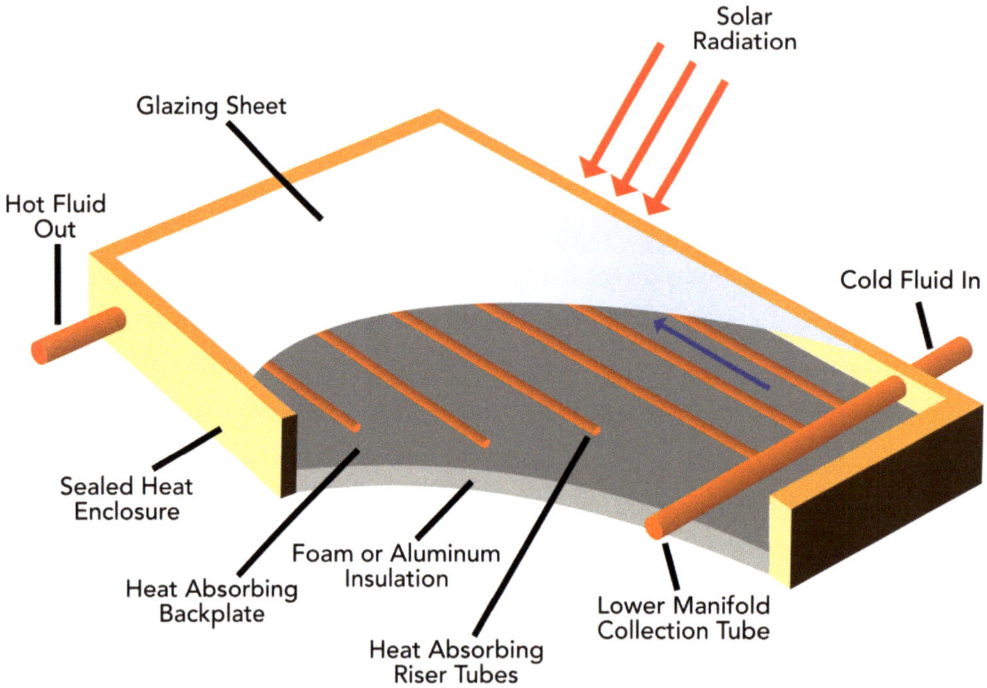


**Figure 12.8** Cut-away section showing the parts of a flat plate collector.

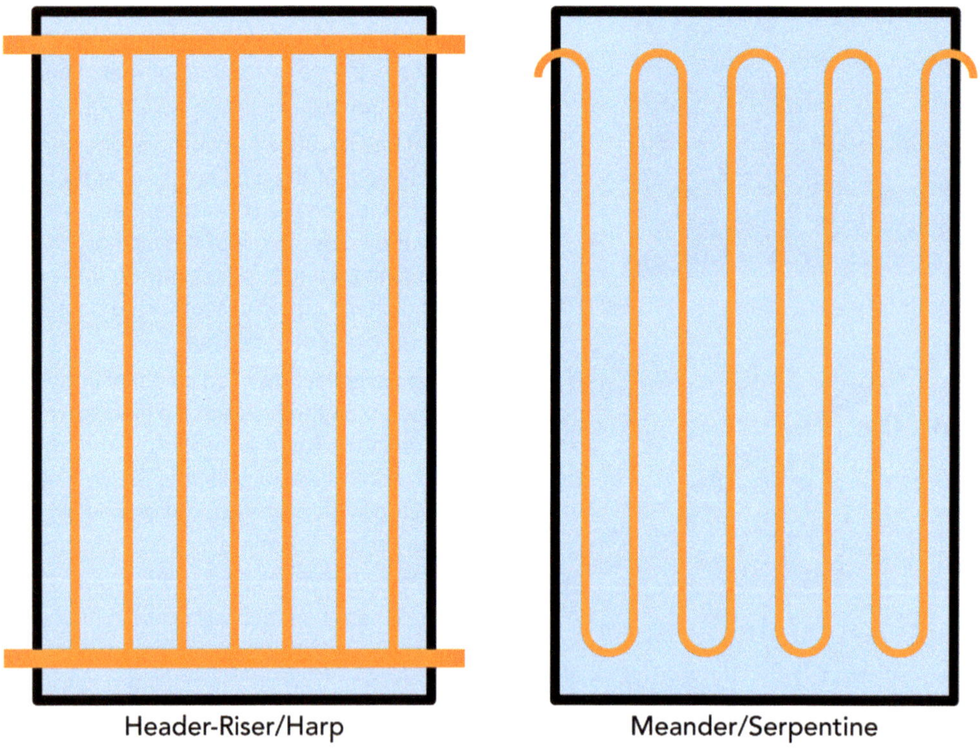


**Figure 12.9** Flat plate piping layouts are either in a header-riser/harp (most common) or meander/serpentine configuration.

very efficient in warmer climates while they may not be so efficient in very cold locations because of having too much heat loss. Flat plate collectors are ideal for producing low target temperatures as they operate to between 86–212°F (30–80°C). If the flat plate is just sitting in the sun without fluid circulation, temperatures can reach as high as 392°F (200°C). There are new developments in flat plate collector design using highly insulated materials and vacuum glazing to increase operational temperatures.

Evacuated tube collectors are made up of a series of transparent glass tubes that are connected into a manifold at the top. Tubes can have diameters from 1″ (25mm) to 3″ (75mm) and are 5′ (1.52m) to 8′ (2.44m)

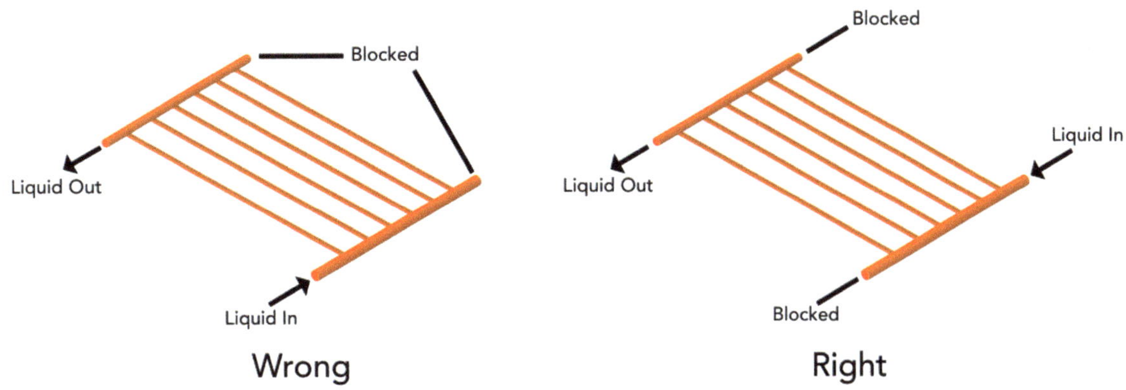


**Figure 12.10** Harp-type layout piping fluid flow needs to be across the unit diagonally from bottom to top.

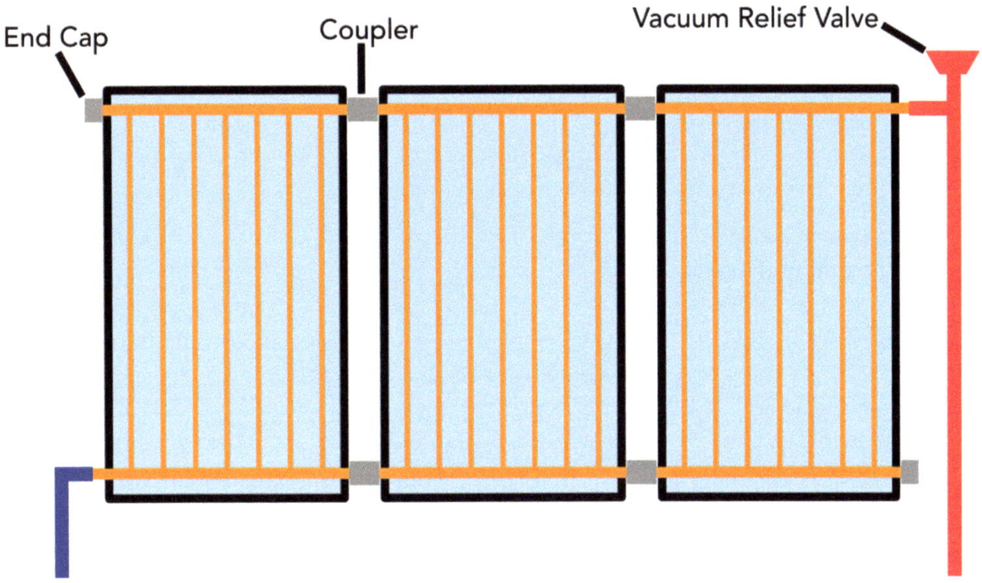


**Figure 12.11** Multiple harp-type collectors are connected together with cold in at the bottom and hot out diagonally across at the top.



long. Tubes have double glass walls and the interior one has coatings to increase solar absorption. The number of tubes in a collector depends on the manufacturer. The configuration of additional heat absorbing plates also depends on the manufacturer.

Basic configurations of evacuated tubes are either an absorber type with direct heat transfer or a heat pipe with a phase change material.

Direct flow evacuated tubes, see figure 12.12, have a U-shaped pipe that allows the fluid to be directly heated inside the tube.

Direct flow tubes can have issues for long-term durability. If a pipe is damaged, the system can lose its fluid and pipes are more difficult to replace/install because they are plumbed into the manifold as shown in figure 12.13. The whole system has to be drained to work on it. The direct connection and U-pipe can be either cumulative or separated in the manifold, or it could be attached directly into a thermosiphon tank. Direct flow evacuated tubes connected to a manifold can be mounted at any angle, even horizontally, because the pump is forcing the water flow through the tubes.

In contrast, heat pipes use a **phase change material (PCM)** to increase the individual tube's absorption

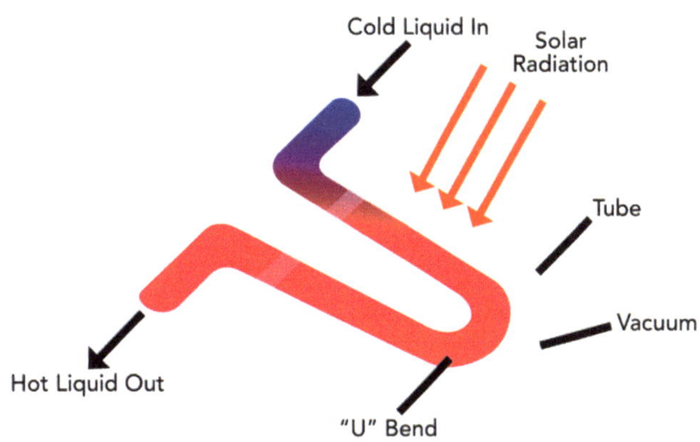


**Figure 12.12** Direct flow evacuated tube directly heats the liquid.

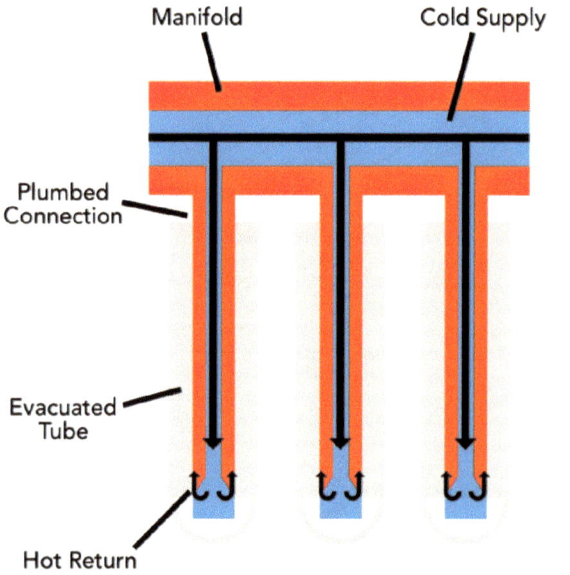


**Figure 12.13** How direct flow tubes are plumbed to the manifold.

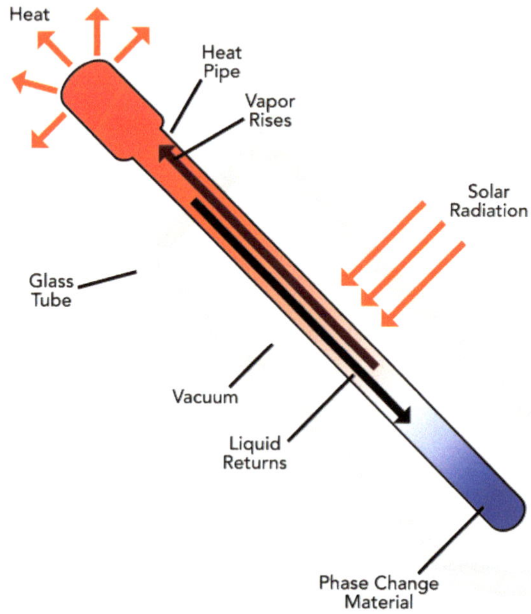


**Figure 12.14** A heat pipe evacuated tube has a pipe with PCM that changes state and heats a bulb at the top of the pipe. The bulb is inserted into the manifold and heats the liquid.

rate. As shown in figure 12.14, when the PCM cools, it turns into a liquid and drops to the bottom of the heat pipe; when heated, the PCM turns into a gas and rises into the condenser bulb at the top of the pipe that is inserted into the manifold.

The condenser transfers the heat into the liquid circulating in the loop through the manifold as seen in figure 12.15. The gas at the top cools off as heat is absorbed, liquifying and flowing to the bottom of the heat pipe. These have had good long-term reliability. If one pipe is damaged, it is easy to replace because it is a dry fit into the manifold.

Evacuated tube collectors are always mounted on a frame that is attached to the manifold. Multiple collectors are attached in series at the manifold. The frame is attached to racks or standoffs as shown in figure 12.16. The frame has threaded hoops at the bottom that line up with holes for the tube bulb tops in the manifold. The exact appearance of the frame and supports varies by manufacturer.

Tubes are slid into place and then caps are screwed into the hoops to keep the tubes in place as seen in figure 12.17. Tubes are lightweight. If one tube is damaged, say in a severe storm by flying debris, then only that tube has to be replaced.

Evacuated tube collectors must be mounted at an angle, not flat, to work properly. These collectors are the most expensive, about twice the cost per area compared to flat plate collectors. One advantage compared to flat plate collectors is that evacuated tubes heat up very quickly because of the round tube, which always has a surface facing the sun perpendicularly, as shown in figure 12.18. Flat plate collectors reflect low angle sun's

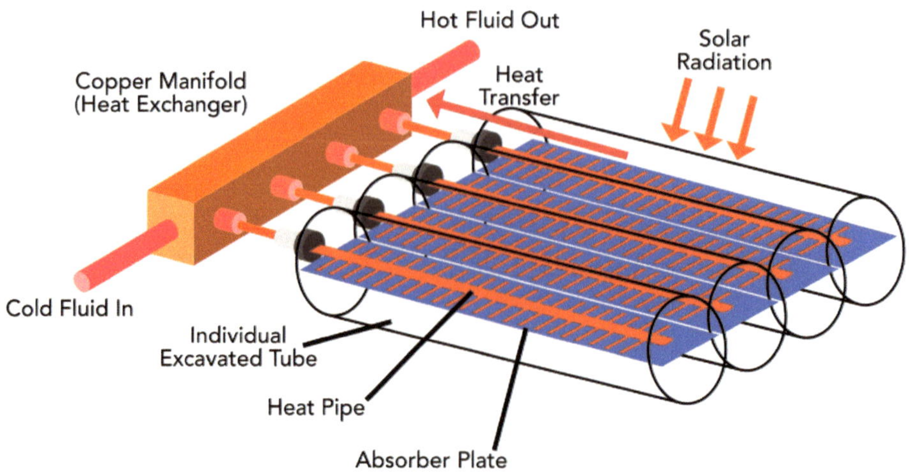


**Figure 12.15** Heat pipe evacuated tubes insert into the manifold to transfer heat.

Figure 12.16 Evacuated tube frames mounted on rails.

Figure 12.17 Evacuated tubes being installed in frame with end caps.

rays. Tube collectors tend to reach higher temperatures than flat plate collectors, which could be important in an industrial application. Evacuated tube collectors operate at temperatures between 170°F (77°C) and 350°F (177°C).

## Other Collector Variations

There are a few other terms and products for solar thermal hot water or air applications that you may see in the marketplace.

Unglazed collectors include the pool heaters mentioned previously but can have other configurations. Unglazed solar collectors do not have a box to trap heat, they just absorb heat. They have dark metal or plastic absorbers and a way for water to flow through. These can be used for preheating water for any industrial or commercial process. Preheating water saves energy because it takes less to heat the water the rest of the way to the desired temperature. Because unglazed collectors don't have insulation, they can lose a lot of heat if the ambient air is cold.

Transpired solar air collectors preheat air for mechanical systems. They are large panels of perforated metal cladding that are a darker color to absorb heat. There is an air space behind the panel. In cold weather air is pulled through the cladding and heated before entering the supply side of the mechanical system. In summer conditions the air space is vented, keeping the building from absorbing as much heat gain through the façade. These panels work well on larger expanses of walls. The air handling units need to be located near the wall with the cladding.

PV thermal (PVT) collectors or hybrid solar collectors combine PV module technology with harp-type solar thermal backing panels. This technology has been around since the 1970s and is quite efficient. When hot, electrical resistance increases in PV cells and wiring relays, so production decreases. Roof temperatures are usually pretty hot. By integrating the solar thermal loop, the PV is cooled down because the solar thermal is transferring the heat away from the collector. PVT has a variety of configurations and can either use water or air to cool them. If water is used, it is hooked up to a storage tank. If air is used, then it is hooked up to the mechanical system (although heat isn't needed in the summertime when the worst PV overheating would occur). PV is always connected to the electrical system. One of the downsides of PVT is that our buildings typically have a much higher electricity demand than hot water demand. It can be very effective for an industrial application that needs a lot of process heat or for a small residential application with a few PV modules.

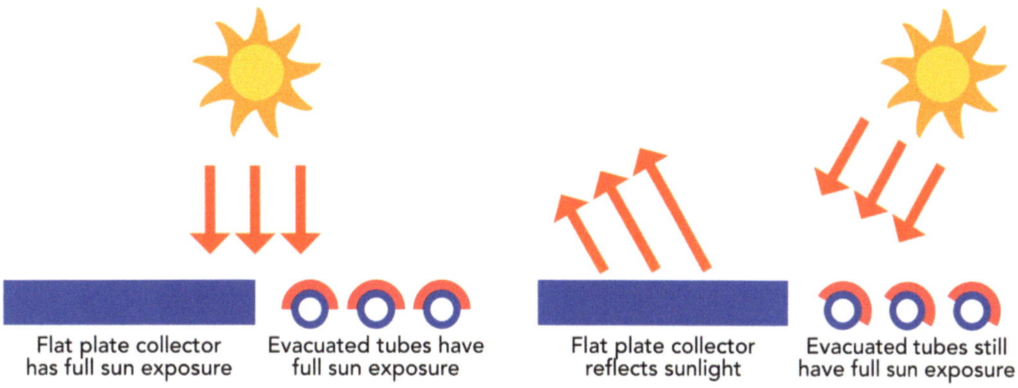


Figure 12.18 Comparison of flat plate and evacuated tube collectors and how well they absorb sunlight at different angles.

## Considerations for Solar Thermal Design

Solar thermal collectors are often located on the roof. There are many other pieces of equipment on the roof, so it is important to coordinate the actual size of the collectors with the amount of available roof space. Solar collectors have to be securely fastened to the building and are subject to wind loads. Flat plate collectors are more likely to have wind loading issues than evacuated tube collectors. Solar thermal and PV modules can share a roof, however, the solar thermal collectors may be at a steeper tilt angle, so care should be taken not to shade the PV. The location of the collectors should be as close as possible to the storage tank, preferably within 150′ (46m) to reduce the amount of heat loss through the circulation piping. That is 150′ (46m) of piping, not necessarily a horizontal plan measurement. In addition to the collectors, space needs to be allocated for the storage tank and other system components. Piping runs need to have chase spaces. Pipes have to penetrate the wall or roof and be well sealed. All piping uses type "L" copper or is pre-insulated flexible stainless steel. Large storage tanks are quite heavy and the building's structure needs to be able to support the weight. A 2000 gallon (7571 liters) tank of water weighs 16,680lb (7566kg).

## Solar Thermal System Configurations

There are two ways that solar thermal systems can be configured: closed loop and drain-back. Both can be used in any climate and any sized system. Designers and installers usually have one type that they prefer; both are proven and viable.

### Closed Loop

Closed loop systems have a single closed loop that circulates a heat transfer liquid. In climates with no frost or freezing potential, water can be used in the loop. In climates where there is frost or freezing potential a glycol–antifreeze mixture, commonly ethylene glycol in a 50% glycol–water solution, is used in the loop. This protects the collectors and exterior portion of the piping from freezing in temperatures down to −25°F (−31°C). Collectors and piping can be damaged from either freezing or bursting pipes. Glycol contracts when freezing while the water expands, keeping the same volume of liquid, so the mixture protects equipment. Water cannot just be added because it would change the glycol mix percentage and could jeopardize the antifreeze. Tap water may be too mineralized or have added chemicals that react poorly with the glycol. Systems with glycol need a 4% higher volume expansion tank. The system needs to be cleaned and flushed as part of regular maintenance to help prevent corrosion. Sometimes additional chemicals are added to the water or glycol mix to help prevent corrosion.

A downside of using glycol is that it doesn't transfer heat as efficiently as water. It also breaks down when it is overheated, when regularly exposed to temperatures at or above 250°F (121°C). Glycol breaks down into glycolic acid, which lowers the overall pH of the mix. Regular pH testing, at least annually, prior to cold weather, is important to be sure that the glycol mix is still effective. Remember that temperatures in collectors can exceed 250°F (121°C), either because power is out and the pump isn't running on a sunny day, or because of the higher operational temperatures of evacuated tubes. If the glycol solution is sitting in a flat plate and overheating, only the amount that is in the collector piping will be affected.

There are other heat transfer products which are used straight from the container without mixing. These kinds of fluids provide frost protection to around −15°F (26°C) and withstand temperatures to around 400°F (204°C). Note that exact numbers depend on the product. These fluids are more expensive than glycol mixes but are often guaranteed for 15–20 years as long as the system isn't exposed to higher temperatures than the product can withstand.

Closed loop systems, as shown in figure 12.19, incorporate a storage tank and collector(s) as well as the **balance of system (BOS)** elements. The collectors can be either flat plate or evacuated tube and are located outside the building. The storage tank, controller, pump, and expansion tank are located inside the building. On the cold side of the loop there is a fill station connection, which consists of two hose bibs that are used to drain and flush the system, as well as recharge the system. Shut-off valves are located on either side of the pump. This allows for work to be done on the pump without having to replace all the liquid in the pipes. There are temperature and pressure gauges to allow for visual monitoring of the system. The controller is wired to temperature sensors that are in the collector and tank. When the collectors are hot and the tank needs heat, the controller turns on the pump. Note that the longest run of piping is the cold liquid and the shortest run of piping is the hot liquid.

### Heat Dump

Heat dumps are important for closed loop systems because overheating the fluid can decompose the glycol. Glycol failure can cause system damage from freeze bursts or overheating DHW system components. Heat dumps are used when no more heat is needed in the storage tank but the collector is still in the sun. Heat dumps can be configured in a number of different ways.

Figure 12.20 shows how water to air heat dissipation is placed on the hot side of the piping and transfers heat to the air through radiant fins that surround the hot fluid filled coils. Finned tube coil heat exchangers can have a number of different shapes from square to rectangular

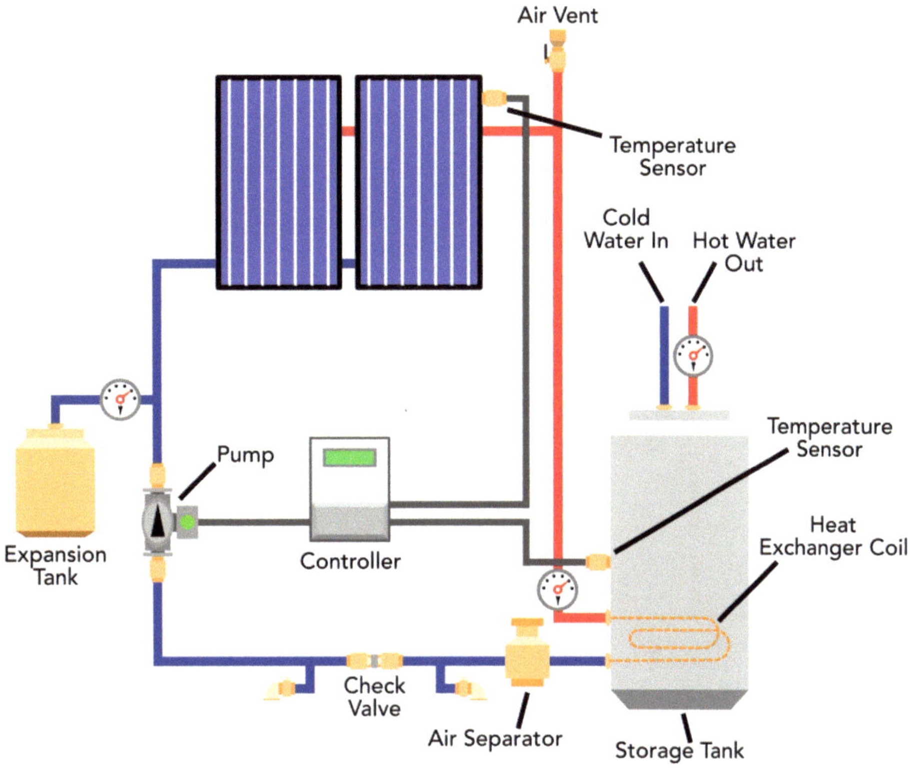


**Figure 12.19** Components of a closed loop system.

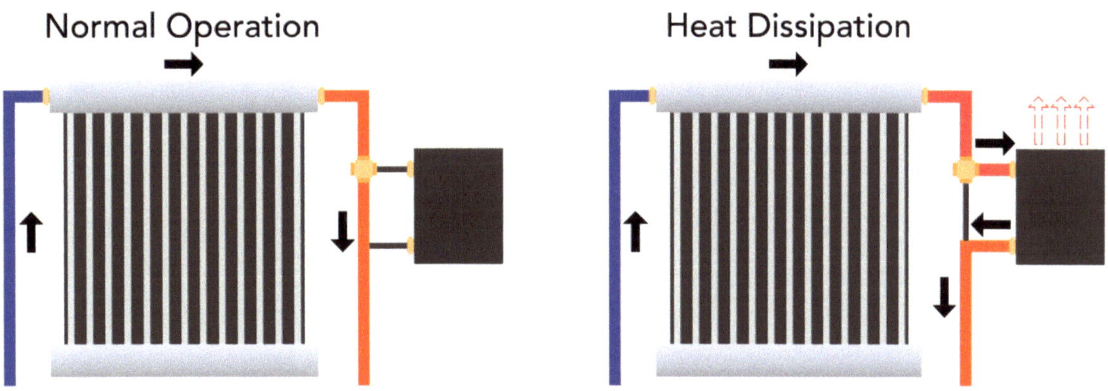


**Figure 12.20** Water to air heat dump for dissipating unwanted heat.



and mount either separately from the collector or on the manifold or frame.

Heat dumps can also be made from loops, or harps, of uninsulated copper piping. All heat dumps should be located in a place where they will not add undesirable heat gain on the interior of the building. Another option is to dump the heat into water, such as a pool or spa.

An appropriate controller should be selected that is able to monitor the system's heat and automatically switch open the heat dump valve when needed.

## Drain-back

Drain-back systems have a circulation loop which pumps water through the collectors. The controller is wired to temperature sensors, one at the collectors and one at the storage tank, and it turns on the pump when the collectors are hot and the storage tank needs heat. In figure 12.21, note that there is a smaller tank, called a drain-back tank, that has a sight glass on the side of it. When the pump is running, as the hot water flows back toward the storage tank, it first passes through the drain-back tank, which is typically mounted just above the storage tank. The water passes through the heat exchanger loop and then is recirculated until either the collectors are no longer hot or the tank doesn't need any more heat. When the pump shuts off, the water drains back through the cold water side and the extra volume of water fills into the drain-back tank. Note that the system has to be mounted or racked at an angle so that all of the water drains out of the collectors and piping that is outside. The standard tilt is ¼" per linear foot (2% slope). The static water line is inside the building.

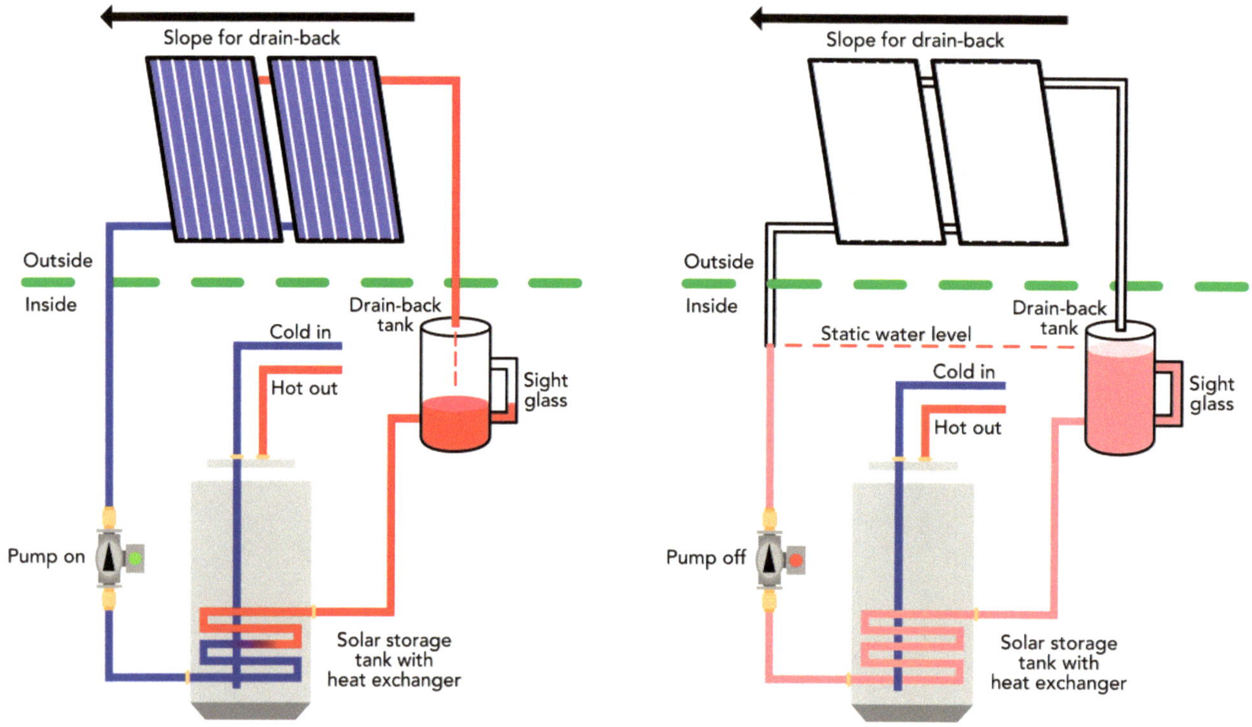


**Figure 12.21** Drain-back system operation.

This means that the water will not freeze or get too hot because it is never just sitting in the collector or exterior piping.

Water has the best heat transfer potential and more water can be added if needed. This figure shows some of the plumbing elements that would be part of the system. The pump's symbol is the circle with the triangle, the triangle point indicates the direction of the water flow. On either side of the pump are shut-off valves. These look like bow ties. There would be a shut-off above the drain-back tank, a drain at the bottom of the system piping near the storage tank, as well as shutoffs and valves for the piping for the potable water into and out of the storage tank but these have not been shown in order to keep the diagram a bit simpler. Notice that drain-back systems do not need an expansion tank or an air vent at the top of the system.

## Parts and Pieces

In this section we'll take a look at the individual pieces of equipment that are part of solar thermal systems. I'll describe what it does and show you what it looks like.

**Controller** (figure 12.22): The controller is an electronic piece of equipment that controls the pump, turning it on and off. It works by comparing the temperature at the collector with the tank temperature. This image shows a controller with the panel cover off on the left, so you see the interior with the ports and labeling of relay attachments for wiring the controller to the temperature sensors and pump. Once everything is hooked up, the cover snaps on, as shown on the right.

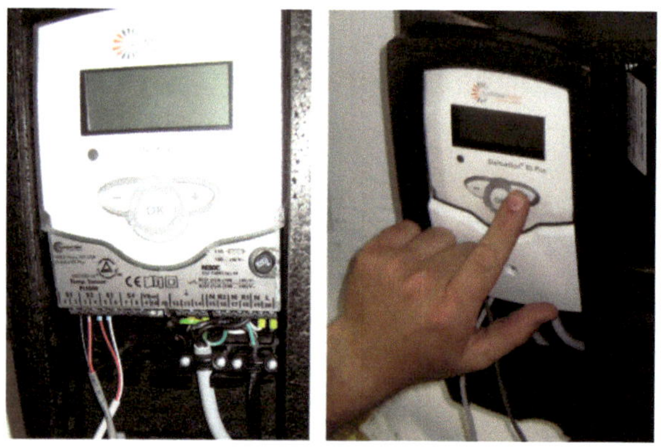

**Figure 12.22** Controller with cover off (left) and cover on (right).

**Pump** (figure 12.23): Pumps are very efficient and appearances vary with manufacturer. Pumps are placed in line on the cold side of the piping loop. If the collectors are vertically located much higher than the pump, two pumps may be placed in line. Sometimes drain-back systems will have two in-line pumps because there is a lot of stress on start-up to get the water moving around the loop. Closed loop systems usually have a check valve above the pump to reduce stress on start-up due to the weight of the water standing in the pipe. To size your pump, you need to know the total length and diameter of the piping loop and the type of heat transfer fluid.

**Pump station** (figure 12.24): Pump stations are often available from manufacturers. They include an insulated box, the pump, temperature and pressure gauges, the controller, and the basic piping fittings that are in the

**Figure 12.23** Circulating pumps: Wilco (left), Taco (right).

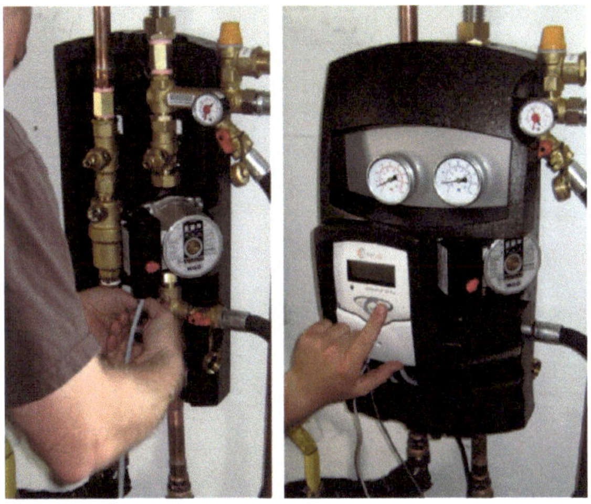

**Figure 12.24** Pump station without cover (left) and with cover (right).

box. Pump stations are nice because they put all the equipment together in one spot and look very tidy.

**Temperature sensor**: The temperature sensor is a long cylinder that inserts into a built-in receiver in the tank (figure 12.25) and the collector manifold (figure 12.26) shows installation in an evacuated tube manifold). The sensor just slides into a sealed gasketed opening in the equipment.

**Figure 12.25** Temperature sensor ready to slide into the storage tank.



**Insulated piping** (figure 12.27): It is very important to insulate piping. Piping can be purchased in rolls with two pre-insulated flexible stainless steel pipes that have sticky tape to keep them together. It also has a wire to hook up to the temperature sensor. The exterior jacket protects the insulation. The piping can be installed both above and below grade. You can also purchase insulation to wrap around piping, which is handy for covering the copper piping on the building's interior. Insulation that is on the exterior of the building needs to be able to withstand the climate and UV rays without degradation.

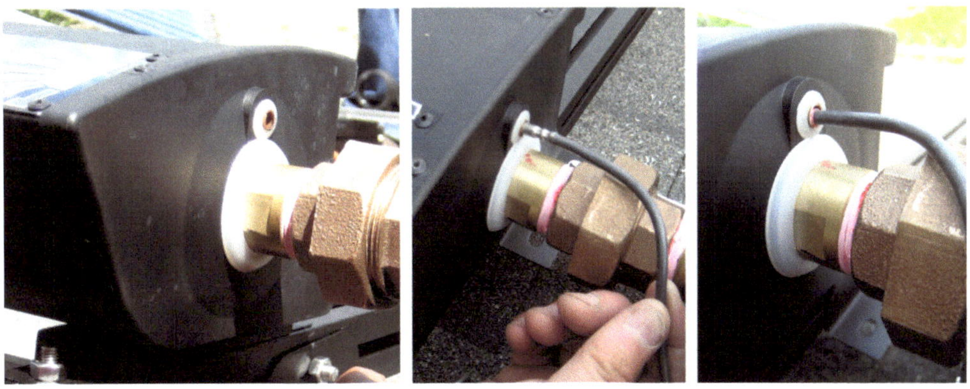

**Figure 12.26** Gasketed sensor slot in manifold (left), installing the temperature sensor (center), sensor in place (right).

**Figure 12.27** Roll of pre-insulated piping.

It is easy to cut the pre-insulated piping to the needed length and attach connectors.

**Expansion tank** (figure 12.28): The expansion tank has a diaphragm on the interior which allows for equalizing pressure in the piping loop. It is installed on the cold side of the loop. On the interior, the tank is divided into two sections by a rubber diaphragm. The side connected to the system piping contains liquid while the other side is dry, filled with pressurized air. As hot water circulates from the collector, the pressure in the piping increases. The diaphragm stretches out, allowing more room for water, which balances the pressure in the system. This prevents damage to pipes or system components due to excessive line pressure. Expansion tanks can be mounted at any angle, facing any direction, and are generally secured with a bracket to a wall or structural element. Expansion tanks on solar thermal collector systems are generally fairly small, not more than 2′ (610mm) in length.

**Figure 12.28** Expansion tank.

**Shut-off valve** (figure 12.29): Shut-off valves are required to stop the flow of liquid through a pipe. They can be quite simple, like the lever in the image below, or have more substantial levers that have handles for easier operation. When the lever is perpendicular to the pipe, the valve is closed and will not let any liquid pass. When the lever is parallel to the pipe, the valve is open and liquid will flow through the pipe. Shut-off valves are shown on a plumbing diagram as a bowtie shape.

**Figure 12.29** Different types of shut off valves in closed (off) position.

**Flow meter** (figure 12.30): The flow meter has a visual readout of either the amount of fluid or how quickly the fluid is flowing in the pipe. The velocity has to be less than 5′/second (1.52m/second), and flow rate below 5gpm (18.9L/min), to prevent erosion in copper piping. Flow meters have different appearances depending on the manufacturer. They are all installed in-line above the pump on the cold side of the piping loop.

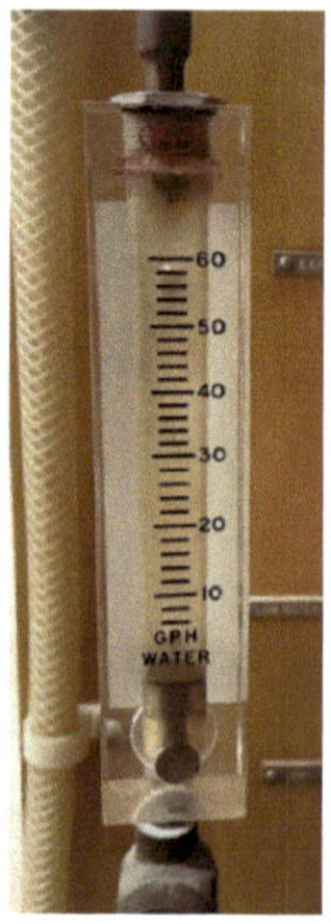


**Figure 12.30** Flow meter.

**Check valve** (figure 12.31): Solar thermal systems typically have very low pressures so the best kind to use are swing check valves. These must be installed in a vertical section of pipe to work properly. The valve swings open when the pump is pushing liquid around the loop. When the pump stops, the liquid tries to fall backwards down the pipe but the valve swings closed and holds the liquid in place. A check valve can be installed above the pump in a closed loop system to reduce start-up drag on the motor. There is usually an arrow on the check valve showing the direction of flow. In a diagram, the check valve is shown as a backwards Z with an arrow next to it pointing in the direction of flow.

**Automatic air vent** (figure 12.32): At the top of a closed loop system, on the hot side of the piping leaving the collector, there is an automatic air vent. This allows any air in the circulating fluid to be released through a valve inside the vent.

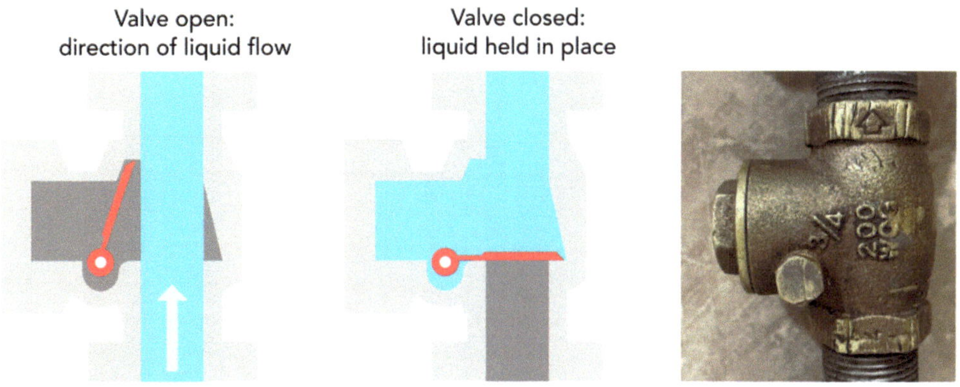


**Figure 12.31** Check valve exterior and cross section of operation.

**Figure 12.32** Automatic air vent.

**Pressurized hot water storage tank** (figure 12.33): These tanks look very much like domestic hot water heaters. They are well insulated and come in various sizes depending on how much water you are storing. They can have as many as three heat exchange coils on the interior. All tanks come with at least one heat exchange coil at the bottom to hook up to the collectors. The plumbing attachments are clearly marked, with cold going out to the collectors at the bottom and hot coming back from the collectors at the top of the coil. Additional coils can be set up for heat exchange to other equipment. Examples would be a fan coil unit that transfers the heat from the storage tank water to air to provide heating in a room, or a loop to a radiant floor heating system. The tank in this image has two heat exchange coils.

**Figure 12.33** Pressurized storage tank.

**Non-pressurized storage tank** (figure 12.34): Larger amounts of storage capacity are often stored in non-pressurized tanks. These tanks can range from 250 gallons (946 liters), or less, to as much as 5000 gallons (18927 liters). The tank's exterior is flexible embossed aluminum which can be crated in a rectangular box that fits easily through standard doors. The tank is insulated and comes with an insulated cover, you can see the cover in back of the tank leaning against the wall. It has a waterproof liner that drapes across the interior of the tank and over the insulation. Copper tubing is used above the water line as tension tie-rods to keep the tank from spreading apart at the top. Heat exchanger coils are placed inside the tank, the number and size of the coils depends on the tank size and the collector configuration. All the hot piping connections are at the top of the tank while sensors and cold piping connections are from the lower part of the tank as seen in figure 12.35.



## Tilt Angle and Sizing Considerations

The best tilt angle for solar thermal hot water is latitude + 15°. This is the best angle because it tilts the collector closer to perpendicular with the winter sun, when there are the fewest hours of sunlight, see figure 12.36. Depending on your latitude and climate, summer insolation may be 2–3× greater than winter, so it is important to maximize winter gain. Because the collector is tilted away from the summer sun, it reduces the amount of heat gain when there are many hours of sunlight. In this way there is a more even amount of solar heat gain all year long and summer overheating is minimized. The collectors should face within 15° of south in the northern hemisphere or 15° of north in the southern hemisphere.

Solar collectors can be partially shaded and still work. The more shaded they are, the less they will be able to heat up since gain is based on heating the entire aperture area. Collectors on large systems may be arranged in rows on a flat roof. Please refer to the chapter 13 (Photovoltaics), in the section on collector spacing, for determining inter-row spacing.

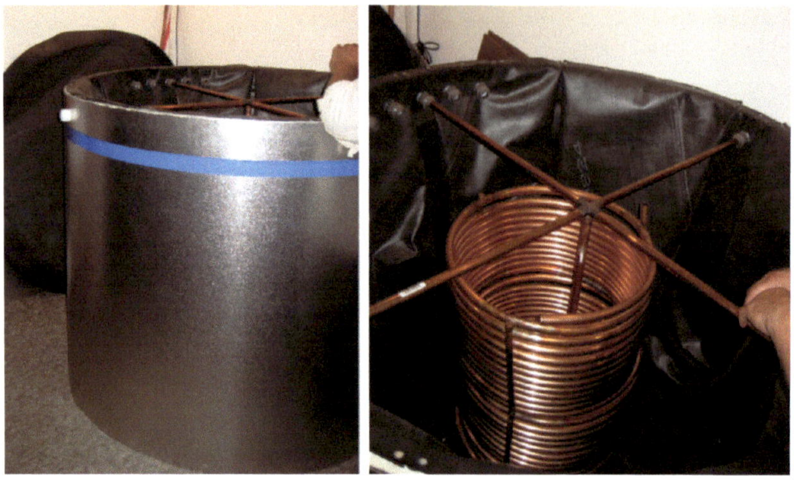

Figure 12.34 Non-pressurized storage tank, exterior on left, with internal heat exchanger coil, interior on right.

## Questions You Need to Answer to Design the Solar Thermal System

**Where will the system be located?** The location or availability of space may be limited and impact the possible system size. Remember that your collectors need to be facing the correct direction and be in the sun for most of the day. You also need to put the collectors within 150′ (46m) of the storage tank. That is the actual length of the piping – not a horizontal distance on a plan. You need to pay attention to where the piping will be located: it needs to be as straightforward as possible and not get in the way of the building's aesthetics or functions.

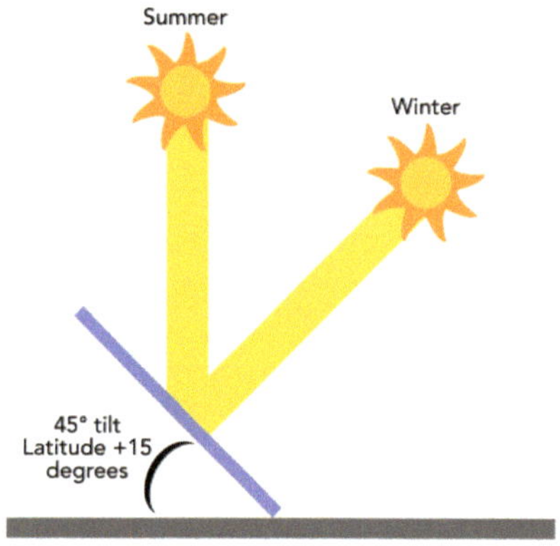


Figure 12.36 The best solar thermal tilt angle is latitude + 15 degrees.

Figure 12.35 Piping and sensors attached to a non-pressurized tank.

**What will the heat be used for?** Domestic hot water or heating or an industrial process – all of these have very different requirements for the type of system, when it will need to make heat, and how hot the water needs to be. If you are using a solar thermal system for winter heating and DHW, it is quite likely that you'll need to cover some of the system's collectors in the summer months to prevent overheating. Special covers are made by manufacturers for collectors that can be securely attached to the collector. Covers prevent damage to collectors when they are not in use. Systems can be designed so that the collectors for heating can be shut off during summer months, while the rest of the system provides the heat needed for DHW. If you are designing something like this, you need to make the collectors easily accessible, both for covering/uncovering and for snow removal in the winter. Evacuated tube collectors do not heat up much on the exterior of the tube, so it can take time for snow to melt off. Snow does let some solar radiation through but not a full amount, so your production will be limited if snow remains on the collectors.

**When is the heat needed and how much demand is there?** This is called a load profile. Load profiles can be tracked daily and seasonally. The better the load profiles match seasonal gain, the easier it is to design an effective system. You may or may not be able to get "real" numbers for exactly how much hot water is going to be needed at any particular time of day or year. That's okay – the idea is to get a sense of what is likely. Some commercial and industrial clients know exactly how much hot water they use. Most clients have no idea. You can't look at the overall water usage of the building as that is for both cold and hot water. Most buildings do not have any metering or measuring system for hot water usage. This means that you'll have to use estimates. Commercial estimates are going to be based on the building's occupancy type (what it is used for) and how many occupants are in the building based on square footage according to code. Some direction for number of handwashes, etc. based on occupancy and use can be found in documentation for third party rating systems as well as manuals used by mechanical engineers. There are computer programs that can estimate hot water usage, such as RETScreen.[1]

There are general assumptions for household hot water usage but care should be taken to question the occupants on how much water they conserve (or don't). In the USA it can be considered to be impolite to ask people about the lengths of their showers and how often they do laundry, or if they leave hot water running while doing dishes by hand. However, these habits have a huge impact on how much hot water will be used every day. In the USA, assuming 20gal (76L) of hot water usage per person per day is reasonable, if people aren't paying attention to how they use hot water and don't really have efficient fixtures. Hopefully you are designing a system for a building that will have water conserving fixtures so you can use a smaller number. If you are using a computer program to model performance, dig around for numbers used for the hot water consumption.

Figure 12.37 shows an imaginary daily load profile for a hotel. This hotel doesn't have a full-service kitchen, so most of the hot water use is for showering and laundry. Across the bottom of the graph is time, morning on the left and evening on the right. The vertical axis is the heat demand load. The curved line centered around the middle of the day is the production curve (note that clock time and local time are often not the same, I've pretended that they are for this diagram). The exact amount of production is going to vary daily, depending on if it is cloudy or clear, as well as having fewer hours of production in the winter compared to summer. However, we can use the annual solar insolation average, based on tilt angle and orientation, to get a daily production amount. The demand is greater in the morning and evening when hotel guests are using more hot water. During the day there is a fairly consistent demand for washing sheets and towels. You can see that the production is happening when the laundry is being done during the day but the guest demand is outside the production curve. That means that the storage tank is going to need to be able to store all the heat (Btu or J) needed for evening and morning use. The system can replenish the heat in the storage tank during the following day's production. Because the majority of the heat demand is outside of the production time, the storage tank is going to have to be larger than if the demand lined up better with production.

The next imaginary daily load profile, shown in figure 12.38, is for a restaurant that serves lunch and dinner. Here you can see that the production profile is again a yearly daily average centered as if clock



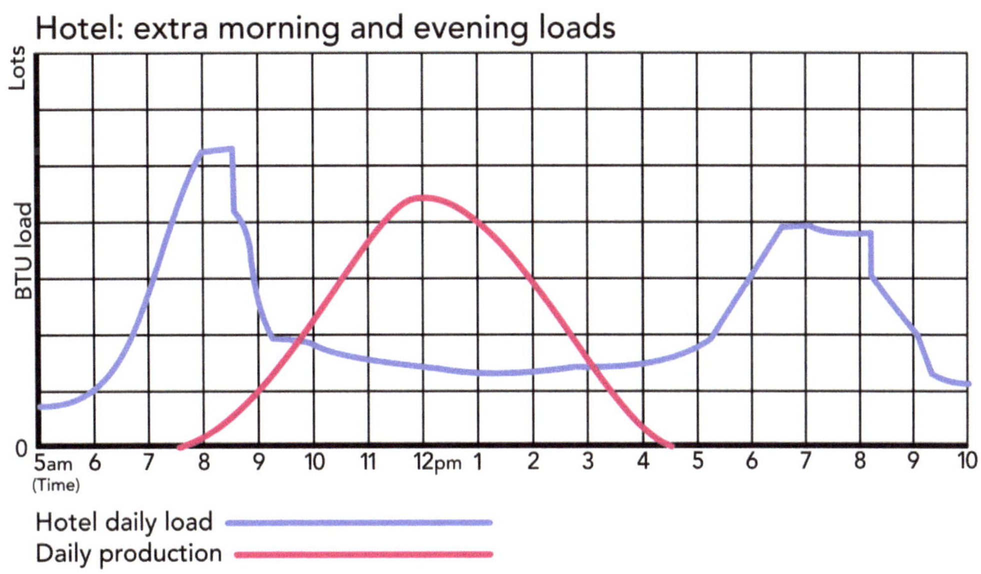


**Figure 12.37** Hotel daily load profile compared to solar production time.

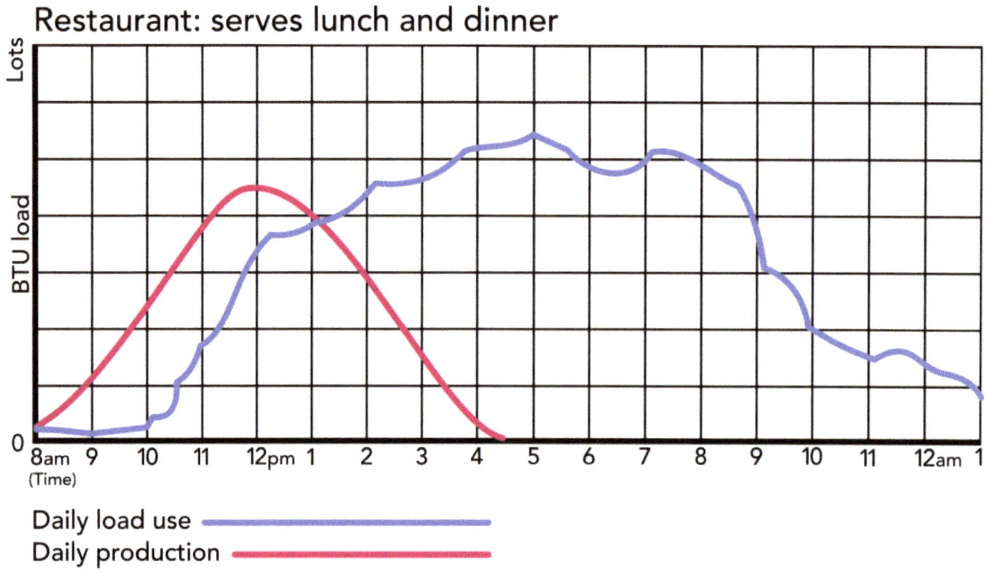


Figure 12.38 Restaurant daily load profile and solar production curve.

time and solar time were both at noon. The demand profile is going to start ramping up as the chefs start prepping food for the day. The profile stays fairly even with demand for washing dishes and linens until the crowd starts dropping off after 9pm. In this scenario you can see that there is a bit of morning production starting before the demand ramps up but about a third of the demand is happening at the same time as the production. There are still two-thirds of the demand outside of production time, so heat would need to be stored to offset need in the later afternoon and evening.

You should also look at the usage over the year. You don't want to have high production when people aren't going to be there to use it. If there isn't a match between usage and production over the year, either part of the system needs to be shut off and covered or the heat needs have a heat dump (like a pool or spa).

Our next imaginary client is a scout camp, located in the northern hemisphere in a higher latitude, that hosts a large number of campers in the summer months but doesn't get much winter use. This is a pretty nice camp that has a camp kitchen and showers. Figure 12.39 has months across the bottom: January on the left, the

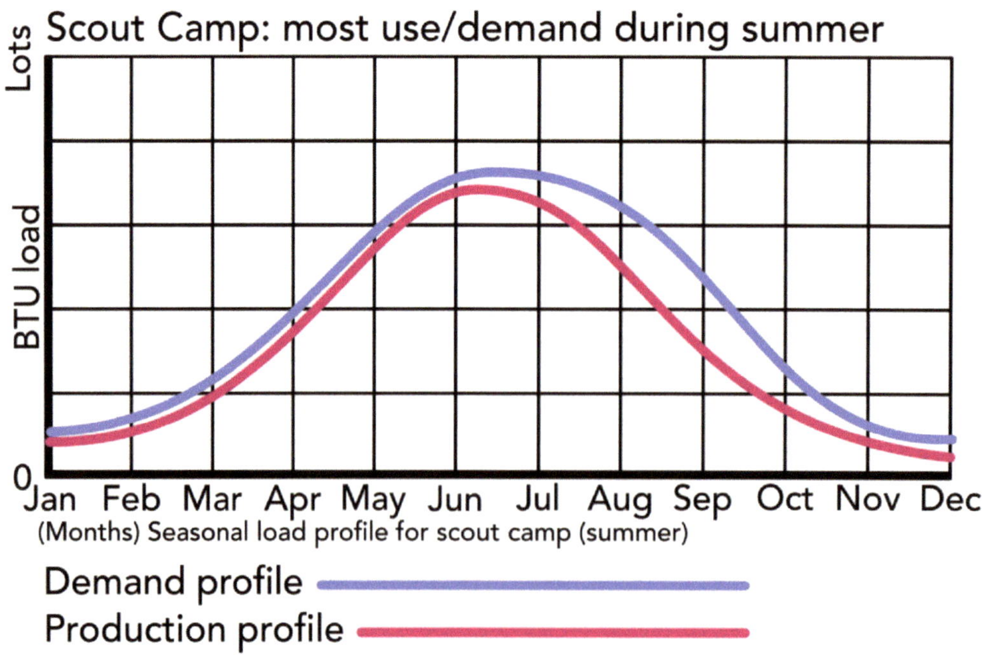


Figure 12.39 Scout camp seasonal load profile compared to production profile.

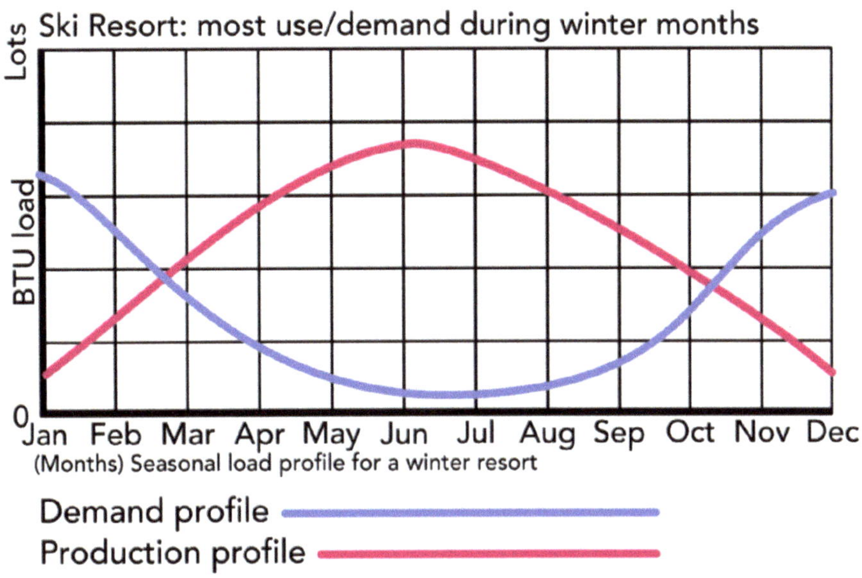


**Figure 12.40** Ski resort seasonal load profile compared to production profile.

summer solstice on June 21, and December on the right. The vertical scale is heat demand with none at the bottom and lots at the top. When I plot the possible solar insolation through the year against the hot water demand of the campers, I can see that there is a fairly good match between the two. It is likely that I could provide a system that is able to offset quite a bit of the heat demand. I can study the tilt angle production effects and perhaps orient the collectors more to the summer sun, maximizing summer production, which will naturally minimize winter production.

For the next seasonal load profile, we are using a ski resort, see figure 12.40. It is also located in a higher latitude of the northern hemisphere. The graph has the same horizontal and vertical scales as the scout camp. This ski resort has not leveraged getting people to their location in the summer. They get "leaf peepers" (people who travel to see the changing colors of the leaves), starting at the end of September. The ski season starts in late October/early November and continues through the middle of February. The plot of the production vs. the heat demand is quite opposite of each other. This shows that it is going to be difficult to design a single system that works well for the whole year. My options include: design a small system that would cover all the summer demand and just a bit of the winter demand; have additional collectors that produce hot water just in the winter (shut off and covered in the summer); possibly wall mounting the collectors to drastically reduce summer production while creating more hot water in the winter – with the added benefit that the collectors will not need snow removal.

**What solar fraction are you aiming for?** Solar fraction is the percentage of the overall heat demand that you intend to offset with collector production. I alluded to this in the ski resort discussion – the solar thermal system would only have a solar fraction of 10% of the total demand – a small amount. Even though that may not seem like much, it is still like getting a 10% discount all year on your hot water heating costs. Remember, these systems are being designed based on averages, and it is important to not exceed 80% solar fraction for DHW demand and 30% for heating demand. There is a diminishing return when using solar radiation to heat up water: the hotter the water gets, the harder it is to transfer the heat into the water. It is easier to transfer heat between two very different temperatures than two similar temperatures.



## Collector and Piping Layouts

Small systems are always laid out in series. There are limits on how many collectors can be connected in series. Look at the manufacturer's datasheet for the numbers, as they change depending on the type and size of collector. The limits relate to maximum flow rates through the collector piping. Flow retarders may be needed in systems with many collectors in series. Collectors connected in series create higher temperatures. Groups of collectors, or large collectors (such as a collector that has 30 evacuated tubes), can be connected in parallel. Parallel collectors provide lower temperatures but more Btu (J) overall.

There is a difference between heat gain in series piping (figure 12.41) and parallel piping (figure 12.42). I haven't provided a set of numbers in these examples because the tank temperature is constantly changing (as hot water is used and cold water is supplied) and the temperature of the collectors is also changing (clouds,

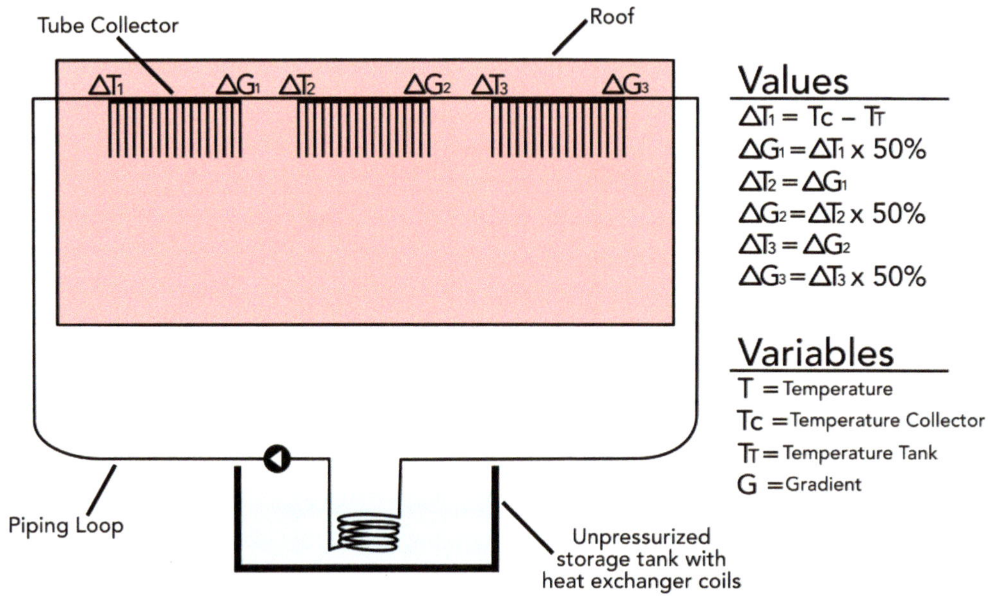


**Figure 12.41** Collectors in series provide hotter temperatures.

time of day). The takeaway is that series achieves higher temperatures and parallel provides more Btu (J) overall. A system's production is going to depend on the available solar insolation and the system size.

The Greek letter delta, Δ, stands for change; in these diagrams there are differences in temperature (T) and gradient (G). The initial temperature difference is going to be between the tank temperature and the collector temperature. Remember that the controller isn't going to turn on the pump unless there is a need for hot water and the collectors are hot. A gradient in this application is the increase of the circulating fluid's temperature from one collector to the next (measured as output). The gradient is always 50% of the ΔT.

For these diagrams, we are saying that the storage tank temperature is denoted by the variable $T_T$ and the collector temperature is $T_c$. Note that the same number of collectors (collector area) is being used for both

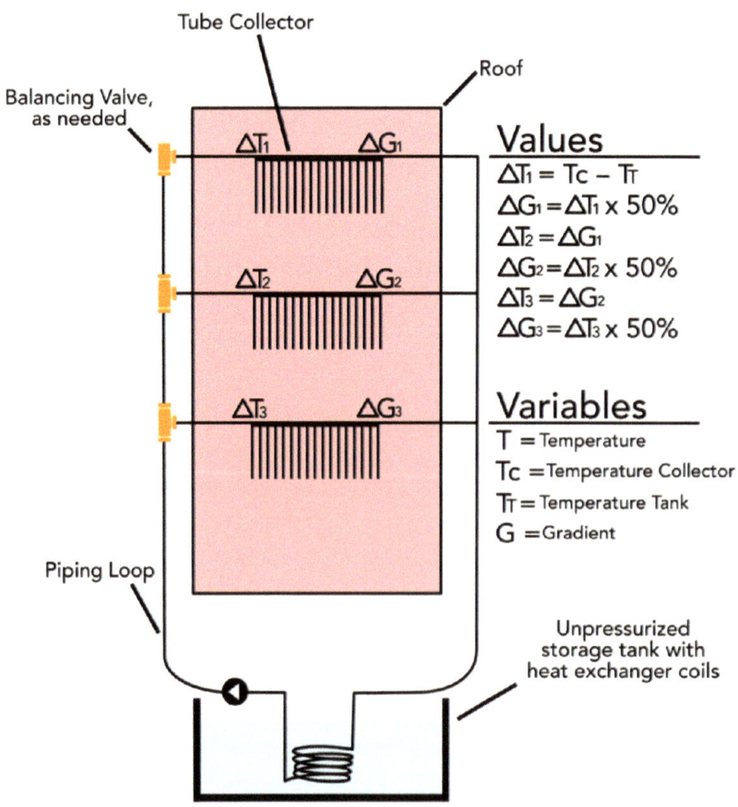


**Figure 12.42** Collectors in parallel provide more overall Btu.

the series and parallel examples. The larger system is likely to have an non-pressurized storage tank, so that is shown here. Also, this diagram shows evacuated tube collectors. Note that all the loop is through the manifold. Remember that the evacuated tubes are their own sealed environment and transfer all the heat to the manifold. What's happening is that the larger the initial temperature difference between the storage tank and the collectors, the larger the initial gradient increase will be. All subsequent heat gains in series are half of the last heat gain because there is a smaller temperature difference and it is harder to transfer the heat.

What's happening in series is that the temperature before the collector is the summation of the tank temperature and all the ΔG up to that point. With more collectors in series, you can heat up the water to higher temperatures. With collectors in parallel, you only get one gradient increase but you get it across each collector, so you get more Btu (J).

If you put collectors in parallel piping, then it is important to balance the system so there is equal pressure through all collectors, see figure 12.43. If the pressure were not equal, the liquid in the loop would only run through the shortest loops and never flow through the longest loop. Pressure will be equal if the piping run difference between the longest and shortest loops is 12% or less. In the diagram below, this is represented by the two sets of arrows. For example, if the short run is 60′ (18.2m) then the long run would have to be within 12% of that distance, 60′ × 0.12 = 7.2′ (18.2m × 0.12 = 2.19m). It is unlikely for our example that the long run could be only 7.2′ (2.19m) longer because that would mean the piping to the longest run collector would be 3.6′ (1.1m) away from the shortest run collector. Collectors are bigger than that and we'd have to fit two in that distance, which isn't going to happen. If the distances aren't within 12% then balancing valves must be located at the cold side branch to each set of collectors in parallel. These are shown as dots on the diagram.

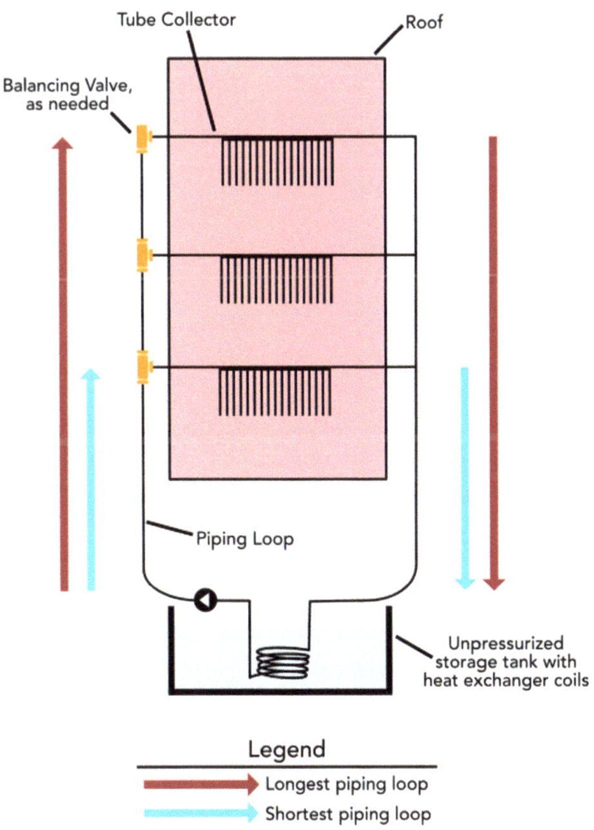


**Figure 12.43** Balancing parallel collector pressure by comparing parallel piping length.

## Calculations: Sizing Components

**Collector sizing**: To size a system you have to estimate the demand, know what temperature the water needs to be, look up deep ground temperature at your project's location, and determine the amount of solar radiation available. Your client may know their hot water usage or have no clue. See chapter 5 on solar insolation to understand how to determine what you have available on your site. Remember, for the most consistent production year-round, the best tilt angle for solar thermal collectors is latitude + 15°.

If the collectors are producing heat that is being used for an industrial process or for heating the building, you can look at the annual or seasonal consumption of the fuel type and convert that demand into Btu (MJ). You can then convert that annual or seasonal demand into a daily average.

If you are working from a quantity of water that is being used, you need it in units of a daily consumption average.

Once you know your daily load, you can multiply it by the solar fraction to determine what percentage of the load will be provided by the solar thermal system. Remember the maximum numbers are 80% solar fraction for DHW demand (or processes) and 30% for heating demand. It can be helpful to start with the maximum and see what the system size would have to be, then compare that to the roof area and/or budget constraints to see what is feasible.

Many manufacturers have DHW kits at a single price that include all the items you need for a functional system. The kits will say that they are suitable for a particular number of people and typically have choices for plates or tubes and closed loop or drain-back. The number of collectors and size of tank increase as the number of people increases.

Here is a general outline of collector sizing process, see Appendix E for specific steps, formulas, and an example.

- Find the gradient: how much are you heating the water? ΔG = design temperature – groundwater temperature

- Determine how much energy it takes to heat the amount of water used every day.
  - Daily heating load = 8.3 × ΔG × total gal/day = Btu/day
  - Daily heating load = 4180J × ΔG × total L/day = J/day (note that you'll need to convert this to MJ/day so it matches units with the solar insolation values)
- Multiply the daily heating load by the solar fraction to determine how much heat energy you need your collectors to produce each day.
- Look up the solar insolation values for annual average and peak (month when there is the highest amount of solar isolation). Convert these values to either Btu/ft$^2$/day or MJ/m$^2$/day. This is how much energy is falling on your site every day in that area. Unfortunately, collectors are only about 73% efficient. So, you'll need to multiply these numbers by 73% to get the solar collector production value.
- Divide the solar fraction load by the solar collector production value of the peak month to get the required area of collectors.
- Compare the area you want with what the manufacturer has available. If you size your solar fraction for the highest amount, round down. Remember that collectors are whatever size they are – you can't change the physical size or have a different tube configuration. You have to use a whole sized collector.
- Compare the area needed for the collectors to the roof size. Remember that you have to keep the collectors vertical and the width and height are fixed. You can't fit part of a collector on a roof – it all has to fit. Most solar thermal collectors are fixed tilt. If you have collectors that are in rows, remember to space them apart so they are not shading each other. Remember that collectors cannot block roof access, see the discussion in chapter 13.
- Once you know what the area of collectors is, then you can determine what the actual output will be. The output will be the area of the collectors multiplied by the annual solar collector production value multiplied by the number of days in a year (365).
- The last steps are to size your tank and piping, figure out where all the system fits, and decide on your racking system.

**Tank sizing**: The size of the storage tank is influenced by the latitude and the temperature desired for the stored hot water. The ratios shown in Table 12.1 are for latitudes 40 and above and need to be increased as you move toward the equator. Equatorial ratio for DHW (low-temperature) should be 5gal/ft$^2$ (203L/m$^2$). Ratios are shown in a range; if the daily load profile is outside of production hours, use the larger number.

How you use the ratio is to take the area of collectors (or number of evacuated tubes) and multiply it by the ratio. This will give you a general size: you'll need to compare the number to what size tanks are available from a manufacturer. Let's say that I have a system with 3 flat plate collectors that each have a net aperture area of 27.7ft$^2$ (2.57m$^2$). The total collector area would be 3 × 27.7ft$^2$ = 83.1 ft$^2$ (3 × 2.57m$^2$ = 7.71m$^2$). I'm going to say this system will work at the low temperature. The 2.5 (101) ratio results in a tank requirement of 83.1ft$^2$ × 2.5gal/ft$^2$ = 208gal (2.57m$^2$ × 101L/m$^2$ = 260L). The 3 (122) ratio gives 83.1ft$^2$ × 3gal/ft$^2$ = 249gal (2.57m$^2$ × 122L/m$^2$ = 314L). I'll have to compare these to what's available depending on if I want a pressurized or non-pressurized tank.

**Piping diameter**: Piping needs to be appropriately sized to maintain a velocity which must be less than 5'/second (1.52m/second), and flow rate below 5gpm (18.9L/min). The flow rate for 1ft$^2$ of flat plate collector or one evacuated tube is 0.028gpm (0.106L/min). Manufacturers will list the flow rate for a particular product on the data sheet. Flow rate of the system = area of collector (or the number of tubes) × 0.028gpm (0.106L/min). Look at a chart for the size for "L" type copper piping needed for that flow rate. (An internet search will provide you with charts using your units and piping sizes.)

## Certification

Certification for both flat plate and evacuated tube collectors is available through the Solar Rating and Certification Corporation (ICC-SRCC™).[2] They are the leading certifier of solar heating and cooling products in North America, performing certification testing for products being used in markets in the USA, Canada,

**Table 12.1** Tank sizing requirements based on storage tank temperature.

| Tank sizing requirements based on storage tank temperature | |
|---|---|
| Required storage tank temperature | Ratio for storage size based on area of collectors, gal/ft$^2$ (L/m$^2$) |
| Low: 95–130°F (35–54.4°C)<br>DHW 120–140°F (49–60°C)<br>Radiant floor heating 120–135°F (49–57°C) | 2.5–3 (101–122) |
| Medium: 130–150°F (54.4–65.5°C) | 2–2.5 (81–101) |
| High: 150–180°F (65.5–82.2°C)<br>Hot water baseboards and radiators 160–180°F (71–82°C) | 1.5–2 (61–81) |

Mexico, and the Caribbean region. They are also accredited under ISO/IEC 17065 to certify renewable energy products. ICC-SRCC™ maintains a list on their website of products and performance certification. The products they certify include solar thermal collectors and system components, solar thermal systems (DHW, pool, and space heating), and PV water heaters. You can search the database by company name or type of product certification.

## Resources


1 RETScreen Innovation Lab. (2021, September 21). *RETScreen* (Version 8.1) [Computer software]. Government of Canada. www.nrcan.gc.ca/maps-tools-and-publications/tools/modelling-tools/retscreen/7465

2 Solar Rating & Certification Corporation (https://solar-rating.org/).


## Study Questions

1. What are the three types of solar thermal collectors and which of the two can heat water to higher temperatures?
2. Draw a diagram of a closed loop solar water heating system that includes the water storage tank with a heat exchanger coil, pump, expansion tank, controller, 2 harp-type collectors, temperature sensors, plumbing loop, and electrical controller connections.
3. Draw a diagram of a drain-back solar water heating system that includes the water storage tank with a heat exchanger coil, pump, controller, 2 tube type collectors, temperature sensors, drain-back tank, plumbing loop, and electrical controller connections.
4. What are the differences between pressurized and non-pressurized water storage tanks? Why would you select one over the other?

CHAPTER 13

# Photovoltaics

**Key Terms**

Amorphous (thin film) solar cell: film of silicon deposited on a flexible substrate.

Commissioning: project systems that are checked after installation for proper operation; rating systems require commissioning to be done by a third party.

Grounding: the electrical system is connected to the earth so that sudden increases in voltage flow do not harm people or equipment.

Monocrystalline solar cell: PV cell cut from a grown crystal ingot, has a solid dark blue, black or gray appearance.

Nominal operating cell temperature (NOCT): PV module electrical production tested at 800W/m$^2$, 20°C (68°F), wind speed of 1m/s (2.2mph), and mounted with an open backside.

Polycrystalline solar cell (also called multicrystalline): PV cell cut from chunks of polycrystalline silicon cast into rectangular ingot; has a mottled appearance varying from medium to darker blues and grays, with an overall blue appearance because of the different sizes colors and angles of the crystals.

Photovoltaic (PV) module: a fairly lightweight rectangular substrate with interconnected PV cells mounted on it that creates electricity from sunlight.

Standard test conditions (STC): PV module electrical production tested at 1,000W/m$^2$ at 25°C (77°F) with no wind; in USA, STC results are required to be printed on a permanent label affixed to the back of the module.

## Concepts for Success

Photovoltaic (PV) cells use solar energy from the sun to produce direct current (DC) electrical energy. The three types of PV cells are **monocrystalline**, **polycrystalline**, and **amorphous (thin film)**. Cells mounted and wired together on a surface are called a module. Modules can be mounted together to create a panel. All the panels in the system are called an array. Arrays are rated according to their DC kW (kilowatt) production under **standard test conditions (STC)**. The actual amount of power that is produced depends on the intensity and duration of sunlight, called solar insolation, and what temperature the module is operating in.

Solar insolation is impacted not only by weather patterns, but also by the orientation and tilt angle of the array. Modules can be mounted so they are installed at a fixed tilt or on tracking systems that follow the sun in one or two axes. Tracking systems are more expensive than fixed tilt racking.

Cells are wired in series in a module so it is important that modules are not shaded. Shading even a small portion of a module can cause the module to drop most of its production. Modules are often wired in series, so if one module is not producing at its peak, the rest of the modules in the series will drop to match the lower production.

To size a PV system, you should first optimize the power system demands for the building. Make the building as energy efficient as possible so the power system doesn't have to be large. Second, you need to know what the actual power demand of the building is, both AC (alternating current) and DC loads, if you are going to have a DC load panel. Third, locate the PV array so you know what the optimal production will be and how much area you have for your array. Fourth, determine the solar insolation. Fifth, size the array. Sixth, if you are including a battery system, size the battery bank. The last steps are to coordinate the system wiring, inverter(s), controller(s) for batteries, and monitoring/metering equipment.

There are fundamentally two different types of PV system inverters. The inverter changes the DC energy produced by the array into AC energy for use in the

DOI: 10.1201/9781003297819-15

building. Microinverters are wired to or integrated with each PV module. The microinverters (AC) are wired together and then to the main service panel. String inverters are used for modules that are wired together in series (DC) and are located near the array, typically at ground level. The string inverter then provides AC to the main service panel. Each module in a string can have a power optimizer (DC–DC) to provide better production and allow the owner to individually track module performance.

All modules are wired together with PV wire which is suited for wet and hot conditions. PV wire has thick insulation. All other wiring needs to be appropriately sized, using correct wiring color types, based on its location and the energy being transported in the wire.

Everything in a system has to be grounded. That means that it is connected to the earth. Grounding protects people from shocks and keeps expensive equipment from being destroyed.

Arrays must be planned so they fit appropriately in the space. Roofs need access to other equipment and must be safe for firefighters on the roof. Escape routes and access points need to be clear and open as well as have rapid shutdown around array equipment so it is safe to walk past.

Certification in various aspects of PV, such as design and installation, are available through the **North American Board of Certified Energy Practitioners (NABCEP)**. To earn certification, you have to qualify for and pass a rigorous exam.

## How PV Works

Solar photovoltaic (PV) cells generate electricity from the sun. Figure 13.1 shows in detail how this process works. Cells are very thin squares, or sometimes disks or films, that are made up of two layers of semiconductor material (n- and p-type) that have been treated differently so that they generate voltage and current in sunlight. When the energy from sunlight (a photon) hits the cell, it activates the electrons on one layer so they break loose and move to the other layer, creating a current as they move (flow). There are wires

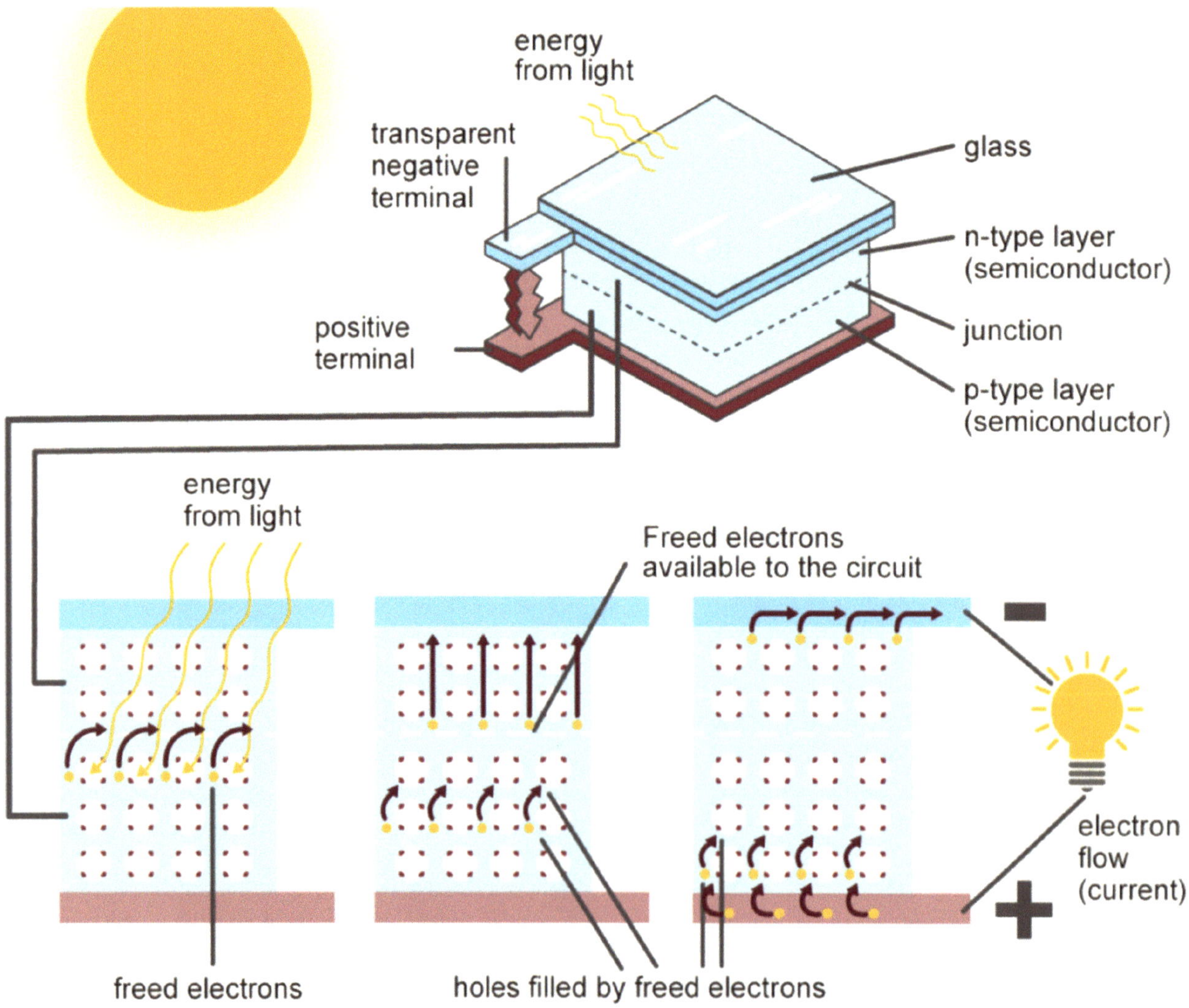


**Figure 13.1** How sunlight creates electricity in a PV cell.[1]

in the cell to capture and move the current around a circuit.

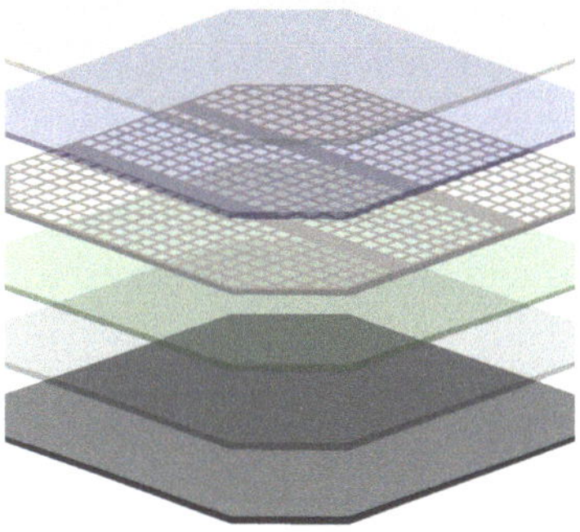


**Figure 13.2** The layers in PV cells are quite thin and are mounted on a backing plate and covered in glass to make them more durable outdoors.

Cells are made up of multiple layers of materials as shown in figure 13.2. They have two layers of semiconducting silicon that have been treated differently so electrons flow easily between them. The silicon is purified and refined to be almost 100% pure. To make it more electrically conductive, it is doped (impurities added) with either boron (p-type) or phosphorous (n-type). This basically creates a hole of one missing electron in the resulting silicon crystal, making it easier for electrons to flow from one silicon atom to another. In figure 13.1 the atoms are shown as white circles with four red dots, while the freed electrons are shown as yellow dots. The arrows show the transfer of electrons across the boundary of the p- and n-type silicon and then around the circuit.

Electrical contacts and wiring are added to each cell in order to collect and conduct the electrons around the circuit. This metal grid is added to the surface of the cell. You can easily see the silver colored wiring that is located on the upper surface of the cell on most modules. The smaller wires that are spaced closely together are called fingers. The thicker busbars, which are spaced farther apart and at 90° to the fingers, can be flat or thin wire. The busbars are acting as positive and negative circuits and are connected across the cells in the module in combinations of series and parallel wiring.

A module is the individual unit that the cells are mounted on and they come in various sizes depending on the number and type of cells. One individual cell can generate 0.5–0.6V. Cells can be used as a unit or cut down. They can be quite small, such as for a calculator, or mounted in groups of 60 on modules as large as 78″ (1.98m) × 39″ (0.99m). Modules are rectangular. The overall appearance of a module depends on their crystalline cell type, how the cells are arranged, and the color of the substrate.

Individual solar cells are arranged on a backing material to create a PV module as shown in figure 13.3. Cells in the module are wired in series to get to higher voltages. Larger modules have both series and parallel wiring. There is typically one weatherproof junction box on the back of the module that connects to the cell wiring. The junction box contains bypass diodes and the positive and negative cables. The cables are about 40″ (1000mm) long and have a standardized connector called MC4. These connectors snap together very securely and require a special tool to take them apart.

Figure 13.4 shows the relationship of the solar cell to a module, panel, and array. Modules can be fastened together (typically on racks) to create a solar panel. The terms module and panel are often interchanged, although they are not the same. A grouping of panels makes up the array. The whole PV system is everything it takes to have the array supply energy to your building.



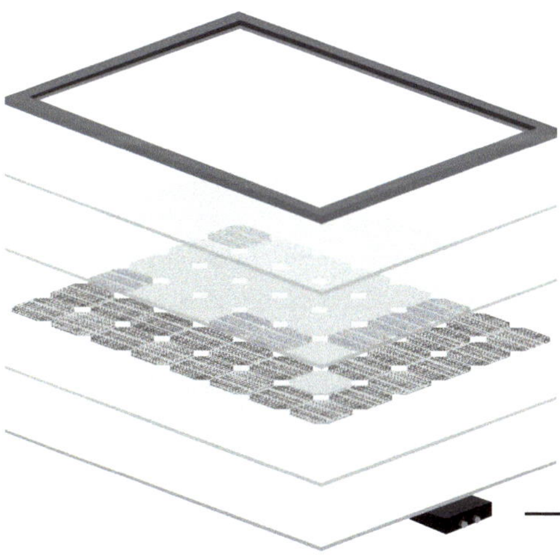


**Figure 13.3** Modules are groups of solar cells wired together, encapsulated and then protected by glass and a metal backing sheet. The aluminum frame protects the edges of the module. The cell circuits are wired to the junction box.

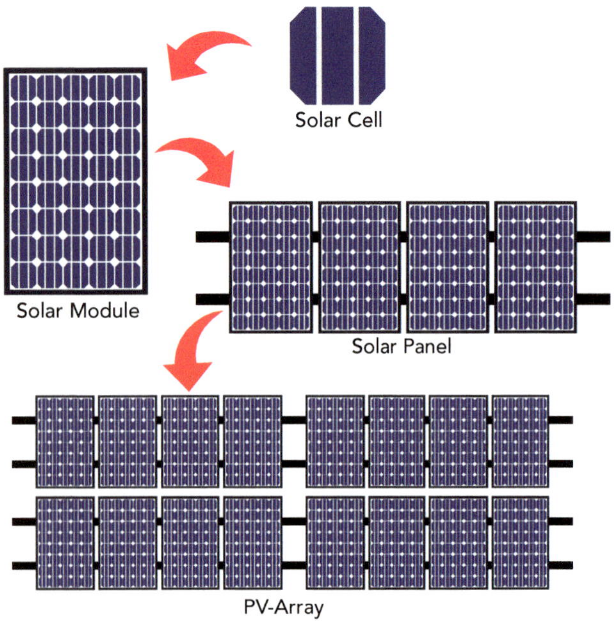


**Figure 13.4** PV terminology from solar cell to array.

## Efficiency of PV

PV are tested for efficiency under standard test conditions. STC are used to evaluate PV module efficiency under lights that provide radiant energy of 1,000W/m$^2$ in a space kept at 25°C (77°F) with no wind. Efficiency is the power rating of the panel divided by the area of the panel multiplied by 1,000W/m$^2$ and then converted to a percentage. Manufacturers provide STC information on all PV module data sheets. This allows for the direct comparison of modules from different manufacturers.

STC is not the same as the actual conditions a PV module will be operating in outside, especially on roofs where operating temperatures are often quite a bit higher than ambient air. Higher temperatures cause resistance in electrical circuits, which means that production is reduced. The sun's radiant energy is variable throughout the day because of weather conditions, air pollution, and location and angle of the sun compared to the collector.

**Nominal operating cell temperature (NOCT)** is a set of testing conditions that are closer to actual conditions. Modules are tested at 800W/m$^2$, 20°C (68°F), wind speed of 1 m/s (2.2mph), and mounted with open backside (air is able to flow around the module and cool it off). Some manufacturers provide both STC and NOCT values on the PV data sheet.

Because PV are located outside, that means that they get dirty. Pollen, dust, snow, or other debris will block sunlight. Locating PV where they can be regularly cleaned or cleared off can keep them operating at their peak.

Premium modules (also called Tier 1) are those that have an efficiency of 19% or more and they generally cost more than standard panels (also called Tier 2) which average about 15% or more.

## Module Performance

The I–V curve shows the output characteristics of a module. Remember that watts = volts × amps. The I–V curve shows the influence of insolation and temperature on module performance. The vertical axis is current (I, in amps) and the horizontal axis is volts (V, in volts). This curve is showing the rated output of the module based on STC conditions. There are two points on both axes and one point along the curved knee of the diagram.

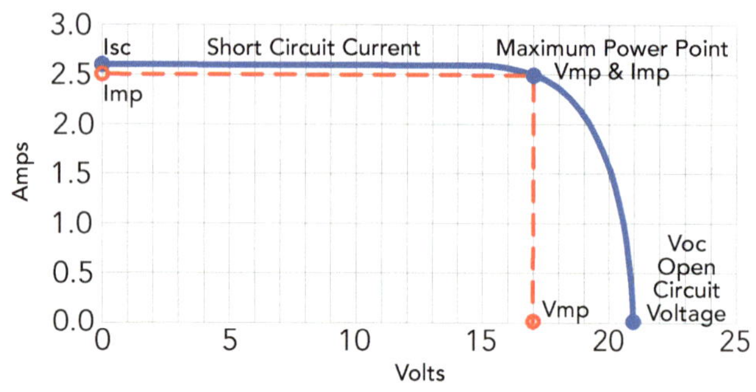


**Figure 13.5** I–V curve showing the terminology.

$\mathbf{I_{sc}}$ is short circuit current. Note that it is located at 0 volts. $I_{sc}$ is how many amps are produced when modules connected in series (positive to negative) are producing but not connected to a load. A safety factor of 1.25 is typically added to array amps in order to size controllers and inverters. If measuring $I_{sc}$, use a multimeter and measure the current between the positive and negative connection between two modules.

$\mathbf{I_{mp}}$ is the maximum power point current and is the best current production for the module when it is producing power.

$\mathbf{V_{mp}}$ is the maximum power point voltage and is the best voltage production for the module when it is producing power.

$\mathbf{V_{oc}}$ is open circuit voltage, which means that this is the voltage that will build up in the module when it is in sunlight but has no load on it. Since there isn't any flow of electrons, there is no current (note it is at 0.0 amps on the vertical scale). This happens in the morning as the sun begins to shine on the modules but the inverter hasn't woken up out of overnight sleep mode. It can also happen if the grid is down and the inverter isn't passing electricity. $V_{oc}$ is used for adding series voltages and the sizing of inverters and controllers. If measuring $V_{oc}$, use a multimeter and measure voltage between the positive and negative leads on the module.

Note that there are dashed lines showing that the $\mathbf{M_{pp}}$ (also seen on data sheets labeled as $\mathbf{W_{pp}}$ and $\mathbf{P_{max}}$) is the maximum power point. This is the best production the module will make under STC. Since this is watts, remember watts = volts × amps, it is the result of the best voltage and current production and shows up on the bending part of the curve. This is where you get the best optimization of power.

Note that the $M_{pp}$ changes value depending on actual insolation and temperature.

**Maximum power point tracking (MPPT)** is programming in either the controller or inverter that works to always provide the best power output no matter what the actual volts and current are. MPPT can increase overall production by 10–40% depending on the season and other factors.

Table 13.1 is the Solaria PowerXT® module data sheet[2] information for the 365R-PD and 370R-P modules. Note that it clearly states the testing conditions and the results. Since there are two modules in this model type, there is a column for each model. Note that modules from any company are typically named with the wattage number for Max Power ($P_{max}$) based on STC test values at the beginning, followed by manufacturer model numbers.

STC uses 1000 W/m², so that will always be the maximum current. However, the intensity of sunlight can be reduced because of clouds and other environmental factors. Having less insolation means that there will be less current, which can lower the maximum power point (still located at the bend of the curve) pretty dramatically. Manufacturers may include a diagram showing the performance of their module under lower insolation. Figure 13.6 is the I–V curve for reduced current due to less intense sunlight for the Solaria PowerXT® 370W module. Note that this diagram assumes the same STC operating temperature, so the voltage isn't changing much. The top line is the STC test number of 1000W/m². Subsequent lines drop by values of 200W/m², down to the lowest current at 200W/m². Note that, as the current decreases, the maximum power point value also decreases.

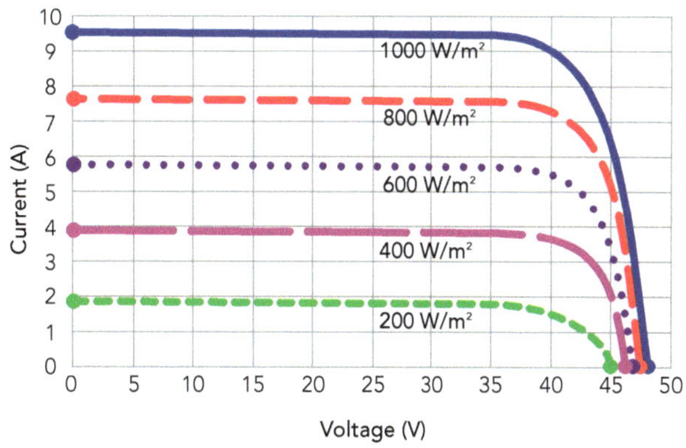


**Figure 13.6** I–V Curve of the Solaria PowerXT® 370W module showing the reduction of current due to lower amounts of insolation. Note that the top line is STC 1000W/m² and the temperature remains the same.[3]

Higher temperatures reduce the voltage because of increased resistance. Note that this diagram assumes the STC insolation for the current is staying the same regardless of the temperature. Days that are, in the summer heat, sunny with little air movement on a darker colored roof can create high temperatures. Most PV systems that are roof mounted are a bit above the roof deck so that air can flow around them and help to cool down the modules. Not all the cell temperature is based on the weather. Cells get hot from operating and from absorbing infrared radiation. Figure 13.7 shows a typical pattern of voltage drop due to operating temperature increases, assuming STC insolation. The outermost curve is STC temperature (77°F, 25°C). Note that, as the temperature increases and the voltage decreases, the overall output is less.

The I–V curves shown in both figures 13.6 and 13.7 are hypothetical because usually both variables of available sunlight and temperature are changing at the same time

13

**Table 13.1** Solaria PowerXT® module manufacturer's data sheet for STC and NOCT performance values for 365R-PD and 370R-PD.[2]

| **Performance at STC (1000W/m², 25°C, AM 1.5)** | | | |
|---|---|---|---|
| **Solaria PowerXT-** | | **365R-PD** | **370R-PD** |
| Max Power ($P_{max}$) | [W] | 365 | 370 |
| Efficiency | [%] | 20.2 | 20.5 |
| Open Circuit Voltage ($V_{oc}$) | [V] | 48.0 | 48.3 |
| Short Circuit Current ($I_{sc}$) | [A] | 9.58 | 9.60 |
| Max Power Voltage ($V_{mp}$) | [V] | 39.9 | 40.2 |
| Max Power Current ($I_{mp}$) | [A] | 9.16 | 9.20 |
| Power Tolerance | [%] | –0/+3 | –0/+3 |
| **Performance at NOCT (800W/m², 20°C Amb, Wind 1 m/s, AM 1.5)** | | | |
| Max Power ($P_{max}$) | [W] | 269 | 272 |
| Open Circuit Voltage ($V_{oc}$) | [V] | 45.1 | 45.4 |
| Short Circuit Current ($I_{sc}$) | [A] | 7.73 | 7.74 |
| Max Power Voltage ($V_{mp}$) | [V] | 36.7 | 37.0 |
| Max Power Current ($I_{mp}$) | [A] | 7.32 | 7.35 |

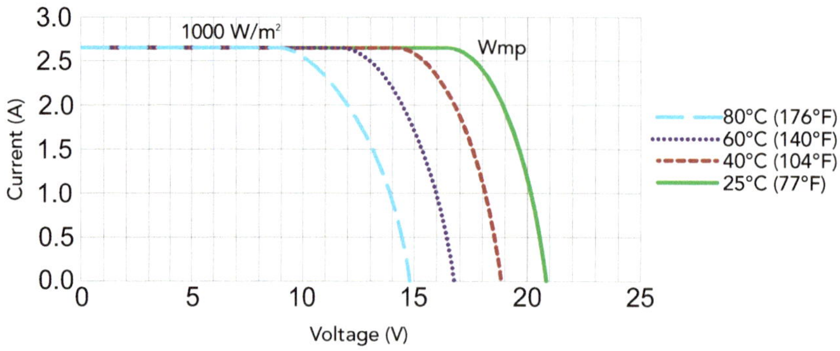


**Figure 13.7** The effect of increased cell temperature reduces the voltage and power output on the I–V curve. Note that the insolation remains at STC value on this chart and the solid line to the right is STC temperature.

over the course of a day. This means that, on a very hot, cloudy day your PV will not be performing at STC values, rather, it will be a lot less.

## Nominal Voltage

Nominal voltage is a category as opposed to the actual voltage of a module or system. Off-grid systems, which have to have battery banks for energy storage, are categorized as either 12V, 24V, or 48V systems. The modules, controller, battery bank and inverter all need to work together to match the desired voltage. Usually, the actual voltages are higher than the nominal voltages. For high voltage arrays with battery backup, the controller or inverter will need to step down the voltage to the desired nominal voltage.

## Types of PV

There are three different types of PV based on the crystalline structure. Monocrystalline and polycrystalline are always mounted on very stiff substrates while thin film is on flexible substrate.

### Monocrystalline Solar Cells

Monocrystalline PV is grown as an ingot of crystalline silicon (mono c-Si). The most common process, the Czochralski method, shown in figure 13.8, creates a crystal by placing a seed crystal in a melted vat of polysilicon, which is then rotated and pulled up as it grows. The resulting ingot is a long cylinder and is one solid crystal.

The round ingots are wire cut to the desired size and shape, usually a mostly square shape with diagonally cut corners to minimize waste. It is then sliced into very thin wafers, about 0.0118in (0.3mm) thick. The thin wafers are polished and treated. Figure 13.9 shows the cutting of an ingot and the typical octagonal shape created from cutting as little as possible off the circle of the ingot to reduce waste material.

Monocrystalline cells have a very consistent color, appearing black, gray, or dark blue, because they are one homogeneous material. When arranged on a

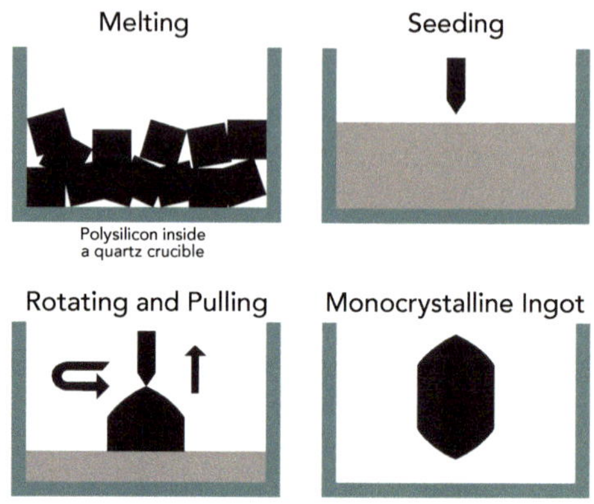


**Figure 13.8** The basic process for growing monocrystalline ingots in a furnace starts with melting polysilicon, then introducing the seed crystal, then rotating and pulling the crystal as it is forming from the melted polysilicon. The final crystal is a cylindrical ingot with pointed ends. The diameter and length of the ingot can vary depending on the manufacturer/furnace and the desired cell size.

module they have a very distinct pattern of diamonds between the cells, as can be seen in figure 13.9. The substrate colors currently available are white or black. The diamond pattern is much less visible on a black substrate. These modules are about 19% efficient, so they deliver about 19W of power per $ft^2$.

Premium modules are all monocrystalline and may use a combination of different technology to increase their efficiency. The following technologies are used to increase the efficiency of monocrystalline cells.

PERC solar cells (Passivated Emitter and Rear Cell or Passivated Emitter and Rear Contact) increase efficiency by about 1% and are easy to add to a cell. This is shown in figure 13.10. It adds a layer of material on the backside of the cells which reflects back the light that has passed though the cell. The layer also helps to keep electrons from recombining, which helps maintain the free movement of electrons within the cell. It also helps to keep cells cool by reflecting light in the infrared wavelength back out of the cell. Otherwise, the rest of the cell and module make-up is the same as usual.

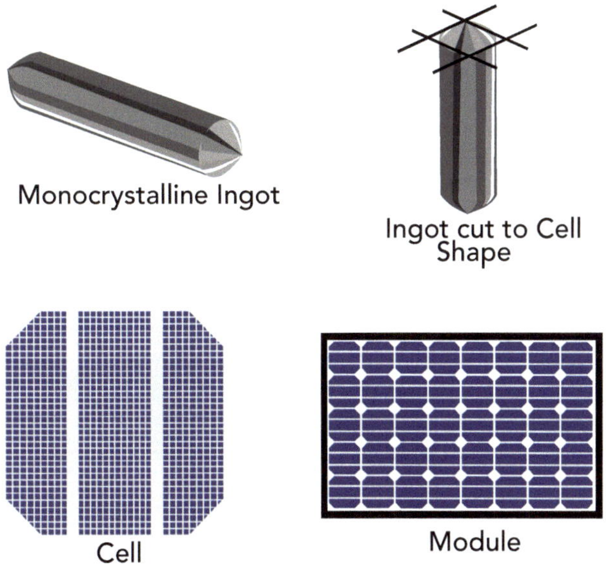


**Figure 13.9** How a cylindrical monocrystalline ingot is cut into an octagonal type shape to reduce waste and arranged onto a rectangular module as tightly as possible. This creates a diamond shape at the edges of the cells where the color of the backing shows through.

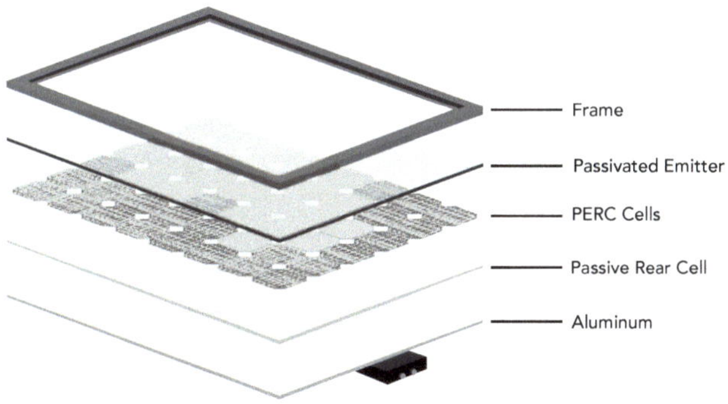


**Figure 13.10** PERC solar cells add layers that reflects infrared light out of the cell and helps electrons to move more freely within the cell.

**HIT** (or **HJT**) solar cells (**Heterojunction with Intrinsic Thin Layer**) add ultra-thin amorphous silicon layers on both sides of the monocrystalline wafer. The technology was developed by SANYO and sold to Panasonic. This technology works well at higher temperatures, producing up to 13% more output on hot sunny days.

**Multi-busbars (MBB)**, shown in figure 13.11, use very thin busbars and wires onto the face of the wafer. Because of their very small size, they use 25–50% less silver, and the round wires reflect light into the cell, which increases performance. There are a number of different processes for printing or integrating MBB.

Split cell technology cuts the traditional wafer in half and then adds busbars into the cell. These modules have a distinctive appearance because of the wider band between the cells across the center of the module as

**Figure 13.11** Multi-busbars are very thin and use substantially less silver than traditional busbars.

shown in the front view of figure 13.12. Split cells have increased efficiency because of lowering the operating temperature and voltage. This results in better module longevity by reducing the negative impacts of thermal cycling. The positive and negative leads (sometimes called split junction boxes) are on either side of the module, as shown in the back view of figure 13.12, rather than the standard single junction box. The voltage drop is due to the module's cell wiring, shown in the circuit view of figure 13.12. The two halves are wired in series and both halves are wired in parallel. Because each half of the module is wired separately, partial shading of the module only impacts half of the module, which keeps production higher than a standard series wired module.



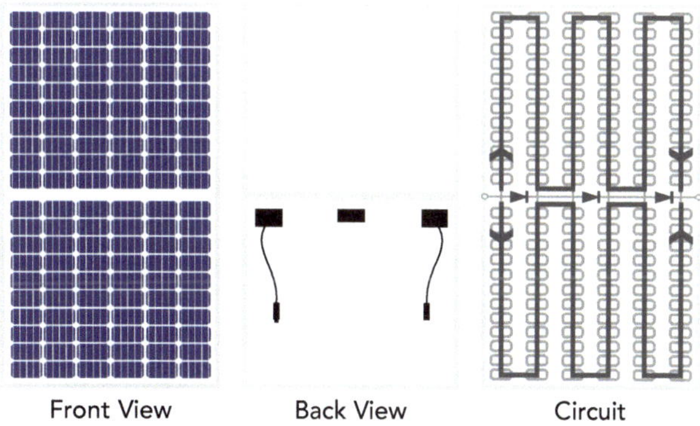


**Figure 13.12** Split cell module showing the front, back, and circuit wiring.

There are other innovations for better performance. For example, Solaria's PowerXT has a patented technology that creates long monocrystalline cells that are slightly overlapped and completely cover the substrate. These are called shingled because they overlap each other like roofing shingles. The cells in the resulting module are wired in both series and parallel. The product doesn't need busbars because of the overlapping of the cells. The final product appears as a solid black color, with more area for production and better performance if the module is partially shaded.

## Polycrystalline Solar Cells

Polycrystalline (sometimes called multicrystalline) solar cells are made from polycrystalline silicon (also called polysilicon, poly-Si). Most polysilicon is created through a chemical distillation process (Siemens process) where large rods which have been broken into chunks are cast into rectangular ingots. These are sized and then sliced into very thin wafers. Because polycrystalline cells are made up of individual crystals that have been melted together, their appearance shows the varying sizes, colors, and angles of the crystalline structure as shown in figure 13.13. The colors vary from medium to darker blues and grays, with an overall blue appearance. The multiple crystalline structure is visible.

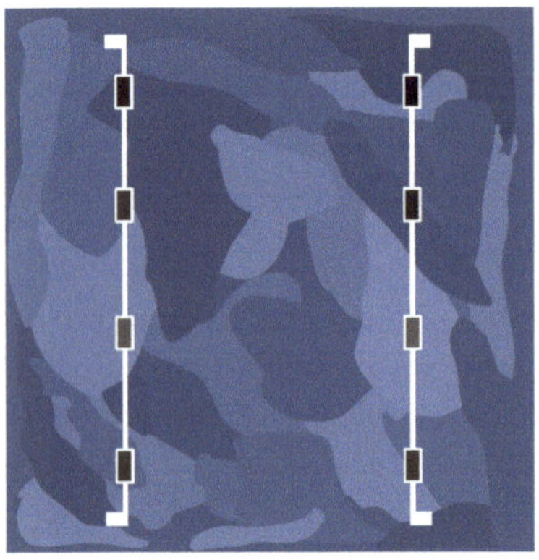

**Figure 13.13** Polycrystalline cell appearance.

Cells are square, so have straight lines between them on the module as shown in figure 13.14. Polycrystalline is cheaper to make than monocrystalline, creates less waste, has higher heat tolerance, and is slightly less efficient (15% or more range) because of smaller crystal areas for the movement of electrons. These modules deliver about 15W per ft$^2$.

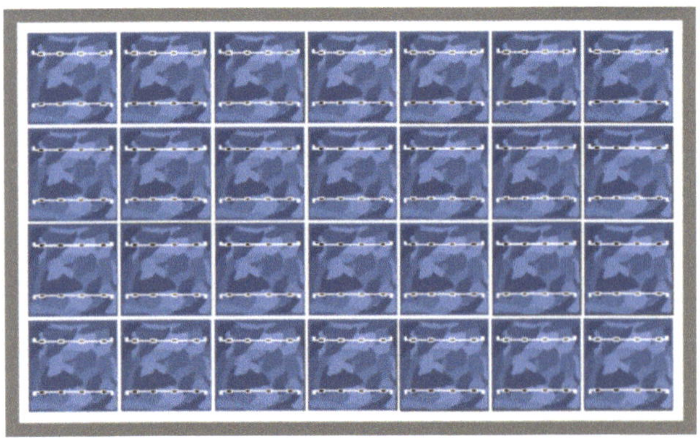

**Figure 13.14** Polycrystalline module has straight lines between the square cells.

## Amorphous (Thin Film) Solar Cells

Amorphous solar cells are much thinner than traditional cells because they are made from depositing thin films of silicon (a-Si) onto flexible substrates rather than being cut from a crystal. An example is shown in figure 13.15. The film is made of very small grains and is less cohesive than grown crystals. This type of cell is not very efficient (5–10%) and can lose significant production within the first half year of operation if it hasn't been treated (annealed) so that it is more durable. There are other technologies using cadmium telluride (CdTe) and **copper indium gallium diselenide (CIGS)** that have higher efficiencies. Depending on the materials used and layer make-up, thin films usually appear as a solid blue or black color.

**Figure 13.15** Amorphous (thin film) module has a flexible substrate.

There is a lot of research in thin film technology and the newest developments include the use of organic materials. They promise less harmful production processes, less expensive production costs, superior production in low light conditions, color options (pretty much any color you want), as well as higher efficiencies and longevity. In labs, organics have been able to quickly produce more energy than it took to make them. These products are not yet commercially available.

The advantages of thin film are that it is inexpensive to make (both time and materials) and it is lightweight and flexible. Typically, thin films do not need to be ventilated, so they can be directly mounted to building elements. The a-Si is a non-toxic material. The disadvantages of thin film are its lack of longevity and lower efficiency.

## Building Integrated Photovoltaics

Building integrated photovoltaics (BIPV) are building materials that incorporate PV. Examples of these products are thin films with sticky backs that adhere to roof surfaces, modules integrated into glazing (windows and skylights) and shading devices, as well as modules that look like (and replace) roofing and façade materials.

Generally, products made with thin film produce about 4–5W/ft$^2$ (43–54W/m$^2$) of PV array area (STC), which is half of that produced by conventional crystalline silicon BIPV at about 10–12W/ft$^2$ (108–129W/m$^2$).

BIPV systems have the same components as traditional rectangular modules and can be used in grid-tied or off-grid applications. Because BIPV is less efficient, you'll need more square feet (square meters) of product than area for a traditional array to create the same amount of electricity. As with any PV, manufacturer's datasheets need to be consulted to determine the actual production based on orientation and insolation. Shading of façade based BIPV can happen easily, so the designer needs to pay attention to the shape and orientation of the building to help boost hours of production.

BIPV are also subject to issues with debris and snow. If angled, snow can melt off relatively quickly. They will need to be washed if in a dusty or polluted environment.

## Orientation and Tilt Angle for Best Energy Production

The best production will happen when the sun's rays are as perpendicular to the PV cells as possible. The challenge is that the sun is a moving target, so it is important to orient your array properly. The best orientation for overall annual energy production is facing south in the northern hemisphere or north in the southern hemisphere. The array should be within 15° of south (northern hemisphere) or 15° of north (southern hemisphere) to get maximum solar insolation. For more about solar insolation, see chapter 5.

There are a few options for how the array can face the sun: fixed tilt, single-axis trackers, and azimuth–altitude dual-axis tracker. Smaller applications are typically fixed tilt as it is the least expensive option. Larger trackers require robust foundations, poles, and mounting equipment, all of which cost more than simple racks. They can double the cost of a project. It is important to compare the value of the additional production vs. the cost of the trackers. Trackers increase production but also add moving parts which can break down and add to maintenance costs over the lifetime of the array. Dual-axis trackers are the most expensive and the panels need to be carefully spaced to avoid shading by other dual-axis trackers in your array.

Trackers can be either passive or active. Passive trackers use a balancing system that has a liquid that changes to a gas at very low temperatures. The liquid boils, turning to gas when exposed to the sun and the weight displacement causes the panels to tilt toward the sun. This type of tracker follows the sun through the day and then stops facing the west when the sun sets. When the sun rises in the east in the morning, once the sun starts heating the liquid, the tracker rotates back to the east. Passive tracking is generally used for simple PV systems as they can lag a bit in following the sun. Active trackers use motors or hydraulics to move the panels to face the sun. Most trackers installed today are active and are used for large-scale, complex commercial applications.

### Fixed Tilt

Fixed tilt means that the array will be mounted so that its angle and orientation do not change. The modules may be racked directly onto a pitched roof, or, on flat roofs, are placed on racks that are set to the desired tilt angle. Racks must be spaced so that none of the modules are shaded. See the discussion of inter-row spacing in the solar insolation chapter (5) and racking in the racking chapter (14) for more information.

Tilt angles are measured from horizontal compared to the angle of the back face of the module. Figure 13.16 shows both PV and solar thermal mounted on a 45° angled roof. Because they are racked directly to the

**Figure 13.16** 45° fixed tilt angle on Middlebury College's 2011 Solar Decathlon house.

**Table 13.2** Comparison of tilt angles of roof slopes facing south and annual average solar insolation.[4]

| **Daily Average Solar Insolation for Williamsport, PA, 41° N Latitude** | | |
|---|---|---|
| **Roof Slope: rise/run** | **Fixed Tilt Angle (degrees)** | **Daily Average Annual Solar Insolation (kWh/m²)** |
| 4/12 | 18.4 | 4.42 |
| 5/12 | 22.6 | 4.49 |
| 6/12 | 26.6 | 4.53 |
| 7/12 | 30.3 | 4.56 |
| 8/12 | 33.7 | 4.58 |
| 9/12 | 36.9 | 4.58 |
| 10/12 | 39.8 | 4.57 |
| 11/12 | 42.5 | 4.56 |
| 12/12 | 45.0 | 4.54 |

roof, they are at a fixed tilt angle of 45°. PV maximizes average production when the tilt angle is the same as the latitude. For example, if you are located at 20° latitude (either northern or southern) the best fixed tilt angle is 20°. If you are located at 52° latitude, the best fixed tilt angle would be 52°.

If the array is being mounted flush with a pitched roof, it may not be at the optimal tilt angle. Depending on how much the difference between the latitude and the roof slope is, there may not be a huge change in the maximum solar insolation if the array is oriented correctly to the south or north. See table 13.2 for roof slope tilt angles facing south for Williamsport, PA, 41° N latitude. Solar insolation values generated from PVWatts.[4] You can see that, for this location, there is very little variance in the solar insolation values between the different tilt angles, although it is slightly better on steeper slopes. This number is per day, for every m² of collector, so even small differences can add up over the course of a year.

## Single-Axis Trackers

Single-axis trackers rotate the panels around one axis, either vertically or horizontally. These can either be installed on the ground or roof, depending on the size and mounting requirements of the unit. Single-axis trackers can improve production 25–35% annually.

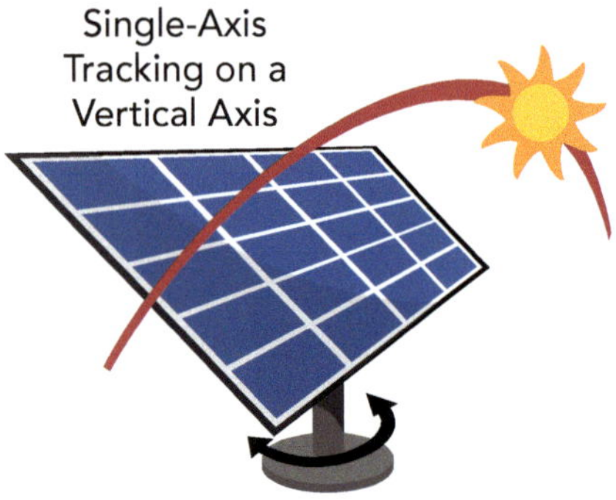


**Figure 13.17** Vertical single-axis tracker. The panels tilt east–west to follow sun through the day.

**Table 13.3** Example of daily average solar insolation for different types of vertical single-axis trackers.[4]

| **Daily Average Solar Insolation for Williamsport, PA, 41° N Latitude** | | |
|---|---|---|
| **Tracker Type** | **Fixed Tilt Angle (degrees)** | **Daily Average Annual Solar Insolation (kWh/m²)** |
| Vertical single-axis | 41 | 5.51 |
| Vertical single-axis with backtracking | 41 | 5.62 |

Vertical single-axis trackers follow the sun from east–west through the day, tracking the sun's azimuth. Panels mounted on single-axis trackers can shade each other in the early morning and late afternoon when they are lined up in a row. Some manufacturers provide a backtracking option which rotates the panel back toward neutral (south or north) to minimize shading effects. The manufacturer's design values for weight sets the limit for the number of modules that can be attached into a panel for a tracker. The smallest trackers can carry 2 or 4 modules, larger ones can carry 12 modules. Smaller trackers can be mounted on flat roofs. The support configuration of these trackers varies greatly between manufacturers. They can look like tripods or poles or even large rotating platforms.

Table 13.3 shows a comparison of values for vertical single-axis trackers. Note that these trackers are providing almost one more hour of insolation each day for this location, which is about a 20% increase. These insolation values were generated using PVWatts.[6]

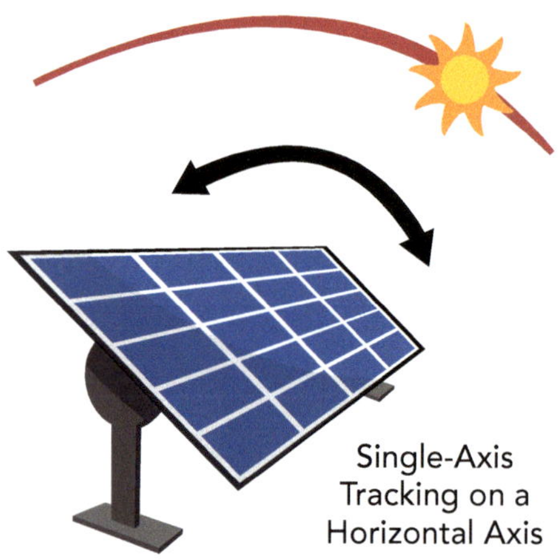


**Figure 13.18** Horizontal single-axis tracker. The panels tilt up and down to follow the sun through the seasons.

Horizontal single-axis tracker follows the sun up and down, tracking the altitude of the sun across the seasons, as shown in figure 13.18. Commercial applications, especially for utility grade production, can use horizontal single-axis trackers because it is easy to mount a large number of modules in a linear array. There is no shading between the units in a line because they are all tilting together. It is important to take into account the distance the rows need to be apart to prevent inter-row shading. This type of tracker is less expensive than the vertical because the panels don't have to be pole mounted and can be closer to the ground.

### Dual-Axis Tracker

Dual-axis trackers follow the sun throughout the day and seasons, tracking both the azimuth and altitude of the sun as shown in figure 13.19. These systems can have as much as a 40–50% increase in production over a year, compared to fixed tilt. These benefits are typically found at higher latitudes because of increased production in the winter months where there are not many hours of daylight. Because of the additional up-front costs and lifetime maintenance, these are usually only found in larger commercial projects or utility production where the added production equals many megawatts of generation. In these types of projects, the array would have to be ground mounted anyway, so it isn't that much more additional cost.

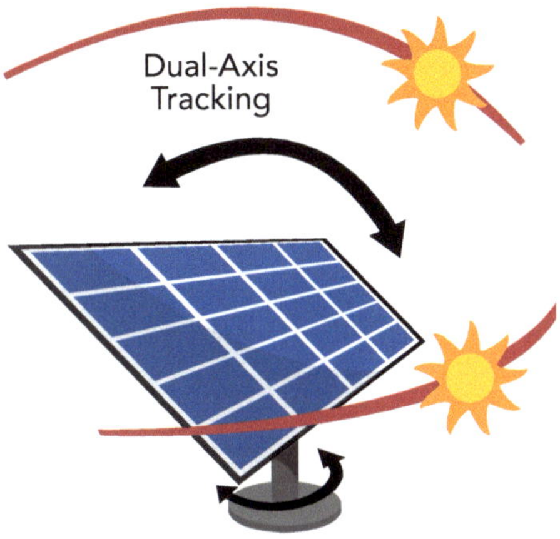


**Figure 13.19** Dual-axis tracking moves both vertically east–west every day and horizontally through the seasons.

Comparing the dual-axis tracker daily average solar insolation in Williamsport, as shown in table 13.4, there is about a 35% increase in production over using a fixed tilt.

Dual-axis trackers have traditionally been pole mounted and are usually installed 12′–20′ above grade. Larger commercial units may be closer to the ground to help reduce torque and wind loading issues. These trackers can have from 12–90 modules mounted into a panel. There are quite a few innovations happening in the dual-axis tracking market which include placing large panels on a tracking frame that rotates on a circular foundation, therefore eliminating poles and reducing maintenance failure issues.

**Table 13.4** Daily average solar insolation for Williamsport using a dual-axis tracker.[4]

| Daily Average Solar Insolation for Williamsport, PA, 41° N Latitude | |
|---|---|
| Tracker Type | Daily Average Annual Solar Insolation ($kWh/m^2$) |
| Fixed Tilt 41° | 4.61 |
| Dual-axis tracking | 6.14 |

Figure 13.20 shows a simulated comparison study of fixed tilt and tracking done by EIA using a location in Los Angeles, CA, USA, 34° N latitude. The lowest production curve results from a fixed array at a 0° tilt angle. It produces much less in every season except the summer, when the sun is basically overhead. The generation created by the fixed array at 20° and single-axis tracker are virtually the same, with the dual-axis producing a bit more midday in the fall and winter. For a small system in this location, it would not be cost-effective to install a dual-axis tracker.

## Effects of Shading on PV Modules

Because the cells of PV modules are in series, shading on even a part of a cell will change the overall production of the module. Because series adds volts and all the amps must match, the output will drop to match the lowest amperage. Shading can happen because of trees, buildings, and other objects that are near enough to cast a shadow onto a portion of your array. PV modules in rows on flat roofs (or the ground), that are mounted at a tilt angle, can shade each other if they are not spaced far enough apart. For information on recording shading on your site and row spacing, see the chapter 5 (Solar Insolation).

Shading can also happen because of snow, dust, dirt, and other debris. Debris such as a wet leaf stuck on a module shades the cell.

Shading is effectively reducing the amount of solar insolation that is available to the cell: remember that less insolation means less current. From 0% to 50% shading the power loss drop for the module is linear. Above 50% shading on one cell, the power loss of the module is a bit less than the direct percentage. This is because manufacturers usually install bypass diodes in parallel across groups of 16–24 cells so that the shaded group is bypassed instead of causing resistance and heat. Depending on what part of the module's cells get shaded there could be some power production or not.

If your modules are wired in series, the shaded panel will cause output from your entire string (any number of modules wired together in series are a string) to drop.

It is important to be sure that all the modules in your array are situated so they receive as much of the solar window (a few hours on either side of solar noon) as possible without shading. For more information, see chapter 5 (Solar Insolation).

## Steps for Sizing a PV System

Sizing a PV system is similar to sizing any renewable energy system. See appendix C for detailed information and worksheets.

13

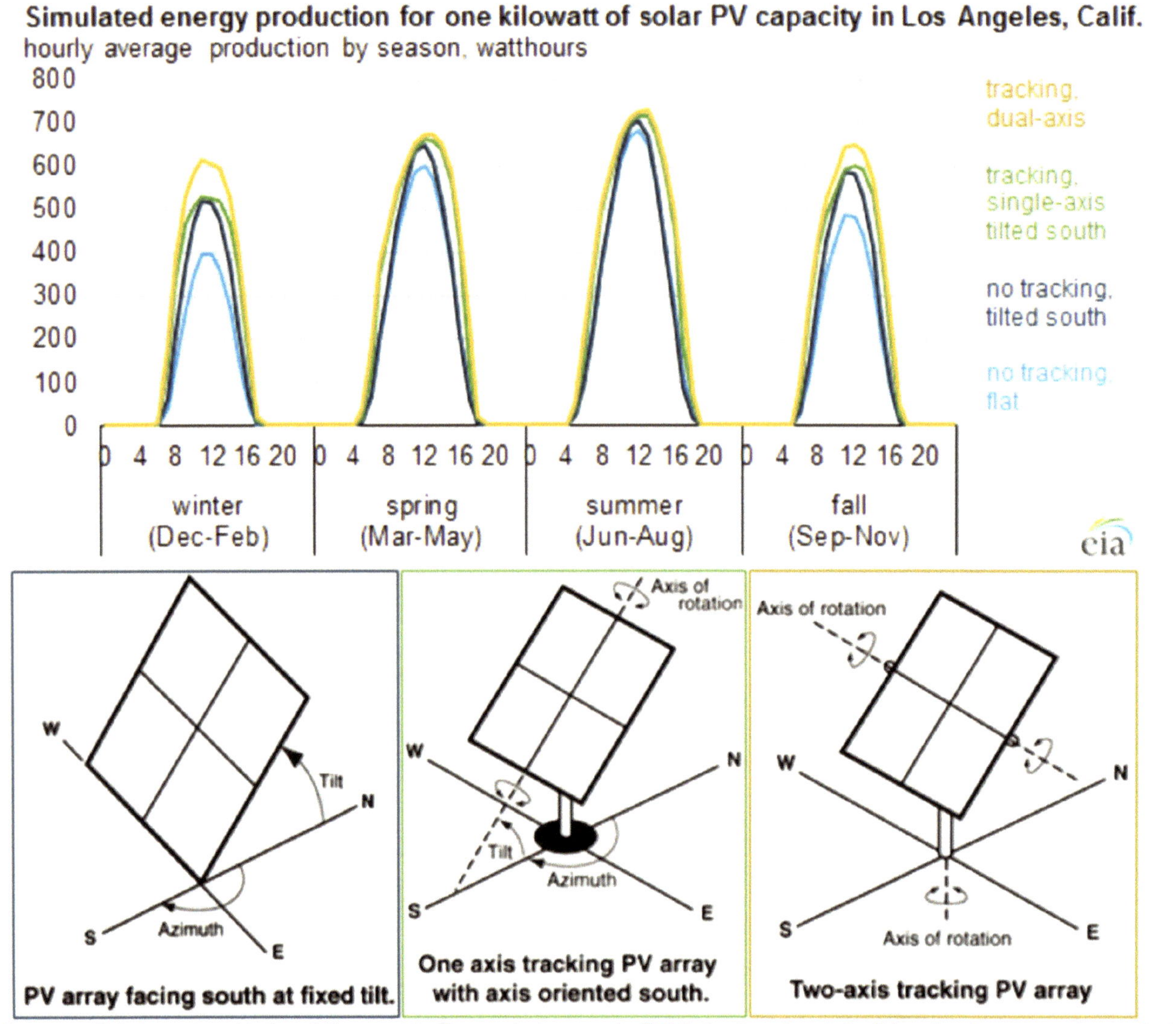


Source: EIA, based on National Renewable Energy Laboratory's PVWatts, using default input values except as noted

**Figure 13.20** Comparison of simulated kW production in Los Angeles, California, USA, between a fixed tilt, one-axis tracking, and two-axis tracking.[5]

## Optimize Power System Demands

The first step is to minimize all loads on the building. This includes overall building performance energy efficiency (refer to various book chapters) as well as making sure that all equipment going into the building is as energy efficient as possible.

All loads should be electrical. All electrical equipment should be as energy efficient as possible. Energy consumption labelling is happening across the world. The United States uses Energy Star which has the EnergyGuide label showing the estimated energy usage. The Energy Star label applies to appliances, HVAC equipment, electronics, lighting, and homes. There are multiple nations that share all or part of the Energy Star rating system including Canada (Energy Star Canada), Japan, Taiwan, and Switzerland. The European Union (EU) has the EU Energy Label **ENERG** which has a scale A–G (A is the best rating). China has the China Energy Label which is similar to the EU label with five categories of efficiency. Australia and New Zealand have an Energy Rating label based on six stars, and India has the BEE Star Label based on five stars. These labelling programs are constantly being updated and more products are being included. It is important to select equipment that is efficient. Customer demand is a huge driver of the development of consumer goods, as are government regulations for minimum performance levels. As manufacturers recognize the demand for more energy efficient equipment, they develop better performing goods.

Another area to pay attention to are phantom loads. These loads are the electrical draw of equipment that is turned off. This is also called standby power. Older equipment sometimes had very high usage of electricity even when it was not being used. Many countries now have regulations in place restricting the amount of standby power to as little 0.5W because phantom loads were using so much power. Even though new equipment may not use much standby energy individually, if you have a lot of items, they still can add up. For instance, a computer that is not turned off uses energy in sleep mode. While the draw of just one computer may not add up to anything, a business running hundreds of sleeping computers can be a big deal. In companies, controlling loads like this becomes policy: directives to turn off computers when leaving for the day and even having programs that will run over the network and shut down computers that have been left on. Tools such as a Kill A Watt meter can be used to measure equipment loads either running or not.

In today's buildings, it is quite common to have lots of electronic equipment that runs on DC power. Anything that has a transformer to plug in is probably converting AC power to DC power. If you are including batteries in a building, it is worthwhile to provide a DC panel from the batteries and your PV array to directly run your DC equipment rather than invert it all and run it through the AC service panel. There are fewer losses, and therefore it is more efficient, to run DC equipment using DC power. For more discussion on this see the chapter 22 (Resiliency and Power).

### Determine the Power Consumption Demands

No matter the size of your building, it is important to have an idea of the power consumption the building will need. Power consumption will cover HVAC equipment, water heating, appliances, and any specialized equipment needed for the building's use. There are a number of computer programs that can be used to do energy modelling. These all require that you appropriately identify all the equipment needed and the periods of building usage. To know the amount of energy needed to heat and cool the building, calculations have to be done that take into account the climate and performance of the building envelope.

You may need to make a list of individual equipment that would be in the building. This list would include both AC and DC loads. If you do this, you need to be sure that you are accounting for all the equipment, have looked up the specifications for its energy use, and understand how long it runs each day or how often it will be used. You can add up usage and then average for a daily, weekly, monthly, and/or an annual number. These numbers can then be used to calculate the needed power generation and storage throughout the day and year.

If you are working with an existing building, the owner has a power bill. If the owner has been at the location for a couple of years, the bill will show current consumption compared to previous consumption over the same time period. This means you have access to both yearly and monthly numbers for kWh used.

### Locate Your PV for Best Production

Determine the most appropriate locations on the project for PV. The important considerations are: facing the array to the best orientation, avoiding shading from the building or anything near the building, determining how you plan to mount the PV system either to the building or on the ground, the importance of ease of access for cleaning dirt/snow and debris from the array, and locating the battery bank near the array to reduce DC losses.

### Determine Insolation

Once you know where you are placing your PV, you'll know what it's orientation and tilt angle will be and you can determine the available insolation. If you have different kinds of PV, for instance building integrated and modules on the roof, you'll have different values for their efficiency and they'll likely have different tilt angles. You may have different orientations and shading conditions for PV panels and arrays on a large commercial project. Each of these has to be considered separately because they will not have the same insolation values.

### Size the Array

The size of your array is dependent on several factors. If your project is off-grid or grid-tied and net zero, then you have to provide a system that is sized to provide your building with all the energy that it is going to use. That's why you need a really good estimate for power consumption. If you are grid-tied and just offsetting some of your power use, then your array size will be limited by the area you have for modules and/or the budget.

If you are adding PV to an existing building, it is still important to encourage the owner to make efficiency upgrades to the building's envelope and to change out older appliances and equipment for new electrical items that have good energy efficiency labelling. In the long run, it is much cheaper to reduce energy use than to purchase a larger array. The least expensive energy is all of it you never have to buy because you didn't need it.

The pieces of information you need to size the array are: daily average generation kWh, daily average solar insolation, inverter efficiency, and system losses.

Daily average generation requires that you know how much energy you want to produce every day. This daily average can be derived from a yearly consumption number. If you have a high annual consumption, it may not be possible to make 100% of your energy, that's where the space you have and the budget will determine your production. It is always interesting to start aiming for 100% and discover how many modules it takes.

Daily average annual solar insolation requires that you know your location, the tilt angle for the modules, and you've determined if your modules will be shaded or not. If modules are shaded, they will not be producing energy, so the hours of production will be reduced. For example, if you have 5.6kWh/m$^2$ theoretically available, but 20% of your production time is shaded, then you really have only 4.48kWh/m$^2$.

Today's inverters are very efficient. Depending on the type and manufacturer they run between 94–98% efficient at inverting DC energy to AC energy. If you have not picked out an inverter you can use 94%.

System losses are general issues with the PV system that keep it from being 100% efficient. PVWatts has the following categories for losses: soiling, shading, snow, mismatch, wiring, connections, light-induced degradation, nameplate rating, age, and availability. Their default numbers, which are explained on the website, add up to estimated system losses of 14.08%. This is with snow being 0%, assuming that there is never snow, or the snow is cleared off right away. If your modules will have snow on them during production hours, then you need to add to the system losses.

Here's an example where each day's production should be 16kWh. The solar insolation is 5.2kWh/m$^2$. The system losses are 14.08% and the inverter is 94% efficient.

Array size = daily average generation ÷ daily solar insolation ÷ system losses ÷ inverter efficiency

Array size = 16kWh ÷ 5.2kWh/m$^2$ ÷ 0.1408 ÷ 0.94 = 23.25kW

Array size = 23.25kW × 1000W/kW = 23,250W

Now you need a module. This example has a big demand, so we'll need a high voltage array. Earlier in the chapter we looked at the Solaria 370W module. The names of modules are generally the STC watt rating of the module.

Number of modules = array size ÷ STC watt rating

Number of Solaria 370W modules = 23,250W ÷ 370W = 62.84 modules.

Since you can't break up a module (have 0.84 of one) this system would need 63 Solaria 370W modules to produce the daily demand for this building.

## Size the Battery Bank

Batteries may be included in a system for several reasons. Remember that if you don't have any batteries and have a grid-tied system, when the grid goes down you have no power. So, you may have a backup battery system that is providing power to essential items in your building. If there is no grid, you probably need batteries to store power when your renewable resource isn't producing energy: PV only makes energy during sunny daylit hours. You could have a series of overcast days where power production is reduced and even after a sunny day you probably want to operate items in your building at night.

In areas with high PV production, batteries are a backup for the grid. This happens because as the sun sets there is an increased demand for power from the grid. It is difficult to start up powerplants to meet big surges in demand for short periods of time. Utilities are using interconnection agreements to allow them to access the energy stored in the batteries in your system to help support grid demand.

Another reason you might want batteries is to keep your building's power demand under a particular number in order to avoid being charged at a higher rate. Utility companies may have tiers for power cost based on power demand at particular times of day and/or year.

Additionally, you might have a battery bank for power resiliency for your building or a neighborhood community in the event of a utility-wide power outage. There's more on this in the chapter 22 (Resiliency and Power).

For specifics on how to size a battery bank, see chapter 18 on Batteries and Energy Storage.

## Inverters

There are three options for PV system inverters. The traditional inverter is a string inverter that processes DC power from the whole array into AC power for your building's main service panel. The second option is to have a string inverter for the array and power optimizers on each module. The third option is to have a microinverter connected to each module in your array, so the DC power is immediately inverted to AC and then sent to your AC service panel.

Modules that are wired in series are called a string and all the strings to the same inverter need to have the same module and number of modules for peak performance. For example, if there were 21 modules in an array, they could be wired in 3 strings of 7 modules. It would not work to have 2 strings where one had 10 and the other 11 because the series of 11 modules would match the production of the 10-module string. If you really wanted 2 strings, they would both have to be either 10 (20 modules total) or 11 (22 modules total). Depending on the $V_{oc}$ rating of the module, you might be able to put all 21 modules into one string, remembering that the maximum is 600V in a string.

Figure 13.21 shows a 12 module array with 2 strings (6 modules in each) connected to a string inverter. You can see the six modules connected in series (positive to negative). The two strings are connected in parallel in a DC combiner box. Disconnects are required on both sides of the inverter and are used to shut off the power to make it safe to repair or replace the inverter. Some inverters have disconnects that are built in; if it does, then you don't need to have external ones. Everything on the array side of the inverter is in DC power. Everything on the other side is AC power. The AC power from the inverter is connected to the main service panel.

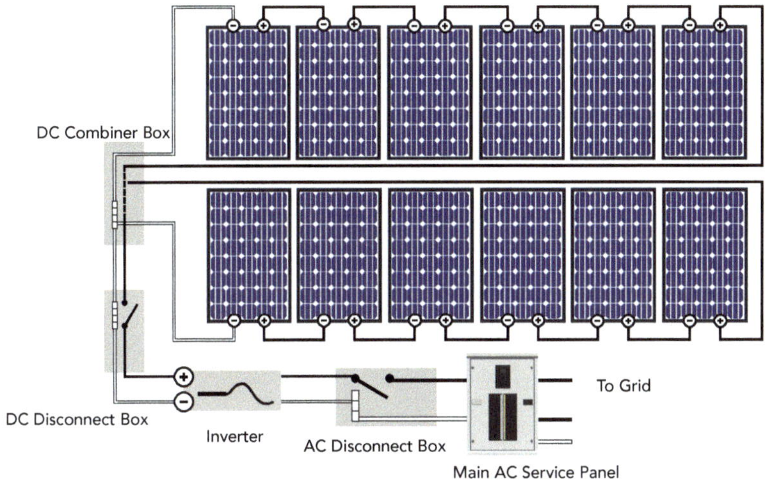


| | Module Rating STC | Inverter AC Rating |
|---|---|---|
| Rated Power | 150 W | 1800 W |
| Open Circuit Voltage | 43.4 V | |
| Max Power Voltage | 34 V | 120 VAC |
| Short Circuit Current | 4.8 A | |
| Max Power Current | 4.4 A | |

**Figure 13.21** Array with 2 strings of 6 modules wired in series. The 2 strings are wired in parallel at the combiner box. DC power flows to the inverter where it is changed to AC power and supplied to the main service panel for the building.

Power optimizers are DC–DC converters and work within the string to maximize production. This can be helpful if there are variations in shading or orientation of the string modules. Power optimizers can be part of the PV system monitoring platform, which allows the owner to individually monitor each module. This can help to identify module issues. Power optimizers use MPPT to optimize the relationship between the voltage and current from each module to produce the most energy. A system with 20 modules would have 20 power optimizers and the strings would be connected to the string inverter.

Figure 13.22 shows what a power optimizer looks like. It measures about 5″ × 6″ × 1″ (127mm × 152mm × 25mm) without the bracket and has cords about 3′ (914mm) long. All the connectors are MC4 compatible. You can see the slot in the bracket plate for bolting it to the rail.

**Figure 13.22** Power optimizer with rack mounting plate and cords with MC4 compatible connectors.

Power optimizers like the one shown are attached to the rack under the module and wired to the positive and negative cords of the module. Some manufacturers sell modules with the power optimizers built in on the back of the module. Figure 13.23 depicts how external power optimizers are wired to modules and each other in a string.

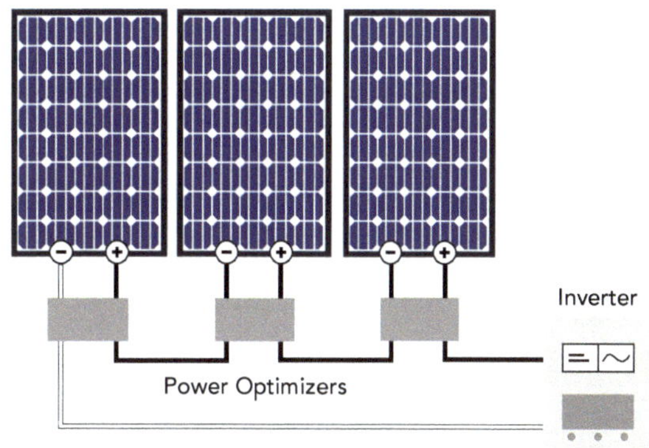


**Figure 13.23** Power optimizer wired into array. Optimizers are DC–DC and allow for individual monitoring and MPPT for each module.

If you have battery storage, you should use a string inverter system because batteries charge using DC power and store DC power.

Microinverters are DC to AC and are installed at the module. They are a similar size to a power optimizer. Some manufacturers build in the microinverter on the back of the module. Figure 13.24 is a drawing representing an ENPHASE IQ7 microinverter which measures about 8″ × 7″ × 1″ (203mm × 178mm × 25mm) without the bracket. You can see the slots that are used to bolt the microinverter to the rail under the module. The MC4 compatible connectors attach to the positive and negative module cords. AC power comes out of the microinverter, so they are wired together and then plugged into the main service panel.

Figure 13.25 shows how microinverters look when installed in an array. Typically, every module has one microinverter, although there are some variations

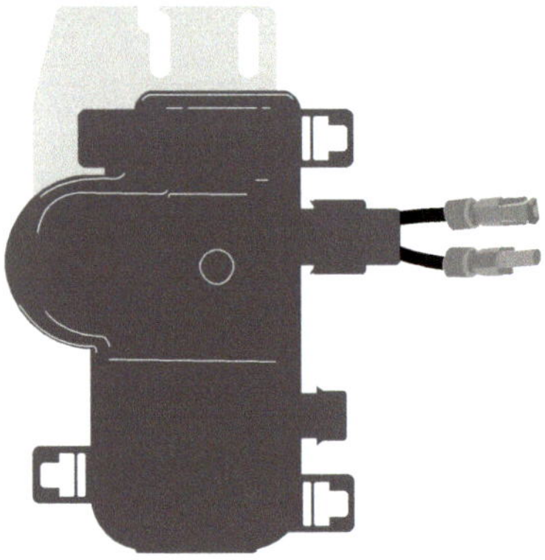

Figure 13.24 ENPHASE IQ7 microinverter with bracket and MC 4 compatible connectors.

depending on the manufacturer. All microinverters have MPPT to optimize array production. The microinverters are inverting the DC power to AC power at the module. Because the power is AC, it can all be wired together and then plugged directly into the main service panel. There will be a gateway connection to a router so that the information about production can be tracked on a computer or phone app. Note that, because the microinverters are DC–AC, no string inverter is needed.

## Wire Types

Codes have basic requirements for the type of wire, its size, and the color coding of the external sheath. Most of the wiring in PV systems is outside and conducting a lot of power. PV wire is used for connecting PV modules and is suited for wet and hot conditions; it has thick insulation to protect the wiring. PV wiring is tested for flammability and sunlight resistance. In the United States, different wiring colors are used for DC and AC power types. In DC wiring, red wire = positive, black = negative. In AC wiring the red wire is often the "live" (negative) wire. White or gray wire is used for the grounded conductor and green or bare copper wire is used for grounding equipment. Bare copper wire is usually used in outdoor conditions. Codes lay out very specific requirements and the National Electric Code (NEC), has sections on PV safety and wire sizing. It is important to realize that the wiring has to be appropriate and correctly installed, otherwise it can become a fire hazard. Overloading any piece of electrical equipment and/or wiring because of improper matching for current, voltage, and power can create overheating and fires.

## Grounding

Grounding is required by code and means that the electrical system is connected to the earth. Grounding is important to protect electronics from damage by lightning strikes, to keep your system from accumulating charge and attracting lightning, reducing shock hazard within the electrical system, and reducing humming from many electrical devices. In the United States, the most common ground is an 8′ (2.4m) long copper-plated metal rod that is driven into the ground. The grounding conductor, bare copper or green wire, is attached to all aspects of the electrical system and equipment and then attached to the ground rod.

Ground fault is when there is accidental contact between a live wire and the ground or metal frame of equipment. This can happen because of water or dirt getting where it shouldn't be, or due to the breakdown of insulation on old wires. If everything is grounded properly, the sudden flow of unrestricted current will be through the grounding conductors rather than through a person. The circuit breaker for the circuit should trip

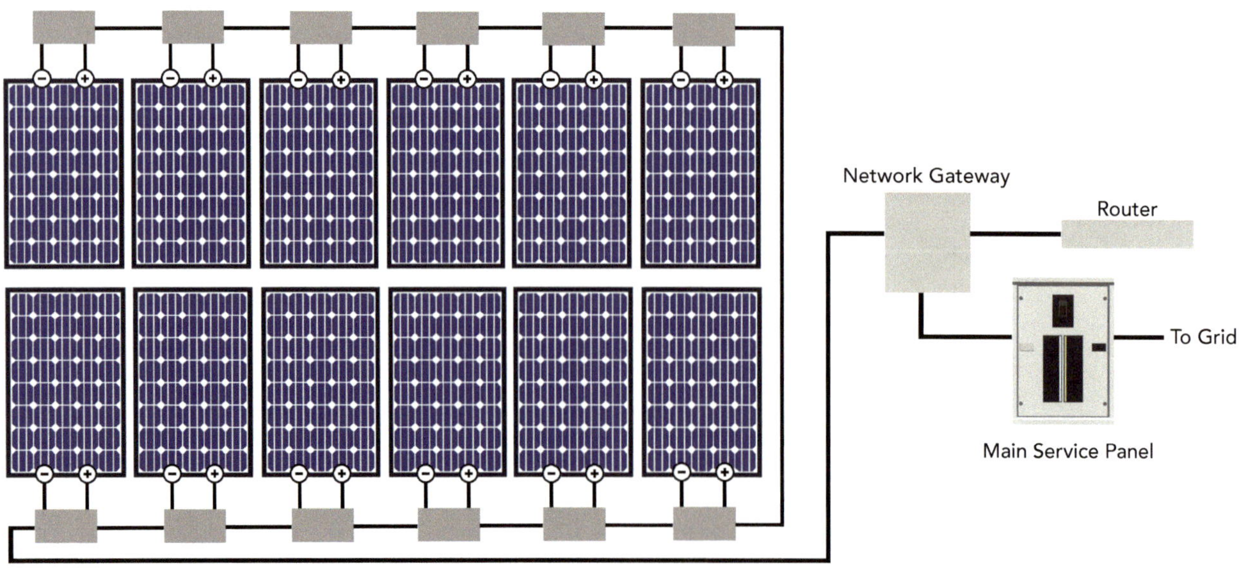


Figure 13.25 Microinverters installed at each module in an array. They are DC–AC, so can be wired together in series and connected to the main service panel.

very quickly when there is a ground fault, which shuts off the power.

Proper grounding ensures that if any part of the electrical system is hit by lightning, the power surge will flow down the grounding conductor rather than through expensive electronic equipment. If a large power surge were to flow through equipment, like the inverter, it could fry the connections in the equipment and render it useless, as well as possibly causing it to overheat and catch fire. Surge protection devices can be added to equipment in areas prone to lightning or with poor grounding conditions.

Often, diagrams of PV systems do not show the grounding conductors or grounding electrode (ground rod) just for clarity. They also often leave out many of the fuses and circuit breakers that are there for safety. That doesn't mean that you can build the system without these features though! Every single piece of equipment must be grounded: every module, every power optimizer or microinverter, the junction boxes, combiner box, disconnects, inverter, and main service panel.

Figure 13.26 shows the typical parts and arrangement of grounding elements for a PV system. No other wiring has been shown for clarity. The dotted line indicates the grounding conductor, typically bare copper wire, that must touch all parts of the system. In the USA, it is required to have a grounding electrode at the main service panel. Note that the inverter is wired directly to the grounding electrode.

Usually there are small connector clamps added to the PV frame that screw down onto the wire to firmly hold it in place. The same connectors are used to attach the grounding conductor to the racking rail. The disconnects, inverter, and main service panel all have built-in slots with screws for attaching the wire. The grounding electrode conductor is also typically bare copper wire which is connected either above or below grade to the grounding electrode with brackets approved for the conditions. Figure 13.27 shows an above grade grounding electrode conductor connected with a bracket to the grounding electrode. The top of the grounding electrode is about 8" (200mm) above grade in this photo.

If the inverter is located a distance away from the main AC service panel, it should have its own grounding electrode as shown in figure 13.28. Note that grounding rods must be wired together. Codes state the size of wire that must be used and the minimum distance rods

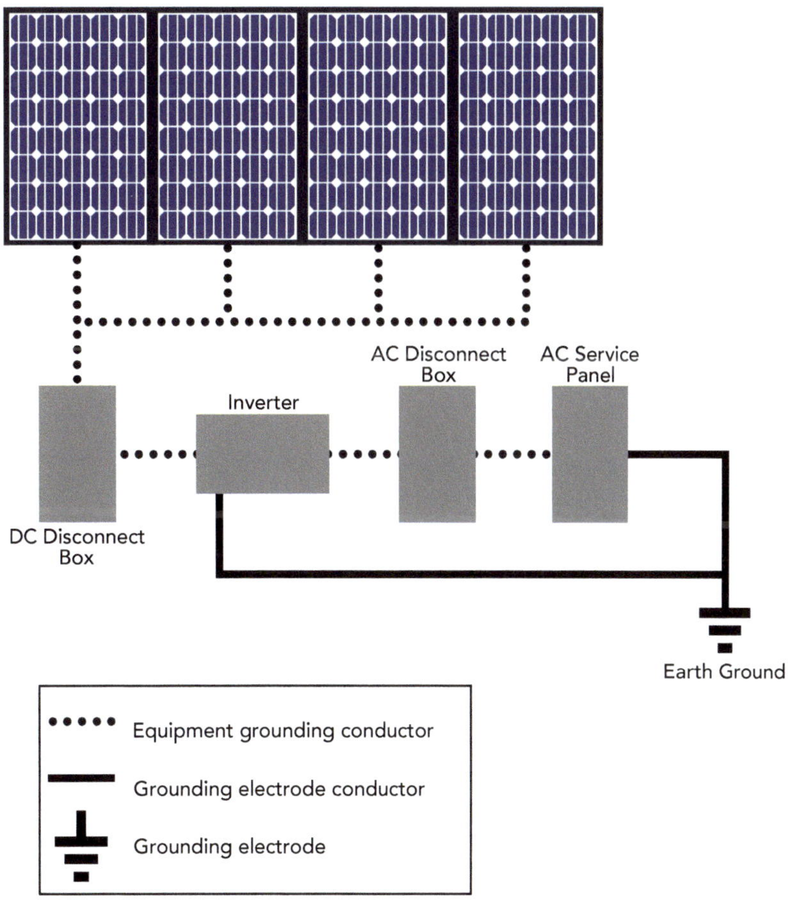


**Figure 13.26** All equipment must be connected with a grounding conductor and then connected by a grounding electrode conductor (typically bare copper wire) to the grounding electrode (8′ copper coated metal rod which has been pounded into the ground).

**Figure 13.27** Above grade grounding electrode conductor connected with a bracket to the grounding electrode.

can be from each other. Generally, they should be at least 10' (3m) apart.

If the array is located on a separate building such as a barn or garage, the inverter should be located near the array rather than the main panel. Figure 13.29 shows how an array that is located away from the main service panel may have its own grounding electrode. Note that the grounding electrode for the array doesn't have to be wired to any other grounding electrode. This diagram shows all the wiring and note that the grounding conductor touches all the equipment. Usually, the grounding conductor is installed along one of the rails supporting the array. The larger dots indicate where the bare copper wire is bonded, attached with a bracket and bolt or screw, to the equipment. Note that the connection is on the array side of all the equipment in order to keep a surge flow of electricity, such as with a lightning strike, from travelling through the equipment and damaging it. Typically, there are also lightning arrestors on grid-tied inverters.

## Roof Coverage: Access and Management

PV is just one of the many pieces of equipment that you are putting on your roof. The array is sharing space with various pieces of mechanical equipment, vents, skylights, chimneys, not to mention solar thermal collectors. Modules may be placed over drain waste vents as long as they can still vent and are protected from snow and nesting animals are prevented from blocking the vent opening.

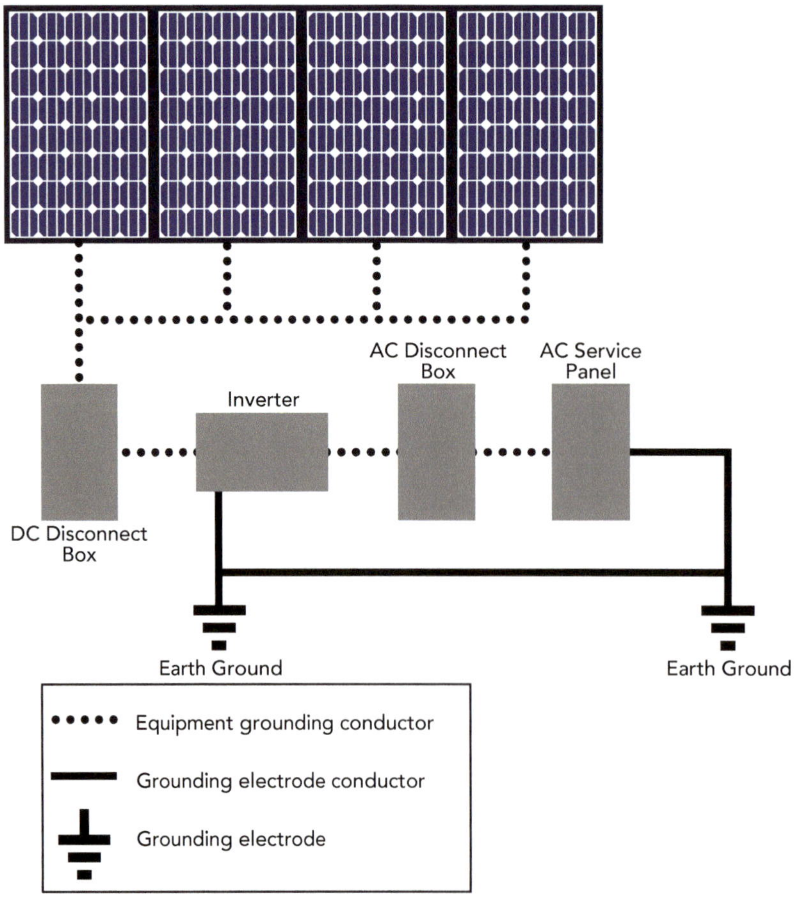


**Figure 13.28** An array grounded using two grounding electrodes.

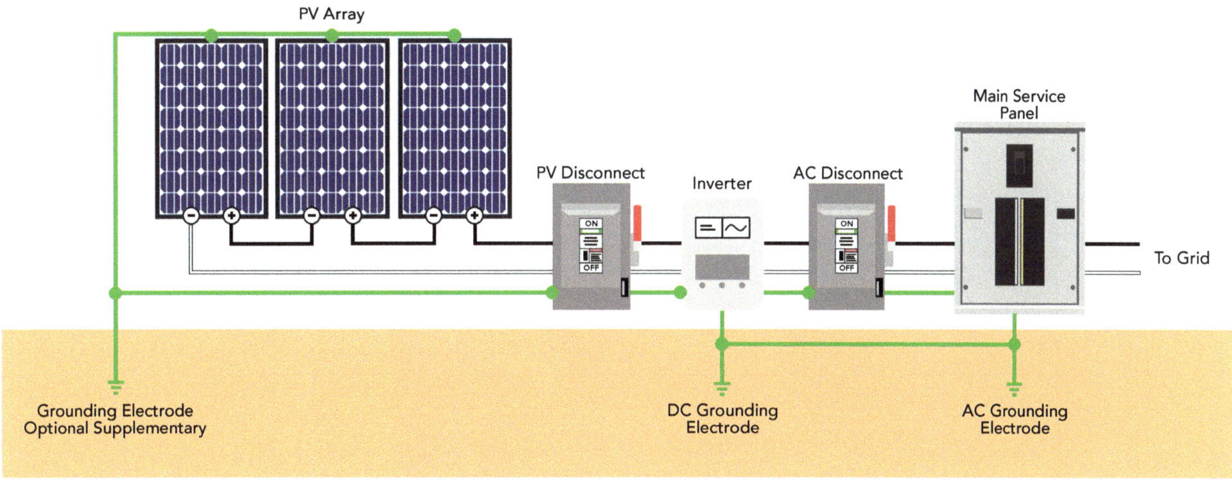


**Figure 13.29** Grounding is required on all electrical components so that people do not receive a shock and equipment is protected.

In the USA, there have been new code requirements to increase fire fighter safety on roofs with PV arrays. The problem is that the PV modules in a string don't stop producing just because the power is shut off at the main service panel. In fact, the voltage can go up, and remain live, because of the module just sitting in the sun ($V_{oc}$).

One requirement is for rapid shutdown of equipment so that when power is lost, or cut, voltage across the system will quickly drop to safer levels. This means that PV circuits located more than 3′ (1m) from the edges of the array must drop to 30V within 30 seconds while conductors inside that boundary are limited to not more than 80V within 30 seconds. If the PV array has no exposed wiring or conductive elements and is installed more than 8′ (2.4m) above the roof or grade then it doesn't have to meet the rapid shutdown requirements. Many new microinverters and power optimizers have rapid shutdown response built in.

Access for firefighters to roofs is important. Figure 13.30 shows the requirements per USA codes which require full access to 3′ (1m) along the entire ridge of a building with pitched roofs, as well as at least one 3′ (1m) clear access up to the ridge on both roof planes. If the modules cover more than 33% of the roof, then both sides of the ridge have to have 3′ (1m) clearances. All these access points must be able to support the weight of the firefighters. On gable roofs that may mean measuring from the bearing wall instead of the edge of the overhang, depending on how the overhang was framed. Modules may extend to the bottom edge of the roof. Modules cannot be placed where windows are used for escape in an emergency: you have to provide at least 3′ (1m) safe path from the window to the roof edge.

The requirements for a gable roof with a large PV array are shown in figure 13.31. Notice that 3′ (1m) access is

13

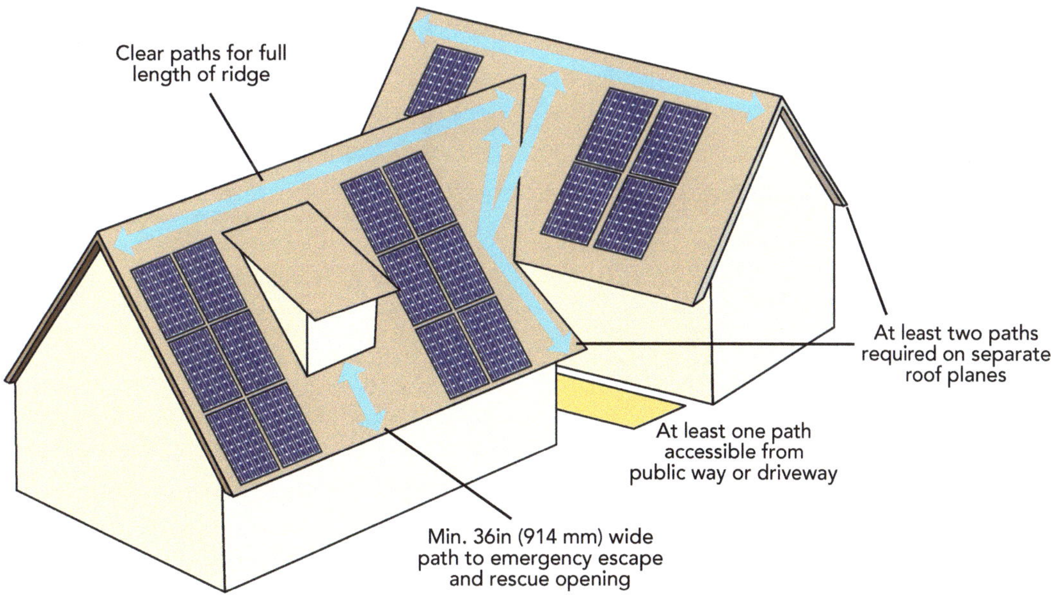


**Figure 13.30** Firefighting access, USA code requirements.

not needed at both gable ends, just one. It is required that access to the ridge is available from both roof planes. It may also be required that the areas under the 3′ wide roof fire access do not have windows or other openings.

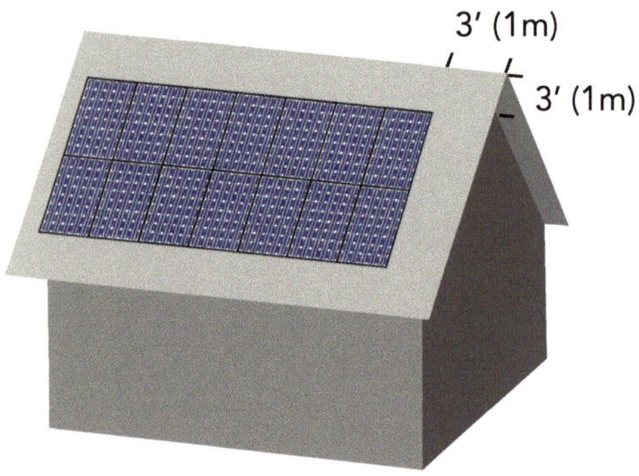


**Figure 13.31** Gable roof requires at least one edge to have 3′ (1m) access as well as 3′ (1m) clear access along the ridge.

Where there are valleys or hips, the 3′ (1m) access can be split in equal parts on either side as shown in figure 13.32 and figure 13.33. Note that the clear access must be maintained along the ridge and no part of the PV or racking system can extend into the accessible walkway.

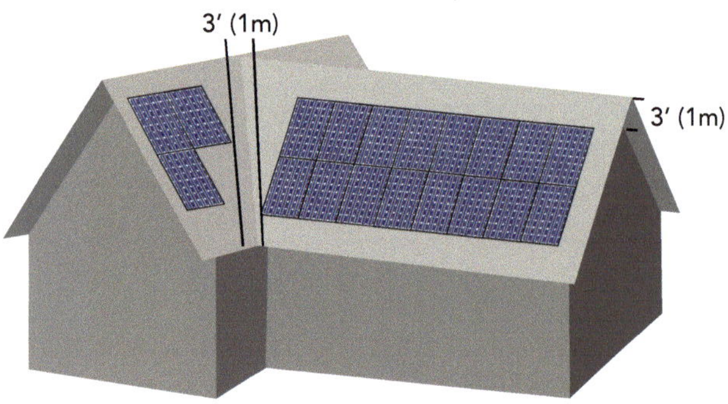


**Figure 13.32** Intersecting gable roofs create a valley and there must be 3′ (1m) clear at the valley, which can be provided on one side or equally split.

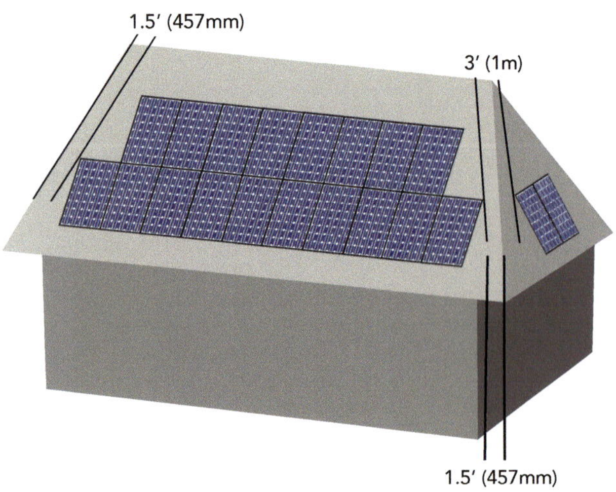


**Figure 13.33** Firefighter access on a hip roof.

For commercial roofs, where there can be great expanses of flat roof, there are basically three types of pathways required under the US Fire Code (2018, section 11.12). First, there has to be access to the roof via the perimeter of the building. Second, the firefighters have to be able to get around the roof once they are up there. These are defined as "other pathways". Third, there have to be pathways that provide access to smoke ventilation areas. All pathways have to be structurally sound to carry the weight of the firefighters.

Perimeter pathways depend on the size of the building. If the sides of the building are 250′ (76m) or less, then you need 4′ (1.2m) wide clear all around the edge of the roof. For buildings with lengths over 250′ (76m) the perimeter pathway has to be 6′ (1.8m) wide. All perimeter pathways have to be continuous around the outside edge of the roof.

"Other pathways" include straight line access to ventilation hatches and/or roof standpipes. These pathways must be at least 48″ (1.2m) wide. If one of these access walkways leads to a roof access hatch, 48″ (1.2m) clear must be provided around the hatch.

The last type of pathway splits the roof into module areas measuring a maximum of 150′ (46m) on either side as shown in figure 13.34. These areas must be separated by an open pathway at least 96″ (2438mm) wide. Depending on the size of your roof compared to the PV modules and how you have them racked, you may not be able to exactly fill 150′ (46m), any additional area would be added into the walkways.

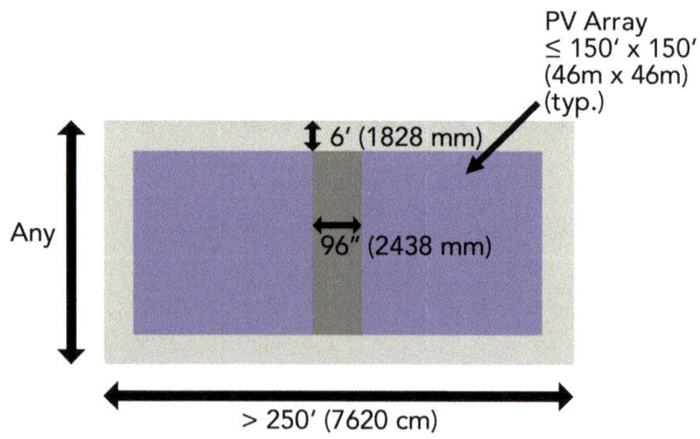


**Figure 13.34** Commercial roof split into 150′ squares for possible array area location.

The requirement for "other pathways" to have 48″ (1219mm) access to and around equipment on the roof might look like figure 13.35.

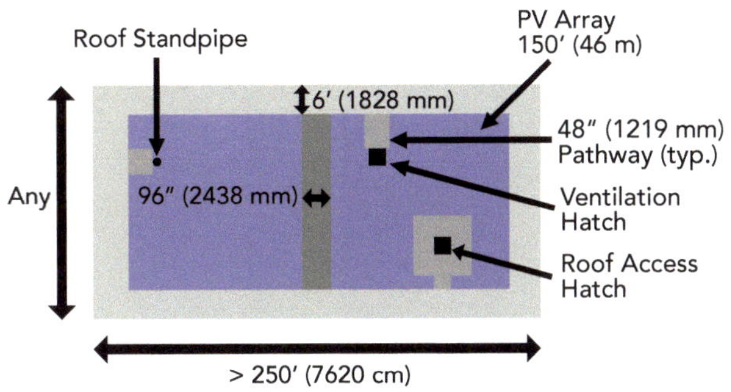


**Figure 13.35** Access pathways of 48" (1219mm) on commercial roofs within the 150'(46m) square array area to vents, hatches and other equipment.

If the designer puts skylights or dropout smoke or heat vents on at least one side of the pathway, then only 48' (1219mm) needs to be provided as shown in figure 13.36.

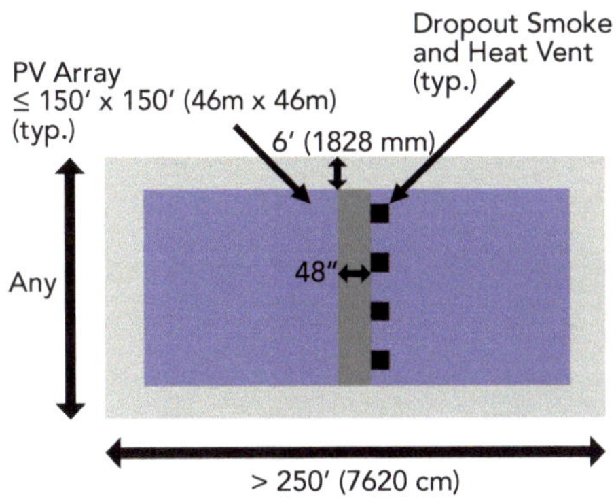


**Figure 13.36** Reduced width of access between array areas when skylights, dropout smoke or heat vents are on at least one side of the pathway.

## Putting the Whole Roof Plan Together

So now you've figured out how many modules need to fit on your roof. You know the size of the modules and the size of the roof. The next thing to do is to determine how much area you have available on the roof for the panels. You need to draw the roof to scale showing the location of mechanical equipment, skylights, and other systems that you need to work around.

If you are working on a sloped residential roof where the array modules will be racked parallel to the roof, be sure that you've drawn the actual size of the roof plane, as it is foreshortened in regular plan view.

Draw the location and width of required access points and pathways, egress from escape windows, maximum array sizes, and other pathways required to access and maintain rooftop equipment. If you are working on a large commercial roof, break it up into 150' × 150' (46m × 46m) maximum squares.

Modules should be aligned in equal rows on a roof. From a performance standard, it doesn't matter if the largest dimension is the width or height. It can make a difference how many modules can fit in an area, so be sure to check what works best. Remember that if the data sheet on a module says it is 19.45ft$^2$ (1.8m$^2$) you can't just divide that into 22,500ft$^2$ (2090m$^2$) (the maximum square area for a commercial roof). Assuming that you are mounting the modules at a 0° tilt angle (horizontal), your erroneous math would say that you can fit 1,156 modules (area of roof/area of module). But you can't change the shape of the module. According to the data sheet for the module you picked, it measures 63.8" × 43.9" (1621mm × 1116mm). Remember that the roof area for the modules is 150' × 150' (46m × 46m): the larger module length of 63.8" (1.62m) would fit 28 full modules (roof length/module length) while the smaller width of 43.9" (1.12m) would allow for 41 modules. That is 1,148 modules. If I have to provide walkways into the area to access roof items or provide fire access requirements, the plan has to remove full modules that are all in line with each other. Remember that the minimum access requirements set forth in codes are a minimum. Modules are the size they are: you can't switch to different sized modules at the edge of an array so they fit the roof better because they wouldn't be compatible with the rest of your array.

For flat roofs you may not want horizontal racking. If you place the modules on tilted racking facing the same direction, you have to take into account the inter-row spacing to figure out roof coverage (see chapter 5, Solar Insolation). If you do racking that alternates tilt, facing to the east and west for instance, there is no need for spacing between the rows because the modules don't shade one another, as you can see in figure 13.37. If you

13

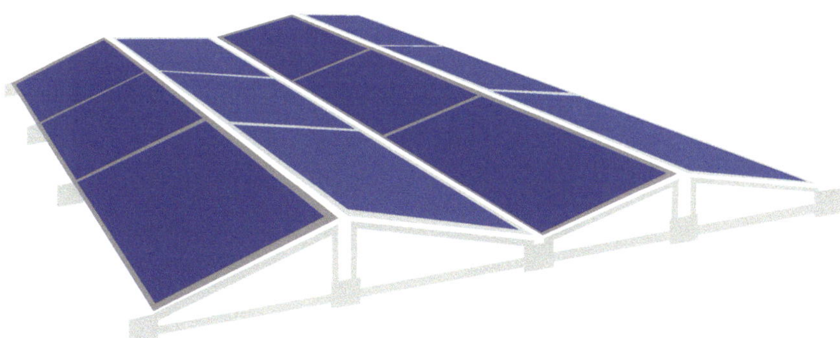

**Figure 13.37** East-west tilted racking on flat roof doesn't require inter-row spacing.

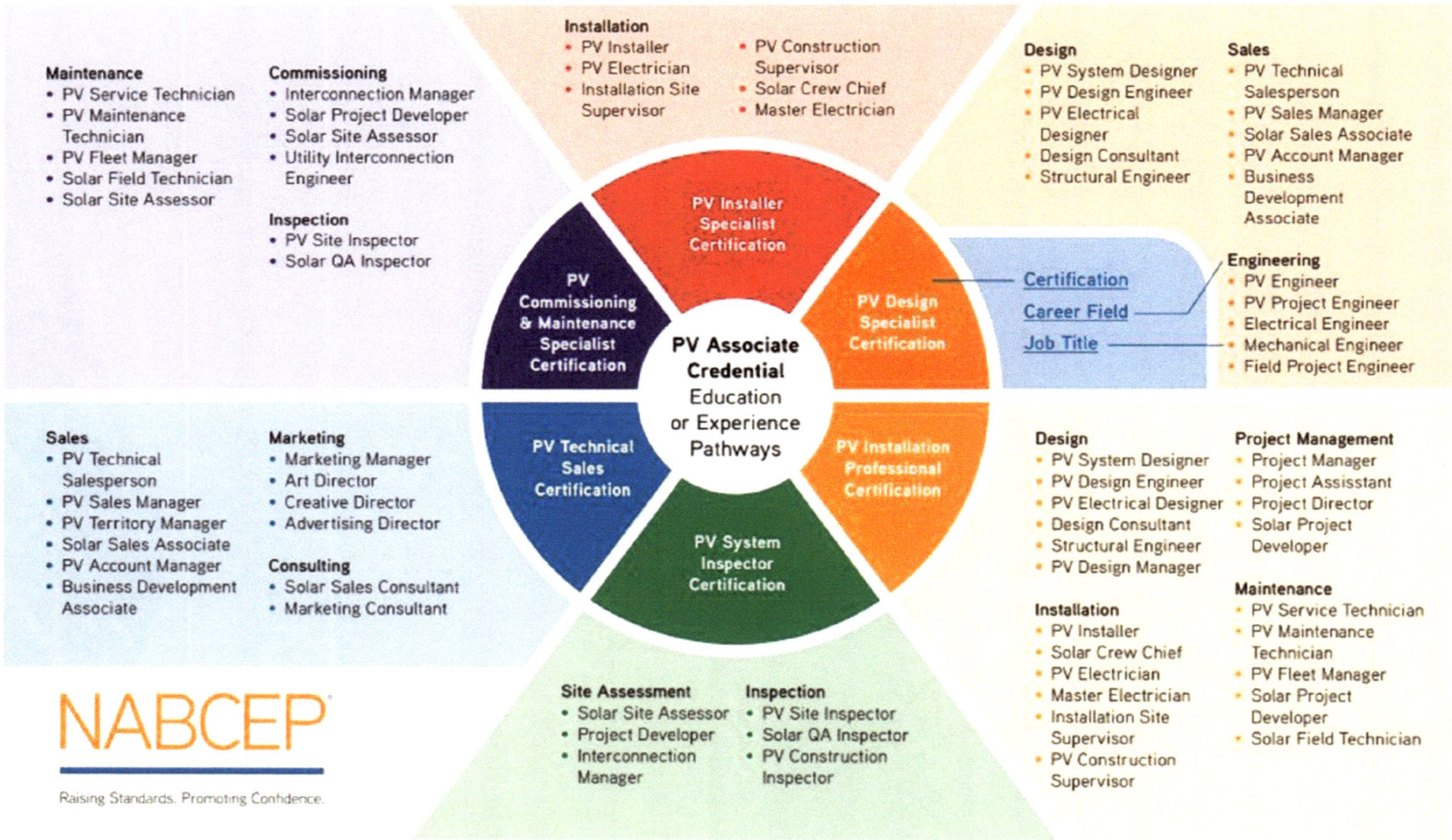


**Figure 13.38** NABCEP infographic showing the types of certifications in the center of the wheel and lists of the career fields and job titles within that specialty. Courtesy of NABCEP.[6]

do this, remember that you have to estimate production correctly based on two different tilt angles and correctly identify the number of modules facing either direction.

## Certification

The North American Board of Certified Energy Practitioners (NABCEP) is a certification program for people working in the PV and solar thermal marketplace. Certifications are available globally although the program started in the United States. They have eight board certifications, six in PV and two in solar thermal. PV certifications include installation, design, commissioning & maintenance, sales, and system inspector.
Figure 13.38 is an infographic from NABCEP that shows their certifications in a pie chart with the career field and job titles within those certifications radiating out from the center.[7] Each of these point to different skill sets within the PV industry. To earn NABCEP certification you must pass a rigorous exam. In order to qualify to take the exam, you have to have appropriate training and experience.

## Resources


**1** From *Photovoltaics and Electricity* [Clip art], by U.S. Energy Information Administration, March 26, 2021 (www.eia.gov/energyexplained/solar/photovoltaics-and-electricity.php).

**2** Solaria PowerXT® Solar Panels (www.solaria.com/powerxt-overview).

**3** Graphic based on the Solaria PowerXT® 370W module as shown in manufacturer data (www.solaria.com/powerxt-overview).

**4** Data generated from PVWatts® Calculator (https://pvwatts.nrel.gov/).

**5** U.S. Energy Information Administration. (2014, November 19). Solar photovoltaic output depends on orientation, tilt, and tracking. *Today in Energy*. www.eia.gov/todayinenergy/detail.php?id=18871

**6** Data generated from PVWatts® Calculator (https://pvwatts.nrel.gov/).

**7** North American Board of Certified Energy Practitioners. (n.d.). *Why get certified?* www.nabcep.org/certifications/why-get-certified/


## Study Questions

1. Describe how a PV cell creates and distributes energy.
2. Describe the five points on the I–V curve and explain what they mean. Answer the following questions:
   a. How are the numbers on the I–V curve determined?
   b. What impact does reducing the amount of solar insolation have on the I–V curve?
   c. What impact does increasing cell temperature have on the I–V curve?
3. Research online and find several modules of each type of PV cell. Compare their appearance, size, and performance information.
4. Describe the tracking options available and explain their pros and cons.
5. Explain why PV module production is impacted by shading.
6. What four items do you need to know in order to size your array? How do you determine the number of modules you'll need?
7. What are the reasons you might add batteries to a system?
8. Explain the difference between a string inverter, a power optimizer, and a microinverter.
9. Why is PV equipment grounded?
10. Describe the access pathways required on a residential roof.
11. Describe the access pathways required on a commercial roof.
12. Explain how you determine the exact number of a particular module that can fit in a roof area.

CHAPTER 14

# Racking for Solar Thermal and PV

**Key Terms**

Racking: elements that securely attach equipment, such as a solar thermal collector or PV module, to a building, pole, or tracker.

Standoff: a mounting element that holds a collector away from the roof or building, typically at least 2" off a roof. Equipment or rails can be attached to the standoff.

## Concepts for Success

Equipment needs to be securely attached to the building. PV and solar thermal collectors are usually mounted to rails. Each collector is supported by two rails. Rails have different profiles depending on the manufacturer and can be supported by a variety of elements. Residential sloped roofs use standoffs, which are usually attached with lag bolts to the rafters, to support racks.

Racking can also be achieved without penetrating a roof by using ballasts. Ballasts are usually solid concrete masonry units (CMUs) and weigh 60lb (27kg) or more. Ballasts are used on flat roofs in locations with winds below 125mph (55.8m/s).

Collectors can also be mounted on the ground using large custom racks or poles. The appearance and size of the mounts varies depending on the size of the array and if it tracks the sun daily and/or seasonally.

Integrating collectors with the building façade and as shading elements has a successful track record.

## Basic Parts

Anything you attach to your building needs to stay attached, no matter the weather or wind. Photovoltaic (PV) modules and solar thermal collectors are no exception. How you attach them will depend on whether they are roof mounted, ground mounted, pole mounted, or otherwise integrated with your building's design.

First and foremost, supports have to be appropriately attached to structurally sound elements. This can be your roof rafters or other structural framing. Anything that penetrates the roof needs to be properly detailed and sealed. Manufacturers make different standoffs for different types of roofing products and slopes.

If your roof framing is wood, then supports will be screwed directly into the rafters through the roof deck with lag bolts. The exact size of the lag bolts and the amount of penetration can vary depending on the wind loading, type of wood, and spacing of standoffs. Installation must meet jurisdictional requirements. Holes through the roofing product need to be properly sealed and flashed so the roof doesn't leak.

Standoffs are the part that is actually attached to the roof. On the top, standoffs can have a single post or have options for L- or U-shaped brackets for attaching single or double rails. The advantage of double rail standoffs is that it reduces the number of standoffs needed on the roof, which means fewer penetrations. Depending on the manufacturer and style, standoffs can have different appearances. Some have cone-shaped flashing plates (figure 14.1).

## Purpose of Racks and Standoffs

Rails are commonly used to mount photovoltaic (PV) modules and solar thermal collectors. PV and collectors are attached to pairs of rails. Rails are long and bar-like, the cross-section varies by manufacturer. Rails should

DOI: 10.1201/9781003297010-16

**Figure 14.1** Cone-shaped single standoffs with rails mounted on a steep slope asphalt shingle roof.

never be directly touching the roof because water and debris (e.g. leaves and dirt) would get caught behind them and create issues for the performance and longevity of the roof. Standoffs hold the racks above the roof and are connected to the building structure. The height of standoffs can vary greatly, with minimum post or bracket heights of as low as 2" (51mm) clear between the rails and roof. Standoffs are sold separately from rails. Note that local jurisdictions may have special requirements due to wind and water issues.

The rails are spaced so that there is one rail near the top and one rail near the bottom of the equipment. PV and collectors have bolted mounting attachments which connect them to the rail. The advantage of using rails is that you can attach multiple pieces of equipment while creating the fewest number of standoff penetrations through the roof.

To further minimize roof penetrations, installers need to balance the module size and the standoff locations. In figure 14.2, you can see that the rail span is measured as the centerline distance between standoffs and the cantilever is from the edge of a module to the standoff.

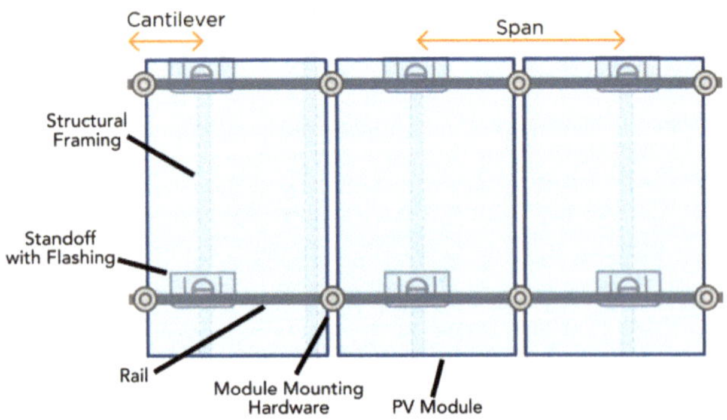


**Figure 14.2** Plan equipment layout for fewest roof penetrations.[1]

Typically, the span is 48" (1219mm) in the USA where wooden residential roof framing is 16" (406mm) or 24" (610mm) on center. Rail system manufacturers have different allowable cantilever lengths based on the design of the rail. These are usually about ⅓ of the span. Modules are all different sizes and can be mounted either vertically (portrait) or horizontally (landscape) on the racks. Modules may have specific information about their mounting requirements.

## Appearance and Installation of Mounts

Standoff mounts are made to be compatible with many different roofing materials for steep, low, and flat roofs. Figure 14.3 shows an example of Quick Mount PV's mounts and flashing products that they make for tile roofs in flat, S, and W shaped tiles.

**Figure 14.3** Tile replacement mounts for flat, S, and W shaped tiles. Quick Mount PV.[2]

The flashing and post attachment are connected to the roof structure and integrated with the roof water barrier, lapping all elements downhill (figure 14.4). For this Quick Mount product, the post is located by sliding a bolt head into the slot and tightening it down.

These same types of mounts can be used to support solar thermal collectors (figure 14.5).

Composition or asphalt shingle roofs are very popular in the USA residential market. The simplest flashing units slide over a course and up under the next courses, leaving about a finger width of space from the flashing to the front edge of the shingle. It is important that flashing doesn't extend in front of the shingle edge as this can trap debris and cause shingles to decompose (figure 14.6). If shingles need to be cut to accommodate

**Figure 14.4** Roof mount is flashed and layered under waterproofing under the roofing tile. Quick Mount PV.[2]

the shape of the standoff, there should be space all around the standoff to allow for proper drainage, as seen in figures 14.1 and 14.6.

Figure 14.7 shows a metal south-facing pitched roof with both PV and solar thermal. This was Middlebury College's 2011 entry in the Solar Decathlon.

**Figure 14.5** Solar thermal collectors installed on steep tile roof in England.

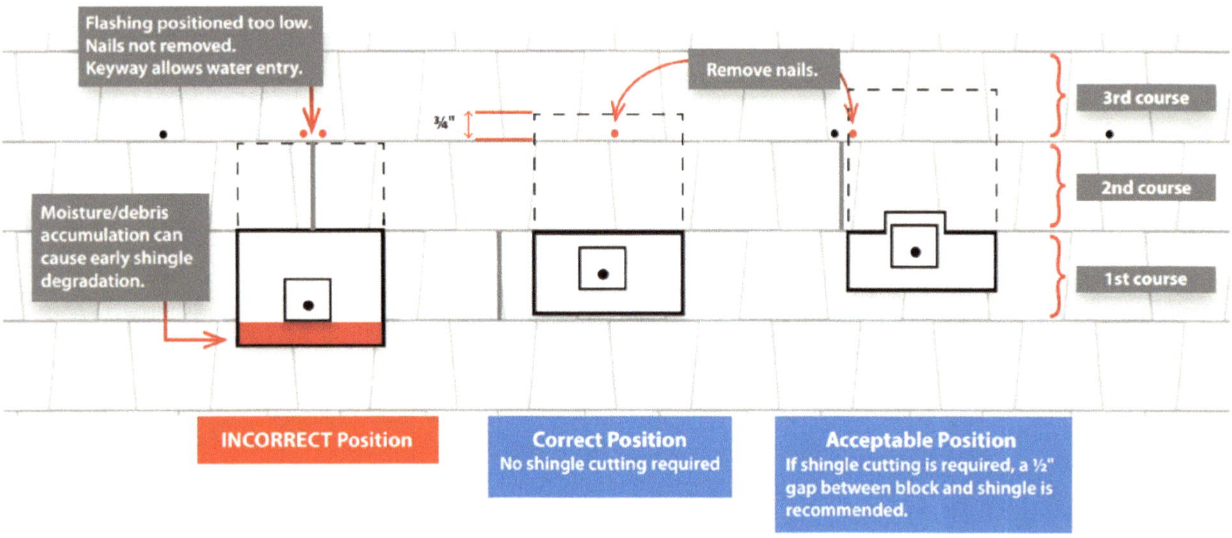


**Figure 14.6** Proper spacing and installation of flashing and standoff on a composition or asphalt shingle roof. Quick Mount PV L-Mount.[3]

**Figure 14.7** PV and solar thermal mounted on a metal roof.

**Figure 14.8** T-shaped rail attached to standoff.

**Figure 14.9** U-shaped rail bolted to standoff through bottom of rail.

## Rails

Rail profiles vary by manufacturer but they all securely attach to the standoff and allow for equipment attachment. Figure 14.8 shows a T-shaped rail bolted with a U-shaped bracket that tightens down with nuts on the standoff bolt. Equipment is secured by bolting through oval slots – one is seen to the upper right of the bolt.

U-shaped rails attach to standoffs by bolting through the bottom of the rail. Equipment is clamped in place with sliding mounts that attach to the top edges of the rail as seen in figure 14.9.

Some rail profiles allow for sliding bolt heads on all four sides of the rail (figure 14.10).

**Figure 14.10** Rail profile that allows for sliding bolt heads to secure equipment.

## Flat Roofs

Flat roofs either need to have standoffs that have been appropriately detailed and sealed for the roofing type or be ballasted. Ballasting is only appropriate for lower wind regions. Solid CMU blocks are used as anchors for bolts or as weight on stands. There must be enough weight on the supports to keep the equipment from being lifted off the roof. Be sure to check your jurisdiction to ensure you are allowed to ballast equipment. Figure 14.11 shows ballasted PV system on sleepers on a flat roof.

**Figure 14.11** Penn College architecture students check out the ballasted PV array on the East Lycoming School District's roof, Hughesville, PA, USA.

In figure 14.12, you can see that the vertical stands have flanges that extend on both sides and the ballasts are sitting on top of the flanges. Note that the array has two levels of rails at 90° to each other and the PV modules are mounted with their widest dimension horizontally. This results in a very low array profile that is hard to see from street level.

Because leaking roofs are a big problem, figuring out a way to integrate the racking system with a flat roof is important. Aim for making the fewest possible holes. If you have a narrow rooftop, you may be able to span it using the parapet structure. You may want to consider creating usable rooftop spaces where the structural support can be integrated with the building (figure 14.13).



## Penetration for Plumbing

There are many types of boots (flashing) that can be used to seal the penetrations through a roof made by

**Figure 14.12** The ballasts are holding the array in place because they rest on top of flanges that come out on both sides of the vertical stands.

**Figure 14.13** Usable space created under a PV array.

**Figure 14.14** Boot for sealing piping penetrating an asphalt roof (uninstalled).

piping. Figure 14.14 shows a boot that is a standard square base pipe flashing for an asphalt roof, prior to installation. This one has stretchy rubber that seals around the pipe. Pipe flashings are available for all types of flat, medium, and steeply pitched roofs for all roofing materials. They need to be installed per the manufacturer's recommendation to be under warranty. Some have flashing similar to standoffs and may require shingles to be trimmed. Solar thermal piping will need to penetrate the building envelope (roof or wall) and be coordinated with racking and other roof equipment.

## Ground Mount

Ground mounting PV and solar thermal collectors can be helpful if access is needed for cleaning off snow. Very large arrays are often ground mounted. Ground mounted arrays need to be located where the general public can't interact with them or the site should be fenced off. All elements supporting the array need to be sturdy and durable – they should last at least as long as the PV modules (20–25 years). Figure 14.15 shows part of a 600kW array mounted to a steel structure. Planting the area with a low height grass helps to keep the array cool, chokes out weeds, and doesn't shade the modules.

Shutoffs and other equipment can be mounted to the structure (figure 14.16). Notice that there is diagonal bracing between some of the posts to provide lateral stability. Posts can be installed on round concrete footings that extend below the frostline. Posts can also be driven into the ground or connected to a ground screw (sometimes also called earth screw). Ground screws are used with round pipes and speed up construction time.

## Pole Mounts and Trackers

Pole mounting is done to reduce equipment footprint, get equipment higher in the air, or have solar tracking.

Fixed tilt arrays often have simple poles and pipes that racks are clamped to (figure 14.17). They can be spaced close together because they don't shade each other.

**Figure 14.15** Ground mounted array with a steel structure.

Figure 14.16 Steel framework for ground mounted array is bolted together and has cross bracing in every other post bay for lateral stability.

Figure 14.17 Pole mounted fixed tilt array.

Figure 14.18 Tracking arrays have to be very sturdy to support changing balance due to array movement.



Pole mounted trackers can be motorized or not and track daily and seasonally. The structure has to be sturdy (figure 14.18). Poles have to be far enough apart that multiple arrays won't shade each other.

Ground mounted trackers can be more economical than pole mounted ones and all the moving parts are at grade (figure 14.19). They always have round foundation pads and turn on rails or wheels.

There are many types of roof mounted trackers, they typically tilt seasonally (vertically) in order to maximize array roof space and prevent shading.

## Solar Thermal

Because the best tilt angle for solar thermal is steeper than PV, they may be mounted on adjustable racks that are attached to rails or other supports. Figure 14.20 shows an adjustable rack for a tube collector. The rack has cross bracing on the tall (back) side. The feet on the bottom can be attached to rails or ballast blocks (figure 14.21). Coordinate the location of solar thermal collectors on adjustable racks so they don't shade PV modules. Remember that foundations need to meet local requirements and that ballasts

**Figure 14.19** Ground mounted trackers rotate the entire structure to follow the sun.

**Figure 14.20** An adjustable angle rack for a solar thermal tube collector.

**Figure 14.21** Adjustable angle rack ground mounted to block ballast.

have to keep the unit from blowing away (ground or roof mount).

## Building Shading

Arrays can also serve as shading devices. Jurisdictional requirements for roof access must be followed, or large arrays must be installed at least 8′ (2.43m) above the roof. Figure 14.22 shows large, sloped arrays extending over building roofs and balcony spaces. Support for the left array is from the building's roof, while the right array also has vertical supports at the balcony.

PV can also provide shading as awnings with proper attention to supporting/attaching the array to the building/glazing system (figure 14.23).

The Heliotrope® designed by Rolf Disch has a large array supported on a central pole by a space frame (figure 14.24). The railings are also plumbed solar thermal collectors.

**Figure 14.22** Architect Rolf Disch, Das Sonnenschiff, large arrays provide shading and produce 4× the energy the building consumes, Freiburg, Germany.

**Figure 14.23** Entry façade shading at Solar-Fabrik Freiburg im Breisgau, Germany.

**Figure 14.24** Architect Rolf Disch, Heliotrope® first Plusenergiehaus® worldwide, Freiburg, Germany.[4]

## Resources


1 IronRidge. (2018). *The advantage of strength: Maximizing cantilevers*. https://files.ironridge.com/education-library/Tech_Tip_Maximizing_Rail_Cantilever.pdf

2 Quick Mount PV. (2020, January). *Tile replacement mount: Installation instructions*. www.quickmountpv.com/downloads/datasheet/QMPV-datasheet-QMTRM-TileReplacement-web.pdf

3 Quick Mount PV. (2019, April). *L-Mount | QMLM / QMLM-ST*. www.quickmountpv.com/downloads/datasheet/QMPV-datasheet-QMLM-LM2.LMount-web.pdf

4 Rolf Disch Solar Architektur. (2021). *The Heliotrope*. www.rolfdisch.de/en/projects/das-heliotrop-2/


## Study Questions

1. Go online and search "PV module racking system". Compare the types of racking systems in the image results for various products: ground mounted, pole mounted, ballasted (flat roof), and mounting for steeper roofs of different materials. Click through the links for a couple of the manufacturers and check out their products in more detail. Explore the versatility of the product and understand how it would be used in a project.
2. Go online and search "PV array building shading system" and "building integrated PV" to see different ideas for how PV arrays can be used on the building façade to both collect and shade.

# CHAPTER 15

# Wind

> **Key Terms**
>
> Anemometer: an instrument that measures wind speed and (sometimes) direction, often using a series of cups that rotate in the wind to gauge its speed.
>
> Nacelle: a streamlined housing that contains the interior workings of a wind turbine.
>
> Turbulence: strong, disruptive, swirling motion of the wind current caused by objects in the environment (buildings, trees, etc.) blocking the wind.
>
> Yaw: rotation around a vertical axis, the nacelle yaws to orient the turbine blades into the wind.

## Concepts for Success

Wind is caused by the uneven heating of the earth's atmosphere and has been used for centuries to power boats and machines. Wind patterns are predictable and are generally stronger in winter than summer months. The units used to measure wind speed are miles per hour (mph) or meters per second (m/s).

Wind turbines generate electricity because the wind speed can cause the turbine blades to rotate, similar to the lift on an airplane wing that raises the plane into the air. For reliable electricity production the turbine needs to be located where there is a regular wind resource of at least 10–12mph (4.47–5.36m/s). Greater wind speeds have the ability to produce more energy. Turbines need to be located above and away from turbulence caused by obstacles such as trees and buildings.

There are generally three categories of sizes for wind turbines, based on how much energy they make and how large the machine is: small wind (for a single property), community (for a small number of properties), and utility (large-scale production).

Most turbines have a tower to get the blades and **nacelle** (where the generator is located) high enough that it is located in a good, steady wind resource. Tower heights for small wind may be as low as 35′–100′ (11–30m) while utility towers can be over 260′ (79m) tall. Towers for smaller wind have a variety of configurations while large wind always have a self-supporting single pole. Maximum tower heights may be regulated by local or regional zoning requirements.

There are two general categories of wind turbines depending on the orientation of the rotor axis. **Vertical axis wind turbines (VAWTs)** have the rotor shaft oriented vertically while **horizontal axis wind turbines (HAWTs)** have the rotor axis oriented horizontally. There are different blade configurations and orientations in both categories. HAWTs have been much more commercially viable than VAWTs.

The amount of power in the wind is based on the wind speed and the efficiency of the turbine to convert the kinetic energy into electricity. The theoretical limit is 59%, known as the Betz limit. Most modern wind turbines operate in the 35–45% efficiency range, however, some VAWTs and small turbines can perform at 15–20%. There is a mathematical formula for determining the power available in the wind based on the wind speed and swept area of the blades (the circle defined by the blades when rotating). Annual energy output (AEO) of a turbine is typically listed on the manufacturer's cut sheet based on annual average wind speeds.

## Capturing Wind Energy

Wind energy has been used for many centuries by humans to power sailboats and windmills. Any kind of wind machine uses the kinetic (moving) energy of the wind and puts it to work. A wind system includes the wind machine, the tower and all the required equipment. Windmills are machines that generate mechanical motion to do work. The wind turns the sails or blades of the windmill and rotates a shaft. Windmills were (and still are) used for pumping water and are

DOI: 10.1201/9781003297819-17

sometimes called windpumps. Windmills ran all kinds of machinery during the industrial revolution, including grist mills, sawmills, and papermills. Wind and water powered all kinds of processes, however, eventually most of these were replaced by the steam engine, which could be built and run anywhere.

Wind turbines produce electricity from the kinetic energy of the wind. Wind turbines were used in the 1930s in the USA to power farms. Investment and development of fossil fuels stifled the use of wind turbines. It wasn't until the 1970s that wind power gained popularity and governments around the world started providing incentives for wind turbine development.

In order to have effective power production, there has to be a steady wind resource that is fairly reliable (blows most of the time) and is not very turbulent. Turbulence is the swirling, unsteady movement of the wind caused downwind by natural land formations, trees, buildings, and other obstacles. Turbulent wind doesn't provide steady, linear pressure to rotate the turbine's blades. Typically, the higher a turbine is located off the ground, the farther away it is from the objects that cause turbulence, so tall towers have access to better, less turbulent wind resources. Wind maps are available worldwide, see chapter 3 on Locating Renewable Resources. Map information is always reported based on a hub height above the ground.

Wind turbines are classified based on the wind speed they are designed for. The classes are I high wind (10m/s, 22mph), II medium wind (8.5m/s, 19mph), and III low wind (7.5m/s, 17mph). Each classification has three letters indicating the turbulence limits at that wind speed: A (16%), B (14%), and C (12%).

Today wind turbines come in a variety of sizes. The larger the turbine, the more power it can produce and the taller tower it is on. See figure 15.1 for a comparison of size. Smaller turbines are used for a single home or farm and are rated at 10kW or less. Turbines for powering small communities are rated between 10–250kW. Utility scale turbines are the largest and can be located on- or offshore and are rated between 250kW–5MW.

The basic parts of a wind turbine are shown in figure 15.2. Community and utility sized turbines always use tubular freestanding towers with gearing to rotate the nacelle so that the nose is always facing into the wind. The rotational movement of the nacelle is called **yaw**, which is the same term used to describe the rotation around a vertical axis for boats and airplanes.

Figure 15.3 shows common components of a wind turbine generator. These vary depending on the size and make of the turbine. The blades catch the wind and turn the rotor. The pitch system rotates the blades to alter their orientation to the wind. The turning of the rotor hub causes the generator to spin and creates

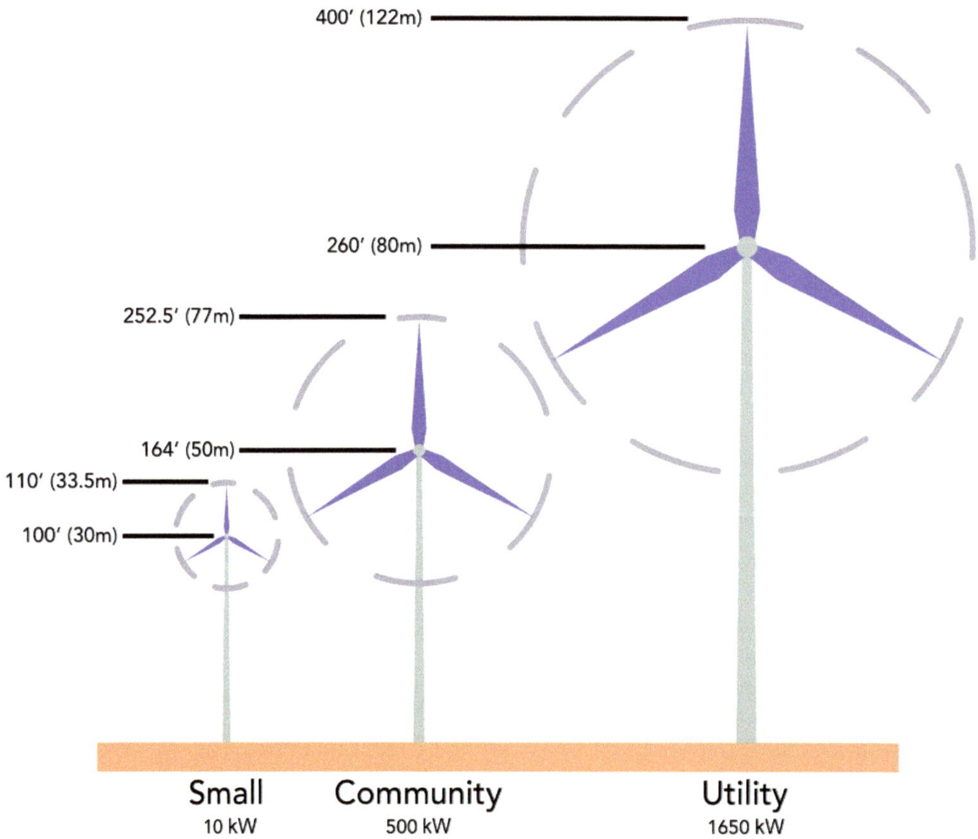


Figure 15.1 Wind turbine size and height comparisons.

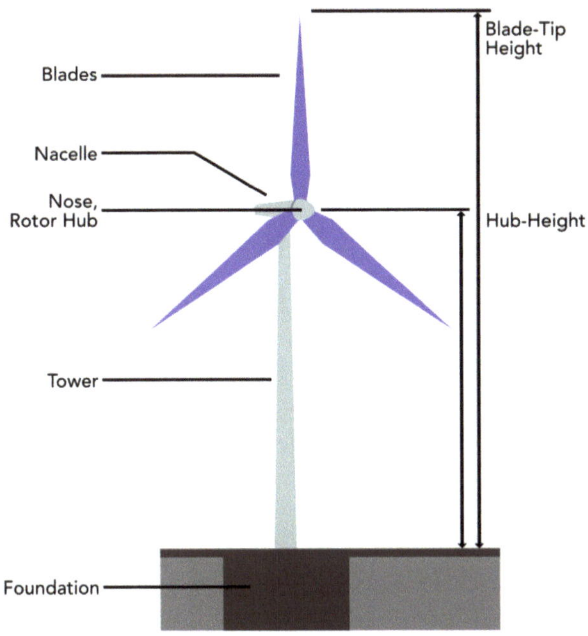


Figure 15.2 Basic parts of a wind turbine.

electricity. To keep the rotor spinning at a relatively slow speed, gearing is used between the rotor drive shaft and the generator to increase the rotational speed of the generator. The brakes stop the rotation of the shafts, which keeps the rotor from spinning. The controller is connected to an **anemometer** which checks the direction and speed of the wind. The controller controls the braking and yaw of the turbine.

Blades are shaped to be aerodynamic and they catch the wind in a similar fashion to airplane wings. The leading edge of the blade is rounder and thicker while the trailing edge is thinner. As the wind passes over the blade, lift is created on the underside. Figure 15.4 shows a cross-sectional cut of an older model Bergey blade. Manufacturers have different designs for blades which vary depending on the type and size of turbine. Blades are designed to increase lift and reduce drag. The lifting action is what causes the blades to move and then turn the rotor hub.

Figure 15.4 Cross section of a blade, leading edge at the bottom of the image.

There can be a pitch system that rotates the blades in place on the rotor hub to change their pitch in order to slow or speed up the overall turning speed of the rotor. The change in pitch increases or decreases the amount of drag. Figure 15.5 shows the pitch mechanism for a Xzeres Skystream turbine.

There is a big difference between small and utility scale wind turbines because of their size. Training and safety are very different, as well as pole and foundation requirements. It takes large crews, heavy equipment, and big cranes to install utility scale turbines because of their very tall tower heights, large blades, and the weight of the nacelle.[2] Smaller turbines have a wider range of options for the type of tower and turbine. They can be installed by small crews using typical residential construction equipment.

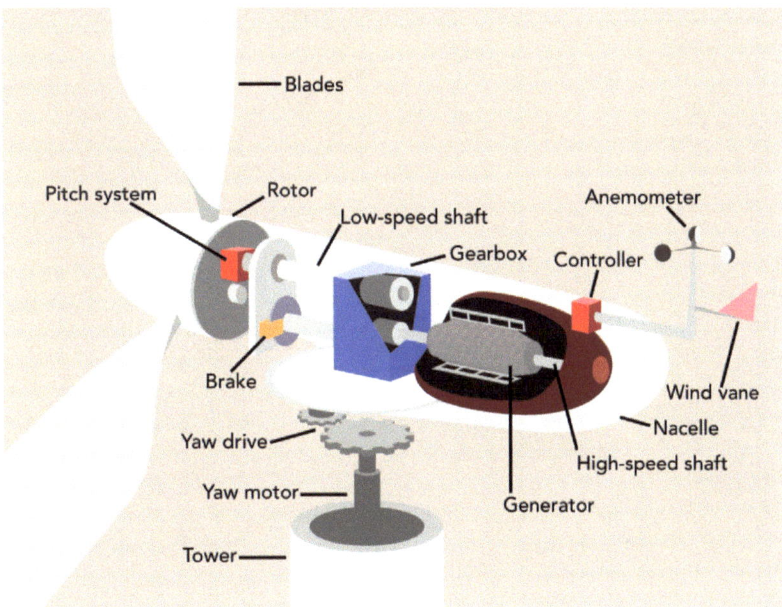


Figure 15.3 Common wind turbine components.[1]

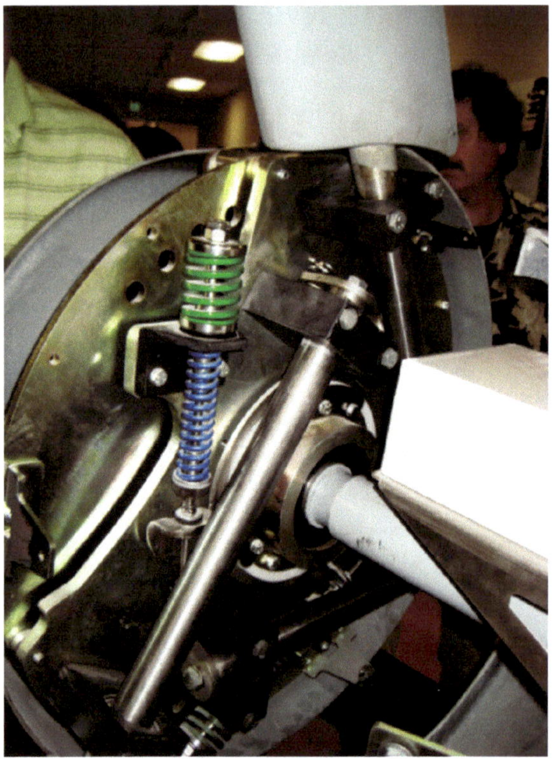

Figure 15.5 Blade pitch mechanism. Photograph by author.

## Categories and Types of Wind Turbines

There are two general categories of wind turbines depending on whether the rotor is orientated vertically or horizontally.

### Vertical Axis Wind Turbines

Vertical axis wind turbines have the rotor shaft oriented vertically. They come in two types: Savonius (drag) and Darrieus (lift).

#### Savonius (Drag)

Drag type VAWTs are named after the inventor Savonius who designed two cups arranged in an S-curve. Today Savonius VAWTs can have S-curved blades with a helical spiral (figure 15.6, left) or two or more parallel cupped blades (figure 15.6, right). These turbines turn because they are being pushed by the wind, as opposed to developing lift, and are therefore called drag devices. At best, these types of turbines can capture about 15% of the available wind energy.

#### Darrieus (Lift)

The other type of VAWT is named after the inventor Darrieus and is a lift device. It relies on blades that are either curved and attached to the top and bottom of the rotor shaft (figure 15.7, left) or in an H configuration extending from the rotor (figure 15.7, right). Blades can be straight or in a vertical helix. The H configuration can have any number of blades. The curved blade Darrieus VAWT has 2 or 3 blades and is generally quite large, while the H-type can vary from small to medium diameters.

#### VAWT Advantages and Disadvantages

The advantage of VAWTs are that they are omnidirectional, so they do not have to turn to face the

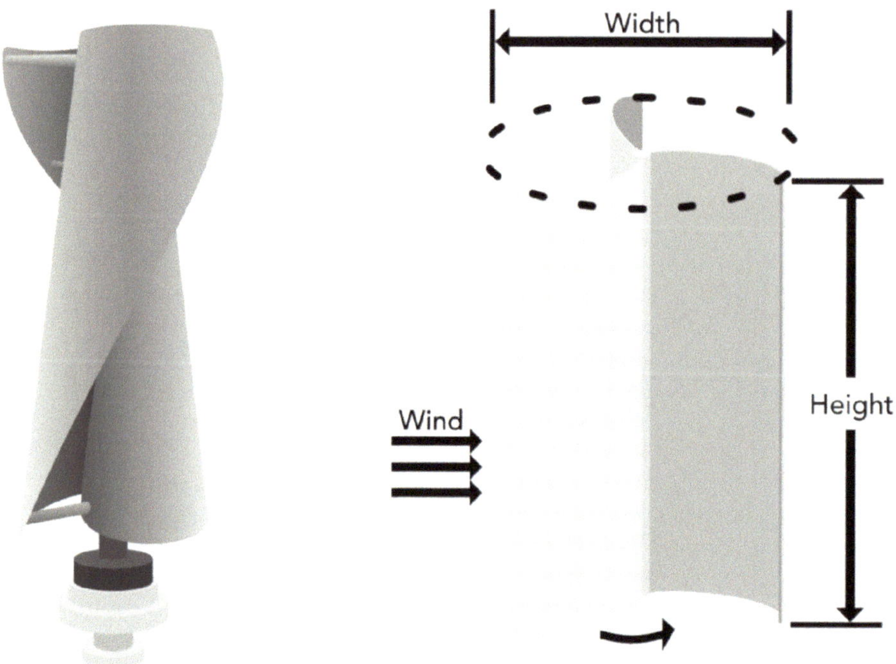


Figure 15.6 Savonius drag type VAWT. Left: 2 S-curved blades with a helical spiral. Right: 3 parallel cupped blades catch the wind, causing rotation.

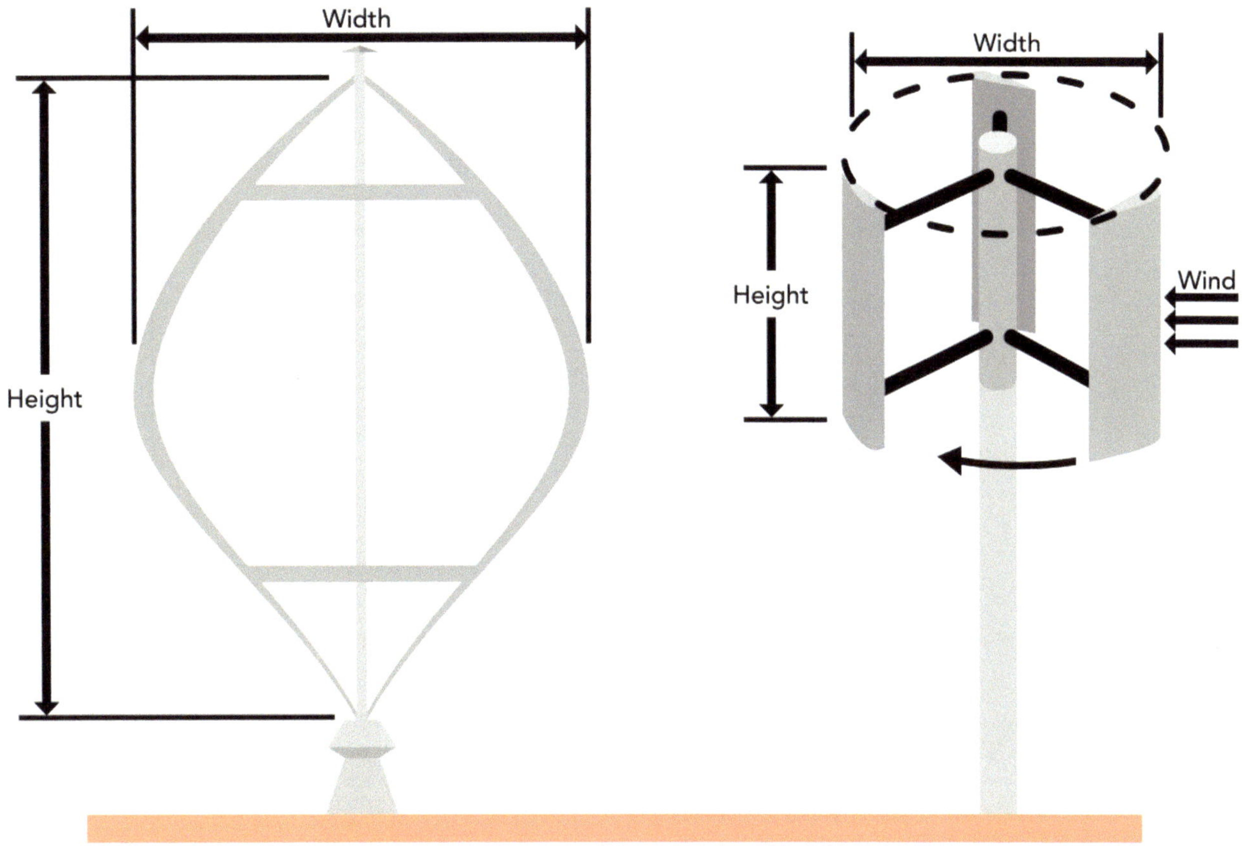


Figure 15.7 Darrieus type VAWTs.

wind. The major components can be located closer to the ground with the Darrieus curved blades, where it is easier to access. Because VAWTs have the rotor oriented vertically, they may be able to have lighter weight towers. Theoretically, VAWTs can generate electricity using less materials than horizontally oriented turbines.

The disadvantages of VAWTs are many. The large Darrieus curved blades have the rotor near the ground where wind has more turbulence; they require bracing at the top; there have been issues with the blades failing because of the centrifugal force stresses; and the entire rotor has to be removed to replace the bearings. All VAWTs have poor self-starting capabilities because of higher drag forces. In general, VAWTs have not been particularly commercially successful and have had issues with poor performance and reliability. Paul Gipe,[3] one of the best-known authors and general gurus of wind in the modern world, has published many articles concerning VAWT performance. There are often new VAWT designs that are touted as a revolutionary breakthrough that outperforms everything else and can be installed on a roof or in a city location where the wind has a lot of turbulence. When evaluating these claims, remember that wind turbines only work well when they have a good resource; they vibrate a lot because they are moving and these vibrations (and noise) can be transferred through the building's structure, which can be irritating to occupants; and finally, if something sounds too good to be true, it probably is.

## Horizontal Axis Wind Turbines

Horizontal axis wind turbines (HAWTs) have the rotor axis oriented horizontally. Figure 15.8 shows the two types of HAWT based on blade orientation: into the wind (upwind, shown on right) or away from the wind (downwind, shown on the left). Smaller upwind HAWTs have a tail vane. The tail acts like a rudder on a boat and keeps the blades facing the wind. Larger upwind HAWTs do not have tails and have yaw motors which are operated by the controller (see figure 15.3). Note that the nose of the upwind HAWT is facing toward the wind direction. Downwind HAWTs use the blades as a rudder to stay oriented to the wind. Only smaller turbines are designed this way. Note that the nose of the downwind HAWT is facing away from the wind direction. HAWTs can have any number of blades, although three is probably most common. Manufacturers have distinctive blade and tail designs.

HAWTs have been commercially successful. The biggest challenges are in how to mount and access the nacelle on the tower, as well as keeping the blades far enough away from the tower that they don't create drag and slow down as they rotate.

It is important that the rotor maintains a consistent turning speed once it has reached rated power. Turbines

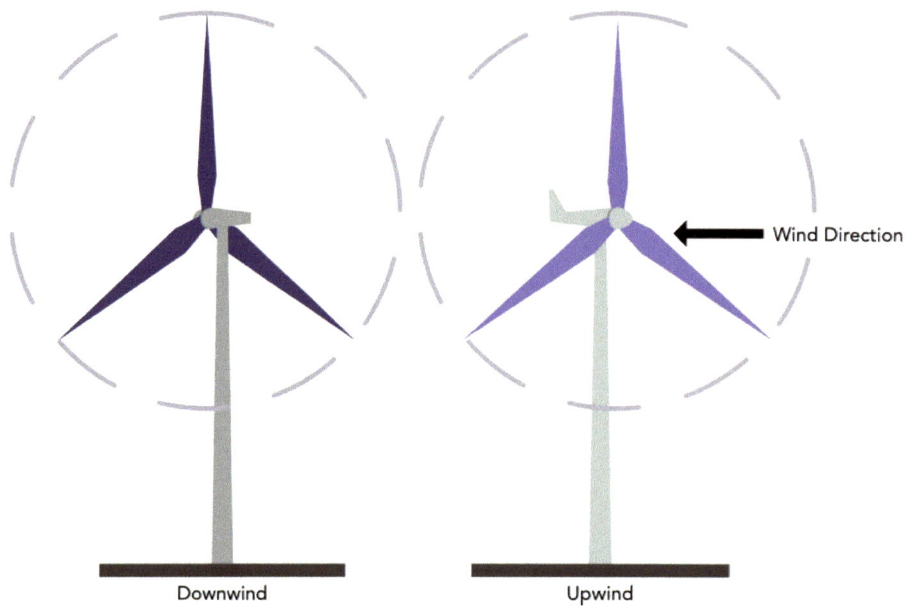


Figure 15.8 HAWT downwind (left) and upwind (right) blade orientation.

may have a maximum wind speed in which they can operate without damage. They will stop in order to protect the machinery. Smaller upwind HAWTs may use furling to slow down and stop blade rotation instead of having a braking system. Furling means that the nacelle folds towards the tail, which moves the blades out of the direct wind and slows them down. Figure 15.9 shows what this looks like. In the left image, the wind is blowing normally and the tail is parallel to the wind, keeping the blades perpendicular to the wind. In the center image, the wind is stronger and the nacelle is beginning to fold toward the tail. This moves the blades at an angle to the wind, which creates drag and slows down their rotation, keeping the blades spinning at a consistent speed. In the right image, the wind has become too strong for the rating of the device and the nacelle has folded against the tail. This completely stops the blades from rotating. The machine can be manually furled in order to stop it so maintenance can be performed.

## Tower Types

Wind towers are either freestanding or guyed, as shown in figure 15.10. All towers are either a lattice or monopole design. Guyed towers are the less expensive than freestanding towers. Lattice towers are made of diagonally braced lattice sections made of pipe or tubing, while monopole towers are a single round shape. Utility scale turbines are always the monopole freestanding type.

For smaller wind turbines, it can be easier to have a tilt-up tower so that maintenance can be done on the ground, rather than climbing up to the nacelle. They can also be lowered to prevent damage from extreme weather events such as hurricanes. Tilt-up towers

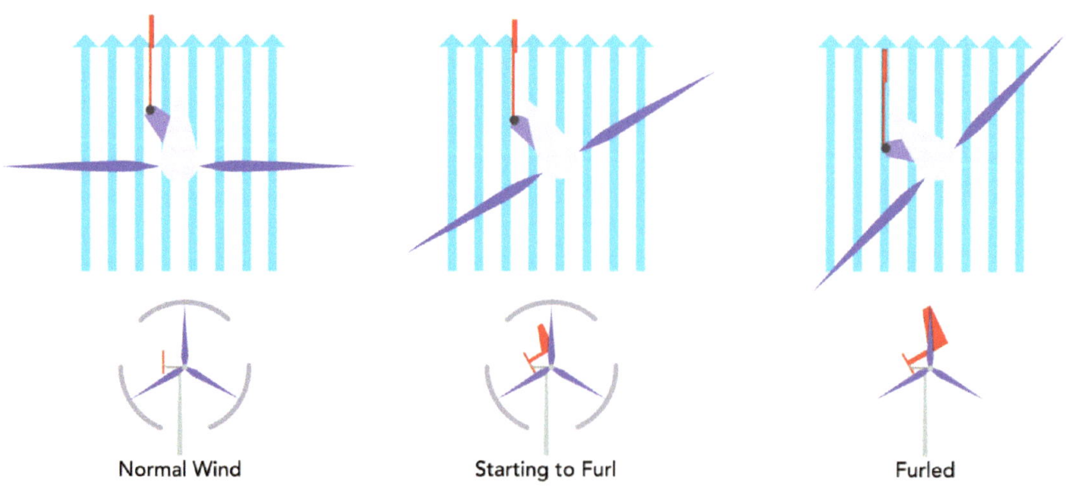


Figure 15.9 HAWT furling to slow and stop blade rotation.

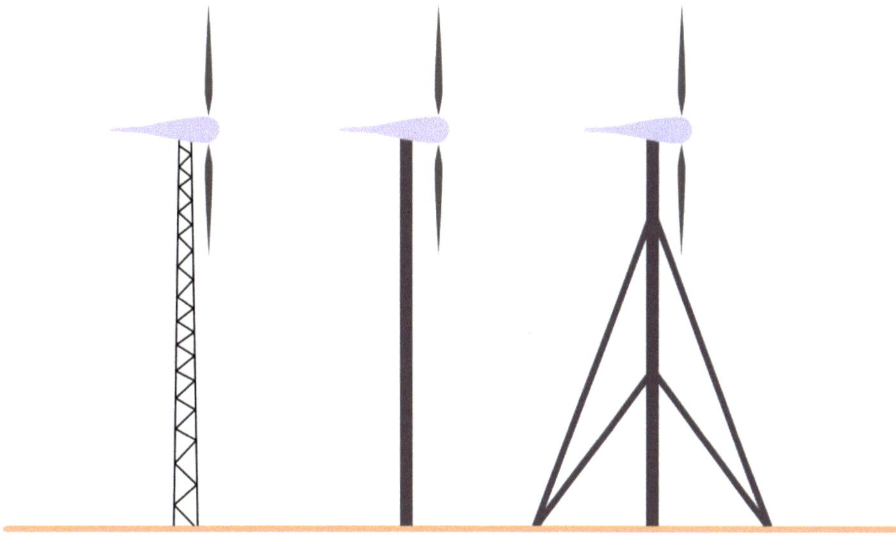


Figure 15.10 Towers are either freestanding or guyed.

can be freestanding or guyed and use either lattice or monopole towers. Figure 15.11 shows a guyed monopole tilt-up tower in elevation: the left image is when it is lowered to the ground; the right image is when it is vertical.

The gin pole, the bar from the hinged base to the winching guy, is always in place. Figure 15.12 shows a photo of a guyed monopole tilt-up downwind HAWT. Note that there are four sets of guy wires connecting the pole to each foundation.

Guy wires require more area around the tower on the ground than a freestanding tower. For tilt-up towers, there must be room for the tower to be lowered onto the ground. Figure 15.13 shows the kite-shaped area requirements for a guyed tilt-up tower, compared to those of a fixed guyed tower and a freestanding tower.

Compare figures 15.11 and 15.13 (left plan) to see the guyed tilt-up tower in both elevation and plan. Note that concrete foundations are required at the tower base and as anchors for the guy wires. Local requirements for depth of foundation need to be followed so foundations are below the frost line and suitable to support the gravity loads of the tower and turbine plus the overturning loads created by the wind. Note that the guy radius is 35–60% of the tower height. The gin pole will be 75–100% of the guy radius and remains in place. Note that the gin pole is parallel to one of the four sets of guy wires and that the full length of the tower height must be open on the other side of the gin pole in order to raise and lower the tower. The pink kite shape defined by the guy anchors and the height of the tower is the minimum area that must be kept clear around the tower.



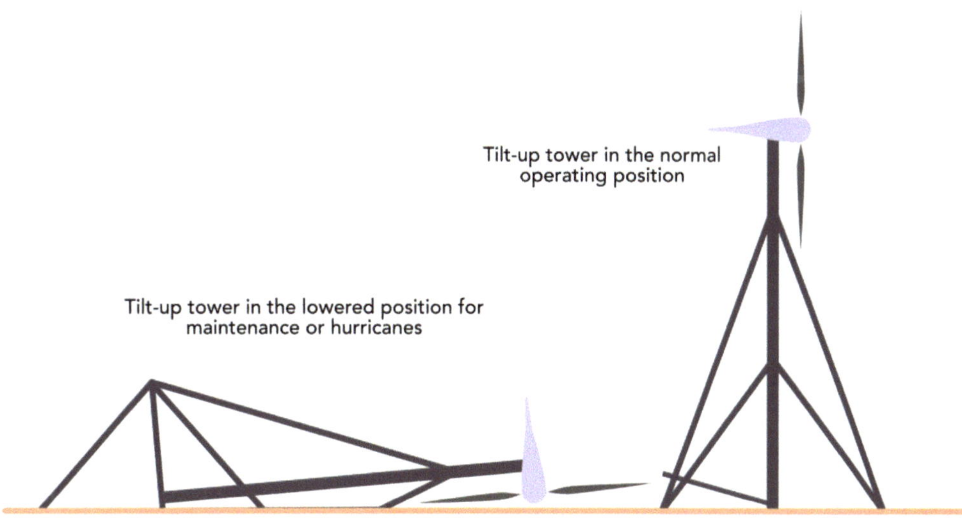


Figure 15.11 Guyed monopole tilt-up tower.[4]

Figure 15.12 Guyed monopole tilt-up downwind HAWT with gin pole to the left.

The fixed guyed tower (figure 15.13, center plan) has a different footprint from a tilt-up tower. Fixed towers have a triangulation of three sets of guys. Foundations must be provided at the tower base and at each guy anchor. The guy radius is 50–80% of the tower height. The minimum cleared area is shown in pink and outlines the location of the guy wires from the tower to the anchors.

Freestanding towers (figure 15.13, right plan) have the smallest footprints. Comparatively, a lattice tower will take up more area than a monopole. The tower base will need a radius of 7–10% of the tower height and be located on a concrete foundation. The area directly around the tower's concrete foundation needs to be kept clear. Note that very large towers for utility sized wind will have foundations that, like a building's foundation, may be backfilled. There will always be an exposed concrete pad around the base of the tower.

Monopoles can be tilt-up as shown in figure 15.14. The gin pole is separate and attached with a guy wire to a pulley motor and the monopole. The footprint requires clear access around the tower and the length of the tower height and gin pole (the vertical dimension shown in figure 15.13, left plan).

All towers will come with a bolt pattern for the concrete foundation so that anchor bolts can be located correctly.

Community and utility sized wind turbines are always mounted on freestanding monopoles. These towers are much taller than small wind turbines and have equipment and access to the nacelle located inside the tower. In figure 15.15 you can see part of a wind farm of Enercon E-66/18.70 turbines located in a forested area near Freiburg, Germany. Because of the tall tower height, the blades are well above the treetops. The turbine is part of a grouping of the same turbines that are located along the top of a mountain ridge.

Figure 15.16 is closer to this smaller utility sized wind turbine which has a rated power of 1,800kW, a cut-in wind speed of 2.5m/s, a rated wind speed of 12.5m/s, a cut-out wind speed of 34m/s, a rotor diameter of 70m, a swept area of 3,848m$^2$, and 22rpm (revolutions per minute) max rotor speed.[6]

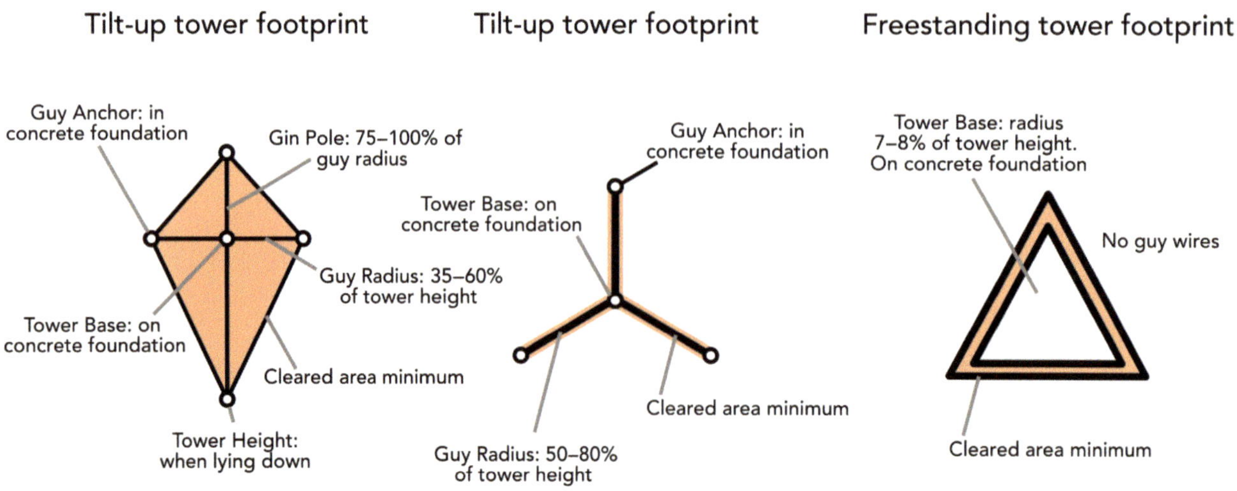


Figure 15.13 Comparative footprints of different tower types.[5]

Figure 15.14 Freestanding tilt-up monopole downwind HAWT (left) lowering to ground (right).

The interior of the monopole is shown in figure 15.17. It has room for equipment and access to the tower. Note the fall net located at ceiling height. Figure 15.18 shows the climb up the tower to reach the nacelle. Workers wear safety harnesses and must be in good physical shape.

## Characteristics of Wind and Good Siting

Wind is created by uneven heating of the atmosphere by the sun. The air of the atmosphere acts in convective currents, hot air rising while cold air drops, which creates wind patterns at both a global scale and in

Figure 15.15 Enercon E-66/18.70 wind turbines on mountain ridges outside of Freiburg, Germany.

**Figure 15.16** Enercon E-66/18.70 wind turbine near Freiburg, Germany.

specific locations. Wind roses are great resources that show what the typical wind patterns are at a particular location. See the chapter 3 (Locating Renewable Resources) for information on wind roses.

Global wind patterns are very stable and move at angles to either side of the equator and were named by sailors as the trade winds, westerlies, and at the earth's poles, the easterlies. Wind also happens because of the earth's rotation: air masses rise and fall in the same atmospheric location, but seem to move on the earth's surface because the world turned. In local areas, wind can be created because of temperature differences between land and water and mountaintops and valleys.

Wind varies all the time. It changes annually, just like rainfall, where there is an average but one year may be a bit more or less windy compared to the average. Wind changes seasonally, generally being stronger in the winter months and less in the summer months. It is important to understand what the seasonal characteristics of the wind at your site are so that you can produce the energy you intended when you need it. Often winter storms dominate wind patterns, rather than convective heating.

Wind also varies during the day, often quieting overnight and being strongest in the mid to late afternoon. Wind also varies instantaneously, from one moment to the next, so turbines may or may not be spinning at exactly the same rate. Because of this, turbine production, which is AC power, can vary widely within a short period of time, which has the most impact on the output of smaller turbines. Varying AC power profile is called "wild AC" and is "tamed" by inverting it to DC power. Larger turbines spin at very steady rates, no matter what the instantaneous change of the wind profile might be, so they are able to maintain a steady AC power profile, which is kept at high voltages (60Hz, 600V) for distribution to and across utility grids.

Utilities will study wind patterns over the course of years prior to installing wind turbines. A group of turbines need to be sited so that they do not produce turbulence for each other and have good wind access. Even small-scale wind installations can benefit from site monitoring to find the best location and verify assumed conditions from wind maps.

The best site location is one that is away from natural and man-made obstacles. Better sites will be located at the top of hilly terrain rather than at the bottom or leeward sides. Valleys and large bodies of water can be excellent locations because of the convective currents created from uneven heating. Both of these types of micro-sites can have strong air currents that change direction over the course of the day.

The wind turbine needs to be located out of the area of turbulent flow created by trees and buildings. Remember that trees will grow, so you may need to look up the full-grown height of the tree type rather than observing the height of newly planted trees. Figure 15.19 shows the rule that the blades should be above twice the height of the obstacle if the turbine is located within 20 times the height of the obstacle. This is sometimes also referred to as the 20/200 rule, 20 referring to being 20′ (6m) taller than an obstacle within 200′ (61m) of the turbine. Because of the 20/200 rule, it is generally difficult to site a turbine away from a house unless the property is at least ½ acre (2023m$^2$).

In the USA, most jurisdictions that have good wind resources have developed zoning applying to both residential and commercial wind projects. Typically, there are different requirements because of the difference in size between residential and commercial turbines and towers. Many zoning ordinances have restrictions about the minimum size of the property, the maximum height of the tower, how close the

Figure 15.17 Interior space at the base of the Enercon wind turbine.

turbine can be located to the property line, how the power is used, and abandonment. In addition to zoning ordinances, there may be covenants on the property that can restrict wind development. There may also be local requirements for maximum tower heights imposed in airspace around airports.

As an example, here is the Lycoming County Zoning Ordinance[7,8] (Pennsylvania, USA) section covering small wind systems:

**3420G small wind energy system**

1. Definition

   A single tower, or multiple towers, situated on a lot to provide energy from a wind turbine source to an individual home, multi-family residential use, office or business and industrial and agricultural uses located on the same lot. The wind energy is not to be provided to others for sale off-site in the power grid. The small wind energy system may follow the rules of net metering under the state policy.

2. Supplemental controls

   Lot size shall not be less than one (1) acre inclusive of the permitted principal use. The tower height inclusive of blade tip shall not exceed one-hundred twenty (120) feet measured from the ground level below the base of the tower. Setbacks from all lot lines shall be a factor of 1.1 times the tower height.

Figure 15.18 Interior view up monopole interior toward the nacelle showing fall net, ladder, wiring, and lighting.

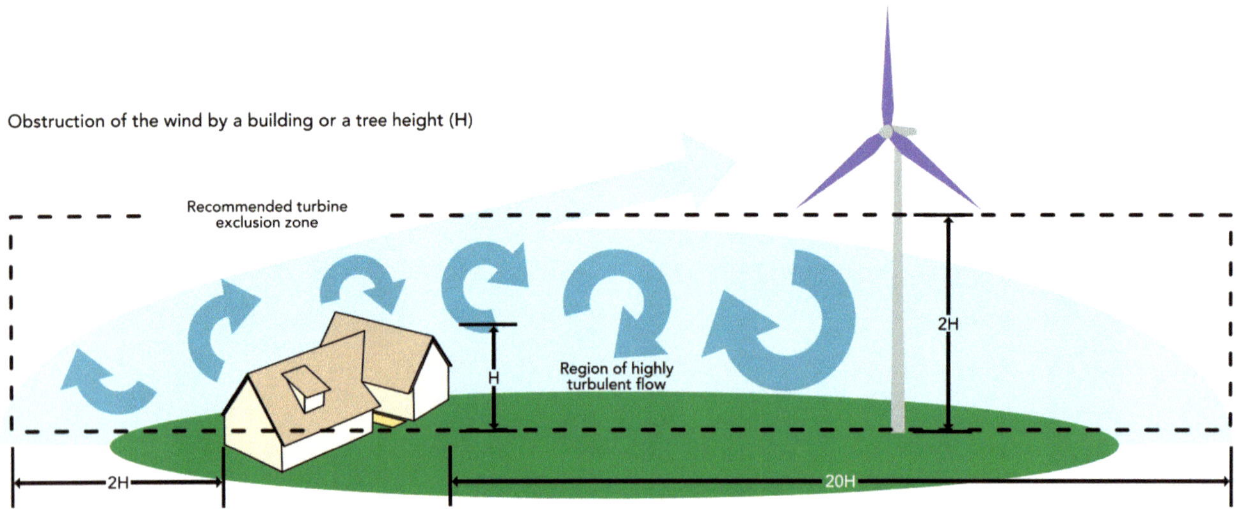


Figure 15.19 Wind turbine placement to avoid turbulence from obstructions.[4]

Towers are prohibited from locating in the front yard. Monopole tower style is encouraged. Guy wires of towers shall be located on the premises of the applicant. Section 3210B of the County Zoning Ordinance on clearing is applicable and Section 5230C of the County Zoning Ordinance need not apply for the removal of trees to have a successful small wind energy system function on a residential lot. The small wind energy system is to operate on the same lot to serve the principal use.

## Power in the Wind

Wind turbine power generation is limited by how fast the wind is, the Betz limit, low rotor rotation energy losses, drag losses from the blade, and generator and drive train inefficiencies. The Betz limit is a theoretical limit that states that the rotor efficiency for removing power from the wind is 59%. Turbines can't completely absorb all the wind's motion; it isn't possible for the air to be still on the backside of the blades. Betz's limit is the amount of power available if nothing else in the system caused power losses. In reality, most modern wind turbines are in the 35–45% range. Remember that Savonius VAWTs are limited to 15%, which is seriously less than all other VAWT and HAWT designs.

When you look at manufacturer's data about a turbine it will list various information about the turbine's performance. Figure 15.20 is an example of cut sheet information for the Bergey Excel 15, a small wind turbine well suited for class II winds.

The American Wind Energy Association (AWEA) has a Small Wind Turbine Performance and Safety Standard[9] which sets wind speeds for specific ratings. Rated power is the maximum power the generator can produce at 11m/s (24.6mph) per the AWEA Standard. Notice that it is in kW while the annual energy is in kWh because it is production over a period of time. Annual energy uses a wind speed of 5m/s (11.2mph) per the AWEA Standard. The purpose of using the AWEA Standard is to make it easier to compare the performance of different turbines. If the manufacturer doesn't list AWEA, then you must be careful to check what wind speed they are using. Peak power is the highest production number on the power curve for the turbine, which will be at a higher wind speed than the rated power. You can see that the peak power is 21.8kW where the rated power is 15.6kW. Cut-in wind speed indicates what wind speed is required to begin to produce electricity. Note that the start-up wind speed of 10mph (4.5m/s) is greater than the cut-in wind speed of 7mph (3m/s). This means that the rotor doesn't start spinning until there is a wind of 10mph (4.5m/s). Once the rotor is spinning, it will continue to produce energy until the wind drops below 7mph (3m/s). Cut-out wind speed will list the speed at which the turbine will stop production. Smaller turbines may have cut-out speeds to prevent damage to the machine in excessive winds. You can see that the Bergey Excel 15 doesn't have a cut-out speed and can produce electricity in winds up to 134mph (60m/s). The rotor speed will depend on the wind speed and varies from 0–140rpm.

Power in the wind can be determined by using this formula: $\frac{1}{2}\rho AV^3$. The variables are: $\rho$ is air density, A is the swept area of the blades when turning, and V is velocity, the effect of wind speed. Note that only the wind velocity is cubed, so the greatest impact on the amount of power available is the wind speed. Figure 15.21 shows the swept area, which is the circular area created by the blades when the rotor is spinning. Note that the swept area is a circle, so $A = \pi r^2$, where r is the radius and $\pi = 3.14$. Usually turbines list the rotor diameter, rather than a radius, which includes the width of the rotor hub and the length of the blades, so you can also use $A = \frac{1}{4}\pi d^2$, where d is the rotor diameter. Swept area may be listed by the manufacturer in units of either square meters ($m^2$) or square feet ($ft^2$).

Figure 15.20 Bergey Excel 15 performance information and specifications.[8]

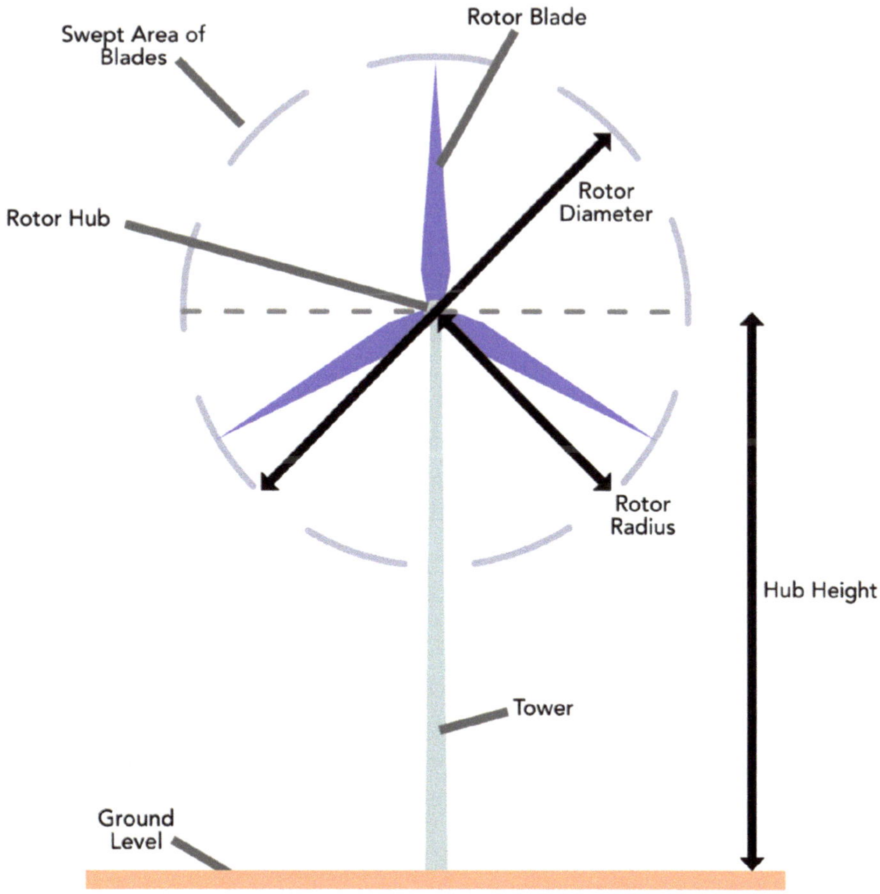


Figure 15.21 Swept area defined by the turning of the blades.

15

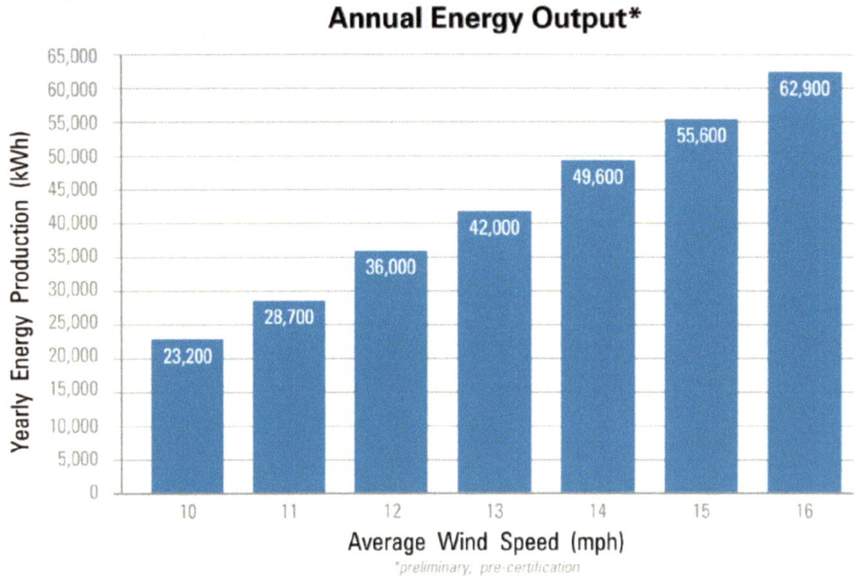


**Figure 15.22** AEO at different wind speeds of the Bergey Excel 15.[8]

The annual energy output (AEO) of a turbine is the amount of energy a wind turbine produces over the course of a year. Manufacturers will typically have a list AEO based on different average annual wind speeds. Figure 15.22 shows the AEO of the Bergey Excel 15 at wind speeds from 10–16mph in 1mph intervals across the bottom and vertical bars showing the annual production in kWh. An average wind speed of 10mph will result in 23,200kWh annual production where a 16mph wind will produce 62,900kWh annually.

You can also manually determine the AEO of a turbine by using this formula: $AEO = ½\rho AV^3 \times 8{,}760$ hours/year $\times 1.9 \times \eta \times$ 1kW/1000W

where 1.9 is the Rayleigh distribution factor and $\eta$ is the overall conversion efficiency of the turbine. Conversion efficiencies can range from 15–45%, and remember the Betz limit is a maximum of 59%. Most HAWTs are 35–45% efficient, although small wind turbines can have efficiencies closer to 20%. Depending on your units, you can substitute the numbers shown in table 15.1 for the variable ½ρ.

**Table 15.1** Value of ½ρ based on units for swept area and wind velocity.

| **Value of ½ρ based on units for swept area and wind velocity** | | |
|---|---|---|
| **Substitution for ½ρ** | **Swept Area Units** | **Wind Velocity Units** |
| 0.6125 | $m^2$ | m/s |
| 0.05472 | $m^2$ | mph |
| 0.00508 | $ft^2$ | mph |

Here is an example problem using a turbine with a swept area of 38.5m² (23′ rotor diameter), an average 12mph annual wind speed, and an assumed efficiency of 30%:

$AEO = 0.5472 \times 38.5m^2 \times 8{,}760hr/yr \times 1.9 \times 0.3 \times 0.001kW$
$= 18{,}177.3kWh/year$

## Certifications and Associations

There isn't a particular certification for people wanting to work installing turbines but there are many training programs for the field. These programs typically focus either on small wind or larger wind because of the big difference in working heights and equipment scale.

There are certification programs for small wind turbines through both AWEA standards and the Small Wind Certification Council (SWCC), which is a program that is part of the International Code Council (ICC). The ICC-SWCC website[10] lists certified turbines and explains their process of certification.

The Global Wind Energy Council[11] is the international trade association for the wind power industry. They produce an annual wind report and have global information about wind.

## Resources

1 Adapted from *The Inside of a Wind Turbine* [Clip art], by Office of Energy Efficiency & Renewable Energy, U.S. Department of Energy (www.energy.gov/eere/inside-wind-turbine).

2 There are a lot of great videos showing the erection of large wind turbines. One of my favorites is a time lapse from MidAmerican Energy Company, posted in YouTube in 2015: MidAmerican Energy Company. (2015, April 22). *From the ground up: Building our energy future, one turbine at a time* [Video]. YouTube. www.youtube.com/watch?v=84BeVq2Jm88

3 Paul Gipe has published many books about wind and his website (www.wind-works.org/cms/) has up-to-date information and articles about issues and products related to wind and renewable energy.

**4** Wind Energy Technologies Office. (n.d.). *What are the basic parts of a small wind electric system?* U.S. Department of Energy. https://windexchange.energy.gov/small-wind-guidebook#parts

**5** From "Wind Generator Tower Basics," by I. Woofenden, February & March, 2005, *Home Power, 105*, p.65 (https://0964fed1-9106-4800-b313-32ef2cf225aa.filesusr.com/ugd/d49ff9_cda2fee8ced34d8e81fa3e034eb14ba8.pdf). Copyright 2005 by Home Power Inc.

**6** Wind Turbine Models. (2017, March 7). *Enercon E-66/18.70*. https://en.wind-turbine-models.com/turbines/135-enercon-e-66-18.70

**7** Strategic Planning Committee. (2011, February 24). *Lycoming County zoning ordinance*. Lycoming County Planning Commission. www.lyco.org/Portals/1/PlanningCommunityDevelopment/Documents/zoning/lycoming_co_zoning_ord_feb2011.pdf

**8** Bergey Windpower. (n.d.). *Bergey Excel 15* [Specifications sheet]. http://bergey.com/wp-content/uploads/Excel-15_Spec-Sheet_Rev.-1_4.7.17.pdf

**9** AWEA certified wind turbines: Interstate Renewable Energy Council. (2019, August). *Certified wind turbines*. https://irecusa.org/programs/certified-wind-turbines/

**10** Small Wind Certification Council (https://smallwindcertification.org/).

**11** Global Wind Energy Council (https://gwec.net/).


## Study Questions

1. Do an online search for "small wind turbine" and look at the image results. Identify the types of turbines (HAWT or VAWT), the size and shape of the tail, the number and appearance of blades, and the type of tower.
2. Do an online search for "vertical axis wind turbine" and look at the image results. Identify the scale of the turbine, if it is a lift or drag device, the number and shape of the blades, how high the turbine is mounted above the ground or building, and the type of tower (if it has one).
3. What is the 20/200 rule and how does it apply to the location and height of a wind tower?
4. Select a turbine and look at the manufacturer's data sheet. Answer the following questions:
   a. What is the swept area of the blades?
   b. What is the cut-in speed?
   c. What is the rated power of the turbine?
   d. Does the turbine have a cut-out speed?
   e. What is the AEO at different wind speeds? If only one AEO is listed, what wind speed was used to determine it?
   f. Does the manufacturer recommend a particular tower?

CHAPTER 16

# Microhydro

**Key Terms**

Dump load (also called diversion or shunt load): a load, usually an electric heating element, that can be used to protect system batteries from overcharging damage and provide a continuous load on a turbine (microhydro or wind).

Head: the vertical drop of water from the top of the penstock to the turbine.

Flow: the volume of water in a stream, microhydro usually diverts 20% or less of the total stream's flow.

Penstock: a pipe that diverted water runs through, located outside of and away from the stream bed.

Riprap: loose stone that is piled in a location to prevent soil erosion due to water runoff.

Weir: a low dam built across a stream or river which reduces the flow and increases the water height upstream.

## Concepts for Success

Hydropower requires that there is a good water resource relatively near your building. Microhydro systems are smaller systems that do not require dams, therefore do not disrupt stream ecology, and are usually easier to gain permission for than larger systems. Microhydro systems typically produce AC power, except for very small systems that produce DC power. Hydropower produces a steady and even amount of energy all day, every day, which means it can either have battery backup or not and is well suited for both off-grid and grid-tied projects.

Microhydro creates power by spinning the turbine runner. There are different types of turbines depending on the type of water resource. A good water resource needs a combination of both head (vertical drop) and flow (quantity of water). Turbines need to always have a load on them so the system should be designed to have a dump (diversion) load.

A microhydro system has an intake at a higher location in the stream which directs water to a penstock (pipe). The penstock is not located in the streambed. The penstock connects to the turbine that is located in a powerhouse to protect the equipment from the elements. The water flows through the turbine and then is returned to the waterway. The whole system is gravity fed.

Head is the vertical distance from the penstock intake to the center of the turbine runner. Because the water is flowing in a pipe downhill over a distance, there will be friction loss. The head with loss is called net head. Head and flow need to be accurately measured at your site. Flow needs to be considered at the lowest water levels during year, because you need to always have water flowing. If your stream is regularly dry, it is not a good resource.

Net head and flow are multiplied together to determine the system's power rating. The power rating is then multiplied by production for the day and month or year. The value of the power depends on what the local cost of electricity is.

## How Hydropower Works

Hydropower works because there is a mass of water that is located at a higher elevation than the turbine. The water mass is potential energy which changes to kinetic energy as the water flows downhill. The water turns the turbine (mechanical energy) and creates electrical power because the turbine is turning a generator. Hydropower is the top producer worldwide for all types of renewable energy. Large facilities (2MW and above) require large-scale dams and can power cities. Microhydro (5kW–100kW) are small-scale installations intended to provide

DOI: 10.1201/9781003297819-18

electricity for anything from a single home to a larger community area. Microhydro has much less expensive component costs compared to other renewables and produces a very reliable amount of energy all the time.

Of general concern is the environmental impact of hydropower on waterways. Small hydro systems are often exempt from costly environmental impact studies. It is important to be sure to have all permits in place prior to installing a system. In the USA, the Hydropower Regulatory Efficiency Act of 2013 (Sec. 3), increased from 5MW to 10MW the size of small hydroelectric power projects which the **Federal Energy Regulatory Commission (FERC)** may exempt from its licensing requirements. The location cannot be on federally owned land nor be located at a dam. If your project meets these requirements, it is categorically exempt from preparing an environmental document. There is a 45-day public review period for comments.

In the USA, the US Army Corps of Engineers has jurisdiction over waterways. Notification of the Corps is required under Section 404 of the Clean Water Act, even for rearranging rocks in a streambed. They can help direct you as to what, if any, permitting is required.

There may be water rights that impact your use of the water to make power. In the USA, individual states control water rights and, even if you own water rights for another use, you may still need a separate water right for power.

Permitting for hydropower can take time depending on where you are located. Microhydro is typically easier to get permits for than larger hydro because microhydro usually has little impact on the natural flow of the river or stream. It is important to start the permitting process early so it doesn't end up delaying your project's completion.

## AC or DC?

Microhydro generators can produce either AC or DC energy. Typically, DC generators are used only in very small systems. AC generators are available in single-phase or three-phase and 120V or 240V (USA typical grid and appliance AC voltage). This makes it easy to directly run typical AC equipment. AC is easy to transform and move over longer distances if your water source is not right next to your building.

## The Importance of Consumption

Let's say the client's household consumption looks like figure 16.1:

The horizontal axis is time from morning (left) to evening (right) and the vertical axis is power with no use at the bottom and more use toward the top. This could be a profile of a family's home where some of the people are home during the day and then power spikes in the evening. The short power spikes are representative of the kind of draw you might see from the compressor motor of a refrigerator or freezer, which runs for a short time and then turns off.

Assuming that you could provide all the power needed, if you don't have batteries and are not grid-tied, your system would have to be sized to meet the maximum power spike. It would be overproducing most of the time, which would be inefficient (figure 16.2). It is unlikely on a small house system that there would be gate controls to open and close to adjust for how much water is flowing through the turbine.

If you size the system to match the average power consumption it would look more like figure 16.3. You can see that there is time early in the morning and overnight where the system is overproducing, and times during

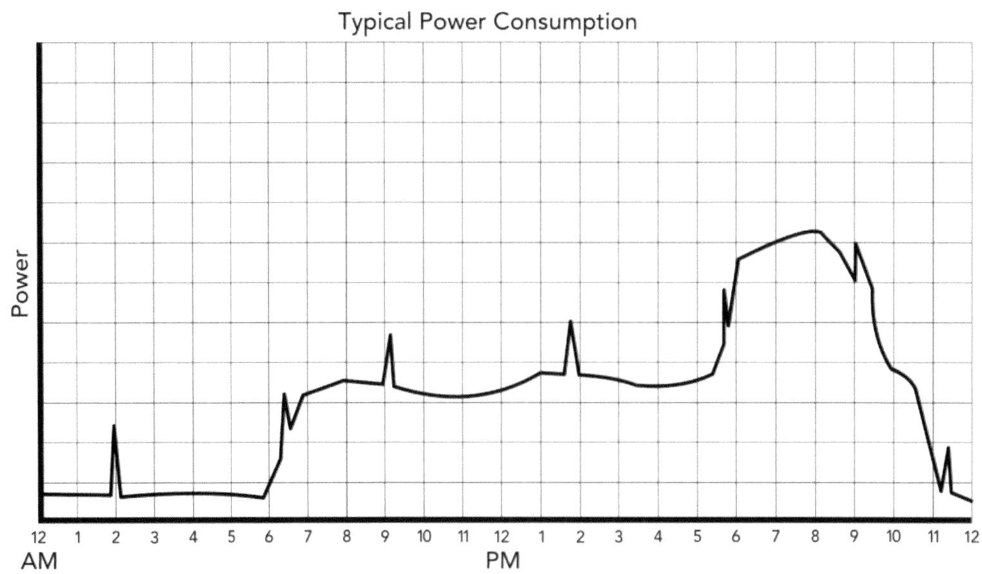


Figure 16.1 Household power consumption profile over a day.

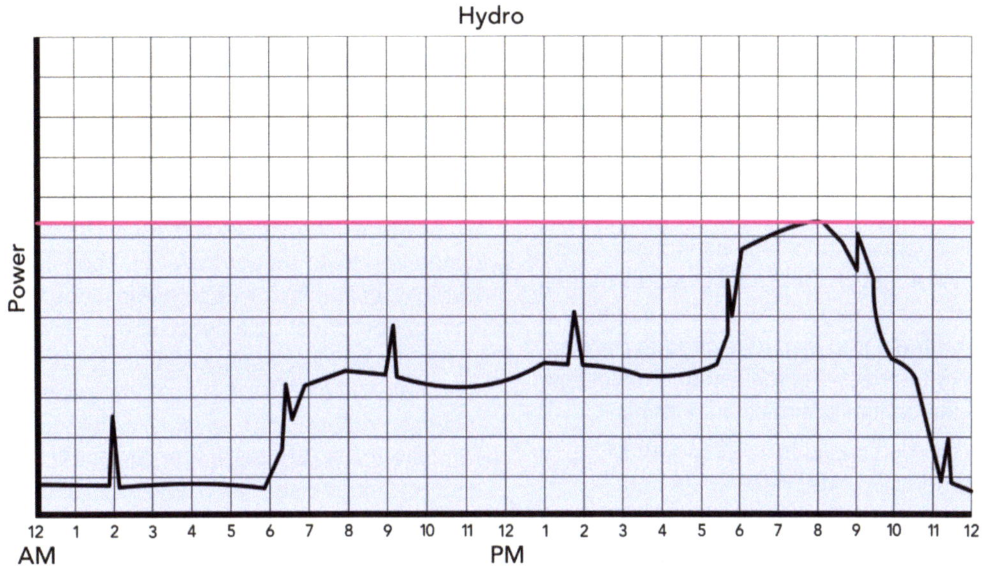


Figure 16.2 Hydropower system sized to match peak load has a lot of over production.

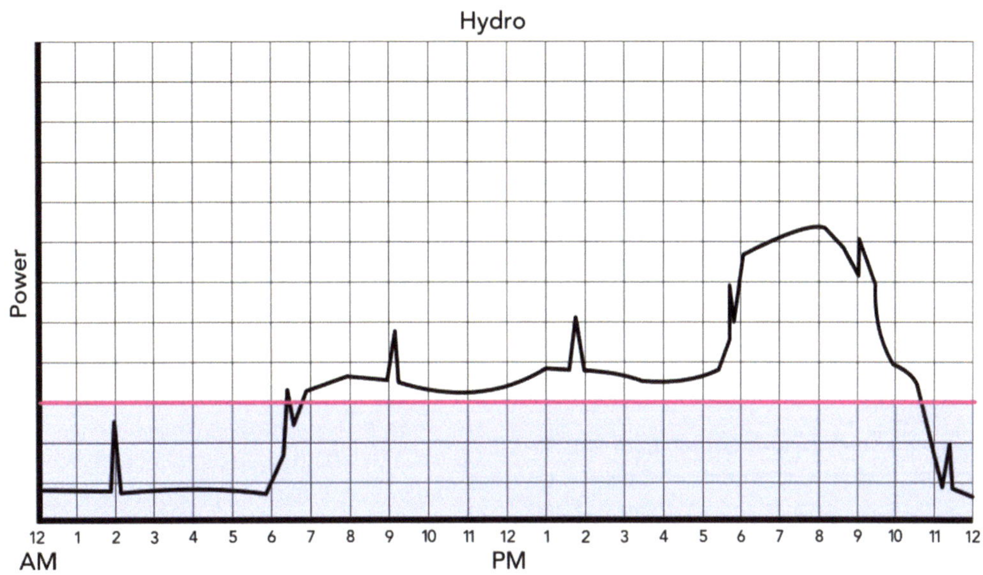


Figure 16.3 Hydropower system sized for average load has some periods of over production and under production.

16

the day where there is unmet demand above the production line.

The first solution would be to work with the client and figure out if it is possible to get rid of the big power spike. If there are still times where demand is exceeding the production of the microhydro system, the user will need to have access to other energy: either the grid or batteries. The spike in power demand happens to be at a time when everyone else is also ramping up demand for grid power so, while the customer will have access to grid power, it is making it difficult for the utility to manage power production. In many markets, utilities charge higher prices for energy at peak demand times, so the customer could be paying extra to use the grid power.

If batteries are going to be used to cover the extra power demand, the designer has to make sure that there is enough "extra" power being produced to charge up the batteries every night and that the battery bank is sized properly to provide the "extra" needed power every day.

## What is a Dump Load?

**Dump loads**, also called diversion loads and shunt loads, are used for systems that have turbines (microhydro and wind). The dump load uses the extra energy produced when the building consumption demand is low. It also provides a load on the turbine at all times. This is important because, if there is no load on the turbine, it could spin very rapidly, causing damage. Also, the turbine can't be disconnected from the battery bank while the turbine is running, so the dump load uses the energy so the batteries aren't damaged from over-voltage.

Dump loads need to match or exceed the required resistance (ohms), current (amps), and power rating (watts) of your system. The resistors can be wired in series or parallel to get the required resistance. The diversion load needs to be able to use more watts than the turbine can generate. The charge controller needs to have a diversion load option in its configuration in order to regulate the flow of energy between the batteries or service panel and the dump load. It is good to have a 25% safety factor between the maximum draw of the dump load and the controller's maximum rating.

The most common dump load is a 12V or 24V DC low voltage submersible water heating element (figure 16.4). The element can be installed in a DC water heater and preheat water being supplied to a water heater.

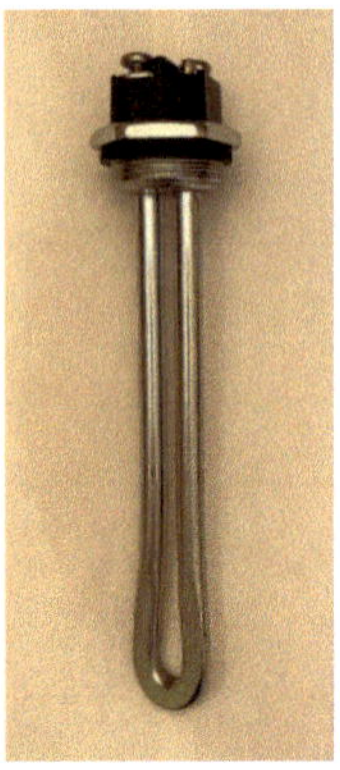

Figure 16.4 Electric water heating element.

The element can also be installed to the air, which then just "burns off" (wastes) the extra energy. If installed this way, you need to provide a cover to keep people from getting accidentally burned by the hot element.

Dump loads can also be used for electric resistance space heating and heating spas and pools. If you plan a space heating dump load, be sure you have an alternate dump load if no heating is needed.

## Off-Grid Microhydro Systems

Because microhydro systems can produce a large, steady amount of energy, you may be able to continuously meet your power demands. If you are able to balance demand fairly evenly across the day you can have a system without batteries that produces AC power as shown in figure 16.5. The turbine is connected to an AC controller that powers the AC main panel and switches to dump loads when needed. This system must be able to produce enough power to run the building at peak load times. It will also need to handle the extra surge draw that happens when induction motors start up (such as the compressor on a refrigerator). It is important to check what the starting load is for items such as power tools and appliances because, while the running load may be fairly low, the start-up can be 3–7 times greater. The system will require an AC controller to monitor usage and send extra power to dump loads.

If the electrical use profile is not a fairly even distribution across the day, the hydro system would have to be oversized to meet the highest power demand. This isn't efficient as the system would regularly be producing much more energy than is being used. A better solution in this case would be to add batteries to the system to even out supply and demand power. The batteries would be charged when the power demand is low and discharged when the power demand exceeds the turbine's production. The generator would have to be large enough to supply demand and charge the batteries every day. The batteries can also be used as backup if the turbine needs to be serviced.

Figure 16.6 shows the components needed for a microhydro off-grid system with battery backup. If the system is small and producing DC power, then it will directly feed the charge controller. If the turbine were generating AC power, there would have to have a converter (not shown in this diagram) before the charge controller. The charge controller monitors the batteries and, if there was a need for the dump load, the diversion charge controller would move the charge to the dump load. Many systems have separate monitors

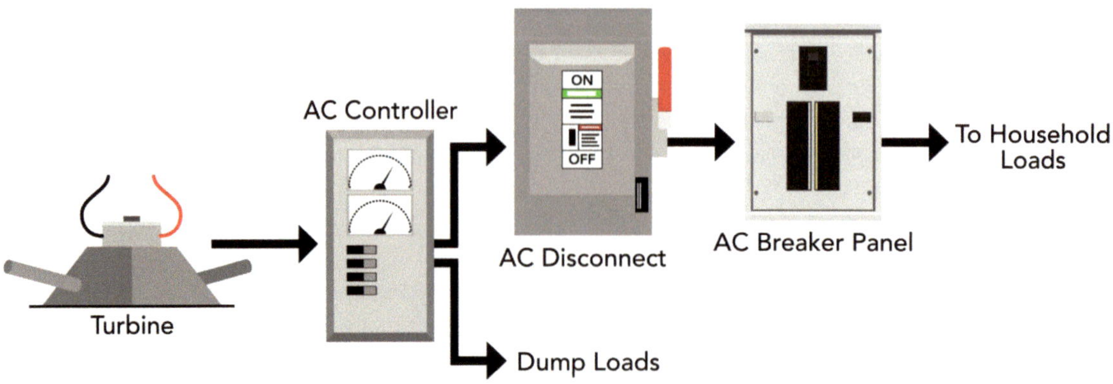


Figure 16.5 Off-grid microhydro without battery backup requires dump loads.

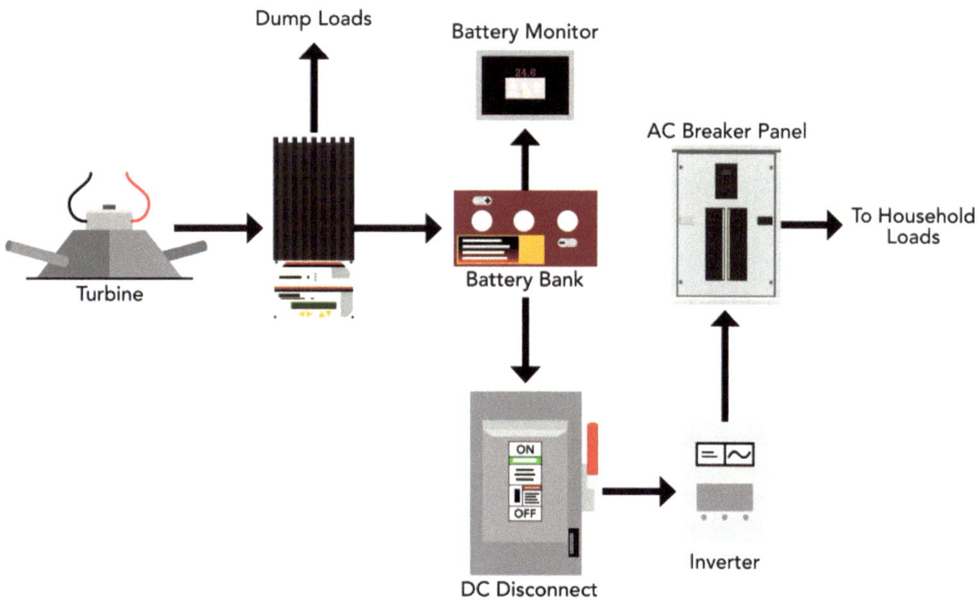


Figure 16.6 Microhydro off-grid with battery backup.

to check the batteries and production. These numbers can be compared to the controller for consistency. The DC and AC disconnects can either be part of the inverter or separate boxes. The inverter changes the DC power to AC power for the main service panel. Note that you could have a DC service panel if you had items running on DC power. This image shows fewer dump loads because your system wouldn't be as oversized as with no battery backup. The battery bank should be sized to cover the difference between the peak load and the power rating of the turbine system.

## Grid-Tied Microhydro

Grid-tied microhydro can have battery backup or not. Figure 16.7 shows a grid tied system with battery backup. Depending on the desired outcomes of your project, the size of the battery bank can vary widely. If it is a backup for when the grid goes down (no power from the grid), then the battery bank will be fairly small and you'll need to include an essential loads panel. Remember, an essential loads panel powers some of the equipment in your building but not everything. If you are designing a net zero building, then it could be sized the same as the previous off-grid system, so that the building isn't really ever drawing power from the grid. You still need the essential loads panel for when the grid fails. "Extra" production is going onto the grid rather than to the dump load. The dump load is there for when the grid goes down.

Batteries aren't necessary if you are grid-tied, as shown in figure 16.8, the controller will cut the power to the main service panel and send it to the dump loads if the grid goes down.

A grid-tied system with an essential loads panel (figure 16.9) will operate the loads tied to the essential loads panel while the main service panel is unpowered in the event of a grid shutdown.

## Microhydro Siting

The most important items for a microhydro site are a reliable water resource that has good flow all year and that the water has vertical drop on site. The vertical drop is called head. The system's power rating is determined by multiplying flow and head. Low head is 32′ (9.75m) or less and high head is 64′ (19.5m) or more. If you have low head, then you need to have a lot of flow. If you have high head, then you can have less flow.

Figure 16.10 shows the features of a microhydro system. At the top is an intake where water from the stream is diverted out of the streambed and into the penstock. The penstock delivers the water to the turbine, which is located in a powerhouse. The energy is then distributed over wiring to the building. Depending on the distance from the powerhouse to the building, the battery bank may be located in the powerhouse or in the building. Remember that low voltages of DC power have high losses over even relatively short distances. If you are thinking of integrating the powerhouse with the building, it is important to consider the wet, damp conditions associated with lots of water! If you are producing high voltage AC power, it can be transmitted up to a mile away.

The small intake structure is not a dam. It is a diversion device. If needed, the intake can connect to a canal that channels the water to the side of the stream and into a forebay tank where sediment can settle out of the water and the water is screened before going into the penstock (figure 16.11). The screen keeps branches, leaves, and other debris or fish from entering the penstock and running through or jamming up the

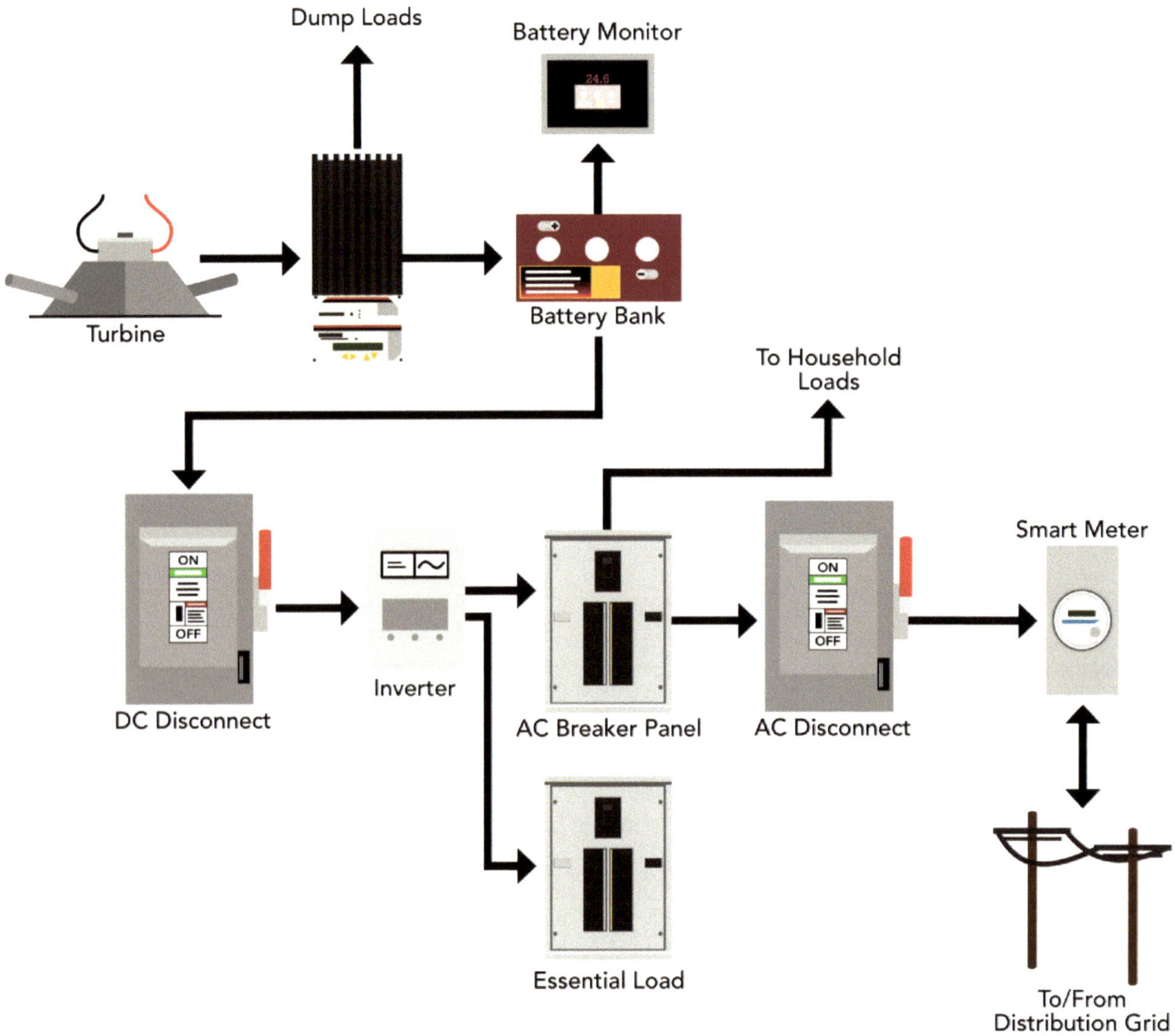


Figure 16.7 Microhydro grid-tied with battery backup.

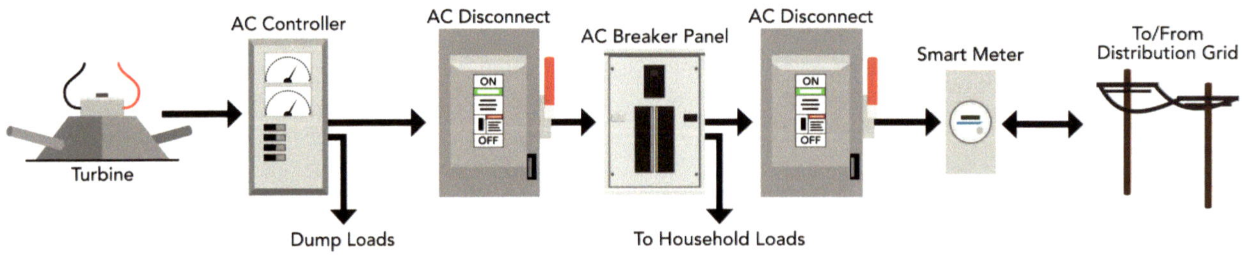


Figure 16.8 Grid-tied microhydro.

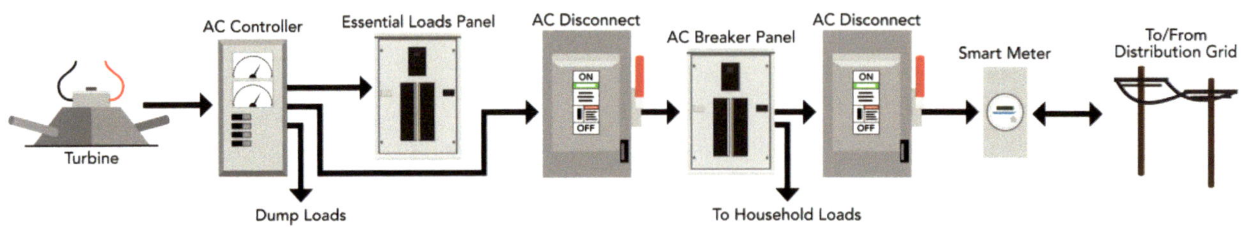


Figure 16.9 Grid-tied microhydro with essential loads panel.

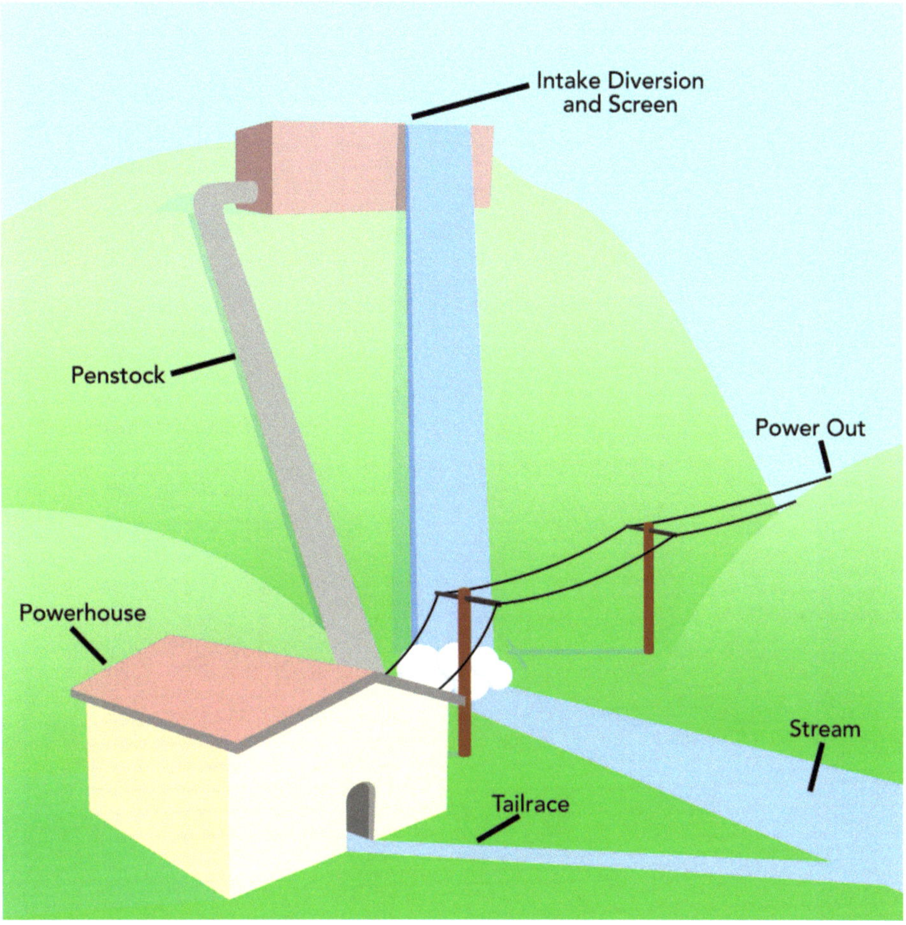


Figure 16.10 Microhydro system features.

turbine. There is a gate that can be closed to stop water from flowing into the penstock. Note that there are no pumps, the water is flowing because of gravity.

Other types of intake structures can be built of concrete or stone and have a settling basin to precipitate sediment out of the water and a screen over the intake (figure 16.12). The intake has a gate or shut-off valve. Note that the intake box height is designed so that, even at its lowest water level during the year, water will flow over the top of the intake structure. The outsides of the intake box are a bit taller than the interior divider which causes sediment to settle out. Note that there is a thin vertical pipe attached to the penstock that prevents a vacuum from developing in the pipe, so you get the most power possible.

The penstock is a pipe that runs alongside the stream, not in the streambed. It may be near the stream or it may cut a different path down the hillside. If the pipe were in the stream, it would be subject to damage by debris from raging flood water. The penstock should be as straight as possible because every bend in it causes a loss in power. If a bend is needed, it should be a very round sweep rather than a sharp bend. The penstock carries water downhill to the turbine. The pipe has to be the correct size to allow the water to flow without causing too much loss due to friction. Friction and pipe bends reduce the effect of the head (drop). Too big a pipe can also cause problems because of emptying too quickly and creating a vacuum. Polyethylene pipe can be used for pressures up to 100psi (7.03kg/cm$^2$). Polyvinyl chloride (PVC) piping has pressure ratings from 160–350psi (11.29–24.6 kg/cm$^2$). Steel can carry up to 1000psi (7.03kg/cm$^2$) or more. Steel is usually used for high head penstocks. Lower heads can use a variety of pipe material, including concrete. Connections at the pipe material joints must be made to meet the expected pressures in the pipe. The penstock can be buried or encased in concrete to help protect from damage by animals (typically chewing damage but could include issues from pasturing cattle or large wildlife such as bears), which can be a problem with smaller plastic type pipes. Trenching can also protect the penstock from vandalism, avalanches, and other types of damage. Consideration for penstock protection depends on the site! Trenches should be 2–3′ (609–914mm) deep and free of sharp objects. All penstocks should be anchored to keep them from moving when filled with flowing water.

The water flowing through the turbine turns the runner (wheel). The spinning of the runner causes the generator to turn, which produces electricity. The turbine and the various electronic equipment are protected from

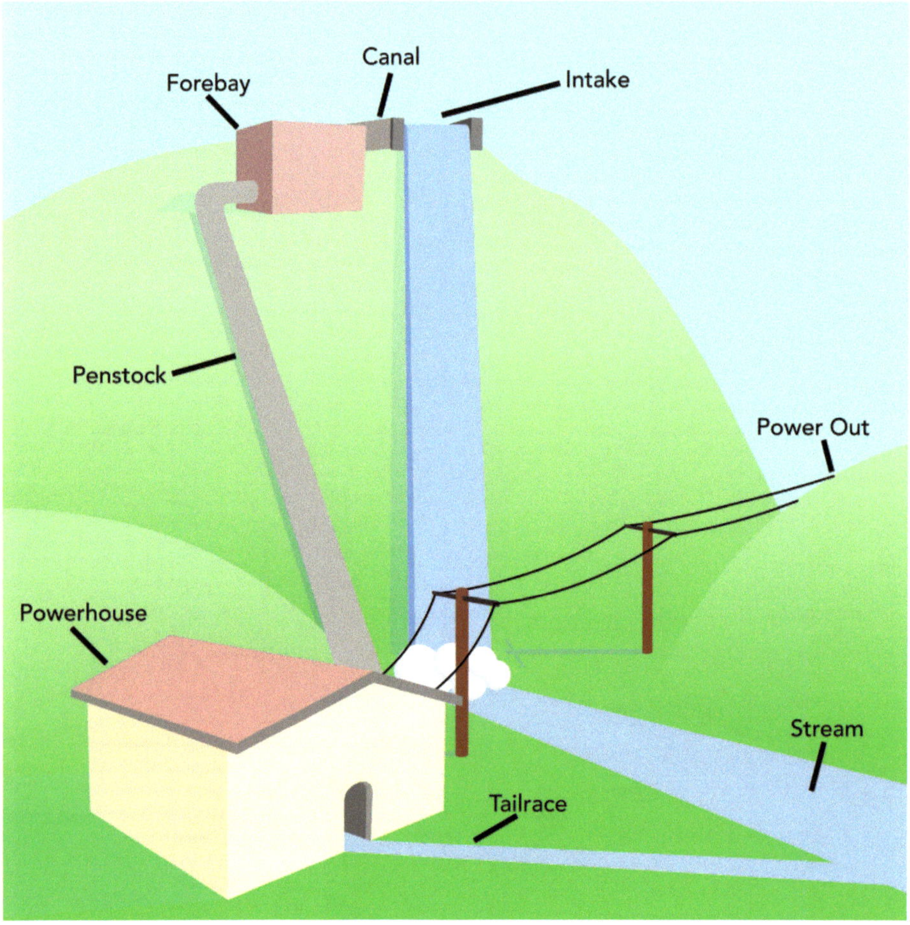


Figure 16.11 Forebay structure includes a canal from the intake.

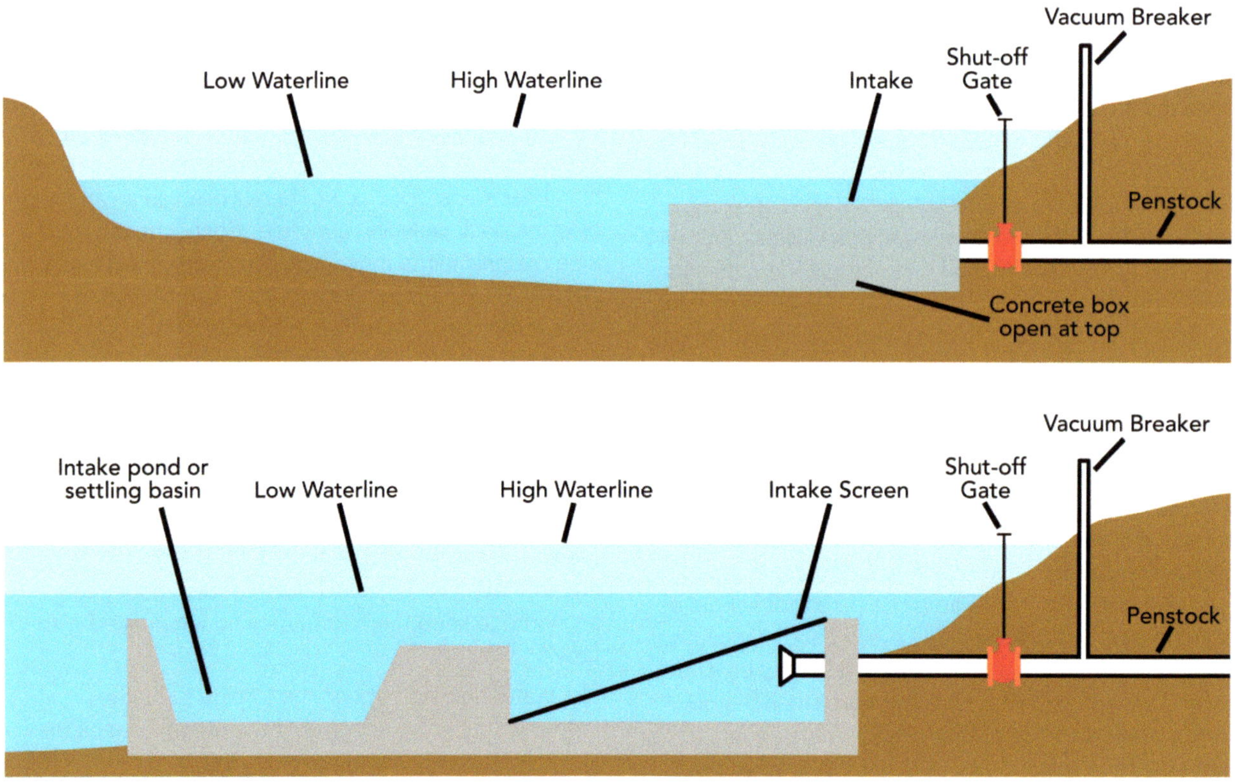


Figure 16.12 Diversion intake structure.

the elements by the powerhouse. The powerhouse is a simple structure and is often not a conditioned space.

After the water flows through the turbine it is called tail water. The tail water is no different from the stream water. It hasn't been processed or heated or used for anything except turning the turbine. Therefore, it can just be returned to the stream. The place where the water runs back into the stream is called the tailrace. The tailrace can just be a portion of ground where the water runs back into the stream. If the stream bank would be eroded by the tailrace water, then it should be stabilized with **riprap**. Mounting platforms for the turbine that include a discharge pipe or spillway are often made from concrete. Everything relies on gravity to get back to the waterway, so the tailrace needs to flow downhill from the turbine.

## Types of Turbines

There are different types of turbines. Most of them are named after their inventor. They are designed to work in different head and flow conditions.

### Pelton turbine

Pelton turbines are used for high head and low flow sites. It works on impulse power. Water hits the impulse blades which look like side-by-side cups or buckets (they can look like spoons) that are mounted around a wheel (figure 16.13). The water hits the center between the buckets and divides into two streams of water, which swirl in the bucket cup and force the runner to rotate. There are usually two jet nozzles but can be more. These turbines can be as small as a few inches (mm) across to 100′ (30.48m) or more. Turgo turbines are a type of Pelton turbine. The jet is smaller and sprays so the water hits three buckets at once to make the runner spins faster.

### Francis Turbine

A Francis turbine is used for medium head and medium flow and is quite versatile (figure 16.14). It has a spiral casing that directs the water flow through vanes on a rotor. The vanes curve out at the bottom which causes both reaction and impulse force from the water on the runner blades. This is a mixed flow turbine because it uses both reaction and impulse force. The stay vanes are the larger vanes which are stationary and help to reduce the swirling of the water to make the turbine more efficient. The second set of vanes to the interior of the stay vanes are the guide vanes which are not stationary, they adjust angle to increase the efficiency of the water striking the runner blades. The runner spins and turns the main generator shaft. The water falls through the runner and out the draft tube. The draft tube is a conical shape (flaring wider in diameter at the bottom compared to the runner diameter). The draft tube is submerged in the tailrace to keep from developing a vacuum.

### Kaplan Turbine

Kaplan turbines are used in low head and high flow conditions. It is a propeller-type turbine with adjustable blades and adjustable wicket gates (figure 16.15). The water drops through the wicket gate, falling on the blades, causing the runner to rotate. This is a reaction

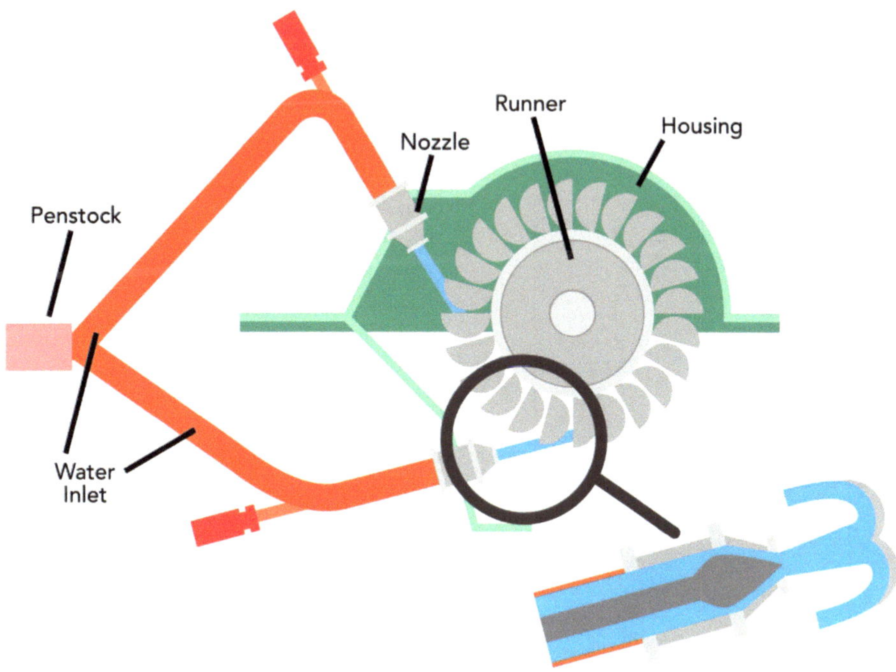


Figure 16.13 Pelton turbine.

16

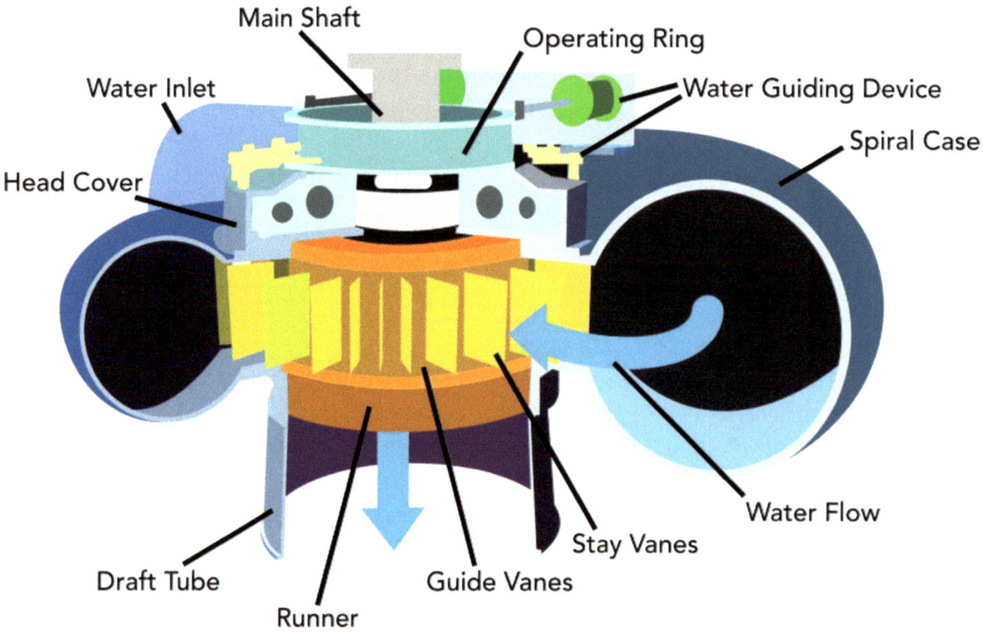


Figure 16.14 Francis turbine.

force. The water exits out the draft tube, which flares and must remain submerged in water.

## Cross-Flow Turbine

Cross-flow (Banki-Mitchell or Ossberger) is another type of low head and high flow turbine (figure 16.16). It has lower efficiency than other types of turbines. The water falls across the runner blades in a fashion similar to a traditional water wheel. The water drops through both the top and the bottom of the wheel to create spinning of the runner. This provides a bit more efficiency than if the water just ran over the top. The water runs out

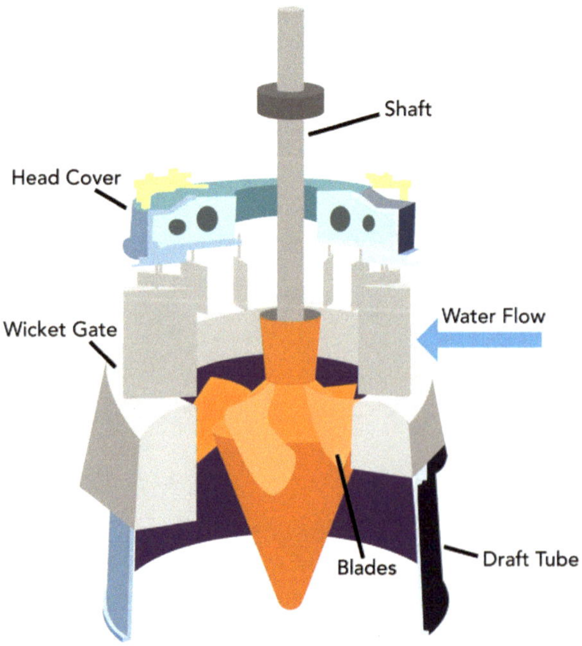


Figure 16.15 Kaplan turbine.

a flared draft tube, which needs to be submerged in water.

## Jack Rabbit Turbine

There is a small drop-in-the-creek turbine called a Jack Rabbit turbine (figure 16.17). It was developed to power aquatic instruments and is also called the Aquair UW Submersible Hydro Generator. It can run in a minimum of 13″ of water and has a maximum output of 100W, which can produce between 1.5–2.4kWh daily, depending on your flow resource; 1.5kWh is based on a flow of 6mph. This turbine doesn't use a penstock. It is available in models rated for nominal 12V, 24V and 48V DC systems. It is quite small, 14.5″ (368mm) long. The generator is 8.375″ (213mm) in diameter and the propeller is 12.5″ (3185mm) in diameter. The turbine needs to be securely mounted so that the propeller faces the water flow. It should be possible to lift the mounting out of the water if there is a flood so as to protect the blades from damage. The propeller is made of aluminum and can be hammered back into shape if damaged by debris in the water. This turbine doesn't have overspeed issues, so dump loads are not required. If the system doesn't have dump loads, then the controller needs to be rated to carry the no-load high voltage capacity of the unit.

Figure 16.18 shows different types of turbines and their application across different amounts of head and flow. Head is the vertical axis, with 1 at the bottom. Flow is the horizontal axis with 1 at the left. The diagonal lines indicate the production at that head and flow in megawatts. You'll note that this chart includes very large numbers that would exceed those of a microhydro application. The larger the power rating of the turbine, the larger diameter the runner will be.

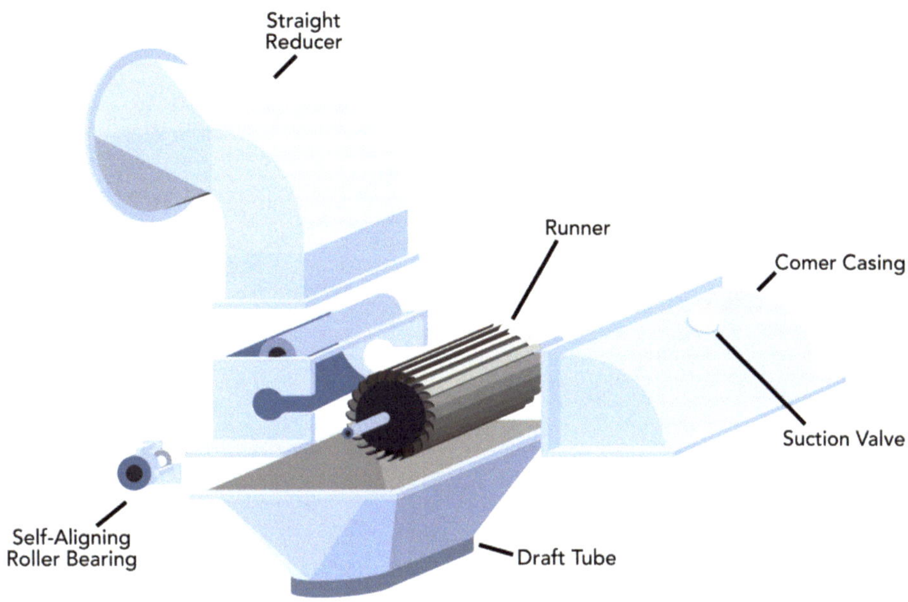


Figure 16.16 Ossberger cross-flow turbine.[1]

## Measuring Head and Flow

Head is water *pressure*, which is created by the difference in elevation between the water intake and the turbine. Head is expressed as the vertical distance (feet or meters), or as pressure in pounds per square inch (psi) or kilograms per square meter (kg/m$^2$). 1psi = 2.31 vertical feet (1kg/m$^2$ = 10 vertical meters).

Head does not include horizontal distance, only vertical drop. Since you probably don't have a waterfall, you'll need to measure the head over a distance. A surveyor can accurately measure the vertical drop available on your site. You may be able to look at topographical maps for the area and get an idea of the vertical drop. The head will be from the center of the penstock intake to the center of the turbine runner. With two people you can get a fairly accurate reading by having one person uphill and one person downhill. The downhill person stands at the location of the turbine and sights horizontally at the bottom of the uphill person's feet. The measurement height will be from the bottom of the sighter's feet to their eye level. The uphill person marks where they were standing and moves uphill again. The downhill person moves to the spot where the uphill person marked they were standing and then horizontally sights the bottom of the uphill person's feet again. This process can be repeated up the length of the hillside, keeping track of how many times the height is measured.

Figure 16.17 Aquair UW submersible generator (Jack Rabbit turbine).

Net head is the actual amount of water pressure that will be creating power. Net head will be less than the head (sometimes called gross head) because of friction loss of the water flowing in the penstock pipe, bends in the pipe, and the length of the pipe. It is good to design for 20% or less loss. Many vendors who sell microhydro systems can help you determine losses and design the proper pipe size and pressures.

Table 16.1 shows typical losses based on 100′ (30.48m) of PVC pipe. The pipe is measured as its actual length. Across the top of the chart is the flow of the waterway. You read the chart by looking in the column that matches your design's flow and then reading down to the amount of loss based on the pipe diameter. Note that PVC pipe is specified by its exterior dimensions, the interior is smaller. A 200gpm (12.62L/s) flow has a loss of 60.6′ (18.47m) (loss of vertical head) in 100′ (30.48m) length of 2″ (18.47m) pipe. Obviously, that is too small a diameter pipe for all that water to fit through. To determine the correct pipe size, we need the head and flow at the site.

Let's say you have 64′ (19.5m) vertical drop (gross head) on your site and the penstock is going to have to be 200′ (60.96m) long. Your flow is 300gpm (18.93L/s). To find the maximum loss, multiply the gross head by 20%: 64′ (19.5m) × 0.2 = 12.8′ (3.9m). Because your penstock is 200′ long and the chart is for 100′ (30.48m) of pipe, the maximum loss will need to be divided into

16

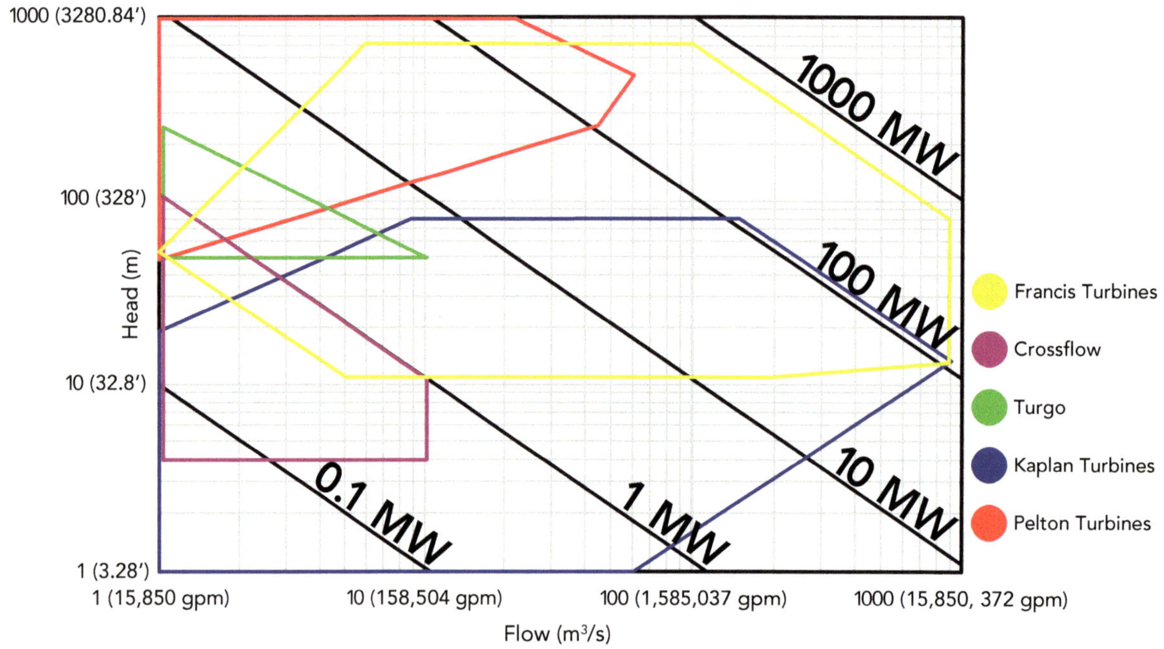


Figure 16.18 Turbine application chart.[2]

100′ (30.48m) lengths, 2 for this example: 12.8′ (3.9m) /2 = 6.4′ (1.95m). That means you want to pick a pipe that has 6.4′ (1.95m) or less loss. Reading down the flow column of 300gpm (18.93L/s), you can see that a 3″ (76mm) pipe has 17.9′ (5.46m) of loss, which is too much: our limit is 6.4′ (1.95m). A 4″ (100mm) pipe has 4.37′ (1.33m) of loss in 100′ (30.48m) of pipe, which is less than the maximum. The actual loss will be in 200′ (60.96m) of pipe, so multiply the loss from the chart by 2: 4.37′ (1.33m) × 2 lengths of pipe = 8.74′ (2.66m). This is the actual loss due to friction in a 4″ (222mm) diameter PVC pipe 200′ (60.96m) long.

Net head = gross head − loss

Net head = 64′ (19.5m) − 8.74′ (2.66m) = 55.26′ (16.84m)

Flow is water *quantity* and is measured in volume of water per second or minute. This could be gallons per minute (gpm), cubic feet per second (cfs), liters per second (L/s), or meters cubed per second ($m^3/s$).

The flow used for your design uses just a fraction of the water in the stream or river at low water levels. When you are measuring the amount of flow, it is important to compare your information with historical records for wet and dry years. Depending on the location of your waterway, there may be a lot of historical data about its flow. It is important to have good measurements on the water flow as it will vary seasonally and annually. Knowing both minimum possible power output as well as peak flooding will help you appropriately design the

Table 16.1 Head loss chart and pipe size for 100′ of pipe. Courtesy of Canyon Hydro.[3]

| **Head Loss Chart** | | | | | | | | | | | | | | |
|---|---|---|---|---|---|---|---|---|---|---|---|---|---|---|
| Design Flow | | | | | | | | | | | | | | |
| GPM | 25 | 50 | 100 | 150 | 200 | 300 | 400 | 500 | 600 | 700 | 800 | 900 | 1000 | 1200 |
| LPS | 1.577 | 3.155 | 6.309 | 9.464 | 12.62 | 18.93 | 25.24 | 31.55 | 37.85 | 44.16 | 50.47 | 56.78 | 63.09 | 75.71 |
| Pipe size and loss per 100 feet (30.48 m) | | | | | | | | | | | | | | |
| 2″ | 1.28′ | 4.65′ | 16.8′ | 35.7′ | 60.6′ | 99.2′ | | | | | | | | |
| 5.08 cm | 0.39 m | 1.42 m | 5.49 m | 10.88 m | 18.47 m | 30.24 m | | | | | | | | |
| 3″ | 0.18′ | 0.65′ | 2.33′ | 4.93′ | 8.36′ | 17.9′ | 30.6′ | 46.1′ | 64.4′ | | | | | |
| 7.62 cm | 0.06 m | 0.20 m | 0.71 m | 1.50 m | 2.55 m | 5.46 m | 9.33 m | 14.05 m | 19.63 m | | | | | |
| 4″ | 0.04′ | 0.16′ | 0.57′ | 1.23′ | 2.02′ | 4.37′ | 7.52′ | 11.3′ | 15.8′ | 21.1′ | 26.8′ | 33.4′ | | |
| 10.16 cm | 0.01 m | 0.05 m | 0.17 m | 0.38 m | 0.62 m | 1.33 m | 2.29 m | 3.44 m | 4.82 m | 6.43 m | 8.17 m | 10.18 m | | |
| 6″ | | 0.02′ | 0.08′ | 0.17′ | 0.29′ | 0.62′ | 1.03′ | 1.36′ | 2.2′ | 2.92′ | 3.74′ | 4.75′ | 5.66′ | 8.04′ |
| 15.24 cm | | 0.01 m | 0.02 m | 0.05 m | 0.09 m | 0.19 m | 0.31 m | 0.42 m | 0.67 m | 0.89 m | 1.14 m | 1.45 m | 1.73 m | 2.45 m |
| 8″ | | | | 0.04′ | 0.07′ | 0.15′ | 0.25′ | 0.39′ | 0.5′ | 0.72′ | 0.89′ | 1.16′ | 1.4′ | 1.96′ |
| 20.32 cm | | | | 0.01 m | 0.02 m | 0.05 m | 0.08 m | 0.12 m | 0.15 m | 0.22 m | 0.27 m | 0.35 m | 0.43 m | 0.60 m |

location of the penstock and powerhouse. You'll need to measure the flow in the stream over the course of the year so you can determine what is a consistent amount of water that is available all year. While there is not a fixed percentage for how much water you might use, you probably shouldn't divert more than 20% of the water to maintain low impact on the stream's ecology. It is not a good resource if the stream regularly goes dry.

If you have to measure flow yourself, if it is a fairly small stream, you can build a temporary dam in the stream so you can divert the flow into a bucket or container. The rate at which the container fills is the flow rate. For example, if you have a 10 gallon (37.9 L) bucket and it fills in 1 minute (60 seconds), your flow rate is 10gpm (2,274L/s).

Another method is to measure with a weighted float (figure 16.19). This method cannot be used in deeper water, more than 3' (914mm), as it can be unsafe (and difficult) to measure the cross section of the waterway. If the water is very fast flowing, it may also be unsafe to wade in it and attempt to measure it. Remember that this should be done when the stream has its least amount of water. Find the straightest and most uniform area along the stream. At the narrowest section, put up a string across the stream at 90 degrees and measure the stream's width. Measure depths at a regular interval across the stream and plot onto graph paper to generate a cross section of the stream. Use the graph areas to determine the area of the water as a cross section. Measure at least 20' (6.1m) upstream from your string and mark that point. Write down how far the points are from each other. You need an assistant to drop the float in at the upstream point as you time in seconds how long it takes for the float to reach the downstream string. The float should float well and not drag on the streambed. Take multiple measurements and average them to achieve accuracy.

To determine the flow velocity, divide the distance between the two points by the average time it took the float to arrive at the downstream point. Multiply the average velocity by the cross-sectional area of the stream. The last step is to multiply your result by a factor to account for the type of surface the stream has. Use 0.8 for sand, 0.7 for small to medium sized stones, and 0.6 for many large stones. Your final number is the flow rate in cubic feet or meters per second.

Building a **weir** is a very accurate way to determine flow. It requires building a temporary dam that has a rectangular opening, called a gate, that allows water to flow over it. You'll measure the depth of the water a good distance upstream from the weir and the width of the gate. You'll need a chart to look up flow based on these numbers.

Remember that you are only going to divert some of this water – if you use 20% then the design flow number is 20% of the flow for the whole stream. In the end, the amount of flow needed will depend on how much power you need to meet the loads in your building.

## Power Equation

The power equation is the production in watts. Net head is the gross head minus losses due to pipe friction and bends. Flow is the quantity of water. Efficiency of

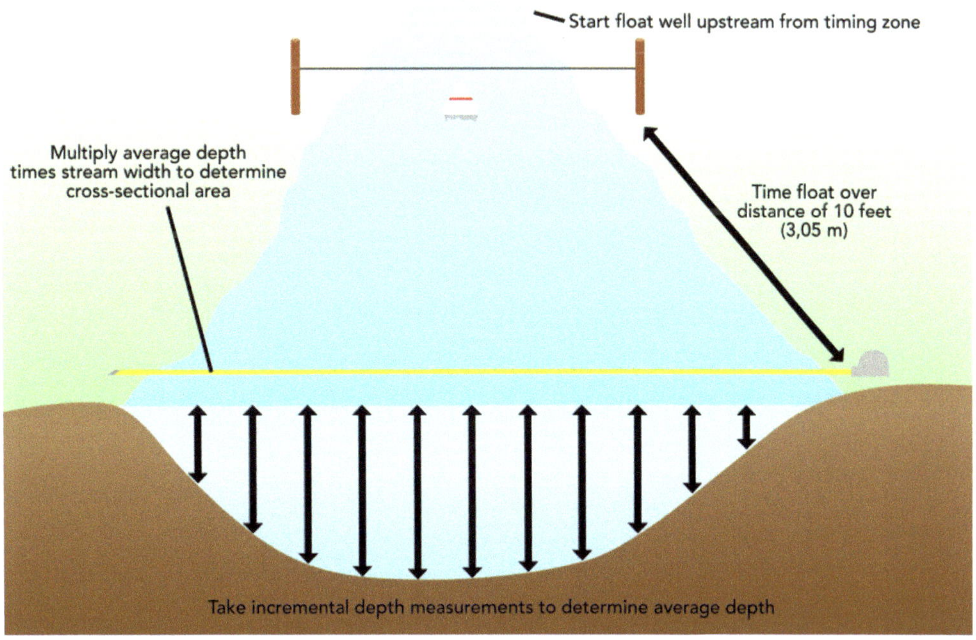


Figure 16.19 Measuring flow with a weighted float. Courtesy of Canyon Hydro.[4]

system is assumed as average of 53% (microhydro is generally between 40–60% efficient).

US units

Watts = net head (feet) × flow (gpm) × 0.1

SI units

Watts = net head (meters) × flow (L/s) × 5.1993

Conversion: $m^3/s \times 1000 = L/s$

From our example earlier, we had a net head of 55.26′ (16.84m) and a flow of 300gpm (18.93lps).

Watts = 55.26′ × 300gpm × 0.1 = 1,657W

Watts = 16.84m × 18.93L/s × 5.1993 = 1,657W

1,657W / 1000kW/W = 1.657kW power rating for this system.

Because this system will produce power all day every day, it is straightforward to determine monthly average or annual production.

Daily production = 1.657kW × 24 hours/day = 39.768kWh/day

Average monthly = daily production × 30 days/month

Average monthly = 39.768kWh/day × 30 days/month = 1,193.04kWh/month

Annual production = daily production × 365 days/year

Annual production = 39.768kWh/day × 365 days/year = 14,515kWh/year

If this is a grid-tied system, look up the cost of electricity in the area and multiply it by the production to find the value.

Value of annual production = annual production × cost of electricity

Presuming that electricity costs US$0.12/kWh:

Value of annual production = 14,515kWh/year × $0.12/kWh = $1,741.80

## Resources


1 Adapted from *Ossberger Turbine Section from Ossberger GmbH + Co in Germany* [Clip art], by PC21, April 4, 2007, Wikimedia Commons (https://commons.wikimedia.org/wiki/File:Ossberger_turbine.jpg). CC BY-SA 2.5 (changes were made: redrawn, background eliminated, and labeled).

2 Adapted from *Water Turbine Application Chart* [Clip art], by Pud, July 11, 2004, Wikimedia Commons (https://commons.wikimedia.org/wiki/File:Water_Turbine_Chart.png). CC BY-SA 3.0 (changes were made: chart was redrawn, gray background removed, labels as legend by color).

3 Head loss table courtesy of Canyon Hydro (www.canyonhydro.com/).

4 Image based on information from Canyon Hydro (www.canyonhydro.com/).


## Study Questions

1. Draw a diagram of a grid-tied microhydro system and label the components.
2. Draw a diagram of a grid-tied microhydro system with battery backup and label the components.
3. Draw a diagram of an off-grid microhydro system without batteries and label the components.
4. Draw a diagram of an off-grid microhydro system with battery backup and label the components.
5. Do an online search for "hydro turbine" and identify the types of turbines shown in the image results (i.e., Pelton, Francis, Kaplan, or cross-flow).

CHAPTER 17

# Marine Renewable Energy

**Key Terms**

Marine renewable energy (MRE): renewable energy that is installed and operated in seawater.
Ocean current: a continuous, predictable, directional movement of sea water.
Tide: the rising and falling of a very large body of water in a particular location caused by the gravitational pull of moon and sun.

## Concepts for Success

About 70% of the earth's surface is covered by water and 97% of all water is in the ocean. About one third of the world's population (2.4 billion people) live within 60 miles of an ocean coastline. The first patent for producing energy from wave power was in 1799.

Tides are caused by the gravitational pull of the moon and sun. **Ocean currents** are created by tides and the earth's rotation. Surface waves are created by winds. In general, water and wind behave similarly, however water has a higher specific gravity so it can produce more electricity at lower velocities than wind.

**Marine renewable energy (MRE)**, also called **marine and hydrokinetic energy (MHK)**, creates energy from wave action and currents.

Costs for MRE are currently high. At this time, the generators are all large, at a community or utility scale.

Devices that capture wave energy are grouped in categories of attenuator, point absorber, pressure differential, and terminators. Each of these types harnesses the wave energy in a different way. Oscillating water column terminator devices have been the most commercially viable to date.

Devices that generate electricity from currents are very similar to wind turbines.

## Creation of Tides, Currents, and Waves

Tides are primarily created by the gravitational pull of the moon and sun on a large, deep volume of water. Tides are long-period waves because the large amounts of water movement can take hours to wash ashore. The moon has more gravitational pull than the sun because it is closer to the earth. The times of greatest pull are when the moon is directly facing or directly opposite a location on earth. Tidal effects between locations vary from no tide to multiple tides in a day due to the moon's orbit. When the sun and moon are in alignment at the full and new moon, they have a stronger gravitational pull which creates high tides. Because the orbits of the sun and moon are cyclical and tides have been recorded for centuries, tides are consistent and predictable. Tides are only noticeable on very large bodies of water such as oceans, seas, and large lakes (like the Great Lakes in the USA). The energy embodied in the rise and fall of tides can be used to create electricity.

As tides flow, they can move through land constrictions, which can create faster water velocity that may be used to produce electricity. Saltstraumen in Norway has one of the strongest tidal currents worldwide. The current is created by large volumes of water trying to enter and exit the Skjerstad Fjord through a narrow inlet. As the tide comes in it has been measured at 6–7 knots (6.9–8 mph or 11–12.9 km/h), and on exiting it reaches over 8 knots (9.2 mph or 14.8 km/h). The average is 6 knots for about 3 hours in each direction. If you look at photos of the inlet, you'll see visible vortices (swirling water) along the banks. These vortices would make harnessing the power of this tidal flow difficult because the water isn't all moving in one direction. This is similar to issues with turbulence in wind, which reduces a turbine's ability to capture power.

DOI: 10.1201/9781003297819-19

There are currents that cross the oceans, caused not only by gravitational pull but also the earth's rotation. These currents are sometimes called submarine rivers and are responsible for moving vast amounts of water around the globe. Many of them circle in areas between landforms, creating five major gyres (large sets of rotating ocean currents). These currents are used (or avoided) by sailors because of their constant velocities and have great potential for future development as an energy resource.

Surface waves are created by wind. Wind is created by unequal heating of the atmosphere from the sun. Wind speed, duration, and fetch (the distance traveled along the water surface) influence how large the waves will be. The wind transfers energy, causing the water to move in a circular motion for long distances without losing much energy. The normal waves you experience at the beach are created by predictable wind patterns.

Hazardous waves can be created by severe storms which have a large storm surge that is much taller and deeper than normal tides. A storm surge is a large sea level rise that arrives all at once, rather than being a big crashing wave. Earthquakes and landslides can create long waves (tsunamis) because they quickly displace water that travels over a great distance. Storm surges and tsunamis have huge destructive potential and cannot be harnessed for creating electrical power.

Production of electrical energy can only use regular, non-turbulent, strong water movement. Like wind, the velocity of water movement is the main factor in how much energy can be produced.

## Marine Renewable Energy Generation

MRE creates energy from wave action and currents. MRE lags behind almost all other types of commercial RE installation. MRE has the advantage of having 24/7 production and documented tidal, wave, and current potential.

Unfortunately, the list holding back development is topped by large price tags. Costs are driven up because of the corrosiveness of salt water, building in durability to withstand the harsh ocean environment and storms, and the large size of most of the current technology. MRE hasn't been a priority as better profits have been found in land-based renewables and improving battery storage to overcome production downtimes for solar and wind.

There are also large environmental hurtles as many jurisdictions require regulatory studies and documentation to gain permission. Environmental concerns include negative impacts on marine animals and changing their habitats, possible changes in water quality and sediment transportation patterns, and impacts on animals that use the earth's magnetic fields to navigate the oceans.

## Wave Energy Power and Categorization

The amount of power available is based on wave height, wave speed, wavelength, and water density. There is a lot of experimentation and development happening for converting wave energy to electricity. There is no particular device that has taken the lead. These categories group similar technology: attenuator; point absorber; pressure differential; and terminators, which include oscillating water columns, overtopping, and oscillating wave surge converter.[1] Each type is located either on the shoreline (where waves are crashing in), nearshore (close to the shore, the shoreline is visible), or offshore (shoreline is not visible, about 10 miles (16 km) out from shore).

### Attenuator

An attenuator has paddles or sections that follow the surface of the water, either parallel or perpendicular to wave motion (figure 17.1). As the paddles or sections move up and down and side-to-side (pitch and yaw) relative to each other, the flexing moves hydraulic pumps which create energy by moving a generator. The location and type of generator depends on the configuration of the device. The energy is cabled to an onshore energy station. Attenuators are located nearshore, either as part of a larger floating or fixed

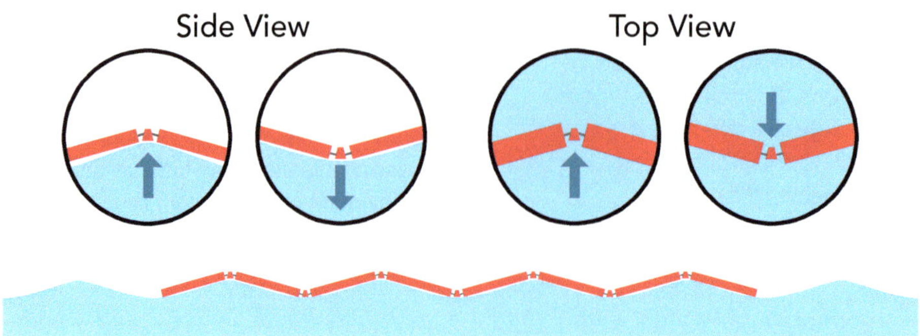


**Figure 17.1** Wave attenuators move up and down or side-to-side based on the water's surface movement.

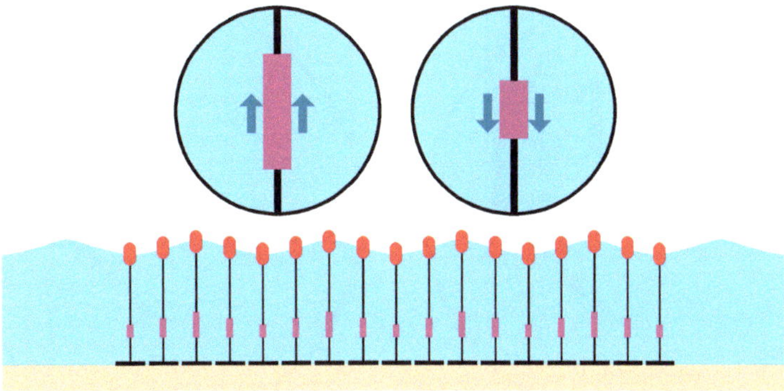

Figure 17.2 Point absorbers are a tethered buoy that moves with the surface waves.

object containing the generator, or they are tethered to the ocean floor.

## Point Absorber

A point absorber has a buoy that moves up and down with the movement of the waves, absorbing the energy from all directions (figure 17.2). The buoy is tethered (offshore locations) or anchored (nearshore locations) so it remains in place. Each buoy has its own generator. The buoy structure may have 8–10′ (2.44–3.05m) showing above the water surface and extend as much as 150′ (45.7m) to the ocean floor. The movement of the buoy powers either a linear or rotary generator. For offshore locations, the electricity is cabled to an undersea substation then to shore. For nearshore locations, the electricity is cabled to an onshore station. Some nearshore point absorbers use the buoy movement to pump fluid through an onshore generator.

## Pressure Differential

With pressure differential the unit is located nearshore on the ocean floor below the waves and uses the pressure difference between sets of waves to compress and expand air in a chamber or set of chambers (figure 17.3). The air movement in or between the chambers powers a turbine and generator and the energy is cabled to an onshore station.

## Terminator Devices

Terminator devices are oriented perpendicular to the waves (i.e., parallel to the shoreline). They have stationary parts that anchor them in place and moving parts that capture the wave energy. Most commonly they are found onshore or nearshore but some types have been developed for offshore locations.

## Oscillating Water Column

Oscillating water column: a partially submerged chamber is pressurized and depressurized because of wave action. Nearshore units float and are tethered in place. Onshore units are a box-like structure with a sloping face toward the incoming waves (figure 17.4). As waves crest, the column of water trapped inside the chamber rises, and when the waves fall, so to does the water column. The changing pressure inside the chamber causes air movement that either rotates a turbine and generator or drives a fluid pump that creates energy in a building nearby. This has been the most commercially successful technology to date.



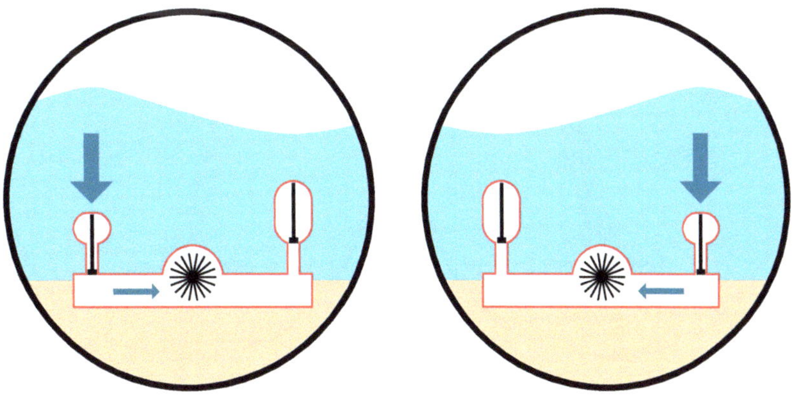

Figure 17.3 Pressure differential devices use the changing pressure of the waves above to expand and contract air in a chamber.

**Figure 17.4** An oscillating water column has a partially submerged chamber where the water moves up and down as waves come ashore, which pressurizes and depressurizes the air pocket, causing it to move and generate electricity.

## Oscillating Wave Surge Converter

An oscillating wave surge converter is usually a large piece of equipment located near shore which is submerged below the water surface (figure 17.5). It has a bottom plate or foundation fixed in place to the ocean floor. The equipment is hinged to the foundation, allowing the oscillator to sway back and forth as the waves move above it. The appearance of the body and type of materials used varies. At the connection to the hinge, the oscillator either generates electricity due to the movement or creates a pressurized flow for hydropower generation in a building nearby onshore.

## Overtopping

With overtopping there is a wide overtopping ramp that is perpendicular to the waves. Waves coming over the ramp fill a reservoir with water (figure 17.6). The water drains, by gravity, through a conventional hydropower turbine back into the ocean. Overtopping devices are installed nearshore and are similar to a large floating pool that is tethered in place. The electricity is cabled to an onshore building.

## Current Generation

Current generation technology parallels wind with horizontal axial flow turbines that act like a downwind HAWT (figure 17.7 left), cross-flow turbines that look like a Derrius VAWT (figure 17.7 center) and anchored kites. There are also oscillating devices that are orientated so that they move in the current similarly to laundry waving in a breeze (figure 17.7 right).

The world's largest tidal farm, a grouping of multiple generators in the same general area, is Project Stroma (MeyGen) in the UK, 2km from the north-east tip of Scotland.[2] It has been in phased development since 2007. At the end of phase 3 it will be a 146MW project. The phase 1 turbines are axial flow, have pitchable blades and a large tripod base that anchors the unit in place. The turbines are connected to a single power export cable to an onshore building where the power is converted to higher voltage for grid distribution. The project placed 13.8GWh of electricity on the grid in 2019.

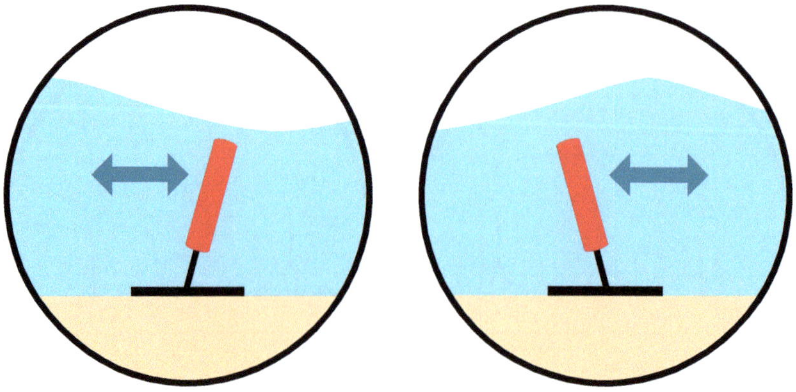

**Figure 17.5** An oscillating wave surge converter moves back-and-forth in the waves.

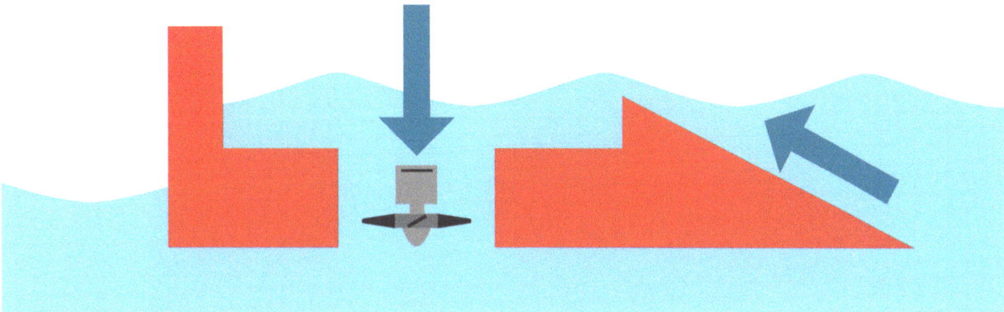

**Figure 17.6** Overtopping devices catch waves over a ramp and catch the water in a reservoir. Water returns to the ocean through a hydro turbine.

**Figure 17.7** Axial, cross-flow and oscillating hydrofoil type of current generation devices.

## Resources


1 Water Power Technologies Office. (n.d.). Marine and hydrokinetic technology glossary. US Department of Energy. www.energy.gov/eere/water/marine-and-hydrokinetic-technology-glossary

2 Adapted from Illustrations based on US Department of Energy images at: SIMEC Atlantis Energy. (n.d.). *MeyGen – Tidal projects*. https://simecatlantis.com/projects/meygen/


## Study Questions

1. List several advantages and disadvantages of MRE.
2. List the categories of MRE technology and provide a brief description of each one.
3. Do an online search of two of the MRE technologies and follow links through to a manufacturer and look at the information they have about the device. Is the device under development or has it been commercially installed?

PART III

# System Implementation, Monitoring, and Payback

CHAPTER 18

# Batteries and Energy Storage

**Key Terms**

Cell (in a battery): the smallest unit making up a battery; single anode (negative) and cathode (positive) metal plates are separated by an electrolyte, which produces voltage and current through chemical reactions.

Days of autonomy: the theoretical number of days that the battery bank would be supplying the load without recharging.

Deep cycle battery: a battery that has been designed to have many cycles of charging and discharging while maintaining its capacity.

Depth of discharge (DOD): the percentage of energy used from a battery; the amount that the battery has discharged, where 50% would be halfway; SOC + DOD = 100%

Islanding: when a building, campus, or community has electrical power during a grid outage.

Phase change material (PCM): stores heat energy through state changes, absorbing and releasing heat when changing between solid and liquid or gaseous state.

State of charge (SOC): the charge level of a battery, where 100% is fully charged.

## Concepts for Success

Batteries are used to store energy that can be used when a renewable is not producing, or the grid fails. They can also be used to mitigate grid demand. The goal when including batteries in a system is to provide power to building loads. Batteries store excess power during production times (charging) and then provide power (discharging). All batteries use a chemical reaction to create the flow of electrons.

The most common batteries used in RE systems today are deep cycle rechargeable lithium-ion or lead-acid. **Flooded lead-acid (FLA)** batteries have been used in systems for years but are being replaced by lithium-ion. Other types of lead-acid batteries are **absorbed glass mat (AGM)** and **gel electrolyte (Gel)**. Both of these batteries were developed to eliminate spills of corrosive battery acid. They are lighter weight than FLA and can be stored in any position.

Batteries have to be stored, charged, and cared for per the manufacturer's recommendations. Batteries lose capacity and life expectancy if they are subjected to cold or hot conditions. Batteries have a three-stage charging process: bulk (85% charged), absorption (100% charged), and float (continuous charge to maintain 100% charge).

The size of the battery bank is primarily determined by the loads you want to power every day, the **days of autonomy**, and the type of battery chosen.

Community and utility scale energy storage systems include pumped storage hydropower (PSH), **molten salt energy storage (MSES)**, and **vanadium redox flow (VRF)** technology.

## Why Use Batteries?

Batteries are used in RE systems for the following reasons:

- Off-grid buildings need to be powered when RE isn't generating;
- Provide backup power for grid-tied buildings in the event of a grid failure (**islanding**);
- Grid-tied buildings can use a battery bank to reduce grid draw at peak times and/or stay below a tier pricing level change for kWh used; and

DOI: 10.1201/9781003297819-21

- Grid-tied storage, accessed through agreement with the power company, used to provide continuity of power as grid usage shifts (particularly an issue in areas of high PV production, which drops off as the sun sets, requiring large ramping up of production).

When you add a battery bank to a system, your goal is to keep building loads powered. The efficiency of RE is typically found by comparing the amount of energy generated to the amount of AC energy that gets sent from the inverter. Each part of the system has different efficiencies; however, batteries tend to lower the overall system AC output by 20–25%. This is because batteries, even if they are not being used daily, self-discharge and need to be recharged and kept topped off with a constant voltage flow. This additional loss needs to be accounted for in systems that are daily or yearly net zero.

## Battery Basics

A battery is a collection of one or more **cells** in a case where the cells store energy which is released as DC electrical power. You are most likely familiar with non-rechargeable batteries, used to power everyday electronics, like those shown in figure 18.1.

**Figure 18.1** Store display of various types of chemical, non-rechargeable batteries.

A battery cell has two different metal plates (varies by battery type) which have different polarity: the cathode (positive) and anode (negative). They do not touch each other within the cell but are separated and submerged in an electrolyte (which can be liquid or paste-like). Electrolytes are made of different solutions based on the battery type. A chemical reaction occurs in the electrolyte that releases the stored energy in the cell to the cathode and anode. The appearance of the cell and the separation between the cathode and anode depends on the battery type.

The batteries in figure 18.1 are single cell and are either zinc-carbon or alkaline in their chemical makeup; they are not used in RE systems. Car batteries, while being rechargeable, are not used in RE systems (although there are developments for using an electric vehicle as backup power).

Rechargeable batteries used in RE systems are labeled deep cycle because they are made to withstand many cycles of being discharged and recharged without damage, while retaining their ability to return to their fully recharged state. **State of charge (SOC)** is used to indicate the charge level in a battery. If a battery is fully charged it has a SOC of 100%. The **depth of discharge (DOD)** indicates how much energy you've used from the battery. SOC + DOD = 100%, so a battery that has 75% SOC is at 25% DOD. The discharge level indicates how much energy must be put in the battery to fully charge it again. A discharge and subsequent recharge of the battery is a cycle. All battery specifications have a recommended DOD. Typically, the greater the DOD, the fewer cycles your battery will be able to handle – this is particularly true for lead-acid batteries.

## Types of Deep Cycle Rechargeable Batteries

There are two types of deep cycle rechargeable batteries: lithium-ion and lead-acid. Lithium-ion battery technology has improved and these types of batteries are quickly overtaking lead-acid in popularity. Lead-acid batteries have multiple lead plates which are submerged in an acidic electrolyte.

### Lithium-Ion

Lithium-ion (Li-ion) batteries have cells that use liquid (most common) or solid ceramic electrolytes (this eliminates leaking and flammability issues). The cathode and anode are separated by a thin layer of insulating material, called a separator, which allows through a flow of lithium ions while blocking electron transfer. The movement of the lithium ions creates voltage. During charging, the lithium ions move from the cathode side to the anode side. During discharging, the lithium ions move from the anode side to the cathode side (figure 18.2).

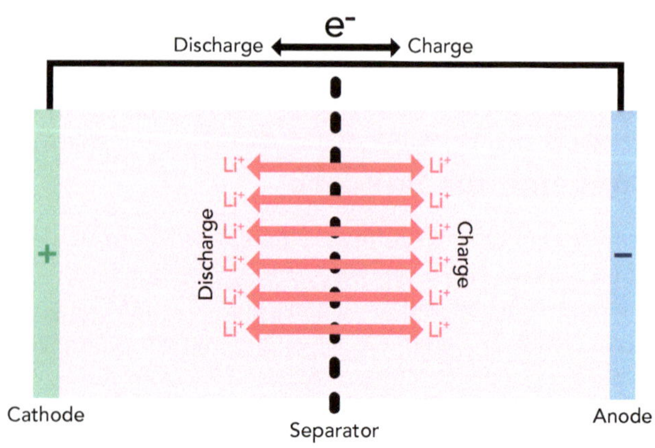


**Figure 18.2** Charging and discharging of a lithium-ion cell.

Lithium-ion batteries used in RE are typically $LiFePO_4$ (lithium ferrophosphate or LFP) and have advantages of longer life, better deep cycling, and cold ratings lower than lead-acid. They come in 12V, 24V, and 48V configurations. Although individual lithium-ion batteries cost more than lead-acid batteries, their number of cycles is much higher. They can also withstand greater discharge – manufacturers recommend 80% DOD or even 100% DOD. This means that you can provide fewer batteries for your system and a lithium-ion battery can last as much as three times longer than a lead-acid battery. This makes the cost of lithium-ion batteries competitive over the long haul. Warranties for lithium-ion batteries are typically longer than for lead-acid types.

Lithium-ion batteries also tend to weigh less than lead-acid batteries when comparing similar voltage and Ah values, making transportation and on-site storage easier and more compact.

## Lead-Acid Batteries

### Flooded Lead-Acid

Flooded lead-acid (FLA) batteries are the original deep cycle battery. They have a well-established track record, are inexpensive, and work well if they are properly maintained. They are typically designed to 50% DOD. These batteries are made of 2V cells. Batteries can have between 1 and 6 cells (figure 18.3 right), the most common are 2V, 6V, and 12V – which can then be wired to create 12V, 24V, or 48V systems. Each battery cell is its own compartment where the chemical reactions for charging and discharging happen (figure 18.3 left). The interior of each cell has positive and negative lead plates which are separated from each other. The positive plates are strapped together and the negative plates are strapped together. The straps are connected to the posts (terminals) by the intercell connector, providing power to the battery's positive and negative terminals and the circuit for electrons to flow through the battery. Terminal posts come in different configurations and must be compatible with the ends on the cables you are using to connect the battery bank together.

The interior of each cell is flooded with an electrolyte which is a mixture of sulfuric acid and water. When a load is connected to the battery, the sulfur molecules from the electrolyte bond with the lead plates, causing electrons in the plates to be released and flow through the circuit. The cell is 100% discharged (dead) (100% DOD) when all the sulfur molecules from the electrolyte are bonded to the lead plates. Recharging the battery causes the sulfur molecules to detach from the plates and return to the electrolyte.

The physical depth of the battery relates to how many amp-hours of energy it can store. Longer lead plates provide more surface area for the sulfur molecules to adhere to, allowing more energy to be stored. Having more cells makes batteries heavier; remember that the plates are lead and the cells are completely flooded with electrolyte. Because of their weight, most battery cases have straps that tie securely to the case. This makes it much easier to lift batteries in and out of a tight-fitting battery box.

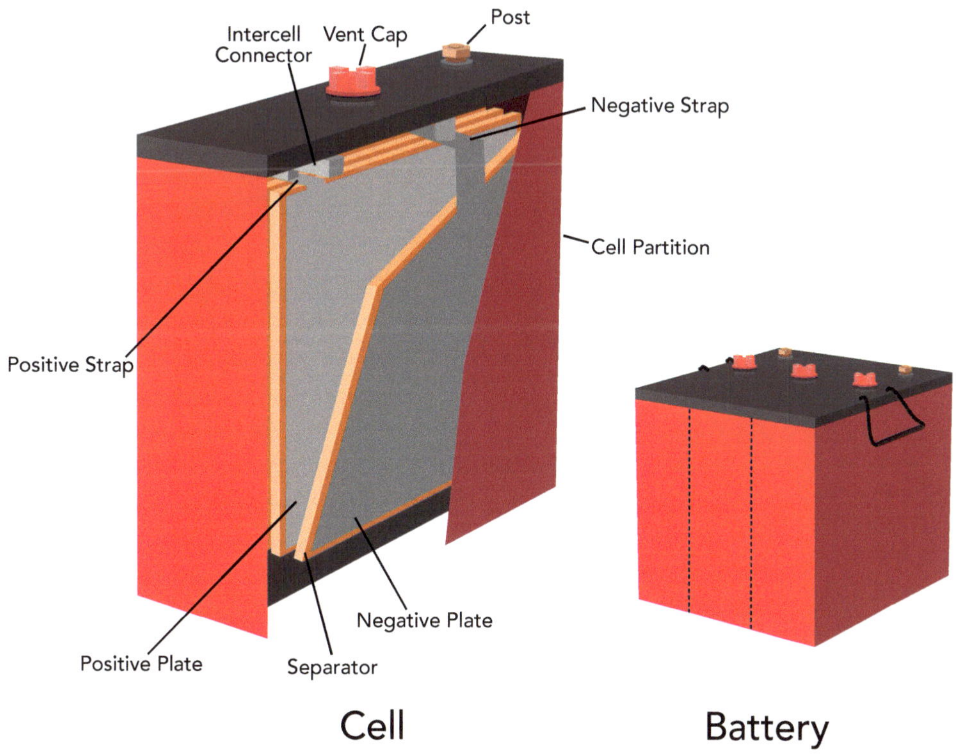


**Figure 18.3** Interior components of a flooded lead-acid battery cell.

### *Watering*

The caps on FLA batteries off-gas hydrogen and oxygen as the electrolyte moves and is heated up from charging and discharging. Batteries should be stored in a sealed battery box that is vented directly outside to avoid potential hydrogen explosions. The off-gassing is from water breaking down in the cell, so the level of the electrolyte needs to be checked periodically and replaced with distilled water if it is low. Heavily used systems may need to be checked weekly or monthly, lesser used systems every 6 months or annually. Distilled water is added to top off each cell: this can be quite time consuming if you have a large battery bank. Watering kits are available to automate watering. This helps to reduce splashing and the release of noxious fumes because the kits have special caps that replace the vent caps on each battery cell. These caps have automatic shut-off valves and connect by tubing to a container of distilled water. All the cells can be watered at the same time from one water storage container located above than the battery bank because the system is gravity fed. Check the manufacturer's information about the watering kit, or any other cap replacement, for compatibility with the equalizing cycle.

### *Specific Gravity*

FLA battery cells are checked at least annually to confirm that they are operating correctly. This is done by verifying the specific gravity of each cell. Specific gravity is the ratio of the weight of the electrolyte solution compared to the same volume of water. The specific gravity is measured when the cells are fully watered and at 100% SOC, and is used to verify that the cell is completely charged. A hydrometer is used to measure specific gravity and the reading can be compared to the charge controller and other system monitors. If a cell is not charging properly, it will affect the battery and battery bank performance.

A desulfator can be used to electrically shake a particular battery in a system to reduce sulfate buildup on the lead plates. There are also battery energy balancers that can be used to equalize the SOC imbalance between batteries in a battery bank by drawing up lower battery currents to match the highest voltage.

## Valve Regulated Lead-Acid Batteries

Sealed **valve regulated lead-acid (VRLA)** batteries were developed to eliminate maintenance, as well as create safer, lighter weight lead-acid batteries (compared to flooded lead-acid). They are sealed so that no off-gassing occurs and use internal valves to capture and remix hydrogen and oxygen back into the electrolyte. There are two types of VRLA batteries: absorbed glass mat (AGM) and gel electrolyte (Gel). VRLA batteries are never equalized.

### *Absorbed Glass Mat Batteries*

AGM batteries use fiberglass mats to absorb the acid electrolyte in the cell. There is no spillable liquid inside the battery because it is contained by the mats. This makes the battery less likely to sulfate and allows for 80% DOD. The cell mats are designed to provide maximum surface area of electrolyte to the plates which helps these batteries to charge faster. The battery can be installed at any angle. Figure 18.4 shows an AGM battery. Note the lack of battery caps (there is a big flat label). AGM batteries typically have a raised portion in the center of the battery top where the capture and remixing valves are located.

**Figure 18.4** AGM battery with no visible cell caps.

Lead-carbon AGM batteries are made with lead-calcium-tin cathode plates and lead-carbon anode plates. They have terminals for each cell as shown in figure 18.5. They provide high energy density, fast charge, rapid discharge, and excellent **deep cycle** life performance. They are suitable for solar, wind and hybrid RE systems and can be used with smart grids as well as islanding.

**Figure 18.5** AGM lead-carbon VRLA battery has terminals for each battery cell.

### *Gel Electrolyte Batteries*

Gel batteries use a silica additive that causes the electrolyte to gel. This prevents acidic stratification inside the cell and, like AGM batteries, can be placed in any orientation (figure 18.6). Gel batteries have a lower power capacity when compared to AGM and need to be recharged slowly or bubbles can form and, although they may eventually work their way to the remix valves, they can inhibit performance and kill the battery before the regular life-cycles are complete. Gel batteries work well in warmer climates and with systems that have slower discharge rates.

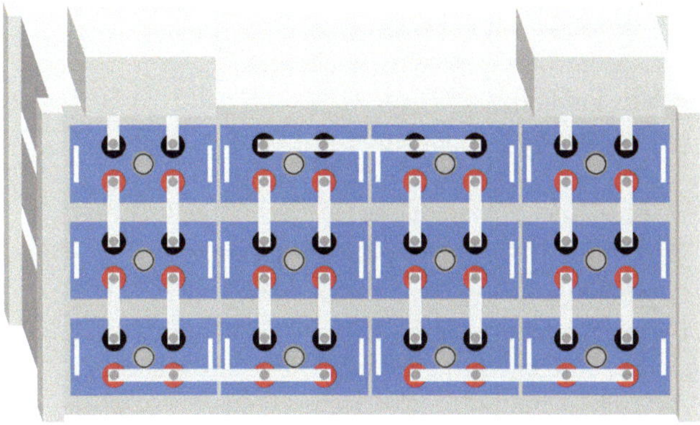

**Figure 18.6** Because their electrolyte is not spillable, and the cells are never opened, AGM and Gel batteries can be installed on their sides in a shelving unit (case) as shown. This can take up less floor space. Note that the shelving cases are specially designed to carry the battery weight and not rack or twist.

Gel batteries can have cathodes made of flat plate or tubular electrodes. Flat plate gel batteries have a flat top on the battery case. Tubular gel batteries usually have dimples or caps that show at each cell location on the battery case top. The tubular plate is a pressure cast lead alloy that is made up of tubes instead of being a flat plate or grid. The tubes have an appearance similar to a comb and are encased, so it looks like a series of tubes along the length of the electrode.

## Charging Process

### Three-Stage Charging

All batteries have a three-stage charging process (figure 18.7). The first stage, bulk, charges the battery to about 85% SOC using a high current. As the battery becomes more charged, higher voltage is needed. In this stage, the sulfite that was deposited on the lead plates is being dislodged. As the battery becomes more charged, the sulfites are harder to remove from the plates. The battery completes charging during the absorption stage. This requires higher voltage than the nominal rating of the battery. The manufacturer's directions for charging settings for the charge controller must be followed in order to prevent damage to batteries. Generally, AGM and FLA batteries have slightly higher voltage for all stages than Gel. In FLA batteries, during bulk charging, the electrolyte bubbles, equalizing the cell and destratifying the electrolyte. This is when cells off-gas because they cannot absorb all the energy being forced into the cell. Note that, during the absorption stage, the voltage remains high while the current drops. In the third stage, a float or trickle charge is constantly run through the battery to maintain 100% SOC.

### Equalization

FLA batteries have an equalization stage which is used periodically where cells are held at or above 2.6V for several hours to deliberately break up sulfation on the plates and stir up (destratify) the electrolyte. Lithium-ion and sealed VRLA batteries are never equalized. Many of these batteries have technology to shut down if subjected to overcharging in order to prevent battery damage and possible explosion.

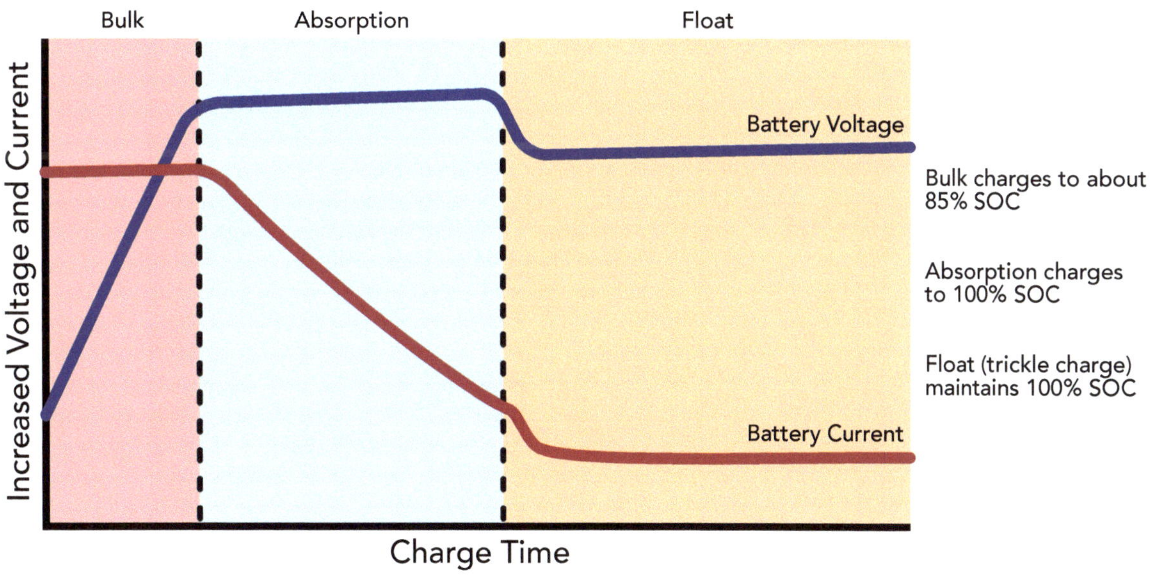


**Figure 18.7** Three-stage charging process for batteries.

18

## DOD, Cycling, and Battery Life

DOD is the measurement for how much energy you take from a battery. The more deeply you discharge a battery, the fewer number of times it will recharge fully. A discharge then recharge is called a cycle. As seen in table 18.1, the FLA battery will complete 1200 cycles at 50% DOD before it will no longer recharge appropriately. Compare this to the Li-ion battery which can complete 2000 cycles at 100% DOD. The AGM battery will complete 3000 cycles at 50% DOD, while the Gel will do 650 cycles at 50% DOD. It is very important to select a battery that will perform for the situation and cycling duty that you expect. Otherwise, the replacement times and costs of the batteries will be greater than expected over the life of the entire system. Table 18.1 compares basic specifications and costs of specific manufacturer's 12V batteries with different chemistry types.

## Operating Temperature

Batteries operate best in a thermally temperate space. There are some batteries that are rated for cold use and these usually have a built-in circuit to warm themselves. The perfect temperature for lead-acid batteries is 77°F (25°C). As temperatures drop, the rated Ah capacity of the battery bank drops. At freezing (32°F, 0°C), FLA batteries will lose around 25% of their capacity. AGM and Gel batteries do better in cold temperatures, perhaps losing only 15% of their capacity. Cycling at very cold temperatures can have a negative impact on battery life, as much as 50%. Some newer batteries will shut down to protect themselves from becoming damaged from cold weather operation. Batteries also do not like to be hot and should never be in direct sunlight. Electrical resistances increase in the heat, causing decreases in battery bank performance. All cut sheets for batteries will list the operating temperatures of the battery and often manufacturers include charging temperature compensation numbers for every 10°F (5.55°C) above or below 77°F (25°C).

## Battery Storage

Batteries need to be kept in a thermally controlled, safe environment. They store large amounts of energy and should never be where people could accidentally interfere with them.

Proper battery storage includes:

- Well insulated area
- Vented (any non-sealed type)
- Locked
- Off the floor
- Protected from damage from anything being dropped on them
- Appropriate warning signage per code (high voltage)
- Appropriate conduits for wiring to/from battery bank per code.

There are many types of battery bank boxes that are oriented either horizontally like a cooler chest or vertically like a refrigerator. Boxes range from very simple to heavy duty heated enclosures. Any kind of box or rack that batteries will be stored in or on must be rated to hold the weight of the batteries. Racks need to have appropriate bracing so that they do not fail. Never use batteries that have been damaged: they need to be appropriately disposed of to prevent harm to people and the environment. Damaged batteries can leak acid, burn skin, and catch fire.

The basic components of a battery bank box are shown in figure 18.8. All holes in the box, except the exhaust vent, should be located below the height of the top of the batteries (lower quarter of box). Highly flammable hydrogen has to be directly vented to the exterior. Use 2" (50mm) or larger pipe for venting. Because hydrogen is lighter than air, it will move upward, so having a sloped, sealed lid with weatherstripping is important. You can tape across the hinge and joint to further seal the lid. You may provide a continuously running in-line vent fan. The height of the box should be a bit taller than the batteries, so it is easy to see the batteries and cabling through the plexiglass lid. The bottom of the box should be insulated with high-density rigid foam

**Table 18.1** Comparison of deep-cycle battery types and their specifications.[1]

| Comparison of Different Battery Types | | | | |
|---|---|---|---|---|
| Battery Type | **Li-ion** | **Flooded Lead-Acid** | **Sealed AGM** | **Sealed Gel** |
| Manufacturer | KiloVaultTM | Crown® Battery | KiloVaultTM | MK/Deka® Solar |
| Model | 1200HLX+UL | CR-215 | 2100PLC | 8G4D |
| DC Voltage Nominal | 12.8V | 12V | 12V | 12V |
| Nominal Ah Capacity | 100Ah | 215Ah | 180Ah | 183Ah |
| Cycle Life | 5000 (80% DOD); 2000 (100% DOD) | 1200 (50% DOD) | 3000 (50% DOD) | 650 (50% DOD); 3800 (10% DOD) |
| Dimensions (L×W×H) | 12.5×6.5×9.25" (318×165×235mm) | 15.5×7×13.52" (394×178×343mm) | 22×4.92×12.6" (559×125×320mm) | 20.75×8.5×10" (527×216×25.4mm) |
| Weight | 25.8lb (11.7kg) | 120lb (54.4kg) | 127lb (57.6kg) | 131lb (58.9kg) |
| List price per battery, USD | $995 | $379.35 | $595 | $715.35 |

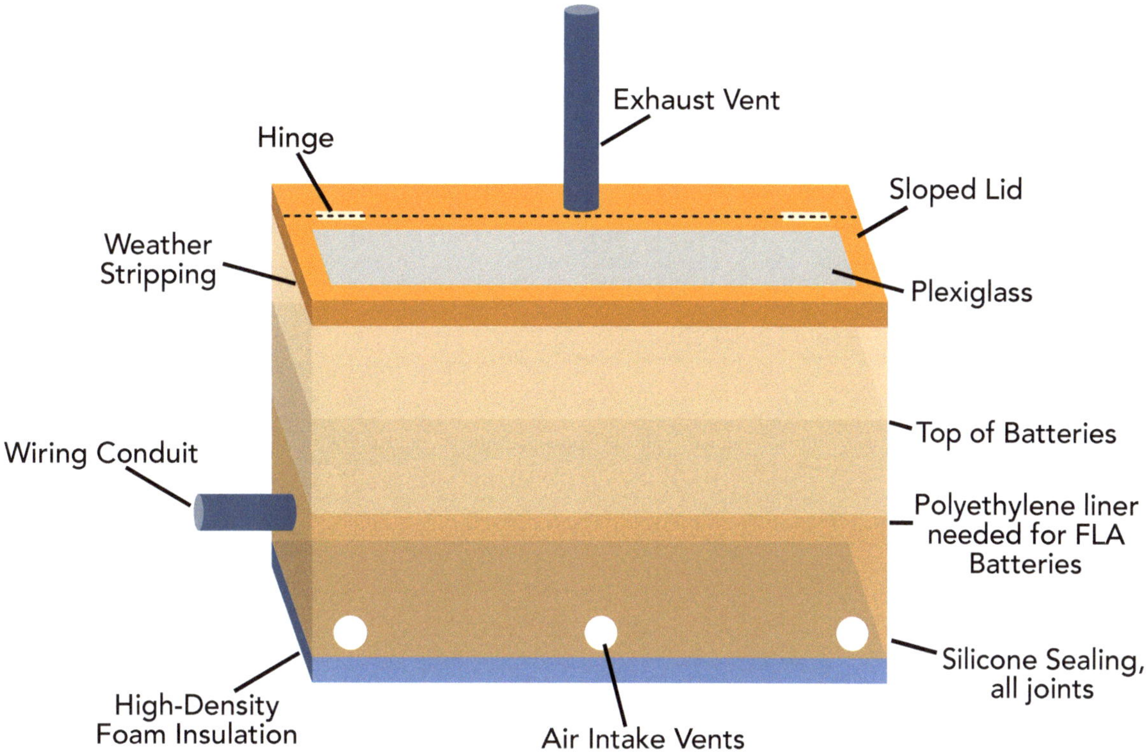


**Figure 18.8** Basic components of a battery bank box.

to create a thermal break with the cold floor. All joints should be sealed with silicone. Provide a polyethylene liner if using FLA batteries.

## Battery Wiring

Batteries follow the same wiring rules as other electrical components. Terminals connected in series add volts (amp-hours stay the same) while terminals connected in parallel add amp-hours (volts stay the same). Remember that every cell in a battery is 2V and they are connected internally to create 2V, 6V, or 12V batteries. Lithium-ion batteries have higher voltages: 12V, 24V, and 48V. Most batteries have one set of positive and negative terminals, although some AGM and Gel batteries provide terminals for each cell. Manufacturer's cut sheets will specify the maximum number of batteries that can be connected in series/parallel in a battery bank.

Battery banks are wired to 12V, 24V or 48V.

For practice wiring batteries see Appendix D.

## Battery Sizing

Battery sizing is based on the:

- Load that you want to power every day (both AC and DC)
- Inverter efficiency
- Days of autonomy
- Discharge limit of the battery
- Ah capacity of the battery
- System DC voltage (12V, 24V or 48V).

Days of autonomy refers to the theoretical number of days that the battery bank would be supplying the load without recharging. The higher the number of days you pick, the larger (and more expensive) the battery bank becomes. Often people pick 2 or possibly 3 days of autonomy. When the system reaches the maximum DOD, the backup generator will come on and charge the batteries.

For steps and examples of battery sizing, see Appendix C.

## Other Energy Storage

Although there has been some experimentation with grid-scale Li-ion battery banks (NREL deployed a 240**MWh** system with a maximum capacity of 60MW),[2] for community and utility power storage needs alternatives include pumped storage hydropower, molten salt energy storage, vanadium redox flow technology, and phase change materials (PCM).

Over 21GW of PSH exists in USA today and it is one of the fastest growing large energy storage types.[3] PSH uses excess energy being produced by RE and uses the energy to pump water to an upper reservoir (figure 18.9). When RE production ceases, the water is gravity fed through the penstock and powers the turbine.

MSES is a successful commercial technology used with concentrated solar power which transfers heat into an insulated tank then is used after sunset to create energy from steam. This storage can provide up to 15 hours of production. See figures 1.24 and 1.25.

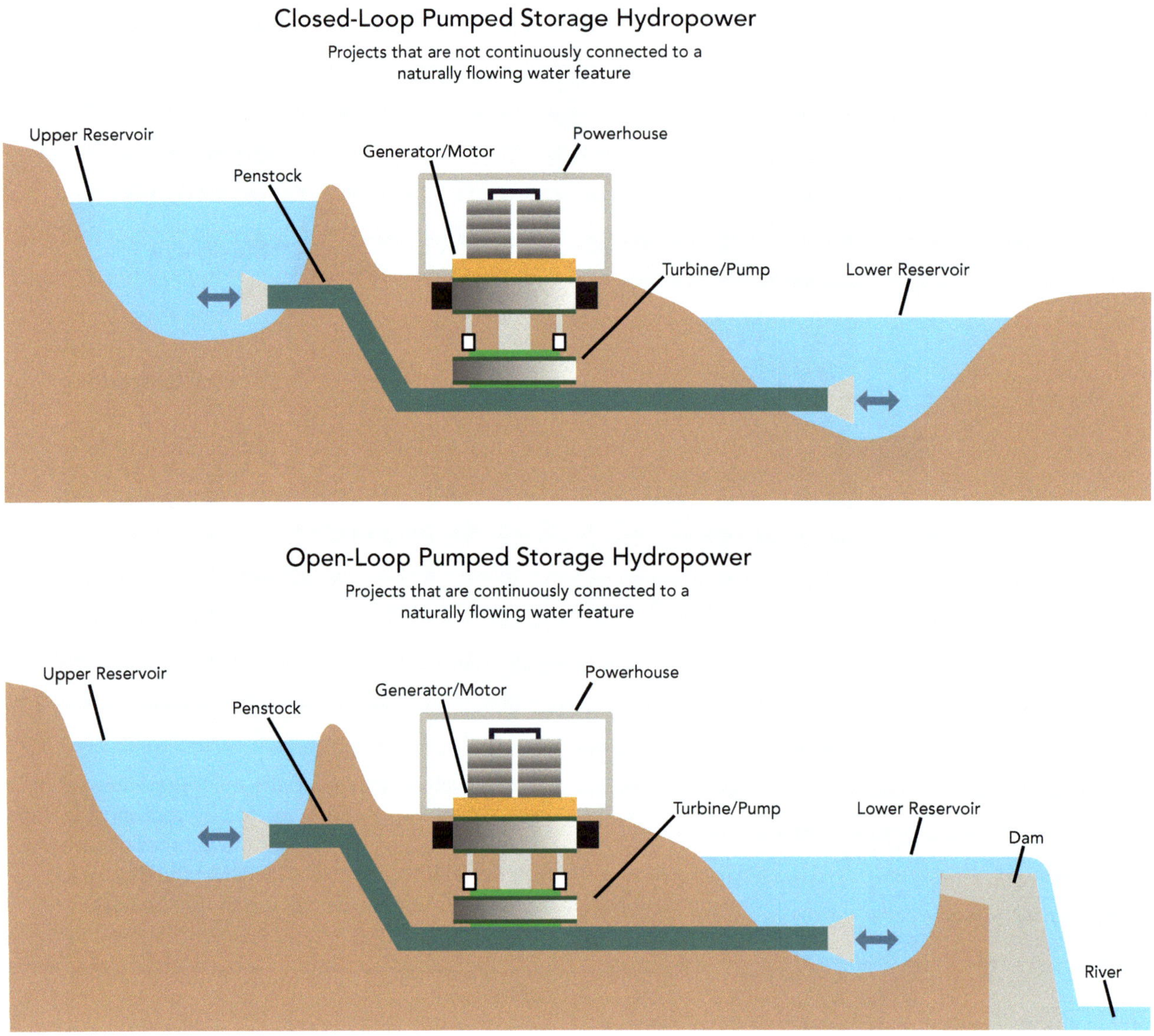


**Figure 18.9** Closed-loop and open-loop pumped storage hydropower use the upper reservoir to store water pumped when there is excess RE generation, recovering the power later by dropping it through a turbine.[3]

VRF was invented in the 1970s by NASA and has been used for large-scale bulk storage. There are companies beginning to offer smaller systems for single homes. VRF works by having two separate tanks containing a vanadium-based electrolyte liquid, one positively charged and the other negatively charged. The electrolyte is pumped into battery cells only when power is needed. The electrolytes are kept separate in the battery by a membrane that allows the transfer of ions. The advantage of this battery is that it doesn't require rare metals and is 100% recyclable. They are non-flammable and they maintain initial capacity after as many as 10,000 cycles.

Phase change materials store heat energy through state changes, absorbing and releasing heat when changing between solid and liquid or gaseous state. These can be included in building materials (see the Rocky Mountain Institute case study, chapter 23A) or separate storage tanks. One example of the use of PCM is for pole mounted PV tracker units that are unpowered. The tracker works because the solar heat on the face of the PV causes a state change and changes the balance of the tracking system. These trackers will follow the sun all day and stop facing west. When they heat up in the morning, they tilt back to the east.

## Resources


1. Pricing and manufacturer's specifications per the altE Store (www.altestore.com/store/).
2. Solar Energy Technologies Office. (2019, March 11). *Solar-plus-storage 101*. US Department of Energy. www.energy.gov/eere/solar/articles/solar-plus-storage-101
3. Water Power Technologies Office. (n.d.). *Pumped storage hydropower.* US Department of Energy. www.energy.gov/eere/water/pumped-storage-hydropower

## Study Questions

1. Answer the following about battery charging:
   a. What is the relationship between SOC and DOD?
   b. What DOD is recommended for Li-ion batteries compared to FLA batteries?
   c. What are the three main steps of charging and what happens during each step?
   d. What is equalization and what kind of batteries are equalized?
2. What are the requirements for proper battery storage?
3. What are the typical DC nominal voltages of a battery bank?
4. What are other types of energy storage besides batteries and how do they work?

CHAPTER 19

# Controllers and Monitors

**Key Terms**

Charge controller: a piece of electronic equipment that limits the rate at which electric current is added or withdrawn from a battery bank.

Hybrid generation: an RE system that combines two generators, typically PV and wind.

Shunt: a low resistance conductor wired in parallel that joins two points of a circuit in order to measure electric current.

## Concepts for Success

Charge controllers protect the battery bank by appropriately controlling how the RE is fed into the battery during charging and float cycles. Wind and hydro turbines need a dump load, diversion controllers can be set to automatically move production to a dump load if necessary.

Smaller controllers match to smaller systems and have fewer features. They will most likely use **pulse width modulation (PWM)** to charge the batteries. Larger controllers match to larger systems and have many possible features such as maximum power point tracking (MPPT) for PV arrays, LED readout panel, and **auto generator start (AGS)**.

Choosing a controller means matching the volts, amps, and watts (including some extra for safety) of your system with a controller that can do the job and has the features you want.

An extra meter for monitoring the system is a good idea: you can verify that the same information is showing up on the meter, controller, and inverter.

## Batteries Use Charge Controllers

Charge controllers are used to monitor and protect the battery. You only need a controller in systems with a battery bank or if you need a dump load for a hydro or wind turbine. A controller is an electrical device that controls the power feed from RE generation to the battery bank. The controller is also connected to the load or inverter. The controller makes sure that the battery is not overcharged (all controllers) or overly discharged (not a feature of very small controllers). On a PV system the controller isolates the battery bank from the array at night to prevent discharge. Diversion controllers are set up with a wind or hydro system to switch the load to a resistance dump load when the batteries are full. A diversion controller can act as either a regular controller or a dump load controller, unless specifically designed as a hybrid controller.

Figure 19.1 shows a small controller for a nominal 12V battery. It has very few functions and a simple set of lights to show information. This type of controller is often called a solar lighting controller because you can hook up a DC light to it. The light will turn on and off based on the PV module production and run the light for the specified times. This is a PWM controller. It controls battery charging by pulsing on and off. PWM works well for smaller systems and is low cost.

Larger charge controllers (figure 19.2) can be nominally 12V, 24V, or 48V. (Note that you'll see 36V systems as well, which are more common for boats and golf carts than RE systems.) Depending on the specifications, these controllers have LED panels on the cover and are more likely to have MPPT rather than PWM. MPPT provides power optimization between PV modules in an array to maximize production (MPPT is also an inverter option). In order for the MPPT to work correctly, the module and controller need to be matched so the MPPT starting voltage is met and will turn on. The system information may be a set display or could have buttons to click through menus. Controllers can be hooked up to a DC load panel or to an inverter. Figure 19.3 shows a controller with a digital display showing battery information.

DOI: 10.1201/9781003297819-22

**Figure 19.1** A small 12V charge controller.

## Controller Options

Options for controllers include: automatically controlling DC-powered loads; system monitoring (via lights, sounds, display); temperature sensor (usually of battery storge area); adjustable voltage set points; low voltage disconnect (shutting off DC load until battery is charged); automatic equalization (only for FLA batteries); and turning on a backup generator (auto generator start, AGS).

Hybrid systems, with both PV and wind generation, need a wind and solar charge controller. These include functions for both MPPT and diversion dump load. If you don't get a hybrid controller, then you'll need separate controllers for battery charging and the diversion load.

Stepdown controllers are used when the PV array is wired to a higher voltage than the battery bank. Voltage stepdown can be useful when adding onto arrays at a later date without having to replace equipment.

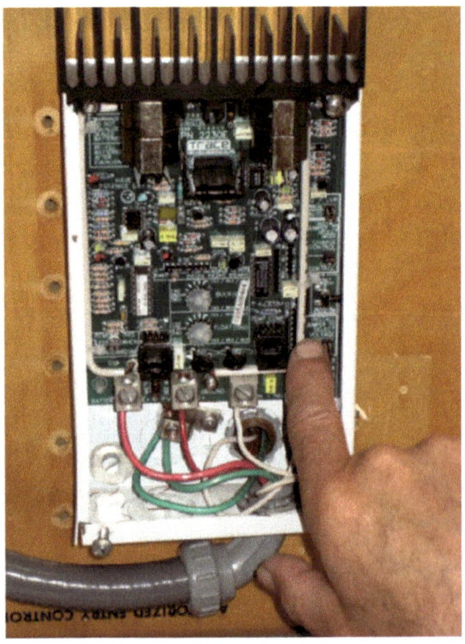

**Figure 19.2** Controller with the cover off.

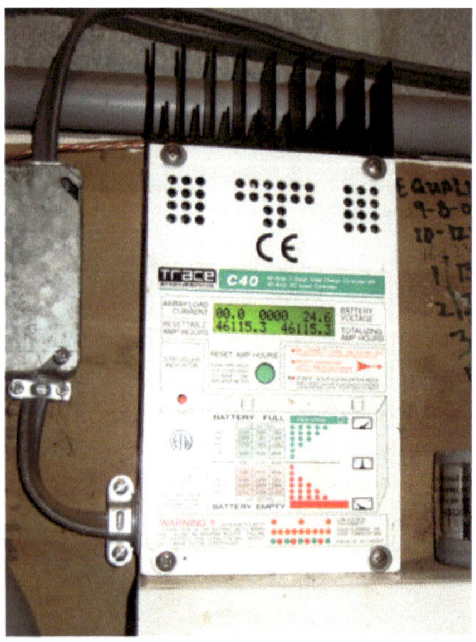


**Figure 19.3** Front of a controller with a digital display, wire in from PV array is on the left.

## Specifying a Controller

It is important when sizing a controller that you make sure it can withstand the array short circuit amps and handle the output amps. The safety factor is 25% for short circuit amps and 16.6% for array amps.

Multiply the short circuit current of the PV module by the number of modules or strings you have in parallel (parallel adds amps) to determine your array short circuit amps. Multiply the result by 1.25 (the safety factor). This is the array short circuit amps the controller has to handle.

To determine the output amps, you need the DC rated watts for your array. Multiply your nominal system voltage by 16.6% to calculate the number you should use: 12V = 14.4V; 24V = 28.8V; 48V = 58.8V. Divide the rated array watts by the adjusted nominal voltage to determine the maximum load amps from the controller.

### 12V example

Consult figure 19.4 for the system layout. The module $I_{sc}$ = 4.68A. All four modules are connected in parallel, so the short circuit amps = 4 × 4.68A = 18.72A total. Add in the safety factor: 18.72A × 1.25 = 23.4A. The array watts = 288W. The system is wired to 12V so, 288W / 14.4V = 20A for the load. Some controllers want array $V_{oc}$; this module has 22$V_{oc}$ (parallel volts stay the same).

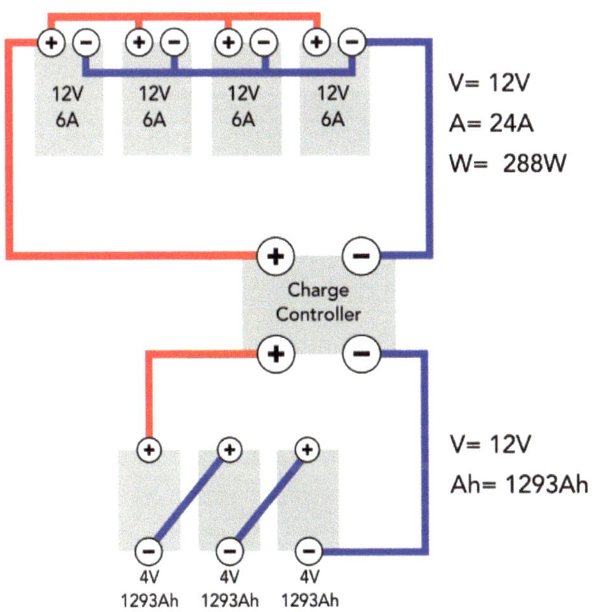


**Figure 19.4** 12V system with 4 modules and 3×4V batteries.

The controller I've selected has the specifications listed in Table 19.1. Compare these features: battery current 40A, load of 30A – both work; $V_{oc}$ is 120V – ours 22V; good for 12V system; handles 550W; can do lights and low voltage; does battery charging. This controller will easily handle the system.

**Table 19.1** Selected controller specifications.

| **Controller Specifications** | |
|---|---|
| Max battery current | 40A |
| Load current rating | 30A |
| Max PV open circuit voltage ($V_{oc}$) | 120V |
| Nominal battery voltage | 12V or 24V |
| 12 Volt battery | 550W |
| LED indications | (1) Status, (3) Battery SOC |
| Transient surge protection | Solar, Battery, Load |
| Ambient operating temperature range | –40°C (–40°F) to +60°C (140°F) |
| Load & Lighting Control | |
| Low voltage disconnect, low voltage reconnect settings 11.4V/12.6V or custom (×2 for 24 volt systems) | |
| Lighting Settings: Dusk-dawn or custom | |
| Mechanical Specifications | |
| Standard Dimensions: 20×17×70mm / 7.9×7.6×2.8 in | |
| Standard Weight: 1.4kg / 3.1lb | |
| Battery Charging | |
| 4.stage charging: Bulk, Absorption, Float, Equalization | |

If you have a larger system, you'll need a controller that will handle the larger loads and voltages. You may need to have two controllers wired in parallel with each other. There are typical break-off points for amp ratings, lots in the 25–40A range, then a jump to around 60A. Manufacturer's list all the specifications, so you just have to dig through them a bit. Some manufacturers have apps to help you select the correct controller.

## Battery and System Meter/Monitor

Controllers and inverters have displays that will tell you what is happening in your system. These are essential for system operation. It is also a good idea to have an additional system monitor: a **shunt** is used to connect the meter to the power line between the controller and the battery bank. This meter will display information about the system. It is not large, usually only the size of your hand. The owner, or technician checking the system, can verify that all of the system displays state the same information about the system. This helps to verify that everything is working properly. Some meters have extra features like AGS and battery equalizing.



## A Whole System

Now that we've covered all the parts, you should be able to recognize them! The PV is on the roof of a barn (figure 19.5). Note that this is an older array, before the 3' access requirement was in place along one edge of the roof – the upper roof layout would not meet USA code today. The rotation of the modules was changed between the upper and lower roofs to maximize the number of modules.

**Figure 19.5** PV modules mounted on a metal barn roof.

**Figure 19.6** Racking and grounding of the PV array.

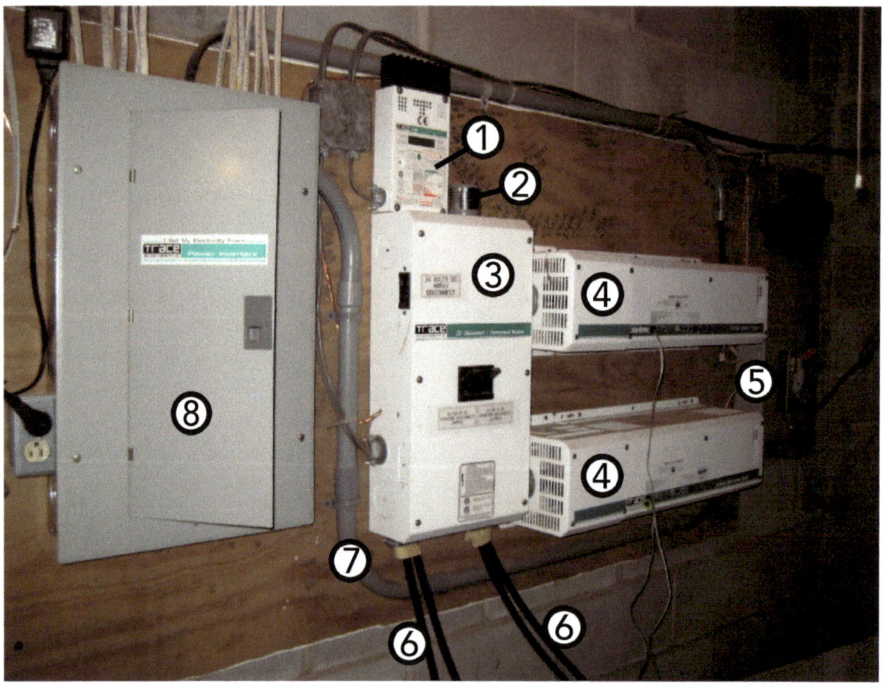


1. Charge controller
2. Lightening arrestor
3. DC disconnect
4. Inverters in parallel
5. AC disconnect
6. Wiring to battery bank
7. Conduit for wiring from AC disconnect
8. AC main service panel

**Figure 19.7** Wall mounted electrical components for a grid-tied PV with battery backup system.

**Figure 19.8** Battery bank on raised platform, 4 groups of 4 batteries wired in series.

Take a look at the racking and grounding – figure 19.6 is looking at the right side of the lower array. The copper grounding wire connects to all the modules and racks. The racks have a tall rectangular profile that the standoffs from the roof slide into on the front edge and the modules slide into on the top.

Mounting the interior electrical system components to plywood is common (figure 19.7). The controller (1) is connected to the PV and the battery bank with wiring (6) through the DC disconnect (3). The controller and batteries are connected to the inverters (4) which are wired in parallel to each other. AC power from the inverters goes into the AC disconnect (5) and through conduit (7) to the AC main service panel (8).

Figure 19.8 shows the battery bank built up off the floor, with distilled water ready to top off the 6V FLA batteries. Note that 4 batteries are wired in series, 6V × 4 batteries = 24V system. These are 246Ah batteries and there are 4 sets of batteries in parallel, therefore 246Ah × 4 = 984Ah.

## Study Questions

1. Go to the altE store (www.altestore.com/store/) and look at the selection of solar charge controllers. Check out the manufacturer's information for two different brands of MPPT charge controllers that handle 30A or more. Compare the specifications and features.
2. Do an online search for "charge controller in a PV system" and compare the system diagrams and features of the results. Pick three that look different from one another. What kind of hookups and displays are shown? What features do they have?

19

CHAPTER 20

# Backup Power

**Key Terms**

Essential load: a building load that has been identified by the owner as needing power 24/7.

Generator: a device that converts mechanical energy into electrical power for use by a circuit.

## Concepts for Success

Backup power provides your building with another power source if there is a grid outage or your off-grid battery bank has reached its maximum depth of discharge.

Grid-tied buildings provide backup power using a battery bank and an essential loads panel. Off-grid buildings use a generator.

Generators burn fuel to create electricity and need to run building loads and recharge your battery bank in the shortest time possible, based on the charging capacity of the inverter. To reduce generator run times, the RE system and battery bank should be sized so that it provides 90% or more of the annual building's power needs.

## What is Backup Power?

If your main power source isn't working, power can be supplied by a backup source. At the most basic level, backup power is supplied by a standby generator (figure 20.1). Under normal operation, the grid's electricity enters through the meter to the main service panel and is then distributed throughout the building. The **automatic transfer switch (ATS)** monitors the grid power and, if the power supply stops, the ATS automatically starts the generator. The power from the generator is fed only to specific circuits so that power only goes to the building functions and not onto the grid. For smaller buildings, like homes, the generator is located in a vented box on a pad outside. Standby generators are not particularly portable and have to be sized large enough to handle the essential building loads – including the larger surge draws for starting up motors and compressors for items like a refrigerator.

Backup power for a **grid-tied RE** building can be a battery bank.

## Why Do You Need Backup Power?

Without backup power your building will be without power in an outage. That means occupants could be in the dark, cold, and without services. If your building has security or special environmental needs, power failure could mean a loss of continuity that could disrupt or destroy the ability to function correctly. These failures could result in harm to occupants and/or property loss. Unfortunately, we have many real-world examples of harm to people because of disabled hospitals and loss of key infrastructure due to power outages.

You need backup power if:

- Your building is off-grid: backup power supplies power when your RE system isn't producing energy and your battery bank is at its maximum depth of discharge.
- Your grid-tied building is regularly without power and the battery bank could reach its maximum depth of discharge: there are many locations around the world where grid power reliability is low or only provided at particular times.
- Your building provides critically needed help and/or services (especially in emergencies), or specialty environments relying on continuous security or environmental control: any function that is critical for emergency services, telecommunications, hospitals, key government offices, air traffic control stations, prisons, museums, etc. that needs to be functioning 24/7.

DOI: 10.1201/9781003297819-23

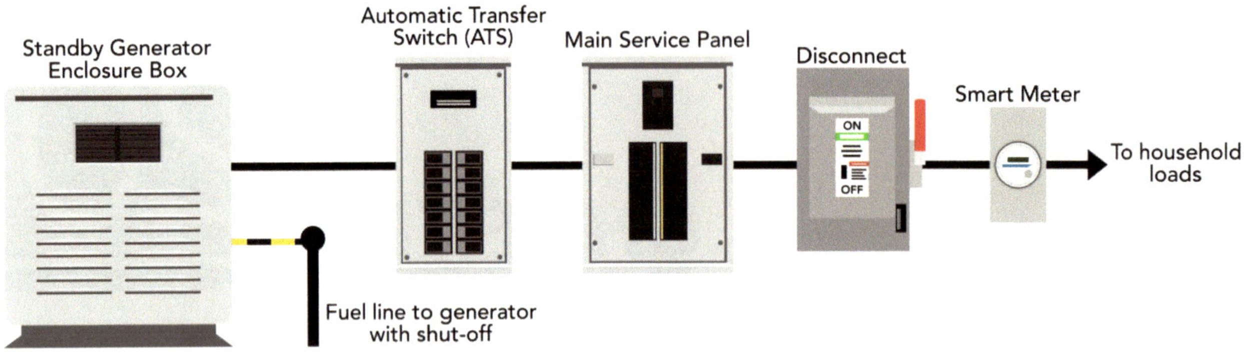


**Figure 20.1** The typical components of backup power provided by a standby generator.

- Your building is located in an area that could suffer catastrophic impacts that disrupt the grid: severe storms (wind, ice, snow), earthquakes, etc. typically cause grid power failure that takes days or weeks to repair.
- Your building is in an area prone to power outages: these can be caused by weather (per previous bullet point), accidents, trees, and grid overload during heat waves or extreme cold temperatures.

## What Building Functions Need Backup Power?

The client and designer must decide what building functions need backup power. Only critical functioning buildings like hospitals, control towers, prisons and the like are going to need whole building operations to continue in a power outage. Most buildings can get by on a lot less and the list will depend on the building use, location, and importance in community infrastructure.

For instance, a house located in the north-eastern USA or Canada, where winters are generally cold, could provide power to the circuits for some lights, outlets, and the heating system. This would be a small percentage of the power generally delivered through the main service panel to the house.

Larger buildings minimally need emergency lighting and elevator power to return passengers to the ground floor. They also require some amount of heating to prevent freezing of water pipes – both potable and fire suppression systems – if the outage is of a longer duration.

A grid-tied building that has an RE system could have a battery bank as the backup power source (figure 20.2). In the event of a grid failure, the inverter will use energy being produced or stored to power the essential load panel, while the standard load (main service) panel is shut down. Only the circuits wired to the essential load panel would be powered during the grid outage.

The battery bank would need to be sized so that it would be able to power the essential loads for a designated period of time. The RE (shown as PV) would have to be sized to be able to produce the power needed to run the essential loads and charge the batteries. There would be a big difference in the amount of power needed, generated, and stored for a house compared to a multi-story building. There may be utility incentives for using your battery bank to reduce demand

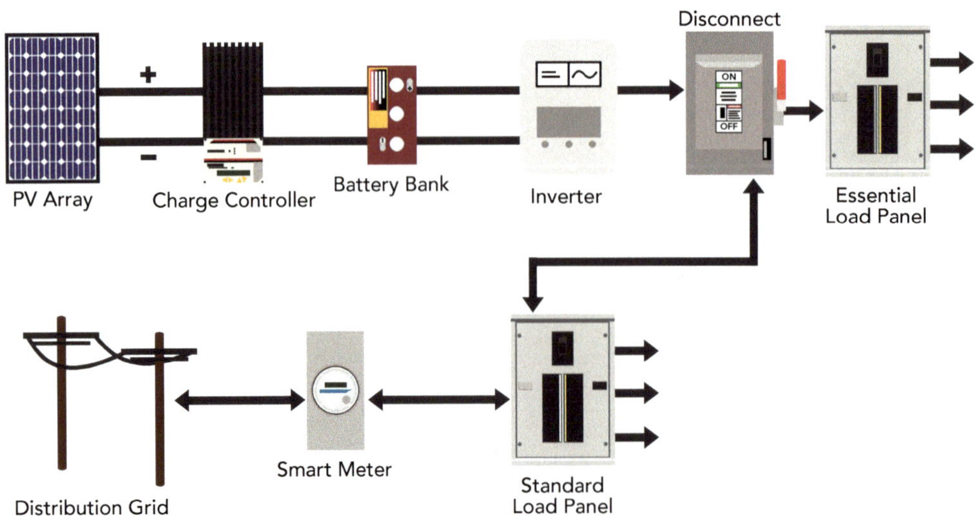


**Figure 20.2** Grid-tied PV system with battery bank backup power to the essential load panel.

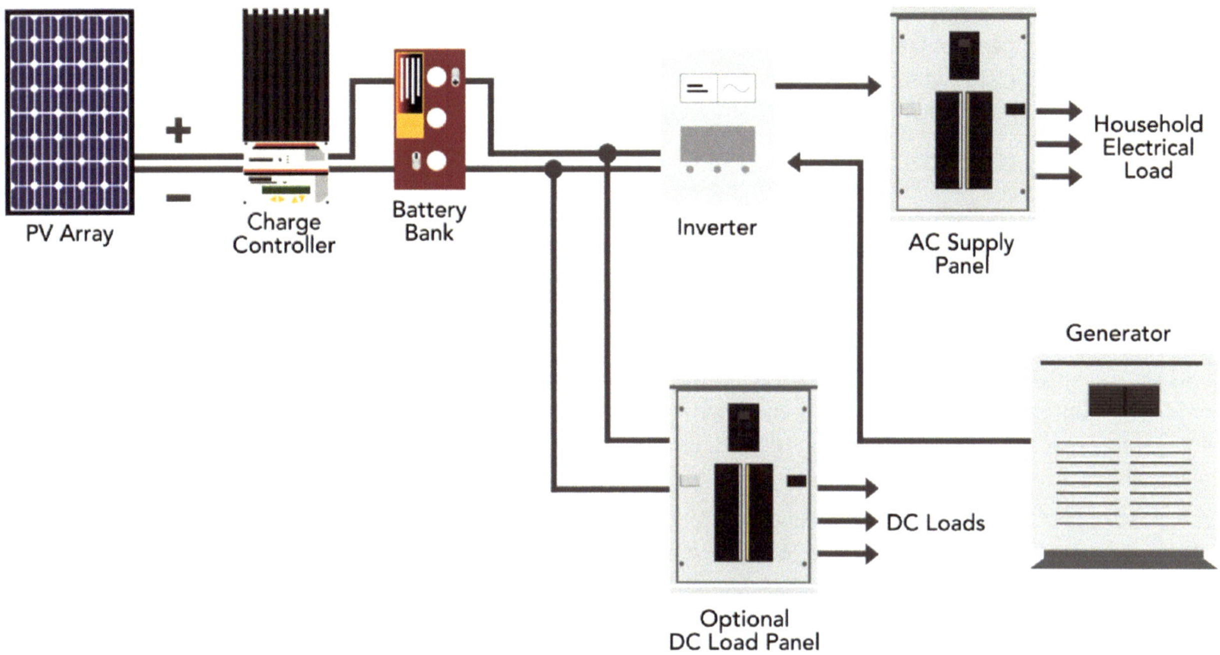


**Figure 20.3** Off-grid PV system with a standby generator.

during peak power periods in addition to working as a backup power system.

**Off-grid RE** systems must have a generator sized to be able to run essential building loads (figure 20.3). The generator should be set up so it will be automatically started by a function of the inverter or a system monitor. The generator will be started when no power is being made by the RE and the battery bank has reached its maximum depth of discharge (DOD) (see chapter 18 for sizing battery banks). The RE system should be sized so that it is providing 90% or more of the annual building power needs without the help of the backup generator. The generator needs to be sized so that it is able to run the building's power loads, handle sudden draws (inrush current) without sputtering and dying, and recharge the battery bank. The generator should be able to recharge the battery bank in the shortest amount of time possible – 7 hours is much better than 24!

## Sizing a Backup Generator to Charge Batteries

A building that has battery storage, either grid-tied or off-grid, may need to be occasionally charged by the generator. The generator runs at one speed: it is burning the same amount of fuel whether it is fully loaded or just powering one light bulb. It should be able to charge the battery bank to a full state of charge and then shut down as soon as possible.

Generators produce AC energy and the battery bank is DC energy. The inverter will invert the energy to DC. As a starting point, the generator should be 1.5–2 times the size of the inverter's rated continuous output in order to both run loads and charge the batteries. The generator and inverter voltages should match. You need a 2.wire start in order to set up automatic start-up. The inverter (or multiple inverters) can be specified to have higher charging rates to reduce charging time.

The estimated time needed to charge your battery bank depends on its overall Ah rating and the maximum DOD. Multiply your Ah by the max DOD to get the total Ah you need to recharge the battery bank. You also need the inverter (or charger) capacity: divide the charge capacity in half (this is because the inverter won't charge at the maximum rate over the entire charge time). Your generator run time = Ah to recharge battery bank / half the inverter charge capacity.

Generators come in either 1800 or 3600rpm; 1800rpm last longer and are more efficient than 3600rpm. Available options will be based on the manufacturer and fuel type.

## Housing a Generator

Remember that generators are loud and give off noxious fumes. They should be located away from fresh air intakes for the building's HVAC system. Smaller generators are usually housed in vented boxes outdoors. These boxes have vents to provide direct combustion air but may or may not have fans to promote ventilation. Typically, boxes are limited to 7500W, 500lb (226kg), and dimensions 28" (711mm) wide × 40" (1016mm) long × 28" (711mm) high.

Larger generators are often located in a shed adjacent to the building or inside the building in a mechanical space. If there is any chance the vibration of the generator would transfer to the building, it should have anti-vibration mounting to the pad or generator support.

These generators always have direct ducting to outside for their combustion gasses. Usually, the combustion air intake is through a large grille with dampers that automatically open and close when the generator starts and stops.

Generators located inside buildings should be located in a compartmentalized mechanical space where no damage to other building systems can happen due to dampers stuck open during cold winter days and nights. Consideration should also be given to how the generator will be replaced: is there a direct path or access with adequate clearance so that the equipment can be easily moved in and out of the space? Remember that all building areas that the equipment moves through have to accommodate its size: corridors, elevators, and turns.

Some jurisdictions have maximum sound and emission limits for generators. You would need to select a generator and install it so that it met these requirements. Depending on the green building rating system and what you are getting points for, you may have to account for all types of energy used and their carbon emissions and carbon equivalents. Remember that the fundamental purpose of an RE system is to reduce carbon emissions: the RE system needs to be appropriately sized so that the generator is used as infrequently as possible.

All generators require annual maintenance and need to be tested regularly. Most generators do a short weekly test which ensures the system is working and charges the start-up battery. Depending on the sophistication of the monitoring system, issues with the test run will be reported so the generator can be inspected and fixed. If the monitoring system isn't automated to report problems, the generator control panel has to be checked manually.

## Generator Fuel

Generators burn gasoline, natural gas, propane, and diesel. Unfortunately, these are all fossil fuels. Appropriately sized tanks must be located close to the generator. The tank size should take into account how much the generator burns, how long it is designed to run for until the batteries are recharged, and how long it is likely that the power outage would last without having RE generation or the ability to refill the fuel tank. Some generators can run on multiple types of fuel while others are specifically for one type.

Diesel generators are very efficient and the generator can last longer but they are initially much more costly (2–3 times) than other fuel type generators. All diesel engines are compatible with biodiesel. Biodiesel does have some special storage requirements as it can easily biodegrade if exposed to heat, oxygen, sunlight, or is stored in nonferrous metals. Additives are needed for cold weather to keep it from gelling. Overall, biodiesel has much lower emissions when compared to regular diesel, except it does have higher NOx emissions. As a greenhouse gas, NOx would have to be accounted for as a carbon equivalent in a rating system or a jurisdiction emission limit.

### Study Questions

1. What is backup power and how could you provide it for your building?
2. What are the reasons you would need backup power for your building?
3. What do you need to know to appropriately size a generator for your project?

CHAPTER 21

# Costing and Payback

**Key Terms**

Incentive: a financial return for investing in an RE system or energy efficiency measure, available in many forms from grants to tax credits.

Return on investment (ROI): system value calculated as a percentage return on the initial investment that compares the current value of your investment in any year to the original investment cost, taking into account the changing value of money and inflation.

Simple payback: system value calculated by dividing the system cost by the annual savings, resulting in number of years until you earned back your initial investment.

## Concepts for Success

Measuring the success of an RE system includes more than just traditional accounting of costs and returns. The owner is investing in an RE system to reduce GHG emissions, have grid independence, and other goals.

System costs include all the major equipment, balance of system (BOS), labor, and permits.

Incentives help to reduce the overall system cost and payback time. Incentives must match the project and apply to the client.

A **renewable energy certificate (REC)** is documentation of each MWh of electricity provided to the grid and is a salable commodity.

The value of production is based on how much the owner would be billed for that amount of energy. Utilities have different pricing structures and you need to figure out how they apply to when you are producing energy.

**Simple payback** and **return on investment (ROI)** are traditional ways to determine how well the money you've invested in your system is doing. Simple payback is calculated by dividing the system cost by the annual savings. The result is how many years until you've earned back the money you spent. ROI is a ratio that compares the current value of your investment in any year to the original investment cost, taking into account the changing value of money and inflation. The result is a percentage return on the initial investment.

## Clients Want to Know about Their System

Building owners want to know how the RE system is working, what it is saving them, and how their investment is doing. Monitoring gives real-time feedback and a summary page or display board that can show production information daily, annually, and since installation.

There are multiple ways to measure the returns on RE generation:

- Carbon saved – tracking offset of GHG emissions saved compared to utility generation;[1]
- Locking in a production value – the RE investment will produce for years, protecting you from energy price increases;
- Independence from the grid – you can run your building even if the grid goes down;
- Simple payback – compares your costs and returns at the current value of money;
- Return on investment (ROI) – compares costs and returns including the effects of depreciation and inflation.

## Total System Costs

Costs for a system include: major system components (PV modules and inverter(s)); BOS items (wiring and electrical disconnect boxes); labor to install the system; building permits; and interest on borrowed money.

DOI: 10.1201/9781003297819-24

## Incentives

**Incentives** shorten payback time because the money is received near the beginning of the project. This can make an RE system investment more affordable and increase its marketplace competitiveness.

There are many types of incentives, each for different applications. Incentives can be aimed at a specific target audience: such as a grant in a low-income neighborhood to help homeowners afford RE or a grant to build to generate megawatts of power that bring permanent jobs to an area. Incentives must match your project and client. Depending on your location, there may be a database available to find incentive listings. The Database of State Incentives for Renewables & Efficiency (DSIRE) lists USA States and Federal Incentives.[2] The EU has listings at their RE website.[3]

When looking for possible incentives, pay attention to details:

- Does it apply to this project?
- Who is it for?
- What counts as system costs?
- How and when does the incentive pay out?

### Types of Incentives

#### Low or No Interest Loans

Low or no interest loans lend money which is paid back with little or no interest.

Advantage: the interest rate is either 0% or well under other lender borrowing rates, saving or eliminating on interest costs; all or most of the project costs are covered.

Disadvantages: the money has to be paid back; the money can be used only for the specific items stipulated in the loan; and there may be liability clauses for non-payment.

#### Grants

Grants are money that is given to the owner for a specific project.

Advantages: grants do not have to be paid back; money is given at the beginning of the project, reducing the amount of money the owner needs to borrow.

Disadvantages: the owner has to apply for the grant money; money can be used only as directed in the grant and the appropriate use must be documented or the grant can be rescinded (has to be paid back).

#### Rebates

Rebates are sums of money given to the owner after showing proof of purchase.

Advantages: owner can send in rebate request as soon as they have a receipt; is a guaranteed amount (at least until the fund runs out).

Disadvantages: owner has to pay for everything up front; rebates are usually fairly small sums of money.

#### Local or National Tax Credits

A tax credit is deducted from the taxes owed by the taxpayer.

Advantages: directly reduces the amount of taxes owed; credit may be carried over to another year if the credit is greater than one year's tax liability.

Disadvantages: only available to individuals or entities liable for taxes.

Clarification: RE expenses may be deductions, which are used to reduce taxable income; the same expense cannot be used to reduce both taxable income and tax liability.

#### Depreciation

Depreciation is an accounting of the decreasing value of an asset over time.

Advantage: counts an expense which reduces tax liability.

Disadvantages: only businesses can depreciate assets; cannot use the same system as a tax incentive and depreciate it.

## REC, SREC, and I-REC

An REC (PV-generated is called solar – **SREC**) is a verified certificate of renewable energy generation that can be sold separately from the energy that was generated by a system. One REC is earned for every MWh of generated electricity placed on the grid. An internationally traded electricity REC is an **I-REC(E)**. Every certificate has a unique number that can only be sold once.

Advantage: additional income for electricity placed on the grid; can make long-term agreements that are better than market rates; can keep for several years before selling – check jurisdiction.

Disadvantage: doesn't count all of your production – only what you didn't use and sent to the grid; market price varies.

## System Cost

The cost of the system includes all the materials, labor, and permits. The materials include all the major pieces of equipment and the BOS. In the USA, a general idea of material costs can be made by pricing the major equipment (i.e., for a PV system the modules, inverter, racks (about ½ the price of inverter)) and adding 10% for BOS. Labor is about 40% of material and permits about 1% of the project. Hypothetical pricing for the example below is in USD.

10 modules @ $388 = $3,880

10 microinverters @ $186 = $1,860
Racking @ ½ inverter cost = $930

Total major equipment cost = $6,670

BOS = $6,670 × 0.10 = $667

Estimated total equipment cost = $7,337

Labor = $7,337 × 0.40 = $2,935

Estimated construction cost = $7,337 + $2,935 = $10,272

Permits = $10,272 × 0.01 = $103

Estimated total system cost = $10,272 + $103 = $10,375

If you are working with an installer, they can give you an exact price. You could also use pricing data such as RSMeans[4] for more accurate labor estimating. You might also add a couple of percent for contractor profit based on equipment cost.

## How to Include Incentives in your Payback Calculations

The type of incentive will help you determine how to deduct it from the system cost. Grants, rebates, and tax incentives are usually deducted from the overall cost. If you have a local tax incentive and a national tax incentive, you have to deduct the local one first. Example is priced in USD.

### Example for Deduction Order

4kW system costs $11,880; local 5% grant; national 26% tax credit

$11,880 × 0.05 = $594 from local grant

$11,880 – $594 = $11,286 system cost with grant

$11,286 × 0.26 = $2,934 tax credit

$11,286 – $2,934.36 = $8,352 system cost after deducting incentives

## Production and Value

In order to determine savings due to generation, you have to determine your production. You probably already know this information because you've designed the system. See chapters 13, 15, and 16 on PV, wind, and microhydro for specifics on system production. These values are based on avoided costs – the owner didn't have to pay for that electricity because it was generated by the system.

The value of your production will depend on how much you would be paying for the energy you are replacing. Utility companies have pricing that can be very simple (1kWh costs US$0.12) or complex (tiered pricing based on kWh used, time of day/week, time of year). Simple pricing is the easiest to figure out. If the tiered pricing is very complex, you may want to create a spreadsheet with all the variables. Examples below use USD.

### Example Value Simple Pricing

System makes 4,994 AC kWh annually. Price of electricity is $0.12 per kWh.

4,994kWh × $0.12/kWh = $599.28 annual value of electricity

### Example Value Tiered Pricing

System makes 4,994kWh AC annually. Pricing tier: summer months pricing is $0.18 per kWh; rest of year is $0.15 per kWh.

kWh production for the three summer months with higher electricity rate is 492kWh + 532kWh + 497kWh = 1,521kWh

kWh production for rest of year is 4,994kWh – 1,521kWh = 3,473kWh

1,521kWh × $0.18 = $274 summer value of electricity

3,473kWh × $0.15 = $521 rest of year value of electricity

Annual savings = $274 + $521 = $795 annual value of electricity

## Simple Payback

Simple payback determines the years to recoup the system cost relative to the annual savings based on the current value of money.

Simple payback = cost of system after deducting incentives/annual savings

4kW system cost after incentives = $8,351.64

Annual savings = $599.28 per year

$8,351.64 / $599.28 per year = 13.9 years to payback

While a lot of investors prefer to have 3–5-year payback period, remember that the equipment will produce for 20–25 years. The owner will continue saving $599.28 every year. At the 20-year mark that is an additional $3,595.68. If the owner sells RECs they would have additional income over the years. It could also shorten the payback period depending on how much energy is supplied to the grid annually.

## Lifetime Costs and Savings

Create a spreadsheet that includes initial system cost, incentives, cost of electricity, interest payments on borrowed money, recurring costs (annual maintenance and inverter replacement at about 10 years), price of electricity, REC sales, and equipment scrap value at end-of-life. This provides the owner a big picture of cost and returns of their system. You could also calculate pounds of GHG emissions offset.

## ROI

ROI is trickier as it considers changes in the value of money. Generally, prices for goods increase over time (inflation) and the value of money is a bit less (discount rate). Actuaries are paid well to guess how this sort of thing turns out. For an RE project, ROI considers the costs and benefits of a system that will continue to produce for over 20 years. I've insulated myself from inflationary costs of energy because, as the cost per kWh increases, the value of my production increases. The money spent has less purchasing power as time passes so it becomes a better investment due to increasing rates of return on it.

In its simplest form, ROI is a ratio of the (current value – investment cost) / investment cost. The change in the current value depends on energy price changes in the future and/or change in the value of money. This example is in USD.

4kW system cost after incentives = $8,351.64

Annual savings = $599.28 per year + 2% inflation each year

Current value year 1 = ($8,351.64 + $599.28) = $8,950.92

ROI year 1 = ($8,950.92 – $8,351.64) / $8,351.64 = 0.0717 or 7% return

Additional year 2 savings due to inflation = $599.28 × 0.02 = $11.98

Year 2 savings = $599.28 + $11.98 = $611.26

Current value year 2 = $8,351.64 + $611.26 = $8,962.90

ROI year 2 = ($8,962.90 – $8,351.64) / $8,351.64 = 0.073 or 7.3% return

ROI is best shown in a spreadsheet – and you can add in the effect of discount rates as well. In this simple example we see that the owner is getting 7% return, a good value in mid-2022 when interest rates are around 1–3%.

## Resources


1 A clear article on the difference between carbon offsets and RECs: Green Power Partnership. (2018, February). *Offsets and RECs: What's the difference?* US Environmental Protection Agency. www.epa.gov/sites/default/files/2018-03/documents/gpp_guide_recs_offsets.pdf

2 Database of State Incentives for Renewables & Efficiency® (DSIRE) (www.dsireusa.org/).

3 European Commission: Energy (https://energy.ec.europa.eu/index_en).

4 RSMeans https://www.rsmeans.com/


## Study Questions

1. What is the difference between simple payback and ROI?
2. What costs do you need to account for in an RE system?
3. How do you know what the value of your RE production is?
4. If you have multiple incentives, what order should they be deducted in?

CHAPTER 22

# Resiliency and Power

**Key Terms**

Microgrid: an area, which could be a single building or a neighborhood, that is independent in regards to power and can operate with or without the grid.

Smart grid: a utility grid that has advanced metering infrastructure which allows for fairly real-time monitoring of customer power use and two-way communication between a building's smart meter and a centralized monitoring system.

Variable generation (VG): RE production from wind and PV that changes daily and seasonally.

## Concepts for Success

Resiliency relating to power means that a building or **microgrid** is able to assist the utility in providing power to the grid and is able to function with or without the grid.

Smart grids are able to manage RE generation, storage, and distribution in real time. This makes it easier to balance demand and generation. Microgrids are defined areas that are able to create, store, and supply electricity whether or not the grid is working. When the grid is down, the microgrid islands, becoming totally self-sufficient.

RE production is variable, making it harder to match with customer demand. In locations with high PV installation, grid demand can be greatly reduced during the day but suddenly increase at sunset. This spike is a problem because it is difficult for the utility to instantly produce more energy.

Energy efficient buildings can greatly reduce spikes in both daily and yearly energy usage. A building that is daily net zero doesn't have demand spikes.

AC energy has been the prevalent grid distributed energy around the world. Today we have many DC devices such as LED lights and smart phones. Because PV produces DC and batteries store DC, buildings with backup energy systems can be designed with a DC loads panel to provide better overall efficient energy usage.

## Smart Grids

Recently in the USA there have been graphic examples of grid failure due to infrastructure age and high electricity demand. In 2020, there were rolling blackouts in California (USA) as large sections of the grid were shut down due to high winds and concerns over live wires starting fires. This happened during a heat wave in mid-August. In a two-day period over 800,000 buildings lost power for hours at a time.

In February of 2021, severe winter storms swept through Texas (USA), and the whole state suffered a power crisis. Demand skyrocketed due to the unusually cold temperatures while at the same time there were equipment problems (due to the freezing temperatures) at some generating plants. Rolling blackouts kept the state grid from being overwhelmed, and many customers lost power for more than three days. Over five million people were affected by power outages.

Many stories like these throughout the world point to the difficulties of keeping power working in extreme weather events and the inefficiencies of older grid distribution systems. This shows the need for buildings and communities to be able to power themselves and for grids to be smarter and able to anticipate demand issues.

A smart grid has a centralized monitoring system that is able to assess fairly real-time power demand through communication with smart meters at buildings. It is able to shift power flows from individual and utility RE generation and access stored power to quickly adjust to demand. Many countries are updating to smart grids.

DOI: 10.1201/9781003297819-25

## Microgrids and Islanding

A microgrid is an area, which could be a single building or a neighborhood, that is independent in regards to power (figure 22.1). It has a defined electrical boundary that can operate with or without the grid. When the grid is down, the microgrid is able to island itself and provide all the power needed within that area. Microgrids operate all the time, adding extra power to the grid and providing backup power to the microgrid in a grid emergency. A controller and switching gear are required at the electrical boundary so that no power is put on the grid when the grid is down. Microgrids are powered by RE and sufficient storage.

Microgrids produce and distribute power in a local area and are therefore less susceptible to failure when compared to a larger utility grid. Within the microgrid there could be different types of RE generation and location of RE systems. This multiple layering of production and storage within a microgrid makes it more resilient and likely to remain powered in an emergency. Microgrids have been used in the USA for hospital campuses and military bases for years but the technology for managing RE production, storage, and monitoring is quite new.

With the ability to create microgrids comes the need for adjusting policy to provide appropriate regulatory oversight. The types of concerns include: defining at what point a microgrid becomes a public utility; defining how or if microgrids can use existing infrastructure; addressing interconnection standards between inverter/monitor controls and communications; and addressing the complexity of shared community power and individual customers and the utility grid, including billing inside the microgrid and opting in or out of the microgrid.

While there isn't yet complete agreement on terms for defining the size of microgrids, here are some general ones: pico-grid only covers a few loads, 1–10W; nano-grids are small, 100W–10kW; and mini-grids are remote, off-grid services. Depending on the application, any of these microgrid types could apply to a single customer. Campuses (multiple buildings in the same area operated by a single owner) located behind a single utility meter can be microgrids. Multi-customer services tend to be larger, 10kW and above.

## RE Production vs. Continuous Power Needs

RE production, with the exception of hydro, is variable. It varies seasonally and daily. This is called **variable generation (VG)**. If a location has more RE production, especially PV, it impacts the amount of electricity the utility needs to produce in order to meet grid demand. At sunset, PV stops production and grid demand ramps up. This effect is called the duck curve and was discovered in 2014. Figure 22.2 shows an example of a day's electrical use and generation in March in California, USA. The top dark blue line is the total load on the grid, without the power generated by RE. Demand is fairly flat until in the evening when it peaks and then drops off. This line is the duck's back and the peak is the duck's head. The lower light blue line is the net load that the utility has to generate. The line dips down during the day because PV generation is being added to the grid from individual building arrays. The net load is offset below the total load and defines the belly of the duck. The RE is being generated from wind (fairly straight red line across the bottom) and PV (green line that is flat until the sun rises, then becomes a tall curve with peak production at midday, then at sunset returns to zero). The effect of the wind generation offsets the net load consistently below the total load, basically parallel in the morning (duck's tail) and evening (duck's neck and head). The dramatic drop of the net load in the middle of the day is because of PV generation (duck's belly).

The problem shown by the duck curve is the very dramatic ramping up of grid power demand as the sun sets and PV ceases production. This ramp is the duck's breast and neck. It is difficult for a utility to suddenly produce more power over a few short hours. It doesn't help to add more PV, in fact, that makes the ramping problem worse. There are many ways to address flattening the duck's belly. One aspect of the puzzle is to store large amounts of energy during peak RE production time so that it can offset the ramping demand. Another aspect is to increase the energy efficiency of buildings so the total loads needed are flatter (no duck head) and lower overall.

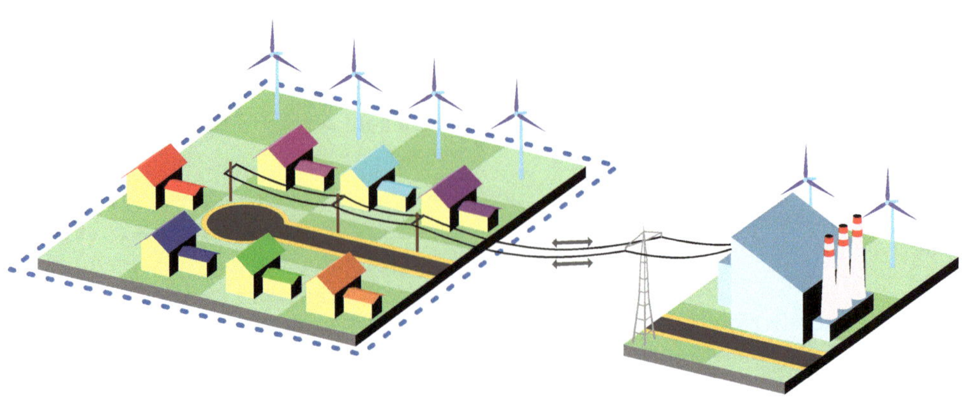

**Figure 22.1** A microgrid is a defined area that has enough power generation and storage that it can operate with or without the grid.

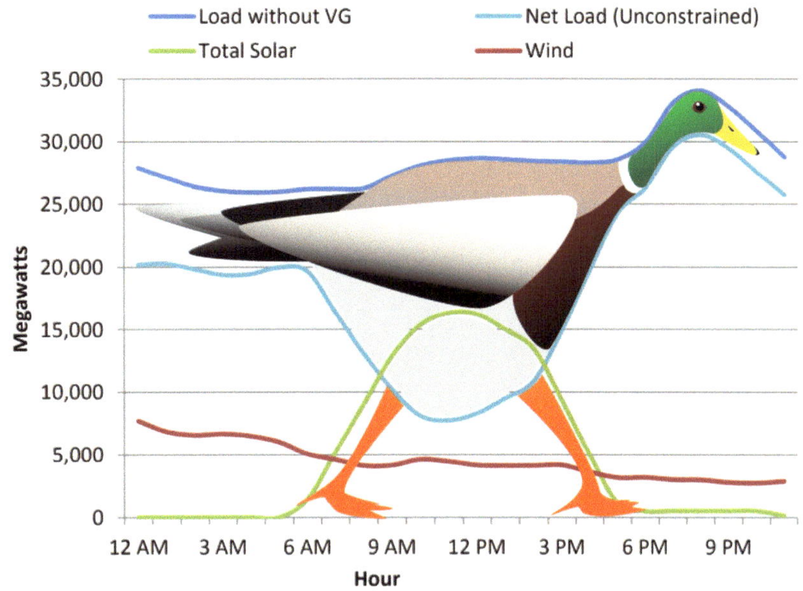


**Figure 22.2** The duck curve is the difference between total load (dark blue line on top of the duck) and the net load on the grid due to solar production (teal line on the bottom of the duck).[1]

## The Importance of Energy Efficiency

Energy efficiency has an impact on both daily and seasonal demand. Figure 22.3 illustrates varying heating demand throughout the year. Most places have little or no summer heating demand and need heat in the winter. This is generic, as there is a great difference in heating degree days (HDD) across the globe. Note that, like our duck curve, this demand has a big belly.

In order to move away from technologies that emit GHG to produce energy, we want to have our buildings run on electricity – preferably RE powered. Figure 22.4 graphs annual PV production over the heating demand curve. Unfortunately, it is not a match, since PV produces the most energy in the summer. This leaves large spikes of heating demand on both sides of our graph, when it is the coldest. This difference will have to be produced by something besides PV. It would take a lot of modules to get the low ends of the PV production curve over the high spikes of the demand curve.

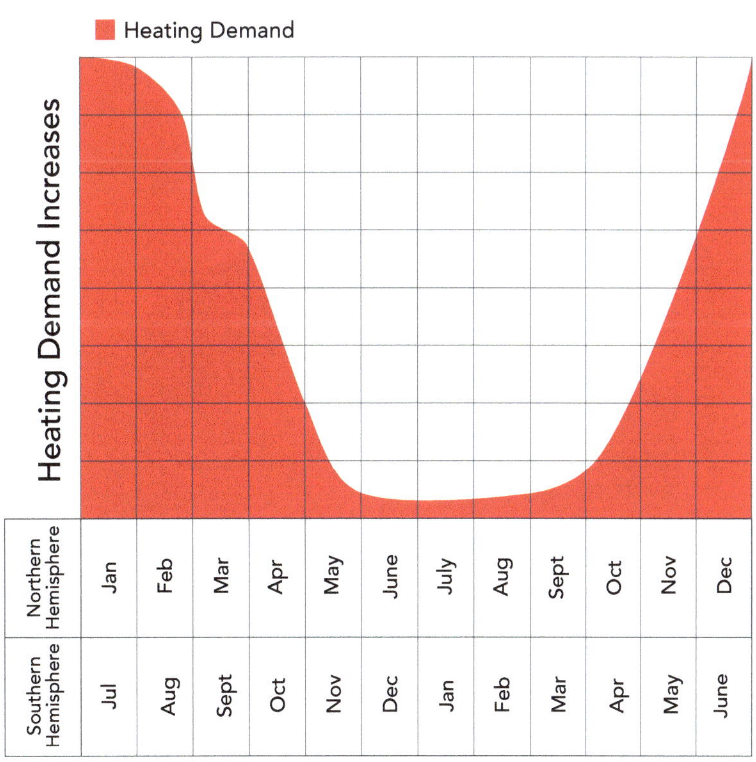


**Figure 22.3** Heating demand is unequal throughout the year, creating a belly curve during the summer months.

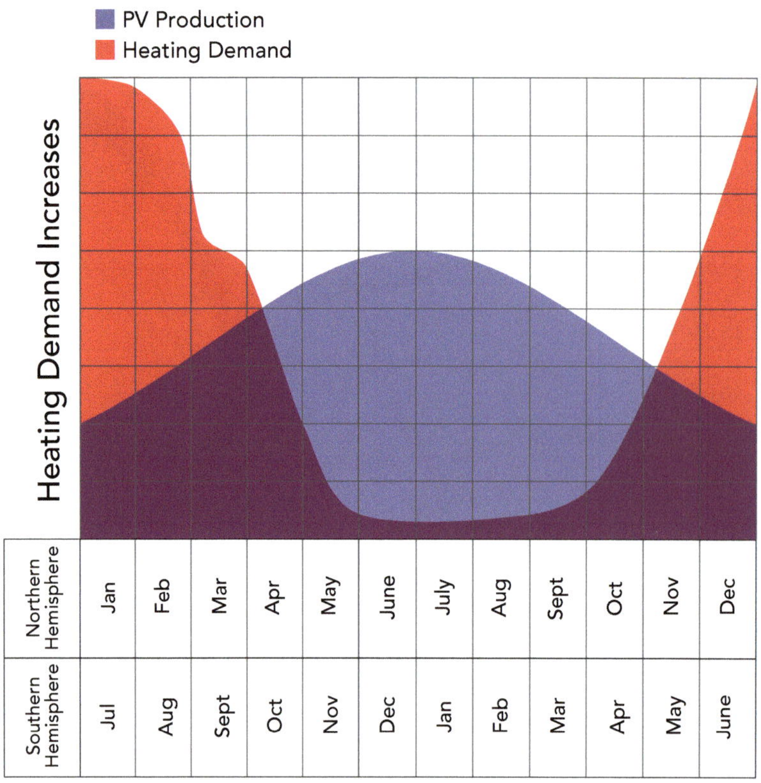


**Figure 22.4** PV electrical production increases in summer months while declining in winter months, leaving peaks of heating demand unmet.

Designing an energy efficient building for the climate with an appropriate high performance envelope and passive solar design dramatically reduces the peak heating demand (figure 22.5). By flattening the demand curve, we are able to install less RE and get closer to meeting all of the demand. If the building were

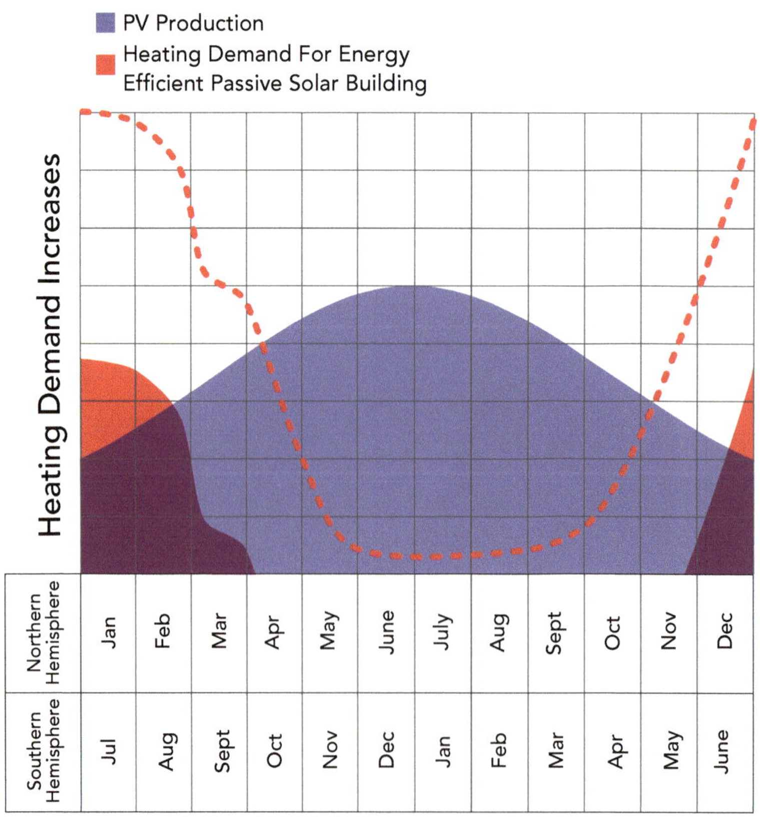


**Figure 22.5** Heating demand for an energy efficient, climate designed, passive solar building is much less, with only small unmet peaks of heating demand.

designed to be daily net zero, there would be no unmet heating demand.

## Current Applications: AC vs. DC

At the birth of the electrical age at the end of the 1800s, Nikola Tesla and Thomas Edison (shown in figure 22.6) were proponents on different sides for AC vs. DC power. There are still no global standards for AC and DC power. Although Edison developed DC, and early electricity was all DC, it was at a low voltage that had high losses over distances and there was no way to step it up and down. Edison's patents were all for DC items, so he wanted everyone to stay on DC power. Tesla believed that AC power could overcome these problems.

**Figure 22.6** Nikola Tesla,[2] circa 1890 on the left and Thomas Edison[3] in 1922 on the right.

The 1893 Chicago World's Fair (USA) was the first big loss for Edison. General Electric's bid for DC was $554,000 while Westinghouse bid $399,000 using Tesla's AC.[4] At the same time Westinghouse started building AC power generation at Niagara Falls. In 1896, when AC power production started at the project, other major electric companies also committed to AC power.

For over 100 years, AC has remained the dominant power type. Even now, if you look at types of energy used worldwide, it is AC (although they are not the same voltage). Interestingly, beyond the advent of personal computing in the 1990s, today's solid-state electronics, LED lighting, smart phones, and the internet of things are all DC-powered. PV and wind produce DC power. Batteries store DC power. Inverting DC to AC loses some efficiency. Today we have technology that can step DC power up and down.

If you have a building that has PV and batteries then it is fairly easy to create a DC panel and run DC loads directly from the PV and batteries, saving inverter losses. DC can be distributed through the building at 380V and fed to power distribution hubs which step down the DC to lower voltages. The distribution hub becomes a plug-and-play center for lighting and other low voltage DC equipment. Providing DC receptacles means you'll never need an AC–DC adaptor to run your DC devices.

## Charging EVs and Powering Buildings

If we make a building so energy efficient that it needs a very small battery to power it, or to power the essential loads panel, it is theoretically possible that **electric vehicles (EV)** could fill the gap instead of using battery backup. The idea is that vehicles are usually parked. They could be charging from PV generation during the daytime, then power your house in the evening. There are some bi-directional EV charging stations. The home's service panels would have to be set up so you could power your building. It is possible that the EV battery could offset ramping demands at the end of the day and then recharge when demand is lower overnight. At this time, however, EV batteries haven't been being designed to power buildings, so they don't necessarily perform like RE deep cycle batteries.

## Resources


1 Denholm, P., O'Connell, M., Brinkman, G., & Jorgenson, J. (2015, November). *Overgeneration from solar energy in California: A field guide to the duck chart* (Tech. Rep. NREL/TP-6A20-65023). National Renewable Energy Laboratory. www.nrel.gov/docs/fy16osti/65023.pdf

2 Sarony, N. (ca. 1890). *Nicola Tesla* [Photograph]. Library of Congress. (www.loc.gov/pictures/item/2014684845/). In the public domain.

3 Bachrach. (1922). *Photograph of Thomas Alva Edison, three-quarter length portrait, seated, facing front.* Library of Congress. (www.loc.gov/pictures/item/92522138/). In the public domain.

4 Lantero, A. (2014, November 18). *The war of the current: AC vs. DC power.* US Department of Energy. www.energy.gov/articles/war-currents-ac-vs-dc-power


## Study Questions

1. What elements define a microgrid?
2. What is the duck curve and why is it a problem?
3. How does energy efficiency help our energy production vs. load time usage issues?
4. Why would it be desirable to provide a DC load panel for a building?

PART IV

# Case Studies

CHAPTER 23

# Case Studies

CHAPTER 23A

# Rocky Mountain Institute, Innovation Center

## Location and Use of the Building

Design for the Rocky Mountain Institute's (RMI) Innovation center began in 2013 and construction was completed in January 2016, see figure 23A.1. It is located in Basalt, Colorado, USA. The office building is 15,600ft$^2$ (1450m$^2$) office with meeting spaces, open office areas, auditorium, and both interior and exterior support spaces. The owner is the Rocky Mountain Institute[1], the architecture firm was ZGF Architects LLP[2], the mechanical, electrical, and plumbing (MEP) engineering firm was PAE[3], the structural engineering firm was KPFF[4], the consultant team was Architectural Applications[5], and Sol Energy[6] designed and installed the PV system. The PV system was funded through a power purchase agreement (PPA) between RMI, Holy Cross Energy and Collective Sun.

Basalt is located at 39.36° N latitude at an elevation of 6,600' (2012m). The climate is very cold, dry, windy, and sunny. The diurnal temperature swings (difference between day and night temperatures) are often as much as 40°F (21°C). There is an average 86" (2184mm) annual snowfall at the site and 16" (406mm) of rain. On average there are 242 sunny days annually, which are fairly evenly spread across the year. Temperatures range from an average high of 83°F (28°C) in the summer an average low of 10°F (–12°C) in the winter. The wind blows along the valley mostly from the south-east and north-west, with higher amounts of strong winds over 20mph from October to April. The average annual solar insolation for 39° tilt angle is 5.73kWh/m$^2$/day, with December having the least availability at 4.08kWh/m$^2$/day (data from PVWatts).[7]

## Benchmarks, Outcomes, and Challenges

RMI wanted a building that set a cutting edge for its innovation in minimizing energy use, incorporating passive design strategies, using durable materials and systems to create a 100-year building, and, most importantly, providing occupant comfort. Figure 23A.2 shows the main lobby and the Amory Lovins Atrium.

The design team was challenged to conceive of the building as a blank slate, starting from nothing: to set aside their assumptions of what the building's systems should be. Figure 23A.3 shows what systems the design team initially thought they would need, based on experience, for a high performance building in this climate. This diagram is conceptual: they didn't ultimately design the building with these systems. They were challenged to start with nothing and add systems rather than assume regular systems and then remove them. This was a big paradigm shift.

The team, which included both the architects and engineers, had multiple design meetings at the beginning of the project, using an integrated process where ideas were suggested and then tested. The team begin with just walls, then added windows, and then tested ideas as an exploration to discover what the project would really need. The focus was on creating interior comfort, using an expanded comfort range of 64–82°F (18–28°C). RMI had very challenging goals for radical reduction of energy use, no cooling, no heat pumps, and delightful workspaces.

Modeling was done using Integrated Environmental Solutions (IES). The first computer models used a simple rectangular box which tested multiple envelopes facing different orientations to discover what impact

DOI: 10.1201/9781003297819-28

Figure 23A.1 RMI southern façade. Photograph by Tim Griffith.

each had on interior comfort. Then windows were added and tested with different R-values. The modeling results showed interior comfort based only on passive design. These results were summarized in an orientation rose shown in figure 23A.4 which clearly shows the fewest hours of interior discomfort were when the longest façade was facing south and included lots well insulated windows. The white area is defined as comfortable, while the dark areas are uncomfortable. This modeling made it possible for the team to optimize the building's performance before putting in HVAC (heating, ventilating, and air conditioning) systems. The final design has R67 roof, R50 walls, and R12 windows (center of glass) with thermal breaks (R5.8 overall) that are tuned to each elevation, which means the amount of glazing and its properties are adjusted based on orientation. The modeled energy use intensity (EUI), without PV, was 17.2 kBtu/ft$^2$/yr with actual energy use at 13.6 EUI. These numbers are well below ASHRAE 2007 and 2030 Challenge targets. Including the PV makes the project net positive, with a predicted EUI of −2.8kBtu/ft$^2$/yr. The infiltration tested at 0.36 ACH50 (0.36 air changes at 50 pascals).

What the team discovered was that they could effectively heat most of the building with about 70% full glass on the southern façade, using increased mass to absorb and radiate heat into the space. Figure 23A.5 shows a building cross section with the final systems. Computer modeling showed that the northern side of the building would get cold with prolonged very cold

Figure 23A.2 Amory Lovins Atrium. Photograph by Tim Griffith.

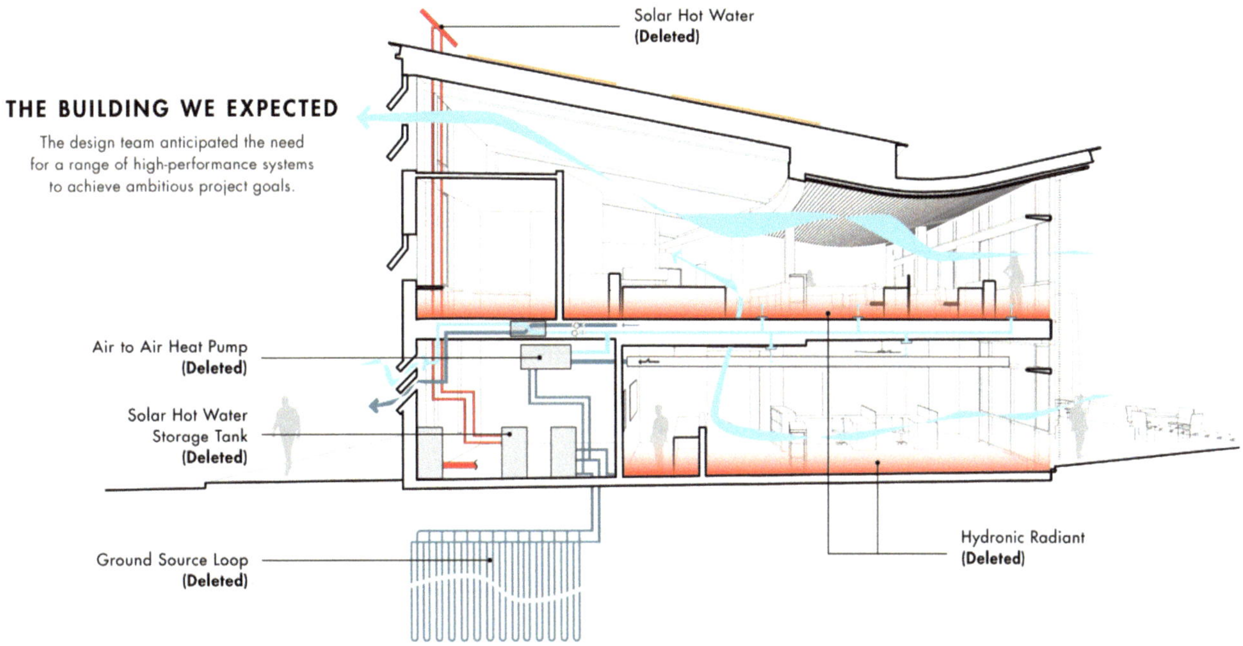


**Figure 23A.3** Building Cross Section of Presumed Mechanical Systems, Courtesy of ZGF Architects.

outdoor temperatures, so a strip of electric radiant heating, 33kW, was added along the wall. The capacity is equivalent to about one hair dryer per 1,000ft$^2$ (92.9m$^2$). Ventilation and heat recovery are provided by a variable air volume (VAV) dedicated outdoor air system (DOAS) which is 90+% effective and eliminates the need for heating in the ventilation system. Cooling could be achieved through night ventilation and thermal mass. This required thermal mass of about two times the area of the floor area, so mass was added to the ceiling and walls. In addition, they used phase change material (PCM) in the southern interior light shelves which freezes (solidifies) at 70°F (21°C) and melts at 75°F (24°C). The building is cooled at night to 65°F (18°C). They did separate energy modeling for the PCM where they changed a mass surface of concrete to have the heat capacity of PCM. In addition to heating and cooling the building, the project uses chairs that assist in heating and cooling the occupants.

One of the largest challenges faced by the team was how to keep the building comfortable in the shoulder seasons. This was solved with an innovative adjustable blind system, shown in Figure 23A.6, that is installed exterior to the windows on the southern façade. Vertical wires are embedded in concrete at grade and remain stationary. The slatted Venetian shades can be fully retracted into the protected housing at each level. They are retracted when full sun is desirable and when winds are over 25mph (1.6kph) to prevent damage. These sunshades, as well as the ones on the east and west façades, are automatically adjusted to the best seasonal angle by the building management system which monitors both interior conditions and external weather conditions. There were security concerns for

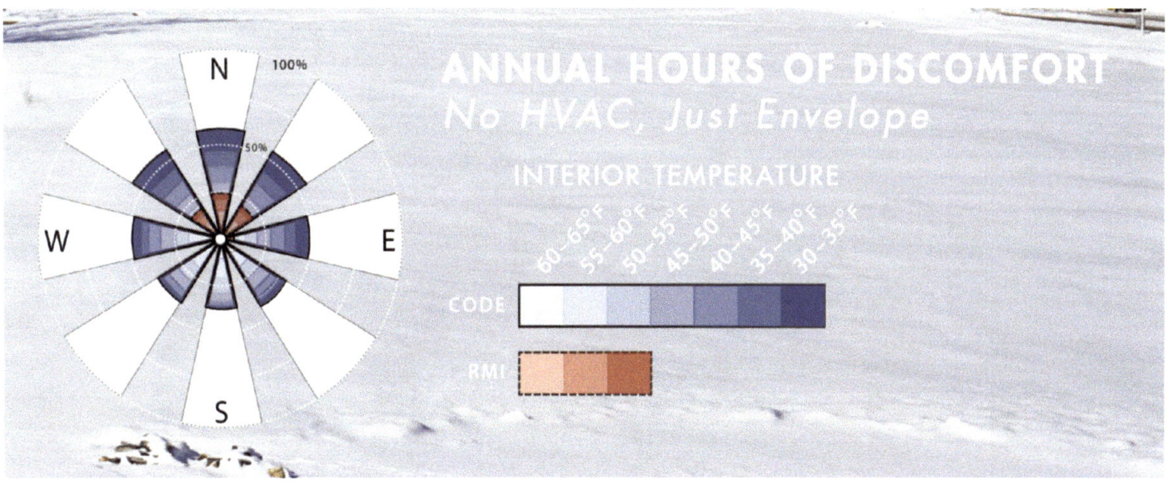


**Figure 23A.4** Envelope Computer Modeling for Interior Comfort, Courtesy of ZGF Architects.

23A

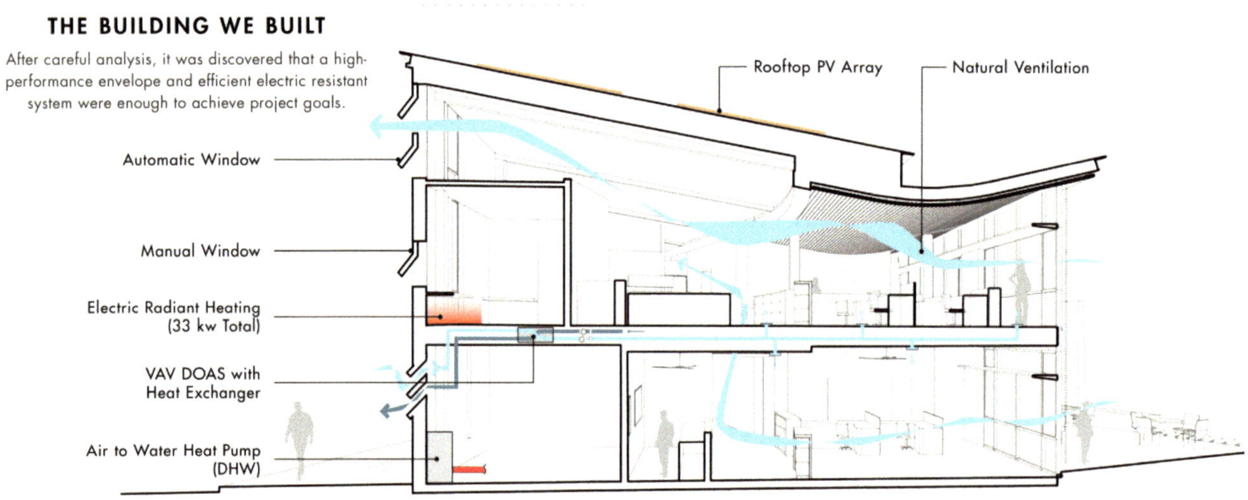


**Figure 23A.5** Building Cross Section of Actual Mechanical Systems, Courtesy of ZGF Architects.

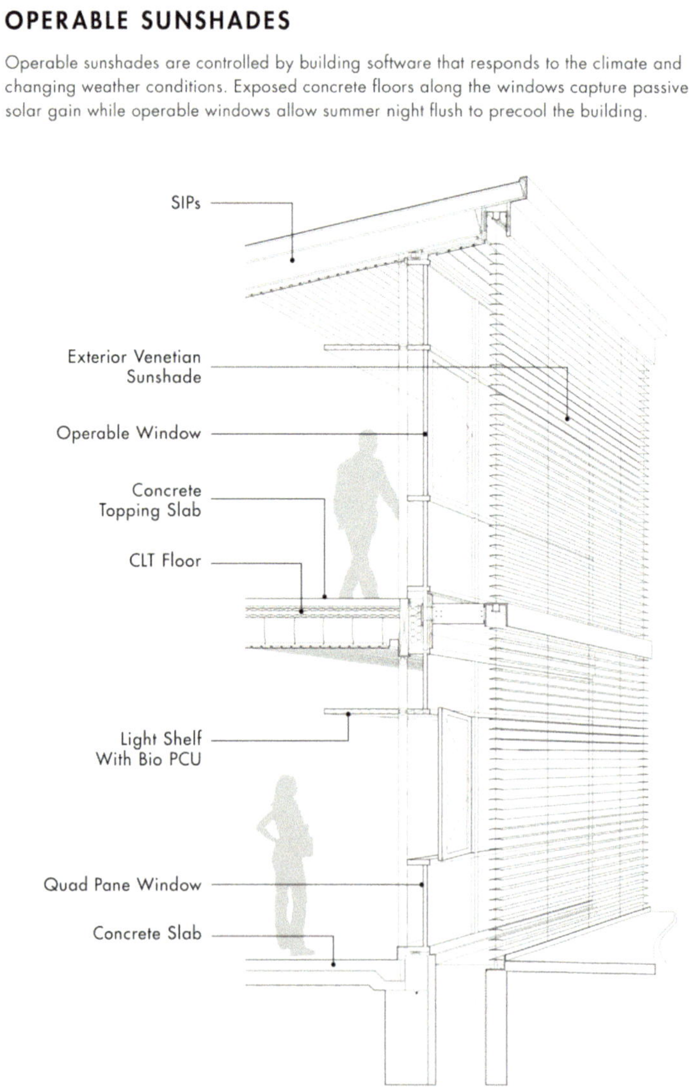


**Figure 23A.6** Southern Façade Operable Sunshades, Courtesy of ZGF Architects.

having windows open overnight as there are a lot of bears in the area. This limited how much the windows could open. The externally mounted shading also limited window operation.

## PV and Interactive Grid Battery System

The PV array is located on the 10° pitched roof and is south facing. It is an 83kW system with a smart grid interactive battery bank. A photograph of the top of the array is shown in figure 23A.7. The building produces 160% more energy than it uses annually. The roof pitch is low in order to maintain city building height restrictions while keeping a comfortable volume of space in the interior open office work areas. The average annual solar insolation for 10° tilt angle is 5.26kWh/m²/day, with December having the least availability at 2.76kWh/m²/day (data from PVWatts[7]). The lower tilt angle reduces the winter production by quite a bit compared to the latitude tilt angle. However, because of the deep winter snowfall, it was determined to be unsafe to try to access the roof to clear snow and the system was designed as if it were going to be completely shaded for three months in the winter.

**Figure 23A.7** PV mounting layout diagram, Courtesy RMI.

The battery bank is a 30kW/45kWh LG lithium-ion battery tower which is intended to provide peak power demand shaving (batteries supplement during maximum building power use). The goal was to keep utility demand levels below 50kW, even when electrical vehicle (EV) charging is occurring, during peak demand times. Staying under 50kW keeps the building in a lower pricing tier. The battery bank is not large enough to provide consistent power for the building in a utility outage. This system is not an islanding microgrid because, at the time, the equipment needed wasn't yet available. Many of the control systems were difficult to acquire at the time (2014) as manufacturers were just developing them. The battery cabinet has integrated controllers for charging and discharging the system as well as providing automatic shutdowns if needed. The inverters are grid interactive and programmable. They control the flow of power to the building and utility.

RMI wanted the system to be adaptable for developments in energy management. This includes providing future functionality for bi-directional EV, DC power, and islanding. Conduit, and possibly existing circuits, to current EV stations could be reused for new bi-directional stations. There was discussion and concerns about having adequate space in the MEPIT (mechanical, electrical, plumbing, and IT) room to allow for additional equipment including disconnects, a second battery tower, and a critical load panel that would be needed for islanding.

The array is roof mounted to a standing seam roof. The racking and array layout are shown in figure 23A.8. There are two rails supporting each module which were bracketed to the roof using low profile attachments, to minimize wind loads.

Each module has a DC–DC power optimizer, shown in figure 23A.9, which disconnects power to the building in a grid outage and provides rapid shut down compliance. The system was designed as 2 arrays, each with 8 strings of modules connected to 4 inverters. One array has 16 modules in each series and the other has 15, for a total of 248 modules.

RMI retains the SRECs (solar renewable energy credits) and has a PPA with the utility for the sale of the power distributed to the grid.

## Lessons Learned

Although this is a passive building, there are a lot of controls. These systems are critical for the proper operation of the building. Commissioning was very important and RMI is interested in doing another commission to check that the building is running as optimally as possible. Careful consideration of the interface between the controllers and the software should be done, keeping the communication and operation as simple as possible. Preferably system communications can be binary on/off in order to keep system firmware updates from being unable to communicate with equipment.

The PV designer was involved early on in the project which was important because optimization of the roof needs to be considered all through the project. The PV is an integral part of the building's systems design. This project aimed for annual net zero, which allowed for the system to be covered in the winter by the deep snow. The battery system and controls were a larger challenge than expected at the time due to new/emerging technologies and interoperability challenges at the time.

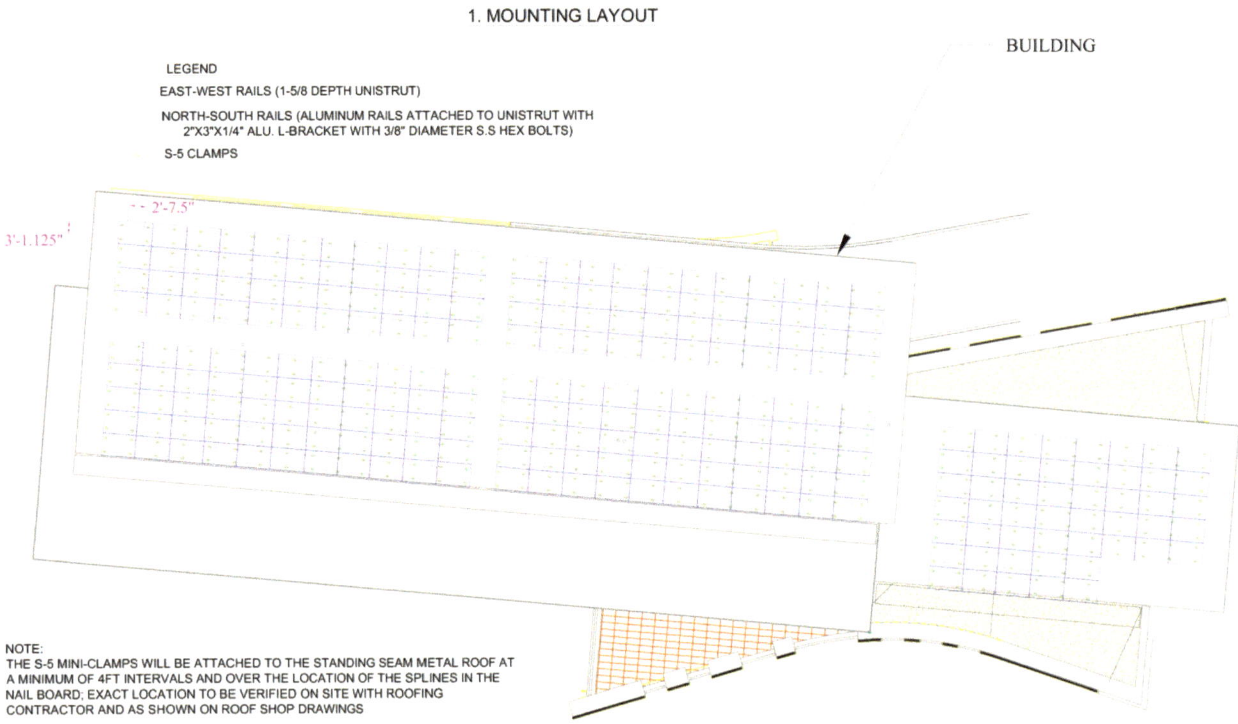


**Figure 23A.8** PV mounting layout diagram, Courtesy RMI.

This project was tremendously successful, both for efficiency and comfort. The current operational EUI is 16kBtu/ft$^2$/year, making it 74% more efficient than an average office in the same climate and 50% more efficient than a LEED V2009 building. It has received certifications of LEED Platinum, PHIUS+2015 & PHIUS+ Source Zero, as well as three Living Building Challenge Petals. The thermal comfort survey has positive scores for 88% of respondents.

The building is an excellent example of Amory Lovins' "Tunneling through the cost barrier", where paying premiums for items allows for reducing or eliminating systems and achieving reduced payback times because of operational savings. Figure 23A.10 illustrates how the project achieved a 4.year payback.

There are many other innovative systems included in this building which are available in other published reports.[8] RMI achieved their intent to have this building be a model to show that it is economically feasible to

**Figure 23A.9** Mounting and power optimizers of rooftop array, Courtesy of PAE.

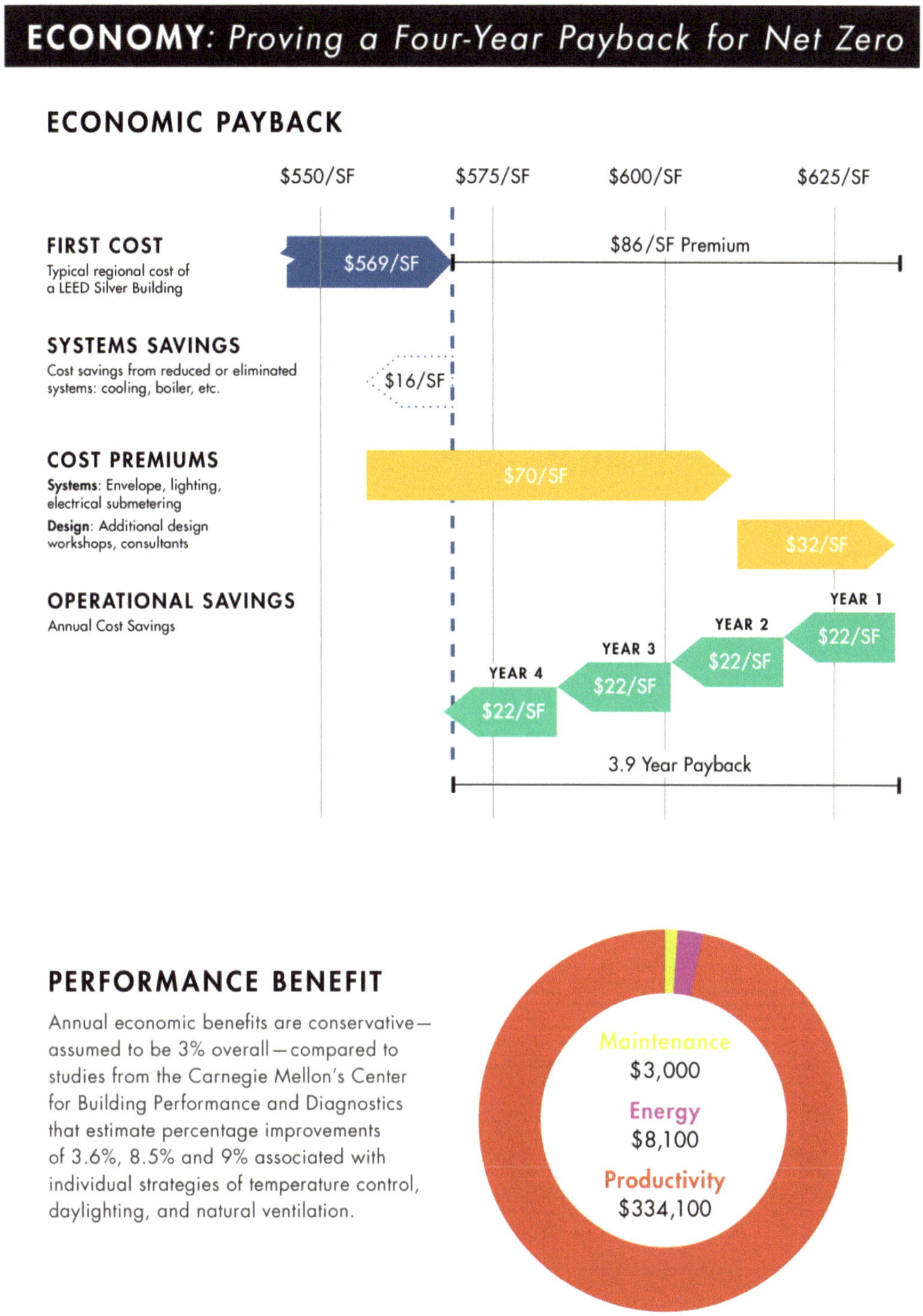


**Figure 23A.10** Economic Returns, Courtesy of ZGF Architects.

achieve a beautiful, high performance building that is comfortable and sustainable.

## Resources

1 Rocky Mountain Institute (https://rmi.org/).
2 ZGF Architects LLP (www.zgf.com/).
3 PAE (www.pae-engineers.com/).
4 KPFF Consulting Engineers (www.kpff.com/).
5 Architectural Applications (www.architecturalapplicati ons.com/).
6 Sol Energy (https://sol-energy.us/).
7 PVWatts® Calculator (https://pvwatts.nrel.gov/).
8 An excellent, very thorough case study can be found at https://cbe.berkeley.edu/wp-content/uploads/2018/12/Livable-Building-2018-RMI.pdf.

CHAPTER 23B

# City of Libby Microhydro PRV

Libby is a city of 2,703 people located Lincoln County, in the northwestern corner of Montana, United States. Similar to most cities in the USA, Libby has a public potable drinking water supply system that is run by the city. It is estimated that in the USA, as much as 1% of all electrical consumption is used for moving and treating public drinking water and can account for 30–40% of a municipality's energy costs.[1]

Supply systems for public drinking water are pressurized. They use **pressure-relief valves (PRVs)** to maintain the correct water pressure in the distribution piping. This protects pipelines and equipment from damage from excess pressure. Energy recovery hydropower, which is hydropower using water that is flowing in existing, pressurized lines, can be installed in parallel with existing PRV, using the pressure and flow of the municipal water to generate electricity. The Hydropower Regulatory Efficiency Act of 2013 (HREA) made it much easier for projects to be developed without environmental impact studies where infrastructure of dams or piping already exist.

The conception of the project began when Jim Hammons went to work for the City of Libby. He saw that the city's reservoir had a dam. He applied for and received a grant to study the viability of a hydro project in 2014. SOAR Technologies completed the study but found that it wouldn't be feasible to use the dam because the reservoir is located 4 miles out of town. SOAR Technologies suggested that the city should consider using the water treatment plant for a hydro system. During this project Canyon Hydro purchased SOAR Technologies, so both companies assisted with project completion.

The entire water system in Libby is gravity fed. The treatment plant is located on a hill and has two large 500,000 gallon (1,892,706 liter) tanks; one is inside and one is outside. The water leaves the treatment plant in a pipe that drops through a vault room located under the tanks. When the city first built the treatment plant, they installed three lines branching off from the main distribution pipe for PRVs but only used two of the pipes. The two existing PRVs were reducing the treated tank water from 116psi to 60–55psi for the distribution in the municipal water system. It was feasible to install a turbine on the unused third pipe and have it act as a PRV. Figure 23B.1 shows the installed turbine with the two other pipes behind it. The head is 150' (45.7m) from the treated water tank to the turbine and the amount of flow depends on water usage in the community. There tends to be more usage in the summer than the winter.

The PRV turbine, shown in figure 12B.2 in plan view on the top and elevation view on the bottom, was a prototype that was designed and built by SOAR Technologies and Canyon Hydro, both located in Washington state in the USA and was manufactured and assembled in the USA. It is a small physical size as the top is about 5' (1.5m) above the floor slab. It is a model ILT06 Francis turbine with fully adjustable wicket gates to allow for variable flow operation. The turbine's controllers monitor the downstream pressure of the water system. If pressure drops, the turbine's wicket gates are opened to allow more flow.

DOI: 10.1201/9781003297819-29

Figure 23B.1 ILT06 in-line hydro turbine. Courtesy of Jim Hammons, City of Libby.

If downstream pressure increases, the wicket gates are closed to reduce flow. There were some necessary adjustments in the beginning of the turbine's operation but since then it has been working well without downtime.

To get approval for the project, Canyon Hydro submitted the required Federal Energy Regulatory Commission (FERC) **Notice of Intent (NOI)** requesting approval for the project on June 10, 2016. FERC published the notice on June 29, 2016, which allows for 45 days for people to file comments and 30 days for people to file motions to intervene.[2] FERC approval for the project was granted within 60 days. The turbine started commercial operation on July 28, 2017.

The flow range of the turbine is 400gpm (1515 L/min) up to 1250gpm (4732 L/min) and has a net head of 150′ (45.7m). The flow range covers approximately 80% of the water flow through the station, allowing the turbine to run continuously all year. When flows are greater than the turbine can carry, the parallel PRVs pick up the remaining flows. If the turbine goes offline, then the PRVs automatically take over. The turbine's rated power output is 24kW. It has an induction generator which produces single-phase 240V AC power. The

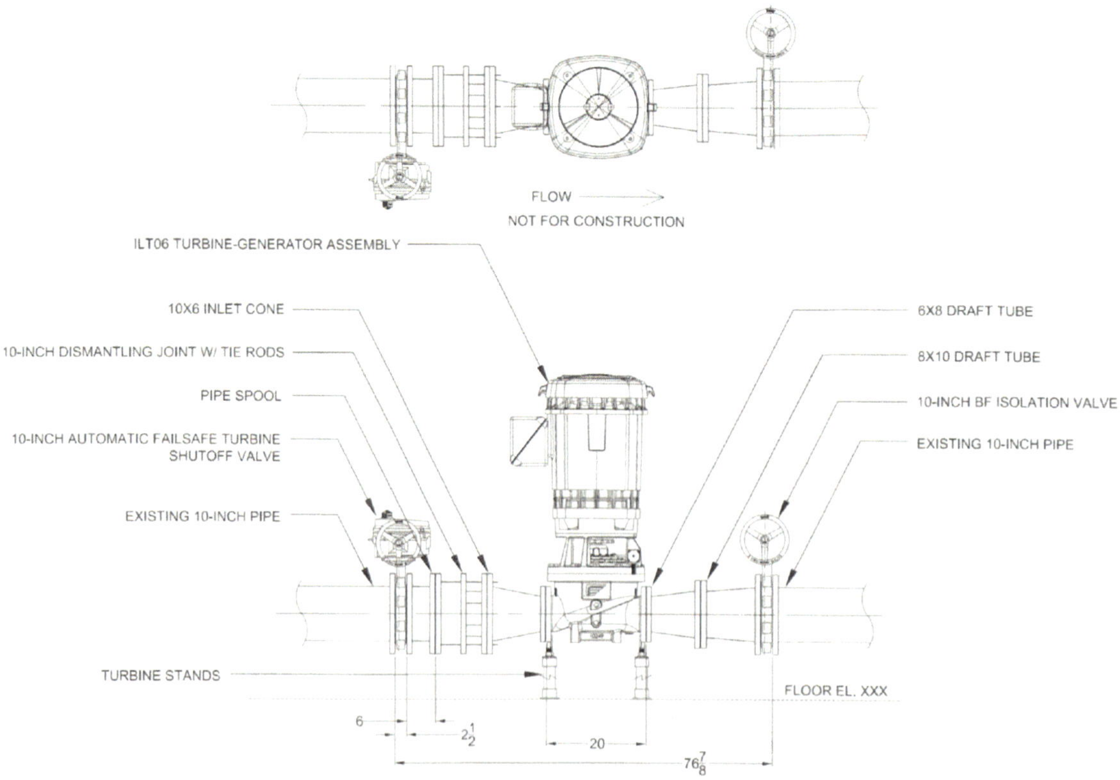


Figure 23B.2 SOAR ILT06 Standard Turbine-Generator Installation. Courtesy Jim Hammons, City of Libby.

panels containing the control equipment are shown in figure 23B.3.

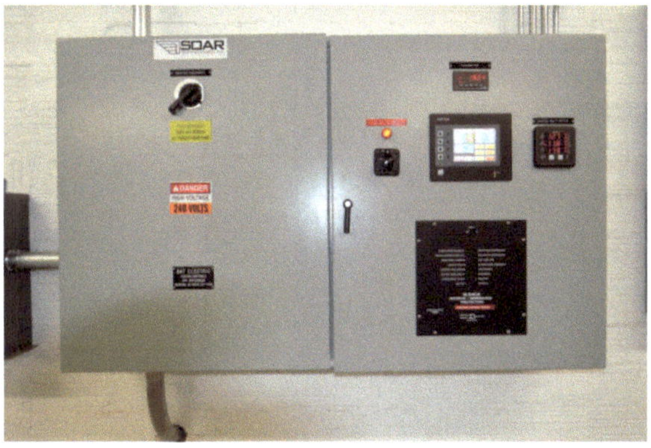


**Figure 23B.3** Control equipment for the turbine. Courtesy of Jim Hammons, City of Libby.

The ILT06 hydro turbine has in-line flanges that allow for bolt installation. City employees were able to install the flanges and fittings themselves so it was ready for the turbine. You can see the circular pipe fittings and bolts in figure 23B.4. Because of the existing third, unused pipe, the turbine and all associated accessories and control system were able to fit into the existing vault in the PRV building. Both the ability of the employees to do part of the installation and not having to build new facilities improved the project's feasibility and economics.

**Figure 23B.4** Turbine and control equipment in vault. Courtesy of Jim Hammons, City of Libby.

The project was funded through an arrangement with the local utility, Flathead Electric Cooperative. They were interested in being able to have a long-term contract for consistent renewable energy generation as part of their portfolio. In the USA, most states have requirements that utilities have renewable energy generation and get customers to reduce power usage. Flathead Electric Cooperative paid for the project upfront in exchange for a contract for the generated power. Power that is produced at the station is net metered. When the project has produced enough power to pay back the costs, then the power will be used by the city to offset power consumption at the nearby water treatment plant. Figure 23B.5 shows an example of the reports generated monthly by the Flathead Electric Cooperative that track production.

You can see from the report from Flathead Electric Cooperative, that the initial cost was counted as a contract for prepaid power. They track the credits and generation each month and show the output. You can also see comparative graphs for each month from 2017–2021. Currently, it appears the prepaid contract will be completed in July 2037. Jim Hammons (personal communication, July 14, 2021) said that this project has been very successful and he would recommend that municipalities look into the feasibility of using their systems to create electricity.

## Resources


1 Johnson, K., Levine, A., & Curtis, T. (2018, January). *Energy recovery hydropower: Prospects for offsetting electricity costs for agricultural, municipal, and industrial water providers and users, July 2017 — September 2017* (Tech. Rep. NREL/TP-6A20-70483). National Renewable Energy Laboratory. www.nrel.gov/docs/fy18osti/70483.pdf

2 City of Libby, MT; Notice of Preliminary Determination of a Qualifying Conduit Hydropower Facility and Soliciting Comments and Motions to Intervene, 81 FR 42346 (June 29, 2016). www.federalregister.gov/documents/2016/06/29/2016-15386/city-of-libby-mt-notice-of-preliminary-determination-of-a-qualifying-conduit-hydropower-facility-and

**FLATHEAD ELECTRIC COOPERATIVE**
Community. Innovation. Reliability.

**City of Libby Hydro Power Purchase & Exchange Agreeement**

**Month End: June 30, 2021**

| Contract Detail | |
|---|---|
| **Dollars** | |
| Contracted Prepaid Power | $ 122,125 |
| Cumulative Credits | $ 23,901 |
| Prepaid Power Balance | $ 98,224 |
| Monthly Average | $ 509 |
| **Kilowatt Hours** | |
| Contracted Prepaid Power | 2,035,417 |
| Cumulative Generation | 398,325 |
| Prepaid Power Balance | 1,637,092 |
| Monthly Average | 8,475 |
| **Estimated Contract Completion Date Based on Current Average:** | |
| July-37 | |

| Production Detail | | | | |
|---|---|---|---|---|
| Period | kWh Generated | Power Cost ($.06/kWh) | Power Balance | kWh Balance |
| Jun-21 | 7,411 | $ 445 | $ 98,224 | 1,637,092 |
| May-21 | 7,681 | $ 461 | $ 98,669 | 1,644,503 |
| Apr-21 | 8,576 | $ 515 | $ 99,130 | 1,652,184 |
| Mar-21 | 10,512 | $ 631 | $ 99,645 | 1,660,760 |
| Feb-21 | 8,061 | $ 484 | $ 100,276 | 1,671,272 |
| Jan-21 | 9,808 | $ 588 | $ 100,760 | 1,679,333 |
| Dec-20 | 9,201 | $ 552 | $ 101,348 | 1,689,141 |
| Nov-20 | 5,275 | $ 317 | $ 101,900 | 1,698,342 |
| Oct-20 | 4,875 | $ 293 | $ 102,217 | 1,703,617 |
| Sep-20 | 6,457 | $ 387 | $ 102,510 | 1,708,492 |
| Aug-20 | 14,767 | $ 886 | $ 102,897 | 1,714,949 |
| Jul-20 | 13,865 | $ 832 | $ 103,783 | 1,729,716 |
| Jun-20 | 8,247 | $ 495 | $ 104,615 | 1,743,581 |
| Previous | 283,589 | $ 17,015 | $ 105,110 | 1,751,828 |
| Total | 398,325 | $ 23,901 | | |

**Monthly Production Summary & Annual Comparison**

| | January | February | March | April | May | June | July | August | September | October | November | December |
|---|---|---|---|---|---|---|---|---|---|---|---|---|
| 2017 | | | | | | | 2,954 | 4,308 | 5,362 | 5,436 | 6,530 | 8,145 |
| 2018 | 8,813 | 5,407 | 9,481 | 7,151 | 11,210 | 11,880 | 8,582 | 8,022 | 9,028 | 9,717 | 8,623 | 9,373 |
| 2019 | 8,413 | 4,660 | 5,153 | 5,048 | 5,188 | 8,286 | 13,219 | 13,060 | 10,797 | 8,560 | 8,204 | 8,641 |
| 2020 | 8,652 | 8,478 | 8,841 | 8,738 | 9,629 | 8,247 | 13,865 | 14,767 | 6,457 | 4,875 | 5,275 | 9,201 |
| 2021 | 9,808 | 8,061 | 10,512 | 8,576 | 7,681 | 7,411 | | | | | | |

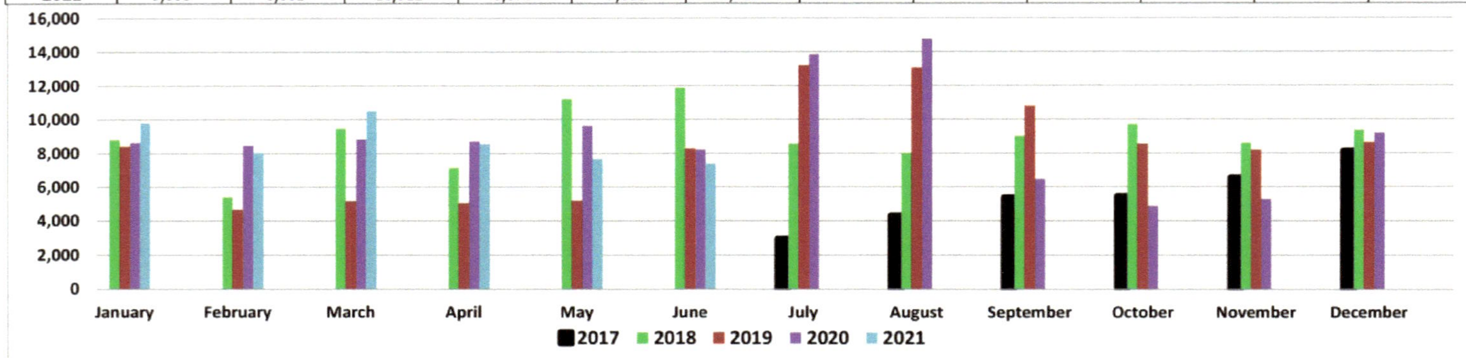


**Figure 23B.5** Copy of update report showing turbine production. Courtesy Jim Hammons, City of Libby.

# CHAPTER 23C

# University of Hawaii FROGs Classrooms

The **net zero energy (NZE) University of Hawaii Manoa (UHM) Flexible Response to Ongoing Growth (FROGs)** are second-generation mixed mode classrooms descended from a line of pre-designed modular/component-based buildings. These were designed and constructed by Project FROG of San Francisco, CA, using an approach called Industrialized Construction (IC). IC uses advanced technology to design and manage off-site pre-fabrication techniques to optimize resource and labor utilization.[1] The first generation of net zero energy FROGs in Hawaii consisted of three prototype classrooms constructed on the islands of Oahu and Kauai in Hawaii between 2009 and 2012, and are seen in figure 23C.1. These initial buildings were used as benchmarks for comparison to the second-generation buildings.

The second-generation FROGs were built on the UHM campus. The Manoa FROGs are 1,440ft$^2$ each and were completed in November of 2016. The entrances and clerestories of the FROGs seen in figure 23C.2 face to the north while the roofs pitch to the south.

The larger roof area is covered with 31 PV modules facing to the south as shown in figure 23C.3. Note that the glazing is located only on the north and south faces of the building.

Key information summarizing the project's location, climate, funding sources, and project statistics is shown in table 23C.1.[2]

To meet the UHM FROGs research objectives, UH researchers and facilities staff tasked the Project FROG design team to customize a pre-designed 1,440ft$^2$ structural shell to become an energy efficient, net zero energy building. While many standard features of Project FROG designs were already energy efficient, such as having well insulated walls and ceilings, high-performance glazing, and sophisticated lighting controls, several other features were customized to meet the net zero energy design objectives. These included increased natural ventilation and using natural daylighting while maintaining a comfortable environment using mixed mode systems.

Net zero energy buildings have energy efficiency as the primary design consideration because the least expensive kilowatt hour (kWh) is the one that is never used. The three key areas for energy efficient building design are: the building envelope, the air conditioning system, and the lighting system – paying attention to the interaction of these systems with each other. Additionally, mixed mode designs prioritize the following:

- optimize building envelope design to minimize heat gain to the occupied space with increased insulation, external shades, and glazing orientation;
- maximize openable windows for cross-ventilation; and
- incorporate ceiling fans to increase air velocity.

## Building Envelope and Glazing

Elements of the building envelope that have the greatest impact on energy consumption include building orientation, window direction and use of shading structures, and operable glazing for ventilation.

The interaction of the windows with other systems required the most attention when customizing the design. The front wall was essentially a curtain wall (figure 23C.2) and although the rear wall of the FROG has nine windows (figure 23C.3), only three were operable in the base design. The primary objective of the re-design was to eliminate direct solar heat gain through these large sets of windows. The first response was to orient the building north-south so that the largest amount of glazing would receive indirect, diffuse northern light which would have minimal direct solar heat gain while providing even, natural daylighting. The original design on the south side featured a deep

DOI: 10.1201/9781003297819-30

**Figure 23C.1** First-generation FROG (left) with locations in Ewa Beach, Oahu and Lihue, Kauai. Second-generation FROG (right) located on the University of Hawaii at Manoa campus.

**Figure 23C.2** FROG 1 (right) and FROG 2 (left) on University of Hawaii at Manoa campus.

**Figure 23C.3** Photovoltaic arrays on FROG 1 (left) and FROG 2 (right). Photo credit: UH Office of Sustainability.

roof overhang that shaded only the upper portion of the window bank, leaving the lower half exposed to direct sunlight. This was changed during re-design by adding a second louvered overhang to block direct solar gain to the lower half of the window bank. This lower overhang can be seen in figure 23C.3.

These modifications minimized heat gain but the windows still needed to be re-designed to maximize natural ventilation. All nine south-facing windows became fully operable. Additionally, the north facing clerestories (figure 23C.2) were modified, adding actuators to open and close the windows. This encouraged natural ventilation enhanced by the stack effect. Upper and lower transom windows were also added to the north curtain wall to increase the amount of operable window area available to encourage successful natural ventilation throughout the building.

## Air Conditioning

As a mixed mode building, natural ventilation is the main way to achieve room comfort. Air conditioning (AC) is provided to mechanically cool the space only when it is too hot outside to be comfortable. The 5-ton AC system is a conventional split system with an outside air conditioning compressor unit (which can be seen in figure 23C.2 near the rear of the FROGs on the west side) and an air handling unit in the mechanical room. Figure 23C.4 is a diagram of the air handling unit (AHU) that provides fresh air to the conditioned spaces through four supply air registers (shown as rectangles) which are located in a single duct. The return air grille is located within the conditioned space (although shown

Table 23C.1 Key building information.

| Key Building Information | |
|---|---|
| Location | University of Hawaii at Manoa, Honolulu, Hawaii, USA, 21.29° N latitude |
| Climate | 2.6HDD and 4753CDD (65°F), high of 87°F (30.5°C) to low of 68°F (20°C) |
| Annual Rainfall | 48.8" (1240mm), 271 sunny days |
| Average annual solar radiation @21° tilt | 5.87kWh/m²/day |
| Project Completed | November 2016 |
| Floor Area | 1,440ft² (134m²) each building |
| PV System Rating | 8.06kW DC |
| Funding | Hawaii Natural Energy Institute |
| | University of Hawaii at Manoa |
| | through the following grants from the Office of Naval Research: |
| | N00014.12-1-0496, N00014.13-1-0463, N00014.14.1-0054 |

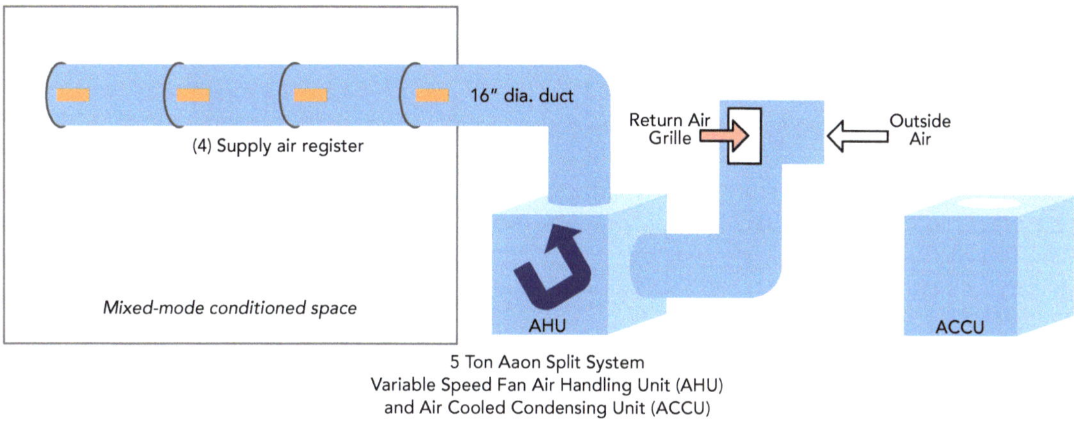


Figure 23C.4 Diagram of FROG mixed-mode air conditioning system.

to the right in the diagram) and the return air duct is tied to the outside air opening (OSA, shown as a blue arrow inwards on the right). The air-cooled condensing unit (ACCU) is shown off to the side in the diagram (because it is located outside) but has a condenser loop tied to the AHU that can cool the supply air when needed.

The controls were designed in a novel manner to reduce system operation in intermittently used classrooms by eliminating cooling of non-occupied classrooms. The AC system was designed with an on-demand thermostat that has to be manually enabled by the classroom users. When started, the on-demand thermostat runs the AC system for one hour, then automatically shuts off. Given 60 minutes to acclimate to the room, the users could then determine whether the room was comfortable or call for another hour of cooling. Because the classrooms were being used intermittently by two different entities, a K-12 charter school and the University, the scheduling of the rooms varied every hour of every day. Rather than automatically operate the AC continually from 7am to 9pm, the on-demand thermostat prevented excessive use during unoccupied periods. An initial study by **Hawaii Natural Energy Institute (HNEI)** documented an estimated 84% savings over a specific time period as shown in figure 23C.5, although the actual savings will vary greatly depending on the room capacity that is scheduled in a particular semester.

## Lighting and Daylighting

Project FROG incorporated into their base design a sophisticated lighting control system that used a Lutron LED ecosystem, Lantana direct-indirect fixtures with daylight harvesting sensors and RDU-12 Drivers; these are shown in figure 23C.6. In the main classroom there are four rows of Lutron Lantana LED fixtures controlled by one occupancy sensor and one daylight sensor. The entry zone has two rows of fixtures tied to one occupancy and one daylight sensor.

The rows of lights were hung parallel to the windows so the rows can be individually controlled to achieve appropriate lighting as the amount of daylighting drops off deeper into the space. The rows in each area are pre-programmed to operate at light levels

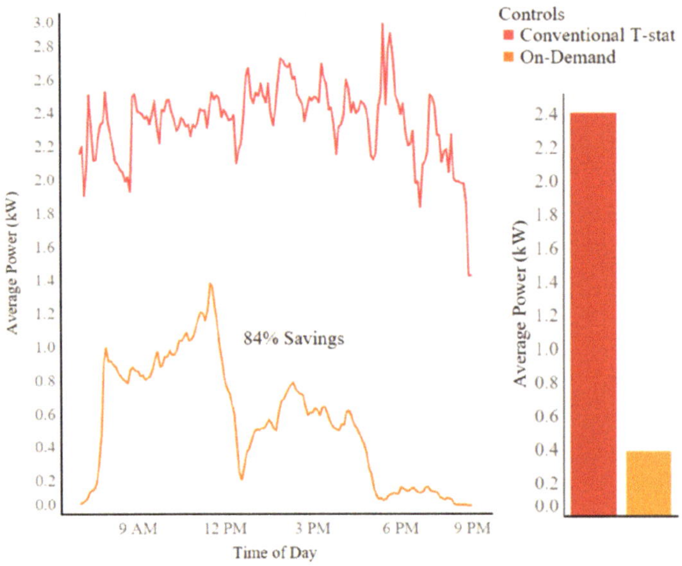


**Figure 23C.5** Comparison of power and energy savings resulting from the on-demand thermostat.

determined by their distance from the windows based on the minimum Illuminating Engineers Society (IES) standards. The occupancy sensor activates the fixtures to turn on and off. Within about 30 seconds of turning on, the lights adjust according to the daylight sensor input. In the main classroom on a typical sunny day the fixtures dim their output so the row closest to the windows are at about 10%, the center row at about 25%, and the two interior rows near the rear of the room at about 40%. On overcast days and in the evenings, the fixtures will ramp up output based on the daylight sensor's readings.

The electrical lighting system interacts with the windows to maximize daylighting in the space. This allows the large banks of windows on the north and south sides to provide natural diffuse daylight for the space during the day. On most days of the year, the lighting system can be completely overridden and shut off, because enough natural daylight is present to accommodate the needs of the space. High-performance glazing (Solarban 70XL) with high visible light transmittance (64%), a low solar heat gain coefficient (0.27), and a light-to-solar gain ratio of 2.37 provides adequate daylight without the associated heat gain.

## Comfort

A challenge common to naturally ventilated and mixed mode buildings is to maintain a comfortable environment with a minimum dependence on mechanical AC. While Hawaii benefits from natural trade winds that can drive natural ventilation, there is a significant amount of time where the trade winds are low or non-existent and/or the air temperatures and humidity are high. For these reasons, three banks of three variable speed ceiling fans were added to augment the original design, providing a source of air movement that cools the skin through evaporation and convection. While the air motion does not reduce the temperature within the space, the users experience a perception of coolth to maintain a satisfactory comfort level. With the ceiling fans on, users frequently choose to not operate the AC.

During the design phase, researchers performed adaptive thermal comfort analysis to determine the effectiveness of increasing ventilation and adding ceiling fans to the base building as shown in figures 23C.7 and 23C.8. The analysis compared the relative comfort produced with and without ceiling fans and adjusting the percentage of operable window area. The figures graph months of the year for the horizontal axis and time of day vertically. Figure 23C.7 shows the results based on no ceiling fans and 25% of the window area is operable. There are massive areas of red indicating very hot conditions for most of the year from 10am until 8pm, all of which are probable hours of occupation for the classroom.

Figure 23C.8 is the annual comfort analysis using ceiling fans and 75% of the window area is operable. Because of increased ventilation and ceiling fans moving the air in the space, overall discomfort is reduced significantly. Almost all the red is eliminated and the darker orange areas indicting hot conditions are limited to smaller areas over the middle of the day and a handful of months. These periods of discomfort are the mid-summer months through October which

Figure 23C.6 The Lantana direct/indirect LED fixtures hung from the classroom ceiling with a fan and the Lutron Remote Driver Unit, RDU-12 Driver (left inset).

are typically the hottest and most humid times of the year.

## Benchmarks and Outcomes

The UH Manoa FROG design is the second-generation of the single zone, mixed mode, net zero energy buildings, designed to be built faster, more economically, and more efficiently than the first three FROGs constructed in Hawaii. While intended to operate more efficiently than the first generation, the second-generation FROGs in fact used about the same amount of energy as seen in figure 23C.9.

The reasons for this are partially due to greater operating hours in the UHM FROGs, as well as a strong cultural commitment to energy efficiency by users in the two Kawaikini FROGs, where the AC was used very little. As a point of interest, a computer server was moved into the Ilima first generation classroom late in the year which created a large air

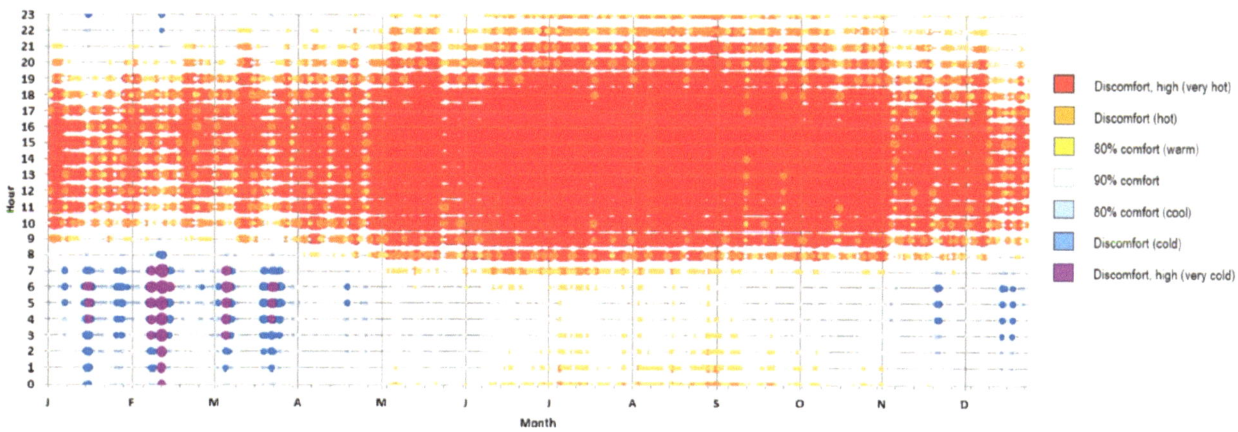


Figure 23C.7 Hourly indoor adaptive thermal comfort for naturally ventilated FROG 1 without ceiling fans and 25% of the window area is operable.

23C

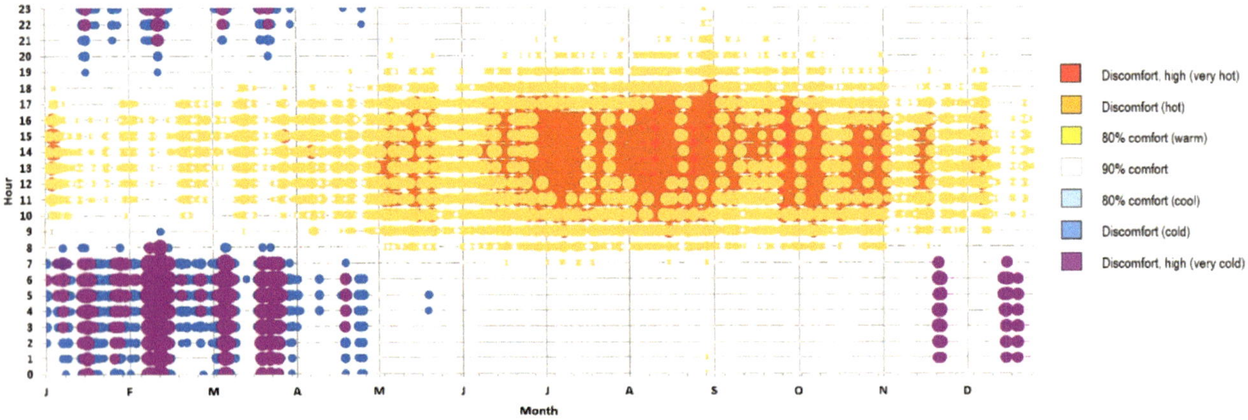


**Figure 23C.8** Hourly indoor adaptive thermal comfort for naturally ventilated FROG 1 with ceiling fans and 75% of the window area is operable.

conditioning load in the building, and increased the annual energy usage.

HNEI has monitored the performance of the UH Manoa FROGs since 2017 and both classrooms have remained consistently net positive, see figures 23C.10 and 11. In both figures the shorter bar on the left is the energy used by the building and the taller line on the right is the energy produced by the PV system on the building's roof. Each data set is stacked for consecutive years, aligning the months vertically. The PV production

based on occupancy and occupant interaction with the control systems.

## Energy Use Intensity

To compare buildings using a common energy metric, site Energy Use Intensity (EUI) was calculated. These compare favorably to the net zero energy targets set by the New Buildings Institute for primary schools.[3] While this is the closest benchmark, it should be noted

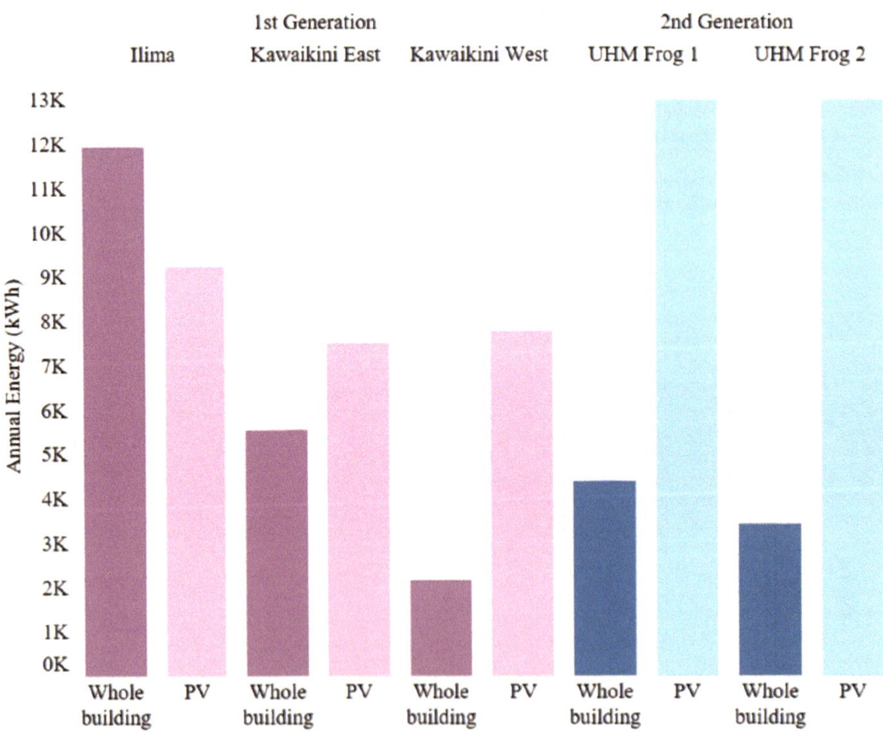


**Figure 23C.9** Overall energy generation and consumption comparison of the three first generation and two second generation FROGs, fall of 2017.

between the two buildings is mostly similar throughout the year, as would be expected. Energy usage varies

that the FROG EUIs are for the classroom only, while the primary school energy accounts for entire schools,

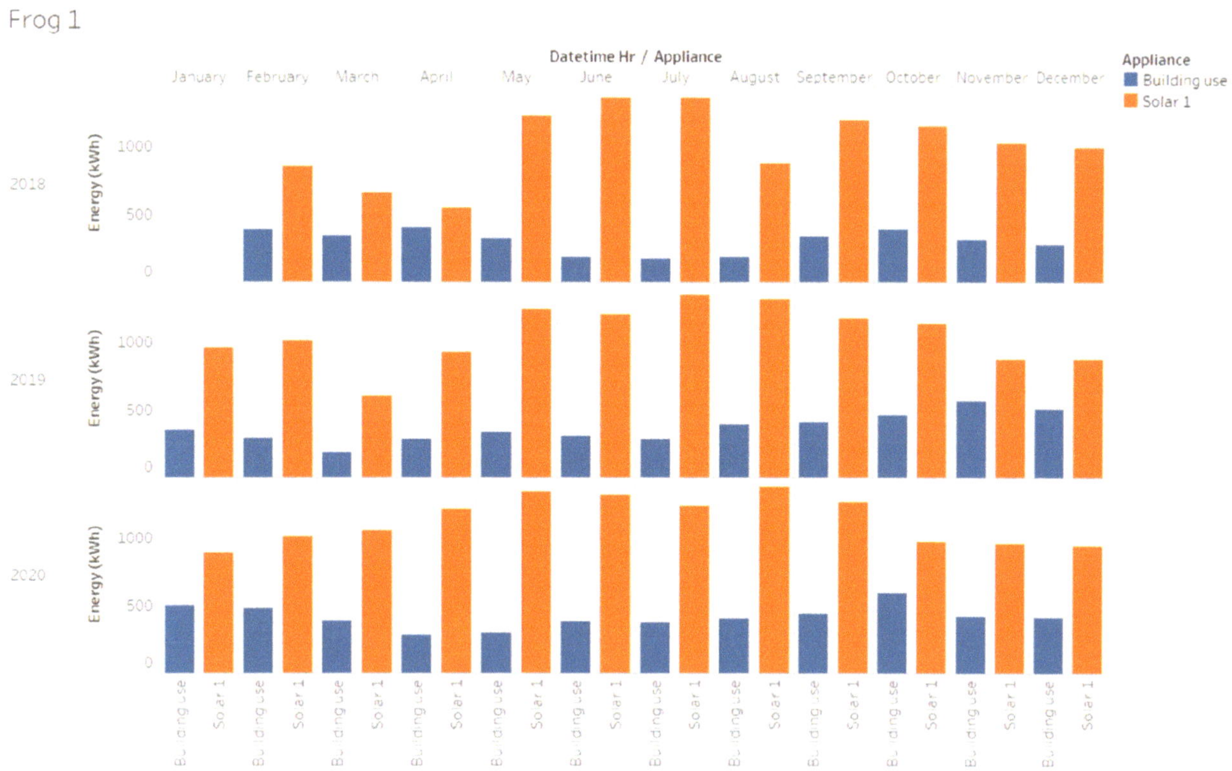


**Figure 23C.10** Generation and consumption for FROG 1 2018–2020.

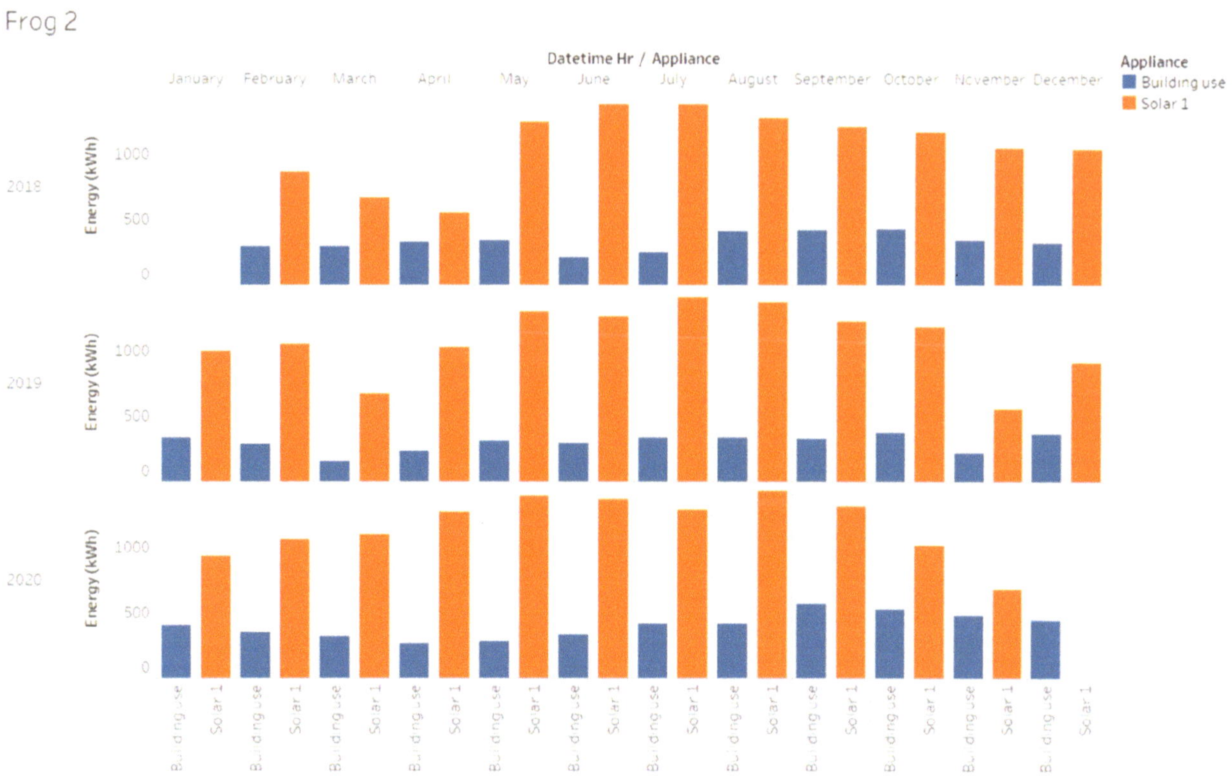


**Figure 23C.11** Generation and consumption for FROG 2 2018–2020.

23C

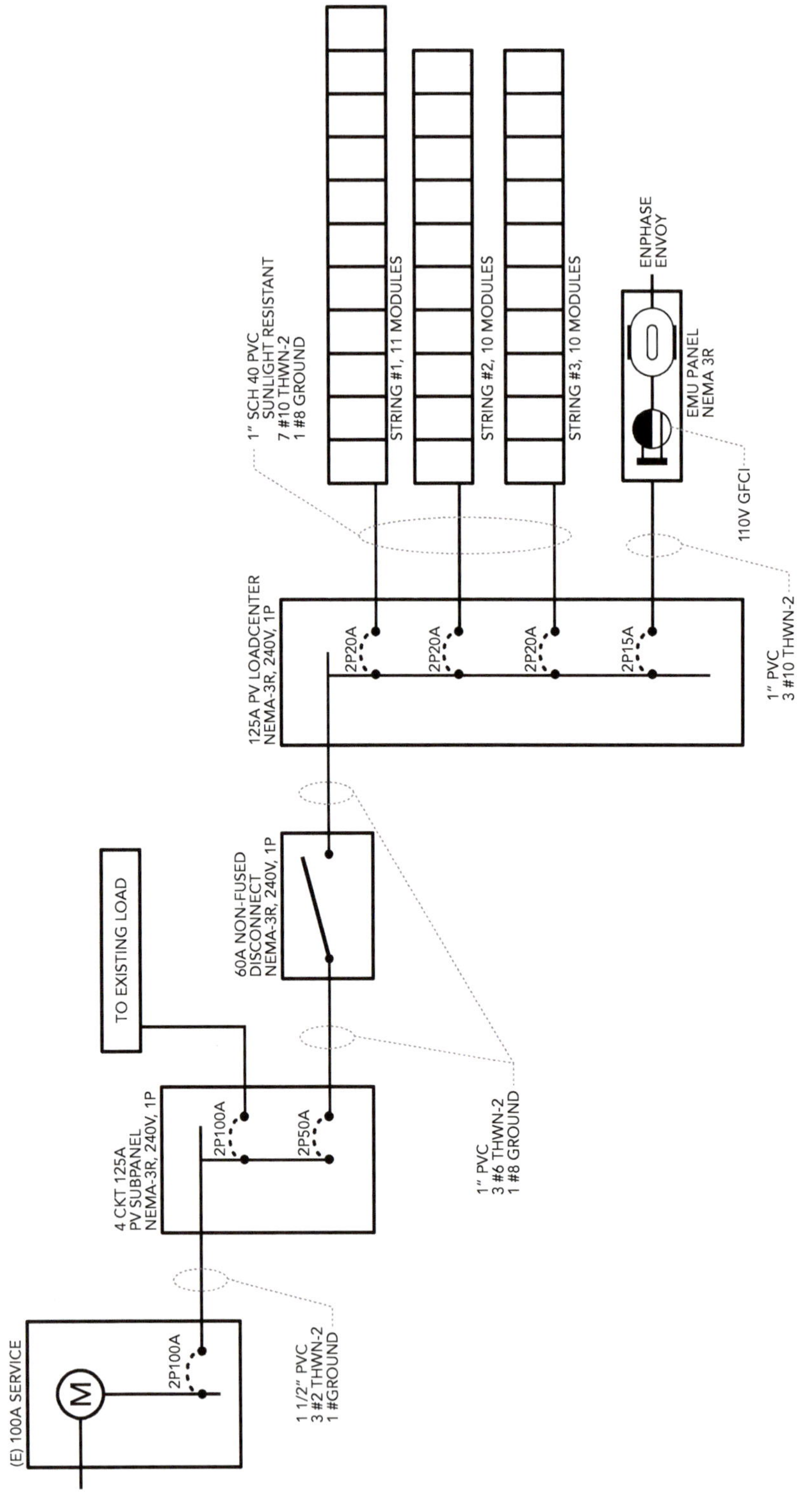


**Figure 23C.12** Single line diagram of UHM FROG PV solar system.

Table 23C.2 Average annual site energy 2018–2020.

| | Site Energy | | |
|---|---|---|---|
| | kBtu/ft²/yr | kWh/ft²/yr | kWh/m²/yr |
| FROG1 | 11.8 | 3.1 | 37.1 |
| FROG2 | 9.9 | 2.9 | 31.2 |

Table 23C.3 NBI NZE Target3 (primary school/CZ1).

| Site Energy | |
|---|---|
| kBtu/ft²/yr | kWh/m²/yr |
| 26 | 82 |

including administrative functions. Note that the EUI is based on the monitoring data usage from 2018–2020 shown in figures 23C.10 and 23C.11.

## Photovoltaic Systems

The renewable energy systems used on the UHM FROGs are each a 31-module array as shown in figure 23C.12. The 260W Hyundai HIS-260RG PV modules are each wired to Enphase microinverters and then wired to a PV loadcenter. The Enphase Envoy provides monitoring for the system and is plugged into a ground fault circuit interrupter (GFCI) outlet connected to the loadcenter. All system parts are grounded. The loadcenter has a disconnect between it and the PV subpanel that powers the building. They both have fuses in them, shown in the line diagram as two dots with an arc in-between. The disconnect is a switch, so it shows two dots with an open angled line between them. The PV subpanel connects to the main service which is part of the UHM campus grid. The UHM campus grid is tied to the utility grid at two substations. Because of their separate grid, the excess energy produced is used to power nearby University buildings.

## Resources


1 Project Frog (www.projectfrog.com/).
2 PVWatts® Calculator (https://pvwatts.nrel.gov/).
3 Carbonnier, K. (2019, September 10). *NBI releases zero energy performance targets for new construction projects*. New Buildings Institute. https://newbuildings.org/nbi-releases-zero-energy-performance-targets-for-new-construction-projects/

CHAPTER 23D

# YWCA Northcentral PA

The **YWCA** building was built in 1928, in the heart of Williamsport's historical district "Millionaire's Row". The front of the building faces north and is shown in figure 23D.1. The YWCA created a safe place for young working women of the early 1900s and today their mission is "to eliminate racism, empower women, and promote peace, justice, freedom, and dignity for all."[1] They maintain meeting rooms and offices for various community services on the first two floors of the building. The upper two floors have 50 small apartments with communal spaces and four large kitchens. In 2011, the YWCA received a grant to install a solar thermal system to provide hot water for the existing building; there were no other additions to the building. The system was designed and installed by Larson Design Group and Stanley C. Bierly. Williamsport is located at 41.29° N latitude, with 4.61 kWh/m$^2$/day solar insolation at latitude.[2] At base 65°F, Williamsport has 5987 heating degree days and 722 cooling degree days.[3]

There were some challenges faced by the designers. The façade of the building is a 4" (100mm) brick wythe that would not be able to support the weight of the collectors. The south face of the building has a clear solar window (a great advantage) and the parking lot asphalt extends directly to the façade (a bit awkward). The boiler room, where the existing gas-fired water heater is located, is about 250' (76m) from the southern façade. Inside the southern end of the building there is a high-ceilinged multipurpose room with a stage. There are also multiple windows on the southern façade. The proposal had to be approved by the Williamsport Historic Architectural Review Board (HARB).

The final installation is shown in figure 23D.2. The project was presented to HARB under the addition of window shades/awnings to the historic structure. You'll note that at the high summer sun angle of the photo, not only are the windows shaded, but also parts of the lower collectors. In the winter the collectors are not shaded. There is a wide variation in the amount of solar insolation available in July (5.82 kWh/m$^2$/day) vs. December (2.68 kWh/m$^2$/day), so the shading helps to keep the production more equal across the year[2]. The location of the collectors vertically on the wall is based on the location of the floor slabs, as the framework supporting the collectors is attached to the underside of the existing slab, as seen below in figure 23D.9.

The system has two separate loops, labeled as east (on the right in figures 23D.2 and 23D.4) and west, each with eight harp-type flat plate collectors. Because the parking is right up to the building, snow and ice guard bars were welded to the lower part of the frame to reduce of sliding snow and ice from shedding from the collectors onto pedestrians and cars. Signage was added to the wall to warn people of the potential for sliding snow and ice, as shown in figure 23D.3.

To keep the storage tank near the collectors, it was placed in the area under the stage. This space was already used for ductwork providing conditioned air to the multipurpose room, as well as storage for various equipment needed for maintaining the parking lot and grounds year-round. In figure 23D.4, you can see the non-pressurized tank which is flanked by panels for the controls for each side of the system. As you can see, the ceiling is low and the tank is tucked in under the ductwork. This is a 500 gallon (1892 liter) non-pressurized tank.

The blue line indicates the maximum height of the water inside the tank. In figure 23D.5 you can see that there are four copper heat exchanger coils that are on the interior of the tank. The coils are in sets, with two connected to the east collectors and two connected to the west collectors.

To organize all the controls, everything was mounted on a painted plywood panel. The panel is braced in place with a tube frame which is mounted to brackets that are bolted to the floor. The side view of the frame is shown in figure 23D.6 while the floor attachment is in figure 23D.7. The front view of the panel shows the piping with two in-line pumps, which is labeled showing the direction of solar fluid flow to the collectors. There are two pumps because of the height between the tank

DOI: 10.1201/9781003297819-31

Figure 23D.1 YWCA Northcentral PA, street façade facing North.

and the collectors. This system uses an external heat exchanger to transfer heat from the collector's closed loop to the coils inside the tank. The collector side piping is filled with a proprietary solar fluid which has a high boiling point that does not break down if it is sitting in the collectors when they are hot. The loop between the external heat exchanger and the copper coils in the tank circulate water from the tank. Water can be used because the piping is all located in conditioned space.

In addition to providing hot water for the building, there is plumbing to a fan coil unit in the multipurpose room to provide heating for the space in the winter. Hot water is circulated through a loop to the fan coil, which is located under the stage behind the right-hand grille and returned to the storage tank. There is a thermostat on the wall of the stage area that is hooked up to a pump controller that turns on if heat is needed and the tank is at an appropriate temperature. The piping loop connection to the tank is shown in figure 23D.7; it is the piping on the left side. This figure also shows the bracing for the plywood control panel as well as the water loop to the heat exchanger to the right. The face of the stage is shown in figure 23D.8.

The stage, shown in figure 23D.8, has three tall windows (seen on the exterior in figure 23D.2) with radiant heating located under each one. There are curtains hung and additional rods (not visible in this photo) for hanging lights and/or a backdrop. Because the piping to the collectors is running directly up the interior wall between the windows, they were boxed over to have a similar appearance to the existing pilasters centered between the windows. These covers extend

Figure 23D.2 YWCA southern façade, photo taken July 27, 2021.

**Figure 23D.3** Snow guard bars on collectors and signage on wall.

**Figure 23D.4** Non-pressurized tank and BOS East and West collectors.

high enough that the piping distribution is not visible from the main multipurpose space. The close-up of the covering, piping and wall brace panel can be seen in the right-hand image.

Figure 23D.9 shows the framing that supports the lower four east collectors. There is an L-shaped bracket that is bolted to the underside of the floor slab at the interior face of the wall. The vertical part of the bracket has diagonal bracing that attaches to an existing steel I-beam. You can also see that the expansion tank is attached to one of the braces. The piping to the collectors penetrates through the southern wall above the windows and runs along the braced frame on the exterior of the building shown in figure 23D.10. In this same figure, you can also see where the exterior frame penetrates through the wall and attaches to the L-shaped bracket. The holes were patched to match the mid-tone color of the brick on the exterior wall.

Back inside the building under the stage, there is a piping loop that runs from the tank to the hot water heater. Figure 23D.11 shows the piping from

**Figure 23D.5** Interior of non-pressurized tank with 4 heat exchanger coils.

**Figure 23D.6** West control panel mounting board front and side view.

**Figure 23D.7** Piping to heat exchanger fan coil in multipurpose room.

the tank and the connection through the wall. The piping runs under the first floor of the building, which is basement area. Because it is about 250′ (76m) from the tank to the hot water heater, there is a circulation pump that runs continuously so that there is always hot water in the pipe. The circulation pump is very small and can be seen to the very right of the photograph. The loop attaches to the cold water supply for the gas-fired hot water heater as shown in figure 23D.12.

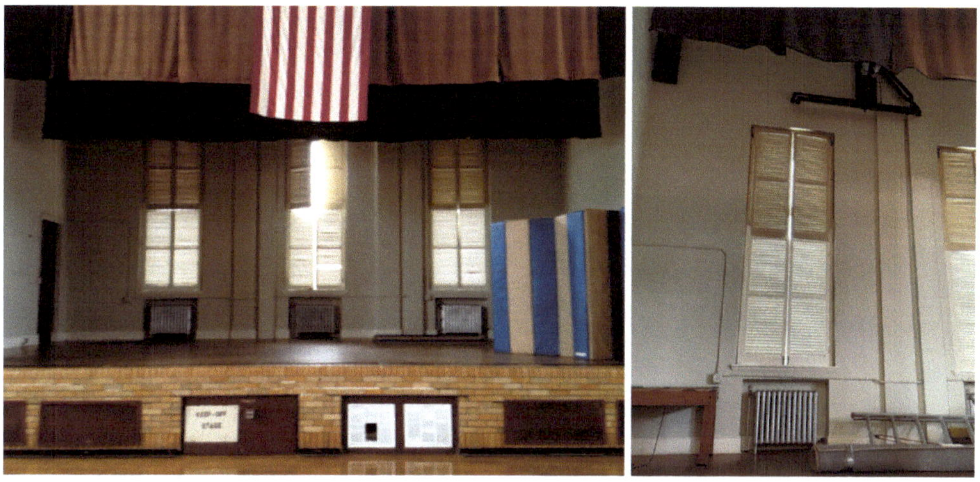

**Figure 23D.8** Stage view of piping from multipurpose room and stage, closeup on right.

**Figure 23D.9** Structural support for east collectors.

The cold water supply connection from the street has been shut off, so all "cold" water to the hot water heater is supplied from the collector storage tank. To help reduce heat loss, all the pipes are insulated and there is just one loop. When there is a big demand draw from the storage tank, the circulation loop pump is overridden and both pipes provide water to the hot water tank. The water heater is set for 140°F (60°C) and only comes on if the water from the storage tank is below this temperature.

This project shows how challenging it can be to integrate new systems into existing buildings. Designers need to apply their creative problem-solving skills to appropriately integrate a system.

## Resources


1. YWCE Northcentral PA (www.ywcawilliamsport.org/).
2. PVWatts® Calculator (https://pvwatts.nrel.gov/).
3. *Williamsport, Pennsylvania (369728) period of record monthly climate summary, period of record: 01/30/1901 to 06/10/2016.* (n.d.). Western Regional Climate Center. https://wrcc.dri.edu/cgi-bin/cliMAIN.pl?pa9728

Figure 23D.10 Exterior attachments collectors to rack, plumbing and racks through wall.

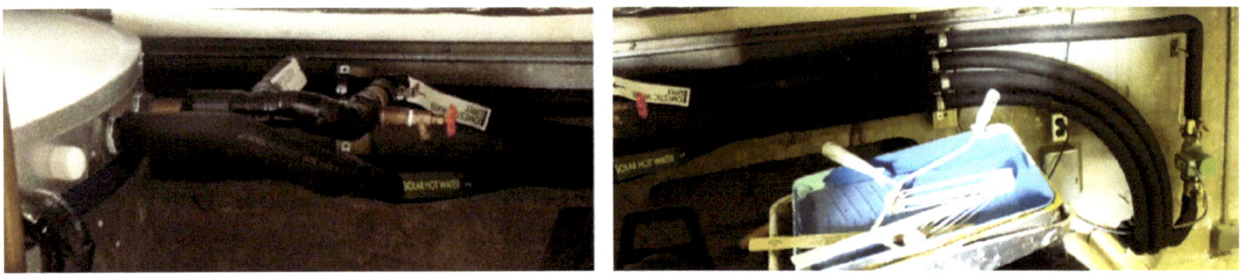

Figure 23D.11 DHW circulation piping, east side of tank.

**Figure 23D.12** DHW gas furnace with solar tie-in.

CHAPTER 23E

# 125 Haus

## Location and Use of the Building

Design for the 125 Haus began in May 2009 and construction was completed from May to October 2011. It is located in Summit Park, a small neighborhood near Park City, Utah, USA. The detached single-family house, shown in figure 23E.1, has 2,444ft$^2$ (178m$^2$) of conditioned space with 3 bedrooms, 2.5 baths, kitchen and dining, living room (shown in figure 23E.2), play area, and studio, as well as a 2-car garage and exterior decks and patios. The architect was Atelier Jörg Rügemer and the General Contractor was Garbett Homes. Structural engineering was provided by LR Nelsons of Salt Lake City and Las Vegas. The original mechanical engineer was Heliocentric, however, their system did not work satisfactorily and was replaced. The current HVAC system was designed as part of Rügemer's research and was installed and has successfully run and been tested over the past 6 years. He also did two years of post-occupancy monitoring on all systems.

Summit Park is located at 40.6461° N latitude at an elevation of 6982 feet (2.100m). It is a cold climate zone, with 8,190HDD and 162CDD. The hottest July day is 81.1°F (27.3°C) with an overnight low of 51.4°F (10.8°C), while winter temperatures drop to highs of 28.6°F (–1.9°C) and lows of 0°F (–18°C). There is an average of 21″ (533mm) of rain and 400″ (10,000mm) of snow annually. It averages 216 sunny days, with the most in June and the least in February. Light snowfall starts in October, with the peak in February and falling off in the beginning of May. Humidity is low in the summer months. The highest wind speeds are just over 6.75mph (3.02m/s) in August, mostly averaging around 5.5mph (2.46m/s), except for calmer times in December and January. Average annual solar radiation at 40° tilt angle is 5.48kWh/m$^2$/day with the least availability in January at 3.64kWh/m$^2$/day.[1,2]

The building received numerous awards including the 2012 AIA Utah Design Award, 2013 Residential Architect Award, 2013 Design Arts Utah Award, and was named as one of 20 best green buildings of 2013 in the USA by Hanley Wood. The house was not PHIUS certified due to complications with its high elevation location but it was designed as close to the PHIUS standard as the budget would allow.

## Benchmarks, Outcomes, and Challenges

The architect used an integrated design process which brought the mechanical and structural engineer and the contractor on the project from the beginning. The first modeling was done with Passive House Planning Package (PHPP) during the design phase. Studies were done on different types of wall construction for both thermal performance and cost. The wall that was built is a double wall with offset 2×6 structural and 2×4 interior studs (11″ cavity with blown-in fiberglass insulation) and exterior 4″ rigid expanded polystyrene insulation with exterior stucco, creating an R58 wall. The roof is R80. At the end of the design phase, the mechanical engineer modeled the building using DOE EnergyPlus software. Results from this model were used to further optimize the design.

A Solar Pathfinder analysis was done and overlaid on the site, as shown in figure 23E.3, to help with the orientation and placement of the building. There was quite a lot of shading in the area because of the hilly, wooded terrain and the nearby houses.

Passive solar gain occurs through the large southern windows and heats the house on sunny winter days. Figure 23E.4 shows a diagram of the current HVAC system. There is a Venmar HE 1.8 88–95% efficient heat recovery ventilator (HRV) used to exchange and preheat ventilation air, as well as to distribute air throughout the home. This can be seen in the diagram in the upper right corner. The blue arrows penetrating the walls represent the air exhaust and intake for the HRV. Note that the intake for the warm, stale air from the interior is shown on the left of the HRV as a red arrow. On the interior of

DOI: 10.1201/9781003297819-32

Figure 23E.1 125 Haus southern façade. Photograph: Paul Richer, Richer Images.

the HRV the heat from the interior air is transferred to the incoming fresh air. When there is not enough sun to heat up the house and/or the exterior temperatures are quite cold, an on-demand gas hot water heater provides warm water to a coil in the HRV. The water heater is a 98% Efficient Navien Condensing Tankless Hot Water Heater that burns natural gas. It is shown on the upper left in the diagram. It is directly supplied with combustion air from the exterior, shown as a blue arrow in, and directly vented to the exterior, shown as a red arrow pointing out. Domestic fresh water is supplied to the bottom of the unit and the hot water piping provides domestic hot water (DHW) for the home and loops through a heat exchanger (shown at the lower left of the diagram by the main loop pump) before returning to the supply line and back to the water heater. The main loop pump runs when heat is called for at the HRV. On the right side of the heat exchanger are two loops, each with their own pump, one going to the HRV heat exchanger coil and the other to the radiant heat loop. The radiant heat loop heats the floors of the lower story, where the studio is, and kitchen's concrete floor. There is also a radiant wall and electric resistance heater (towel bar) in the master bath, which can be chilly as

Figure 23E.2 Interior view of living room. Photograph: Scott Zimmerman.

Table 23E.1 Key building information.

| 125 Haus | |
|---|---|
| Location | Summit Park, 40.6461° N latitude, elevation of 6982 feet (2.100m) |
| Climate | 8,190HDD and 162CDD (65°F)<br>high of 81.1°F (27.3°C) to low of 0°F (−18°C) |
| Annual Rain/Snowfall | 21" (533mm) of rain / 400" (10,000mm) of snow |
| Average annual solar radiation @40° tilt | 5.48kWh/m²/day |
| Project Completed | October 2011 |
| Floor Area | 2,444ft² (178m²) |
| PV System Rating | 2.25kW DC, provides 80% of annual electricity use |

it faces north. The heated towel bar is shown near the center of the diagram.

The garage was designed so that warm cars could be enclosed inside during the winter and the heat from them radiates upwards through the floor assembly for additional warmth on the main floor. The 22kBtu/h direct-vent gas fireplace, shown near the center of figure 23E.4, is located in the living room on the main floor (as seen in figure 23E.2) and can be used for backup heat.

Night ventilation is used to cool the building through operable windows and the "thermal chimney" which is an open riser staircase located in the center of the building, see figure 23E.5. This is effective because of the large summer diurnal temperature swings.

## PV Grid-Tied System

The PV system has 8 Sunmodule Plus SW 275W modules each with an Enphase M250 microinverter, rated at 2.25 kWp (peak) DC and is shown in figure 23E.6. Annual production is about 3,000 kWh. The system has been online since January 2018 and produces 80% of the building's annual electricity consumption.

The system is frequently cleared of snow during the winter months, not only to provide additional electrical yield, but also to keep the panels from breaking under the 3–4' of regular snow accumulation. The location of the panels was determined by using the original Solar Pathfinder diagram. Only the roof's east half is structurally sound to accommodate for the additional wind and snow load developed because of the panels. The east roof has a shorter beam span compared to the west half of the roof. The panels are installed at a 10° tilt angle, oriented directly south, and are ballast-mounted on the thermoplastic polyolefin (TPO) roof to avoid any roof penetrations. The building was originally designed to be solar-ready, which included an empty 2" chase through the roof and down to the breaker panel in the basement.

## Lessons Learned

Computer modeling and the integrated design process were integral to producing a beautiful building that is about 80% more energy efficient compared to the

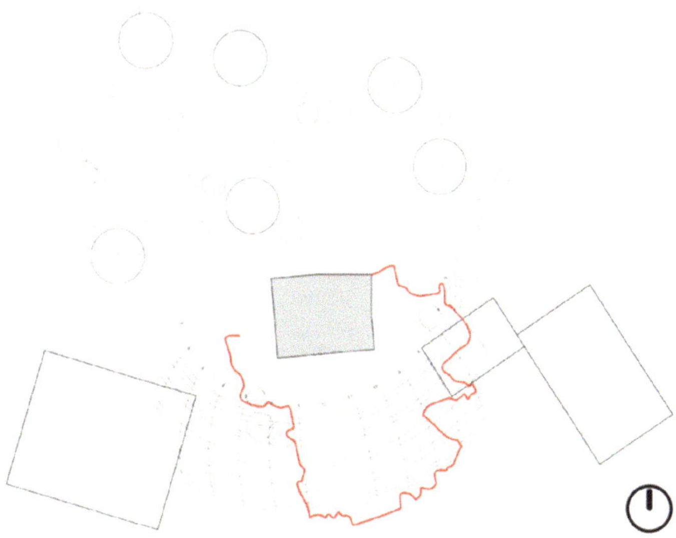

Figure 23E.3 Solar Pathfinder site analysis as decision making tool and solar window site overlay.

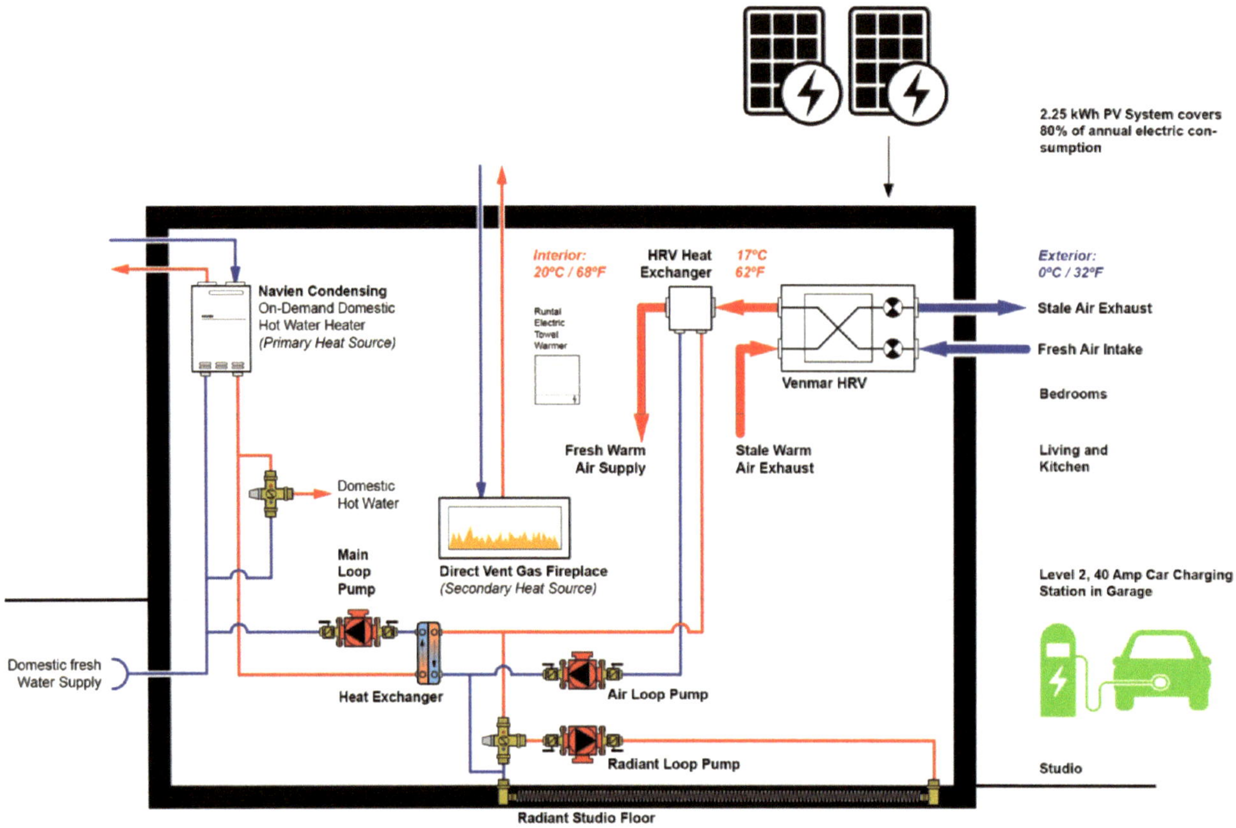


**Figure 23E.4** HVAC system diagram.

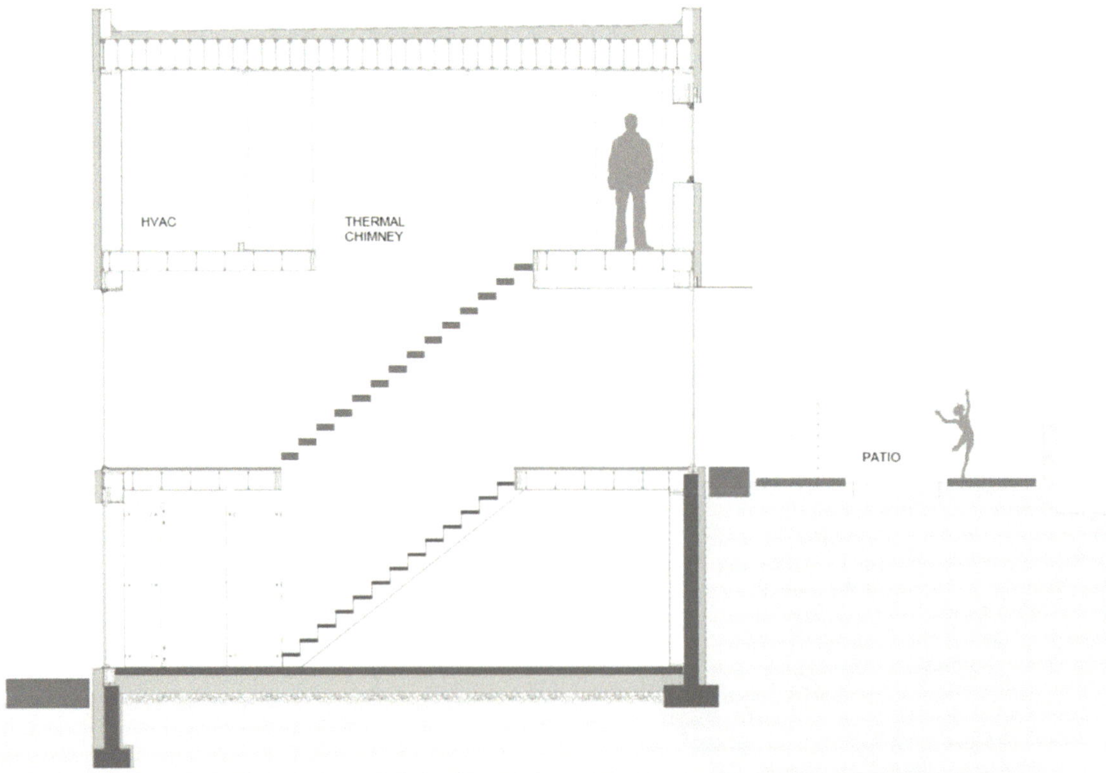


**Figure 23E.5** South-North full building section showing the open thermal chimney at the center of the building.

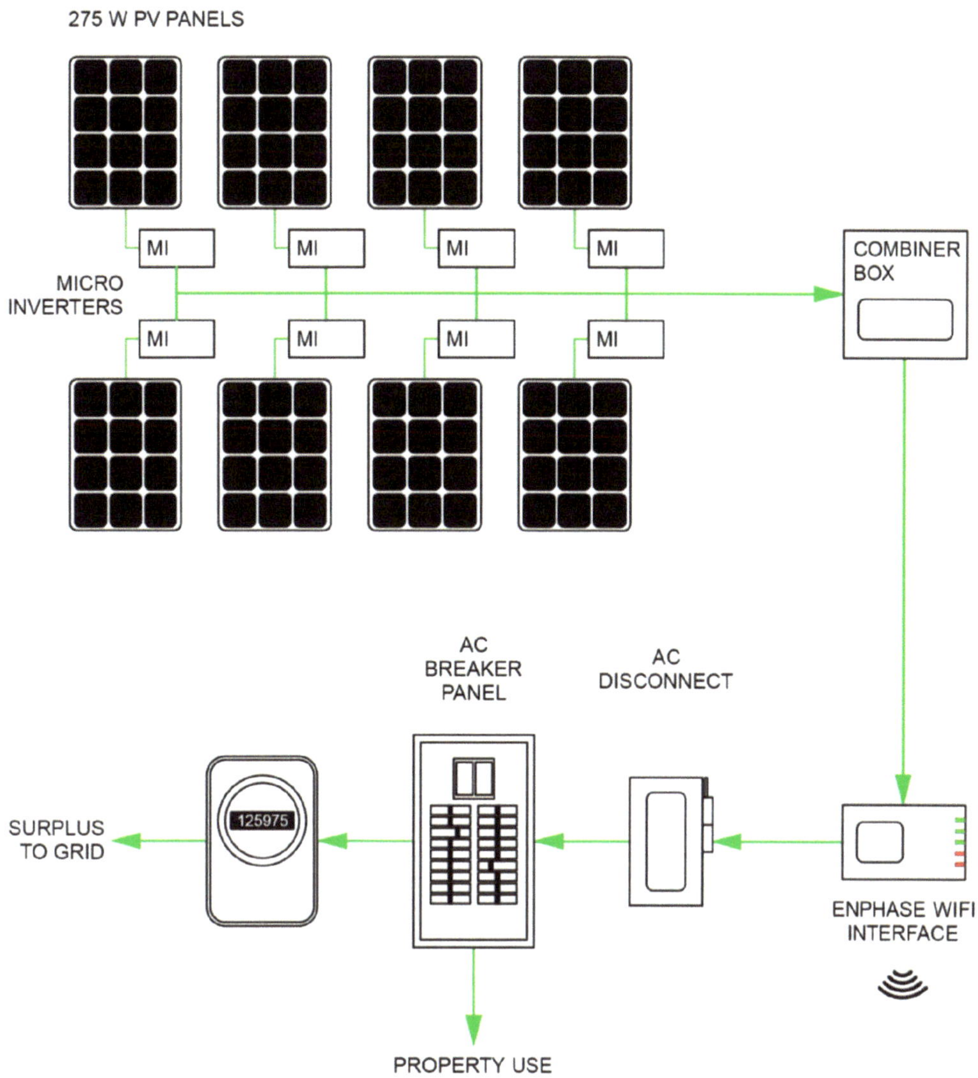


**Figure 23E.6** Diagram of 2.25kW DC rated PV system.

code IECC 2006 (Utah code at time of occupation) standard benchmark. The owner performed 26 months of post-occupancy monitoring to discover the actual energy used. The overall energy consumption cost was $654.01 annually, prior to the installation of the PV system. This is for a family of four. This cost averages out to 30.40kBtu/ft²/year or 80.33kWh/m²/year overall energy. The annual cost to heat and cool the house was US$265.02. This cost averages out to 17.45kBtu/ft²/year or 40.97kWh/m²/year for conditioned net area. This includes all the natural gas used for heating purposes and electricity for the HRV and radiant loop pumps. When these numbers are compared to energy usage according to local area data, 125 Haus is about 80% more efficient than the average housing stock.

The air tightness control layer was the oriented strand board (OSB) sheathing on the exterior of the structural framing. The blower door tested air infiltration rate was 212 CFM50/0.8 ACH50.

The architect had difficulty finding a builder interested in working on an integrated design process. However, by making decisions about construction based on the direct input of the builder throughout design and computer modeling, the building was built for cost of US$117.50/ft², which excluded land cost, impact fees and architectural fees. This cost was very low compared to typical building costs in the region at time of construction.

## Resources


1 PVWatts® Calculator (https://pvwatts.nrel.gov/).
2 Rüegemer, J. (2013). Developing super efficient passive residential buildings: Experiences from building Utah's most energy efficient and cost-effective house. *The International Journal of Architectonic, Spatial, and Environmental Design* 6(3), 77–93. https://doi.org/10.18848/2325-1662/CGP/v06i03/38345

Appendix A

# Glossary of Terms and Abbreviations

24/7: 24 hours a day, 7 days a week – a measure of usage.

A: amperage or amps, also sometimes use I – electrical current

AC: alternating current (electricity); also air conditioning (mechanical system)

ACCA: Air Conditioning Contractors of America

AGM: absorbed glass mat – a type of sealed VRLA battery.

AGS: auto generator start

AIA: American Institute of Architects

Alternating current (AC): voltage (current) is switching polarity (direction) at regular intervals over time – graphs as a sine wave having equal positive and negative values.

Altitude: the vertical height of the sun in the sky measured above the horizon.

Amorphous (thin film) solar cell: film of silicon deposited on a flexible substrate.

Anemometer: an instrument that measures wind speed and (sometimes) direction, often using a series of cups that rotate in the wind to gauge its speed.

ANSI: American National Standards Institute

ASHRAE: American Society of Heating, Refrigerating and Air-Conditioning Engineers, USA

ATS: Automatic transfer switch – provides automated switching from the grid to the standby generator.

Azimuth: the angled location of the sun based on the horizon/cardinal directions.

BEM: building energy modeling

Biofuel: biomass that is turned into a liquid fuel such as ethanol and biodiesel.

Biomass: organic matter from plants and animals that are used as fuel to generate electricity, power vehicles, heat buildings, and cook.

BOS: balance of system – all the "other" equipment needed for a renewable energy system besides the equipment that produces the energy.

BPI: Building Performance Institute

Btu: British thermal unit – amount of heat needed to raise one pound of water one degree Fahrenheit.

Carcinogen: a substance or radioactive material that can cause the formation of cancer in living tissues.

CDD: cooling degree day – number of degrees that a day's average temperature is above a particular temperature.

Cell (in a battery): the smallest unit making up a battery; single anode (negative) and cathode (positive) metal plates are separated by an electrolyte, which produces voltage and current through chemical reactions.

$CH_4$: methane – a greenhouse gas often produced by cattle and decaying landfill materials

Charge controller: a piece of electronic equipment that limits the rate at which electric current is added or withdrawn from a battery bank.

CHP: combined heat and power

CIGS: copper indium gallium diselenide – thin film technology with higher efficiency.

CO: carbon monoxide

$CO_2$: carbon dioxide – a greenhouse gas that is often used as the basis for measuring negative environmental impacts of all greenhouse gasses through equivalents $CO_2E$.

Combined heat and power (CHP): concurrent generation of both heat and power (from burning a single resource) which is located at or near the site where the heat and power is used. This is also called cogeneration. It captures the heat that would normally be lost in the production of electricity and is generally used for industrial and large commercial applications.

Commissioning: project systems that are checked after installation for proper operation; rating systems require commissioning to be done by a third party.

COP: coefficient of performance

COPD: chronic obstructive pulmonary disease

CSP: concentrated solar power – creating electricity at a merchant (utility) scale by concentrating the sun's rays on a collector.

Days of autonomy: the theoretical number of days that the battery bank would be supplying the load without recharging.

DC: direct current

Deep cycle battery: a battery that has been designed to have many cycles of charging and discharging while maintaining its capacity.

Degree days: heating (HDD) and cooling degree days (CDD) are a measurement of how many hours in a day (24 hours) are colder than a reference temperature (so need heat to be comfortable) or are hotter than a reference temperature (so need cooling to be comfortable).

Depth of discharge: the percentage of energy used from a battery; the amount that the battery has discharged, where 50% would be halfway; SOC + DOD = 100%

DHW: domestic hot water

Direct current: voltage (current) which flows in one direction only – graphs as a horizontal line over time.

DOD: depth of discharge

DOE: Department of Energy, USA

Dump load (also called diversion or shunt load): a load, usually an electric heating element, that can be used to protect system batteries from overcharging damage and provide a continuous load on a turbine (microhydro or wind).

EEM: energy efficiency measure

EER: energy efficiency ratio

EIA: Energy Information Administration

ENERG: EU Energy Label

Energy efficiency measure: any practice that reduces energy usage – this can be a system, an approach, a particular piece of equipment or building material – the combination of which will result in the best synergies.

Energy: power expended over a period of time, available in different forms, electrical power is measured in Wh or kWh.

Equinox: when the plane of the earth's equator passes through the geometric center of the sun (the center of the sun's annual path through the sky) where there are equal hours of sunlight and darkness, generally around March 21 and September 21.

ERV: energy recovery ventilator

Essential load: a building load that has been identified by the owner as needing power 24/7.

EUI: energy use intensity

EV: electric vehicle

Expansion tank: a small tank installed in closed loop hot water systems to protect the piping and equipment from excessive pressure.

FERC: Federal Energy Regulatory Commission (USA)

FLA: flooded lead-acid battery – a deep cycle rechargeable battery that uses lead plates submerged in a sulfuric acid electrolyte to store power

Flow: the volume of water in a stream, microhydro usually diverts 20% or less of the total stream's flow.

FROGs: Flexible Response to Ongoing Growth – see UHM case study

Gel: gel electrolyte – a type of sealed VRLA battery

Generator: a device that converts mechanical energy into electrical power for use by a circuit.

GHG: greenhouse gasses

Green building certifications: rating systems that are used to indicate the performance and sustainability achieved by a building project, usually with ranked levels based on the outcomes and number of issues addressed.

Greenhouse gasses: any gas that is able to absorb infrared radiation, therefore trapping heat in the earth's atmosphere.

Grid-tied RE: the building is connected to the grid and supplies surplus power to the grid.

Grounding: the electrical system is connected to the earth so that sudden increases in voltage flow do not harm people or equipment.

GW: gigawatt, equal to one billion watts

HAWT: horizontal axis wind turbine – rotor is oriented horizontally.

HDD: heating degree day – number of degrees that a day's average temperature is below a particular temperature.

Head: the vertical drop of water from the top of the penstock to the turbine.

Heat pump: a device that uses mechanical energy to transfer heat from one location to another.

Heat sink: a mass that absorbs or dissipates heat without much change of its own temperature – for heat pumps used in building HVAC this is typically air, water, or earth.

HERS: Home Energy Rating System

HIT: Heterojunction with Intrinsic Thin Layer solar cells (also HJT)

HJT: Heterojunction with Intrinsic Thin Layer solar cells (also HIT)

HNEI: Hawaii Natural Energy Institute

HPWH: heat pump water heater

HRV: heat recovery ventilator

HSPF: heating season performance factor

HVAC: heating, ventilating, and air conditioning – mechanical means of providing fresh air, heat and cooling to the building's interior.

Hybrid generation: an RE system that combines two generators, typically PV and wind.

I: amperage or amps, also A, electrical current

IAQ: indoor air quality

IEA: International Energy Agency

IEC Class: International Electrotechnical Commission Class for wind speeds and turbulence.

IECC: International Energy Conservation Code

$I_{mp}$: maximum power point current

Incentive: a financial return for investing in an RE system or energy efficiency measure, available in many forms from grants to tax credits.

Infiltration: air that passes through all the holes and discontinuities in the building envelope, i.e., air leakage.

Inverter: a piece of electrical equipment that changes DC energy to AC energy.

I-REC(E): internationally traded electrical REC

$I_{sc}$: short circuit current

Islanding: when a building, campus, or community has electrical power during a grid outage.

J: joules

kW: kilowatt, equal to 1,000W

kWh: kilowatt hour, equal to 1,000Wh

Latitude: imaginary lines that run east–west around the earth, where 0° is at the equator and 90° is at the poles.

LBC: Living Building Challenge

LED: light emitting diode – a DC light that uses very little power.

LEED: Leadership in Energy and Environmental Design

Magnetic declination: the variation, measured in a horizontal plane, between magnetic north and true north.

Marine renewable energy: renewable energy that is installed and operated in seawater.

MBB: multi-busbars used on a solar cell

MHK: marine and hydrokinetic energy

Microgrid: an area, which could be a single building or a neighborhood, that is independent in regards to power and can operate with or without the grid.

Molten salt energy storage: a technology used with concentrated solar power which transfers heat into an insulated tank then is used after sunset to create energy from steam

Monocrystalline solar cell: PV cell cut from a grown crystal ingot, has a solid dark blue, black or gray appearance.

$M_{pp}$: maximum power point (also $W_{pp}$ and $P_{max}$)

MPPT: maximum power point tracking

MRE: marine renewable energy

MSES: molten salt energy storage

MW: megawatts, equal to 1,000,000kWh

MWh: megawatt hour, equal to 1,000kWh

$N_2O$: nitrous oxide – a greenhouse gas that increases air pollution issues

NABCEP: North American Board of Certified Energy Practitioners

Nacelle: a streamlined housing that contains the interior workings of a wind turbine.

NEEP: Northeast Energy Efficiency Partnerships

NOCT: nominal operating cell temperature

NOI: Notice of Intent – see City of Libby case study

Nominal operating cell temperature: PV module electrical production tested at 800W/m$^2$, 20°C (68°F), wind speed of 1 m/s (2.2mph), and mounted with an open backside.

NZE: net zero energy

Ocean current: a continuous, predictable, directional movement of sea water.

Off-grid RE: the building is not connected to a grid and must provide all of its own power.

Parallel circuit: has more than one path for the flow of electrons, adds amps while volts stay the same, positive terminals are connected to positive terminals while negative are connected to negative.

Passive design: strategies that reduce energy usage of building (mechanical, lighting) without using any energy.

PCM: phase change material

Penstock: a pipe that diverted water runs through, located outside of and away from the stream bed.

Phase change material: stores heat energy through state changes, absorbing and releasing heat when changing between solid and liquid or gaseous state.

Photovoltaic (PV) module: a fairly lightweight rectangular substrate with interconnected PV cells mounted on it that creates electricity from sunlight.

$P_{max}$: maximum power point (also $M_{pp}$ and $W_{pp}$)

Polycrystalline solar cell (also called multicrystalline): PV cell cut from chunks of polycrystalline silicon cast into rectangular ingot, has a mottled appearance varying from medium to darker blues and grays, with an overall blue appearance because of the different sizes colors and angles of the crystals.

Power: how fast energy is produced or used, electricity measured in W or kW.

PRV: pressure-relief valve

PSH: pumped storage hydropower

Psychrometric chart: a chart that shows the physical and thermal properties of moisture in the air.

Pumped storage hydropower: during times of excess RE production, water is pumped to a higher elevation reservoir which is later dropped through a turbine when RE is not producing.

PV: photovoltaic – creating electricity from sunlight.

PWM: pulse width modulation – a feature of a controller.

Racking: elements that securely attach equipment, such as a solar thermal collector or PV module, to a building, pole, or tracker.

RE: renewable energy

REC: renewable energy certificate

Relative humidity: the amount of water vapor in the air expressed as a percentage compared to saturation (100%) at the same temperature.

Renewable energy: any energy source that comes from naturally renewing sources or processes.

Return on investment: system value calculated as a percentage return on the initial investment that compares the current value of your investment in any year to the original investment cost, taking into account the changing value of money and inflation.

Riprap: loose stone that is piled in a location to prevent soil erosion due to water runoff.

ROI: return on investment

RSI: R SI-value – R-value in SI units.

Scale (water chemistry): a white, chalk-like substance of minerals that is deposited on surfaces, left behind after water evaporates.

SEER: seasonal energy efficiency ratio

Series circuit: has one path for the flow of electrons, adds volts while amps stay the same, the positive terminal is connected to the negative terminal.

Service panel: an electrical box that has power service input that is then distributed through circuits to all the building loads.

Shunt: a low resistance conductor wired in parallel that joins two points of a circuit in order to measure electric current.

Simple payback: system value calculated by dividing the system cost by the annual savings, resulting in number of years until you earned back your initial investment.

Smart grid: a utility grid that has advanced metering infrastructure which allows for fairly real-time monitoring of customer power use and two-way communication between a building's smart meter and a centralized monitoring system.

SOC: state of charge

Solar insolation: the amount of radiant energy available at a particular location due to latitude and weather.

Solstice: when the sun reaches maximum and minimum declination (the extreme edges of the sun's annual path through the sky). Summer solstice occurs at the maximum declination (the greatest number of daylit hours), around June 21 in the northern hemisphere or December 21 in the southern hemisphere. Winter solstice occurs at the minimum declination (the least number of daylit hours), around December 21 in the northern hemisphere or June 21 in the southern hemisphere.

SREC: solar renewable energy certificate

Standard test conditions: PV module electrical production tested at 1,000W/m$^2$ at 25°C (77°F) with no wind; in USA STC results are required to be printed on a permanent label affixed to the back of the module.

Standoff: a mounting element that holds a collector away from the roof or building, typically at least 2" off a roof. Equipment or rails can be attached to the standoff.

State of charge: the charge level of a battery, where 100% is fully charged.

STC: standard test conditions

Thermography: use of thermal imaging, which visually displays infrared energy emitted, transmitted, or reflected by an object.

Tide: the rising and falling of a very large body of water in a particular location caused by the gravitational pull of moon and sun.

Turbulence: strong, disruptive, swirling motion of the wind current caused by objects in the environment (buildings, trees, etc.) blocking the wind.

UHM: University of Hawaii Manoa

UNHCR: United Nations Refugee Agency

UV: ultraviolet – a wavelength not visible to the human eye.

V: volt or volts, electrical pressure or force

Valve regulated lead-acid batteries: these are either absorbed glass mat (AGM) or gel electrolyte (Gel).

Vanadium redox flow: a battery type that uses vanadium-based electrolyte.

Variable generation: RE production from wind and PV that changes daily and seasonally.

VAWT: vertical axis wind turbine – rotor is oriented vertically

Ventilation: fresh air introduced into the building.

VG: variable generation

$V_{mp}$: maximum power point voltage

$V_{oc}$: open circuit voltage

VOC: volatile organic compound

Volatile organic compound: gasses that are emitted into the air at room temperature from an organic (carbon) compound – many are human-made chemicals used for a wide variety of processes and emitted by a wide variety of products – and are often found in much larger concentrations inside buildings.

VRF: vanadium redox flow

VRLA: valve regulated lead-acid batteries

W: watt or watts

Weir: a low dam built across a stream or river which reduces the flow and increases the water height upstream.

Wh: watt hours

WHO: World Health Organization

Wind rose: a circular graph aligned with the cardinal directions that has bars that indicate the velocity, amount, and direction of wind in a particular location.

$W_{pp}$: maximum power point wattage (also $M_{pp}$ and $P_{max}$)

Yaw: rotation around a vertical axis, the nacelle yaws to orient the turbine blades into the wind.

YWCA: Young Women's Christian Association – see case study 23D

# Appendix B

# Designing an RE System

## Planning for an RE System

All buildings use energy. No matter how efficient you make a building, it is still going to use energy. To get to net zero (building produces what it uses) you'll need a renewable energy (RE) system.

RE systems make electricity from a consistent and reliable energy resource: sun, wind, or water. Your building must have availability and access to the resource.

Concerns for sun:

- How much solar insolation is available annually at your building's location?
- How much solar insolation is available in winter? (If you need daily net zero this is critical – if there isn't enough sun, then you'll need to supplement with something else.)
- What options do you have for tilt angle/orientation/collector mounting that could impact production?
- Are there large elements (buildings, trees, hills) near your site that will shade your collectors?
- What balance of system (BOS) items do you need to plan for so your solar collectors work optimally?
- Do you need to have safe, easy access to the collectors to clear them of snow?

Concerns for wind:

- How much wind, and at what height, is available annually at your building's location?
- How much wind is available in the summer? (If you need daily net zero this is critical – if there isn't enough wind then you'll need to supplement with something else.)
- Are there zoning restrictions that will impact your project?
- Do you have enough site space to locate the turbine away from the building and other obstacles that cause turbulence (nearby buildings, trees, hills)?
- What BOS items do you need to plan for so your wind turbine works optimally?

Concerns for water:

- Do you have a reliable amount of water/tide/wave power all year at your building's location?
- Do you need special permitting to access and/or divert the water?
- What BOS items do you need to plan for so your water power production works optimally?
- Remember that flowing water creates a lot of humidity and potential for leaks: how are you designing and protecting the system and building so as to control humidity and possible mold and water damage?

### Optimize Power Demands

Buildings use energy to create a comfortable interior temperature and humidity level as well as to power equipment. To create energy efficient buildings, you need to design for the climate and have a clearly defined and consistent building envelope (refer to chapters 6 and 7 on these subjects). Power demand for HVAC will be dependent on the interaction between the envelope and the climate, based on heat gain and loss compared to the conditions desired indoors, and is most often modelled by computer. The model considers the loads and the type of HVAC equipment.

Other loads to optimize include the equipment used by the occupants for lighting, cooking, clothes washing and drying, computers, and everything people want to operate to make their lives better and run their businesses. Items to consider: efficient officially rated appliances (such as Energy Star[1]); elimination of phantom loads; use of a DC loads panel if there's a battery bank.

### Determine Equipment Power Consumption Demands

Many computer modelling programs have lists of equipment you can pick from (or add your own) so that you can accurately determine the equipment power consumption. This is impacted by the building's occupancy and usage pattern (for instance hospitals are used 24/7 while a school may not be used in the summer months).

Existing buildings will have power bills that you can look at. Have the client consider building efficiency and upgrading appliances to reduce demand.

If you want to determine these loads by hand (perhaps you have a very small system such as an RV or tiny house), you must list all the loads and add them up to get a daily load profile. Listing all the loads means listing everything electrical that would be used in the building: lighting, water heater, exhaust fans, toaster, hair dryer, etc.

Identifying equipment loads can be achieved by:

- Look at the electrical labelling on the piece of equipment (assuming you own it already);
- Look at the manufacturer's cut sheet information about the product; or
- Use a meter, such as a Kill A Watt[2], to measure energy use of the equipment over a given time period.

Information you need to know about each piece of equipment:

- How many you have of them
- AC or DC power
- Volts and amps or watt rating
- How often it is used (every day or what number of days in a week)
- How many hours it operates every time (day) it is used (includes cycling loads such as a refrigerator)

Make a list of all the loads, how many of each there are, and multiply it by the volts and amps or watt rating to get the total AC or DC watts needed. For each item, multiply the watts by the hours used each day and how many days it is used per week. This will give you the total Wh needed to run that piece of equipment during a week. Take the weekly Wh and divide by 7 days/week to get the average Wh needed every day. Add up all the AC loads and all the DC loads. It is easy to set up a spreadsheet, like the one shown in Table B.1, to do the math for you.

## What are Power Production and Storage Goals?

You (and your team) need to be very clear about how much and when your RE system is making and supplying power. The decisions you make have a big impact on what components you'll need as part of your RE system, its sizing, and backup power needs.

### Daily net zero

If you are off-grid you are aiming for daily net zero: your RE system needs to provide all of your building's power

Table B.1 Example Excel spreadsheet content to calculate total daily watt load of all AC and DC powered equipment.

| **Estimated Equipment Power Consumption** | | | | | | | | | | |
|---|---|---|---|---|---|---|---|---|---|---|
| Equipment name | Quantity | AC Volts | AC Amps | or AC Watts | Total AC Watts | Hours of use per day | Number days used per week | Total Wh per week | 7 days in week | Average AC daily load W |
| toaster | 1 | | | 850 | 850 | 0.0333 | 7 | 198.135 | 7 | 28.305 |
| | | | | | | | | | | |
| | | | | | | | | | | |
| | | | | | | | | | | |
| | | | | | | | | | | |
| | | | | | | | | | | |
| | | | | | | | | | | |
| Total AC daily load W | | | | | | | | | | 28.305 |
| Equipment name | Quantity | DC Volts | DC Amps | or DC Watts | Total DC Watts | Hours of use per day | Number days used per week | Total Wh per week | 7 days in week | Average DC daily load W |
| | | | | | | | | | | |
| | | | | | | | | | | |
| | | | | | | | | | | |
| | | | | | | | | | | |
| | | | | | | | | | | |
| | | | | | | | | | | |
| Total DC daily load W | | | | | | | | | | |

needs 24/7. Only water-based systems inherently have the possibility of 24/7 production. That means that you'll probably need to have storage. You'll also need to consider how you can get backup power that isn't fossil fuel based when your RE system isn't producing and your battery storage is used up. It is critical that you know exactly how much production potential is available when there is the least amount of resource: i.e., winter sun, summer wind. If your site has potential for both sun and wind production, you'll want to consider a hybrid system as they tend to have opposite high/low production seasons.

### Annual net zero

Annual net zero is easier to achieve than daily net zero because your RE system just has to make what it uses in a year. Say you are designing a PV system: if your winter production is low, that could be okay because you can make up for it with "extra" summer production. Note that this choice requires that you are attached to the grid so you can power your building in the winter. This option may or may not have a battery bank since it is grid-tied. PV systems that have no battery backup create a demand surge as the sun sets (see duck curve in chapter 22, Resiliency and Power). This is very difficult for power companies to accommodate as a region becomes saturated with PV production.

### Storage

Energy storage can impact your RE system and building design for several reasons:

- Mitigating demand peaks on the grid (i.e., PV production ceasing as sun sets);
- Allowing utility to access your building's stored energy to power the grid as demand increases;
- Providing power continuity in event of blackouts (grid failure) or brownouts (intermittent or rolling times when there is power to your building's location);
- Providing DC power center for DC building loads;
- Keeping grid energy draw of building below a particular threshold (reduced cost of electricity in a tiered pricing system); and
- Providing a building, neighborhood, or community islanded microgrid.

## Resources

**1** ENERGY STAR (www.energystar.gov/).
**2** Kill A Watt® (www.p3international.com/products/energy-savers.html).

Appendix C

# Sizing a PV System

## What Size PV System?

The size of the system you need is determined by the building's loads. You need to know how much power your building will be using daily and annually and how much of the load you intend to provide with PV. See Appendix B (Designing an RE System) and chapter 13 (Photovoltaics) for more detail.

### Your Optimal System Size can be Determined by One (Or More) of These Factors:

- Building is to be net zero daily or annually so you have to cover all of the building's power demand.
- Client has a set budget to spend on the system.
- There is a limited amount of space where you can mount the PV system.

### Know your site:

- What is the available insolation for the climate? Understand when there is the least available solar resource and what orientation(s) and tilt angles have reasonable production.
- What shading will happen at your site? Can you fix it (cut down trees) or not (the tall building is going to shade yours)? Remember that your site shading characteristics reduce the theoretical available climate insolation and need to be accounted for. Be sure that new landscaping will not turn into future shading.
- Will there be standing snow, sand or dust, thick pollen, pollution, or similar coating agent on your PV modules? If so, this will negatively impact production as these will act as shading. You may wish to discuss with the client the viability of creating access to the PV and providing cleaning to minimize this impact (consider cleaning every 6 months as part of basic maintenance).

### Determine Daily Average Generation – How Much Electrical Production is Needed Every Day:

- Off-grid: must be daily net zero;
- Grid-tied: you can pick any amount of production based on project goals and site limitations.

### Equipment and Performance Efficiency:

- System losses: PVWatts[1] has a base assumption of 14.08% (soiling, shading, snow, mismatch, wiring, connections, light-induced degradation, nameplate rating, age, and availability).
- Inverter efficiency: most inverters today are 94% efficient (or better).
- Batteries: off-grid or a larger battery bank that is constantly being used and recharged will drop system efficiencies by 20–25%.[2] It may be that a smaller, rarely used grid-tied backup battery bank would have less loss but all batteries discharge even when not being used. The batteries would need to be charged after a power outage, as well as constantly kept at a full state of charge.

### Array Size Formula:

DC rating of array = daily average generation / daily solar insolation / system losses / inverter efficiency

### Choosing a Module:

Modules come in various sizes, materials, and efficiencies – all of which change the power rating of the module. Here are some general tips:

- A standard 60-cell module is 3.25′ × 5.5′ (990mm × 1676mm), this is a common size for residential rooftop PV modules. Commercial solar can use 72-cell modules.

- Building integrated PV (BIPV): mono- or polycrystalline products average 10–12W/ft$^2$ (0.093m$^2$); thin film averages 4–5W/ft$^2$ (0.093m$^2$).[3]
- Monocrystalline PV are typically 20% + efficient and are considered premium modules. A 60-cell module produces between 310–350W.
- Polycrystalline PV are typically 15–17% efficient and are considered standard modules. A 60-cell module produces between 250–300W.
- Thin film modules have a big range of efficiency, 8–13%, depending on what they are made of.

### Number of Modules in the Array:

Number of modules = DC rating of array / STC watt rating of module

- DC rating of array is determined from the array size formula (above);
- STC watt rating of chosen module is from the manufacturer's data sheet. If you haven't selected a particular module, then use general numbers from choosing a module (above).

## Inverters and Power Optimizers

There is more detail on this subject in chapter 13 (Photovoltaics).

### Types of Inverters

- String inverters are used if off-grid, when a large battery bank is included in the system, there is a DC power system in the building, or the array is quite large. You may need more than one string inverter on an array.
- Power optimizers, DC–DC installed at each module, are used with string inverters so that individual modules can be monitored and power production equalized/optimized across the string.
- Microinverters are typical for grid-tied systems. Microinverters are installed at each module, are DC-AC, and allow for individual monitoring of array modules. Check manufacturer for limit to number of microinverters that can be connected together. These are always used on grid-tied systems without battery backup.

### Sizing String Inverter:

Note that some inverter manufacturers have web apps for sizing/selecting string inverters. These will verify the voltage needed (number of modules in series) for a particular inverter or suggest the correct inverter(s) based on the array information you input.

#### Grid-Tied Inverter

- Determine the maximum number of watts the inverter will need to pass through. Multiply the number of modules in your array by their STC watt rating.
- Select the inverter you'd like to use by verifying:
  - DC watt continuous power rating;
  - DC input voltage range, pay attention to where the inverter "wakes up" and starts passing power;
  - All strings connecting to inverter have the same voltage;
  - Total array amps and short circuit amps (may need string as well);
  - If one inverter will not handle the whole array, based on the DC STC array wattage, then determine how many inverters you need. This can be a great option for arrays that are oriented in different directions (a problem because the higher generation would be dropping to match the lower generation) or have a different number of modules connected in series (volts) in your strings.

#### Grid-tied Inverter Example

Given the array shown in figure C.1, let's select an appropriate inverter.

Maximum DC watts to inverter = 12 modules × 150W = 1800W or 1.8kW

String voltage = 6 modules × 34V = 204V DC passing through inverter

Note that the string voltage is the maximum production at STC values – your inverter should "wake up" below this number as there are many times (cloudy weather) that the array will be producing at levels a lot lower than 204V.

Total array amps = 2 strings × 4.4A = 8.8A

Total short circuit amps = 2 strings × 4.8A = 9.6A short circuit

Table C.1 shows SMA's Sunny Boy-US string inverter data.

The Sunny Boy 3.0-US 240V looks like a good match for our system. The following data is from the column 240V in the input (DC) section rows. It will pass 4800W (we have 1800W); the max DC voltage is 600V (ours is 204V); it has MPPT that works 100–550V (our max is 204V); it starts (wakes up) at 125V and then shuts off at 100V (our max is 204V – that means this inverter will work when the solar radiation is less than STC values); it has 2 MPPT trackers that will connect 1 per string (we have 2 strings); each MPPT can operate at 10A maximum (ours is 4.4 per string) and 18A short circuit (ours is 4.8 per string).

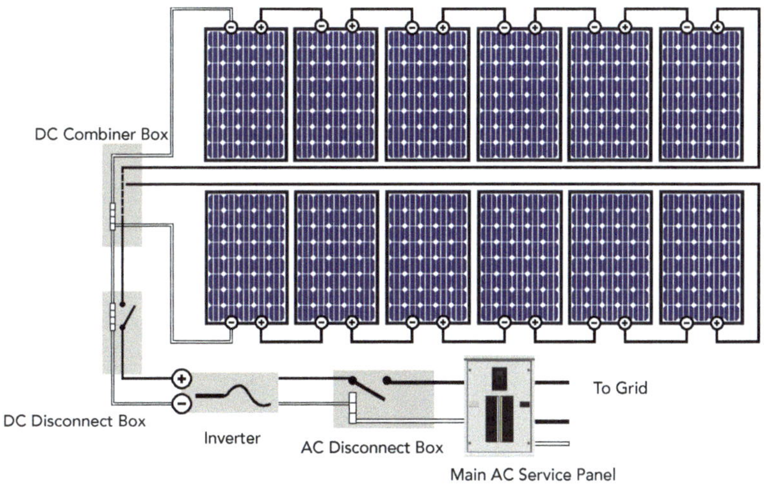


| | Module Rating STC | Inverter AC Rating |
|---|---|---|
| Rated Power | 150 W | 1800 W |
| Open Circuit Voltage | 43.4 V | |
| Max Power Voltage | 34 V | 120 VAC |
| Short Circuit Current | 4.8 A | |
| Max Power Current | 4.4 A | |

Figure C.1 Array with 2 strings of 6 modules wired in series. The 2 strings are wired in parallel at the combiner box. DC power flows to the inverter where it is changed to AC power and supplied to the main service panel for the building.

Table C.1 SMA's Sunny Boy-US string inverter data.[4]

| **Technical data** | **Sunny Boy 3.0-US** | | **Sunny Boy 3.8-US** | |
|---|---|---|---|---|
| | **208V** | **240V** | **208V** | **240V** |
| **Input (DC)** | | | | |
| Max. PV power | 4800Wp | | 6144 Wp | |
| Max. DC voltage | | | 600V | |
| Rated MPP voltage range | 155–480 V | | 195–480 V | |
| MPPT operating voltage range | | | 100–150 V | |
| Min. DC voltage /start voltage | | | 100 V/125 V | |
| Max. operating input current per MPPT | | | 10A | |
| Max. short circuit current per MPPT | | | 18A | |
| Number of MPPT tracker/ string per MPPT tracker | 2/1 | | | |
| **Ouput (AC)** | | | | |
| AC nominal power | 3000 W | 3000 W | 3330 W | 3840 W |
| Max. AC apparent power | 3000 VA | 3000 VA | 3330 VA | 3840 VA |
| Nominal voltage/adjustable | 208 V/• | 240 V/• | 208 V/• | 240 V/• |
| AC voltage range | 183–229 V | 211–264 V | 183–229 V | 211–264 V |
| AC grid frequency | | | 60 Hz/50 Hz | |
| Max. output current | 14.5 A | 12.5 A | 16.0 A | 16.0 A |
| Power factor (cos ϕ)/ harmonics | | | 1/<4 % | |
| Output phases/line connections | | | 1/2 | |
| **Efficiency** | | | | |
| Max. efficiency | 97.2% | 97.6% | 97.3% | 97.6% |
| CEC efficiency | 96.0% | 96.5% | 96.5% | 96.5% |

## Off-Grid Inverter

Off-grid inverters are selected based on:

- DC voltage of system (typically 12V, 24V, or 48V);
- Output AC watts that will be needed to supply loads under standard operation;
- Output AC voltage – USA most off-grid inverters supply either 120 or 240V (can use two 120V in series to get to 240V);
- Desired frequency – USA is 60Hz;
- Surge capacity of about 3x the AC load to cover motor start-up;
- Waveform – check that loads will run on the sine wave: most likely to see modified (has a sine wave that is stepped in profile) or pure (has a curving sine wave as typical for AC power).

## Microinverter Sizing

Microinverter sizing is a matter of comparing the manufacturer's information to your module and verifying that it is compatible with your module. All manufacturers list these specifications in a table similar to the one shown in table C-1. You'll need to compare the number of cells (60, 72, or 96), desired output voltage, the input STC power from the module (usually there is a range the inverter can handle), the maximum input DC voltage from the module, peak power tracking voltage, the operating range, minimum/maximum start voltage, maximum DC short circuit current, and maximum input current.

## Battery Bank

Battery sizing is based on:

- The load that you want to power every day (both AC and DC)
- The inverter efficiency
- The days of autonomy
- The discharge limit of the battery
- The Ah capacity of the battery
- The system DC voltage (12V, 24V or 48V).

Days of autonomy refers to the theoretical number of days that the battery bank would be supplying the load without recharging. The higher the number of days you pick, the larger the battery bank becomes – which means it gets more expensive. Often people pick 2 or 3 days of autonomy. After that point their backup generator will come on and charge the batteries.

### Steps of Battery Bank Design

Determine the inverter losses: divide the AC daily load by the inverter efficiency.

Total AC daily load = AC daily load / inverter efficiency

Total daily load = total AC daily load + DC daily load

Average Ah per day = total daily load / DC system voltage

Number of batteries in parallel = average Ah per day × days of autonomy × discharge limit × battery Ah capacity

Number of batteries in series = DC system voltage / battery voltage

Total number of batteries = batteries in series × batteries in parallel

## Controller

Document system size: Array Watts; Array $V_{oc}$; Array $I_{sc}$. A safety factor of 25% is added for short circuit amps.

Determine the load amps by dividing the rated array watts by the adjusted nominal voltage to determine the maximum load amps from the controller: 12V = 14.4V; 24V = 28.8V; 48V = 58.8V.

Decide what options the controller needs and match the requirements to a controller.

## Fitting Modules to a Roof

All code access requirements (USA) for roofs are covered in chapter 13 (Photovoltaics). Modules on pitched roofs are usually racked parallel to the roof. For flat roofs, modules are typically laid out in rows. Information on determining appropriate inter-row spacing to keep modules from shading each other is chapter 5 (Solar Insolation).

Otherwise, you have to fit the actual size of the module to the actual size of the roof – so figure out the actual size you have to work with. Modules can be oriented either horizontally or vertically. Look at the manufacturer's data sheet for the frame size of the module: you can round up a little as the mounting clips add a finger's width of space between the modules.

My module is nominal 24V, 374W and has the dimensions 79.69 × 39.53 × 1.38 inches (2024 × 1004 × 35 mm).

I have a sloped roof that I could put 21′ (6400mm) of modules across (leaving space for roof access) and 17′ (5181mm) of vertical roof height.

### Horizontal module orientation

79.69 = 80″ across and 39.53 = 40″ high

2024 = 2030mm across and 1004 = 1010mm high

Across the roof:

21′ × 12″/ft = 252″; 252″ / 80″ = 3.15 modules = 3 modules

6400mm/2030mm = 3.15 modules = 3 modules

Vertical height roof:

17′ × 12″/ft = 204″; 204″ / 40″ = 5.1 modules = 5 modules

5181mm / 1010mm = 5.1 modules = 5 modules

Total modules that fit on roof horizontally is 3 × 5 = 15 modules.

### Vertical module orientation

39.53 = 40″ across and 79.69 = 80″ high

1004 = 1010mm across and 2024 = 2030mm high

Across the roof:

252″ / 40″ = 6.3 modules = 6 modules

6400mm / 1010mm = 6.3 modules = 6 modules

Vertical height of roof:

204″ / 80″ = 2.55 modules = 2 modules

5181mm / 2030mm = 2.55 = 2 modules

Total modules that fit on roof vertically = 6 × 2 = 12 modules

For this roof, more modules fit in a horizontal orientation. You have to run the math to find out.

## Resources

1 PVWatts® Calculator (https://pvwatts.nrel.gov/).

2 This blog has an in-depth analysis of all the system losses for an off-grid system: Murphy, T. (2012, September 18). Blow-by-blow PV system efficiency: A case study for storage. *Do the Math*. https://dothemath.ucsd.edu/2012/09/blow-by-blow-pv-system-efficiency/

3 Strong, S. (2016, October 19). *Building integrated photovoltaics (BIPV)*. Whole Building Design Guide. www.wbdg.org/resources/building-integrated-photovoltaics-bipv

4 SUNNY BOY 3.0-US / 3.8-US / 5.0-US / 6.0-US / 7.0-US / 7.7-US (www.sma-america.com/products/solarinverters/sunny-boy-30-us-41-77-us-41.html).

# Appendix D

# PV Wiring Workshop

Review chapter 4 (Basic Concepts about Electricity) prior to trying these wiring exercises. I suggest placing trace paper over the diagrams to try out wiring layouts – or draw a quick representation on paper. Pay attention to the rules for wiring: series adds volts (amps stay the same); parallel adds amps (volts stay the same); you have to have everything wired in series first – then the "left over" terminals get wired in parallel. All the series wiring has to be the same: you have to have the same number of groupings for all the elements. For example, if you want 2 PV modules in series to get to a voltage, then all groups of PV modules have to be in pairs for that array.

Off-grid systems rely on a battery bank to provide continuity of power. Battery banks are wired to nominal 12V, 24V, or 48V. PV modules need to be wired to match the battery bank voltage or require a stepdown controller. PV modules used in these types of systems are labeled as nominal 12V, 24V, or 48V even though they usually have a bit higher voltage rating.

Time to apply this to wiring up some equipment! Remember, series wiring is positive to negative (+ to –) terminals and adds volts, while parallel wiring is positive to positive and negative to negative (+ to + and – to –) and adds amps.

In a project, all the batteries would be the same make and model, and the same for all the PV modules. Batteries are commonly 2V, 4V, 6V, or 12V. PV modules have a wide range of voltages from 12V to 40V. All the specifications for the batteries and modules shown in these exercises are based on actual items available in the marketplace in 2021. These first handful of wiring problems are simplified to help you get a feel for how wiring in series or parallel works.

**Problem 1: Wire 4×PV (12V, 6A) in figure D.1 to 12V and to the charge controller.** The 4×PV modules are shown on the left (with their rated voltage and amps) and the controller is shown on the right (with terminals for wiring). Once you have the modules wired together and to the controller, fill in the DC voltage (adds in series), the amperage (adds in parallel), and watts that the controller will need to pass from the array to the batteries.

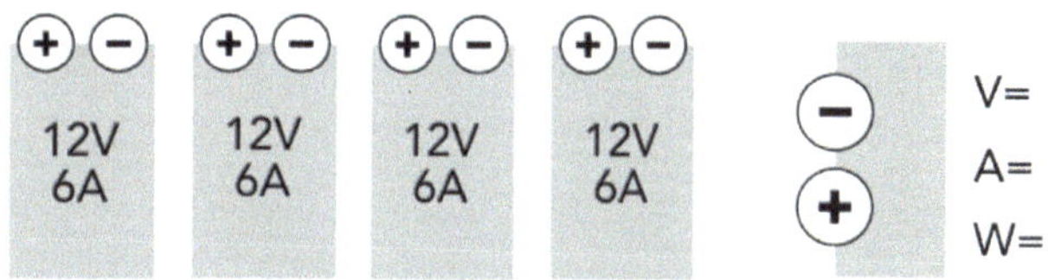


Figure D.1 Wire to be a 12V system with 4×12V, 6A PV wired to charge controller.

These are nominal 12V modules, and the system is 12V, they are all going to be wired in parallel. Your wiring connects all the positive terminals and then there is a separate wire connecting all the negative terminals. Because this is part of a circuit, the terminals will then be wired to the next piece of equipment, either a controller or inverter, in parallel. Note that the wiring to the circuit comes from the terminals that weren't connected and are located at opposite ends of the array to balance the resistance. Answer per figure D.22.

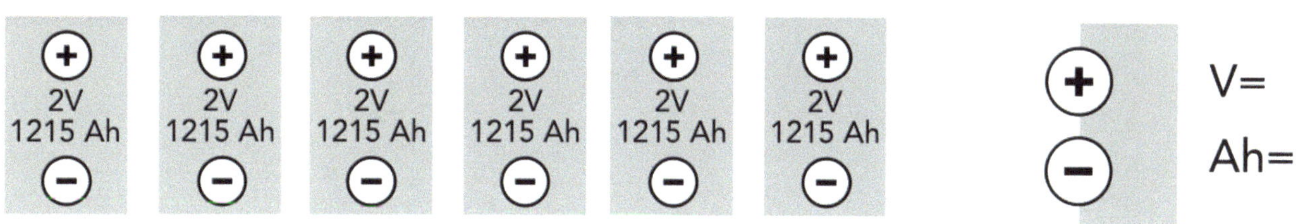


Figure D.2 Problem 2: Wire 6×2V, 1215Ah AGM batteries to 12V and wire to charge controller (Answer figure D.23).

**Problem 2: Wire 6xbatteries (2V, 1215Ah) in figure D.2 to 12V and to the charge controller**. Start by wiring the number of batteries needed in series to reach 12V. All the groups of batteries must add up to 12V. Then wire the groups of batteries in parallel with each other and the charge controller.

Because each of the batteries is 2V, it takes all 6 in series to get 12V. It doesn't matter if you wire the negative to the positive terminal or the positive to the negative terminal as long as you are consistent within the particular diagram – either way the connections are in series. Note that since you added the volts, the amps stay the same. Batteries store amp-hours (Ah), which add only when wired in parallel. Answer per figure D.23.

**Problem 3: Wire 3xbatteries (4V, 1293Ah) in figure D.3 to 12V and to the charge controller.**

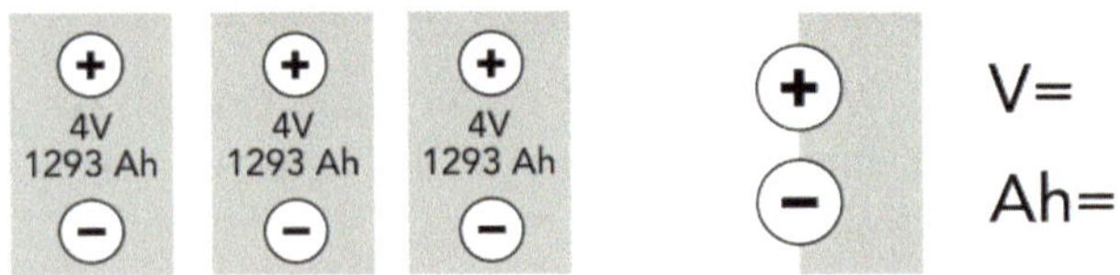


Figure D.3 Problem 3: Wire 3×4V,1293Ah FLA batteries to 12V and wire to charge controller (Answer figure D.24).

All three batteries need to be wired in series to get to 12V. Remember, since you added the volts the amp-hours stay the same. Answer per figure D.24.

**Problem 4: Wire 2xbatteries (6V, 305Ah) in figure D.4 to 12V and to the charge controller.**

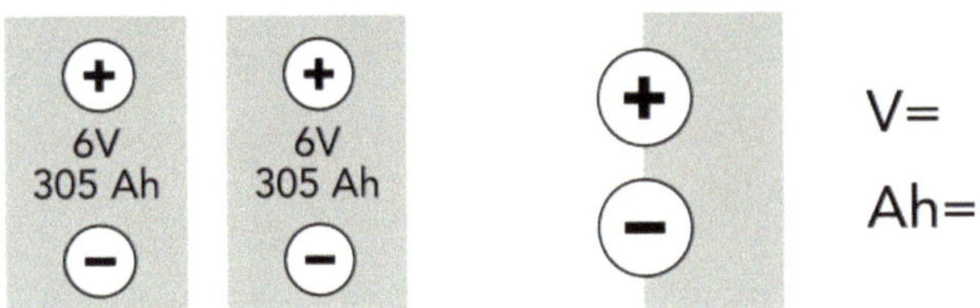


Figure D.4 Problem 4: Wire 2×6V, 305Ah AGM batteries to 12V and wire to charge controller (Answer figure D.25).

These two batteries need to be wired in series to get to 12V and the amp-hours will stay the same. Answer per figure D.25.

**Problem 5: Wire 2xlithium batteries (12V, 100Ah) in figure D.5 to 12V and to the charge controller.**

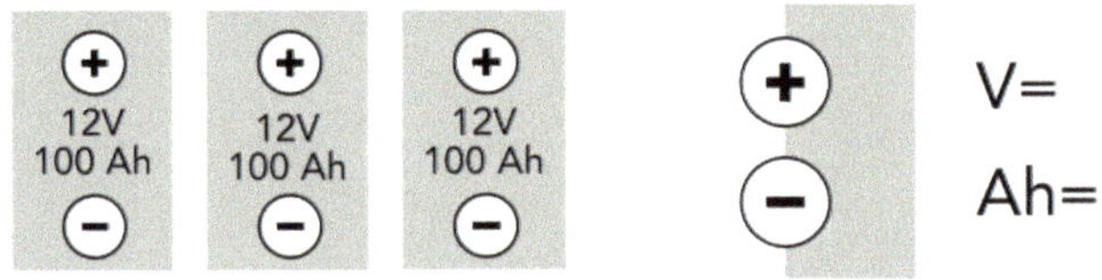


Figure D.5 Problem 5: Wire 3×12V, 100Ah lithium batteries to 12V and wire to charge controller (Answer figure D.26).

Because these batteries are already 12V, they will be wired in parallel. Remember, parallel wiring connects positive to positive and negative to negative. Parallel wiring adds amps and the volts stay the same. Since these are batteries, the parallel wiring adds the amp-hours and the volts stay the same. Answer per figure D.26.

So far, all our examples either used series or parallel wring. Often wiring in a system is a combination of both series and parallel. Always start by wiring in series (positive to negative) to get to the desired voltage and then connect the unused terminals in parallel (positive to positive and negative to negative). In this example, shown in figure D.6, of a 24V PV array, note that there are 4 modules that are 12V each.

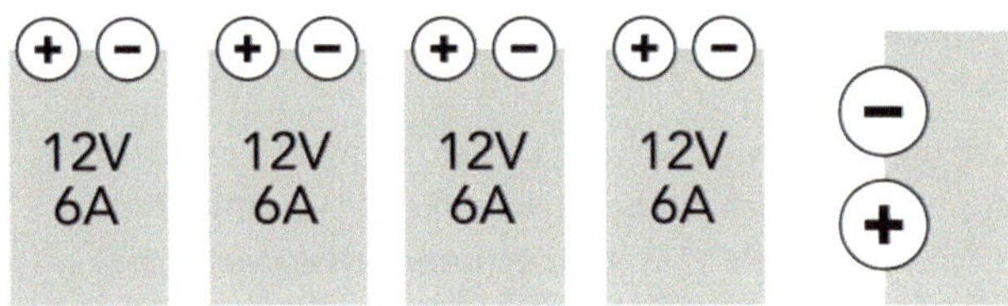


Figure D.6 Wire to 24V: 4×PV (12V, 6A) and wire to charge controller.

Two modules have to be wired in series to get to 24V. If we wanted to have more modules in our array, they would have to be added in groups of two. There could never be a single 12V module with everything else at 24V because the 12V module would pull down the voltage of the 24V modules to match it at 12V.

Step 1, shown in figure D.7, is to connect the modules in series: 2 sets of 2 modules wired in series.

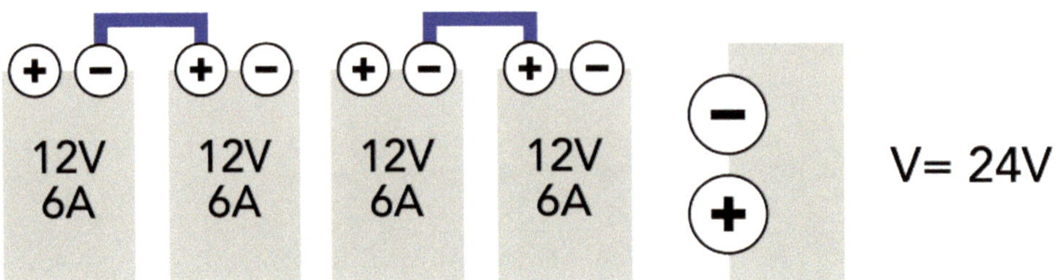


Figure D.7 Step 1 is to wire the modules in series to reach the desired 24V, making sure all sets of wiring have the same number of modules.

Step 2, shown in figure D.8, is to connect the unused terminals in parallel.

Step 3, shown in figure D.9, is to wire the array into a circuit, remembering to take the wiring from the opposite ends to balance circuit resistance.

It can be a bit tricky to add the volts and amps: remember series adds only the volts and parallel adds only the amps. For this array, see figure D.10, you'll have 2 sets of 2 modules, each of which is 24V, 6A. When modules are wired in series, they are called strings. This array has two strings. The next step is to add together however many strings that are wired in parallel. Remember the volts will stay the same, so whatever the voltage is that you've wired the system to is what the array voltage will be. The amps that add are the total amps for each series. This example has 2 strings at 6A. The whole array will have 2 × 6A = 12A.

Figure D.11 shows a PV system with batteries. The charge controller monitors the charging and discharging of the batteries. The PV system will be wired in parallel to the charge controller and the battery bank will be

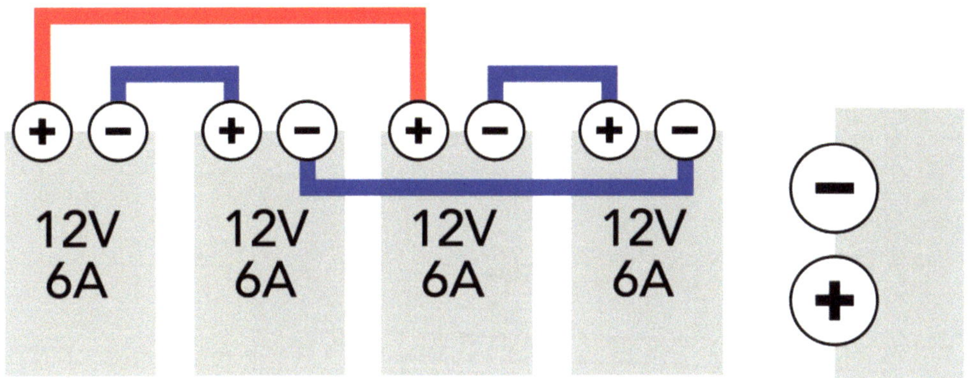


Figure D.8 Step 2 is to wire the unused terminals in parallel to each other.

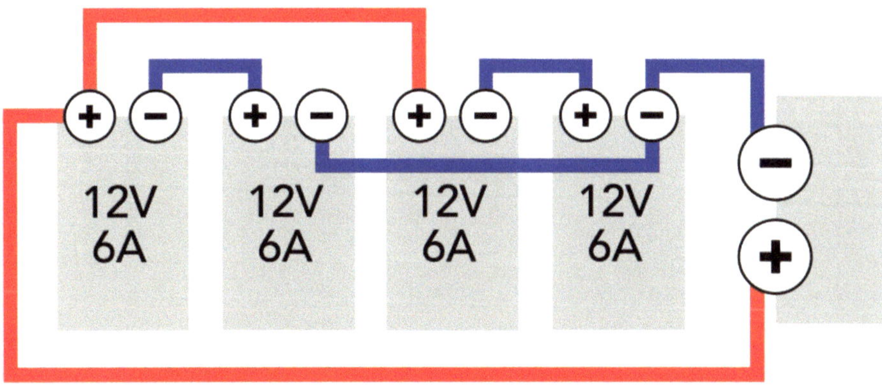


Figure D.9 Step 3 is to wire the positive and negative ends of the array to the charge controller, wiring in parallel from opposite ends of the array.

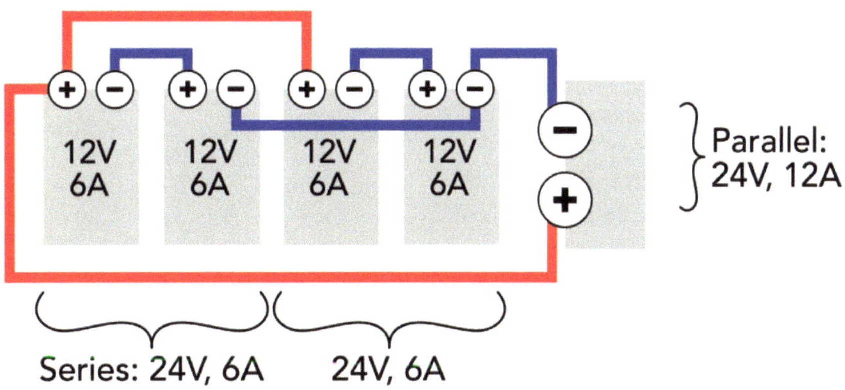


Figure D.10 Groupings of modules in series add the voltage, then the overall system in parallel adds the amperage.

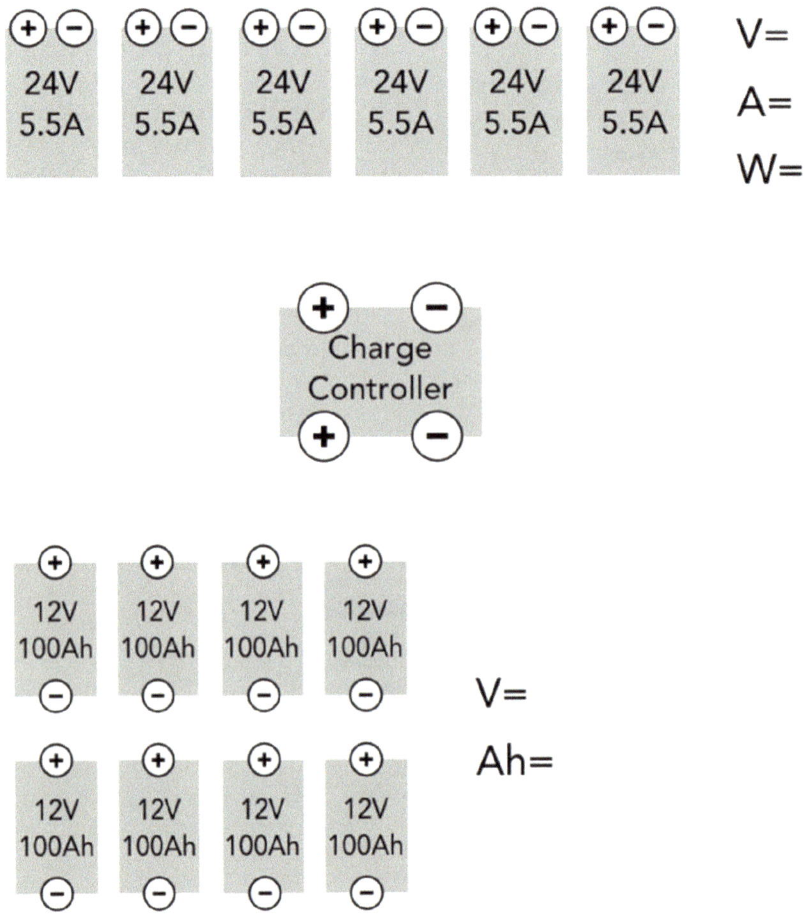


Figure D.11 A 6-module PV (24V, 5.5A) array with 8×batteries (12V, 100Ah) wired to charge controller.

wired in parallel to the charge controller. The whole system will be wired to 48V.

First, wire all equipment to 48V as shown in figure D.12. In this example, the PV modules are 24V so 2 need to be wired in series. That will be 3 strings, each at 48V, 5.5A. The batteries are 12V so 4 need to be wired in series. That will be 2 sets in series, each at 48V, 100Ah.

Step 2, shown in figure D.13, is to add parallel wiring to all the unused terminals: positives to positives and negatives to negatives. The PV modules add 3 strings, 3 × 5.5A = 16.5A. Remember the volts stay the same at 48V. The batteries have 2 sets in series, 2 × 100Ah = 200Ah, volts staying at 48V.

Step 3, shown in figure D.14, is to wire everything into the circuit. Both the array and the battery bank connect in parallel to the charge controller. Note that the wiring to the charge controller extends from the "end" of the parallel wiring you drew in step 2, not from a terminal. In "real life" the parallel wiring is attaching in a combiner box (PV) or busbars (battery) which are then being connected to the controller. The reason these diagrams have been simplified is because all the "extra" wiring can be very confusing when you are first learning.

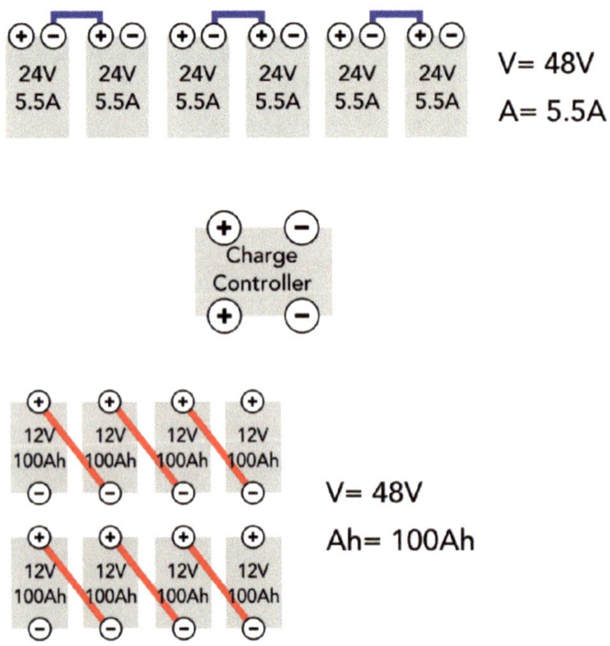


Figure D.12 Step 1: series wiring the PV array and battery bank to 48V.

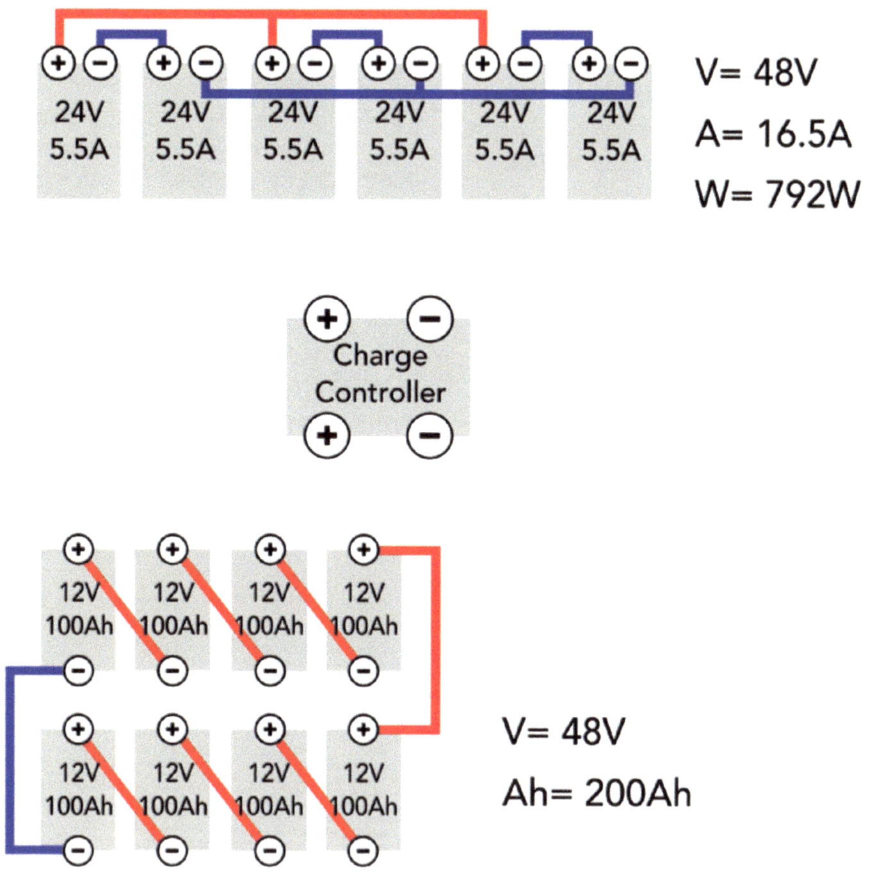


**Figure D.13** Step 2: wire groups of series in parallel, connecting the unused terminals.

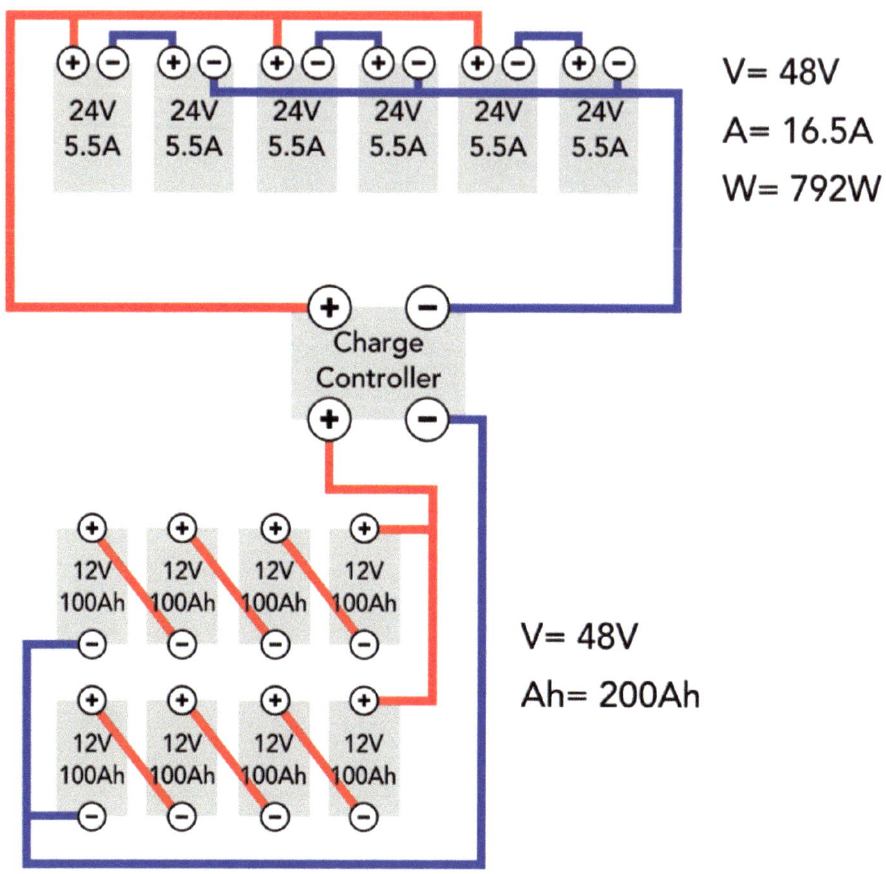


**Figure D.14** Step 3: connect the unused terminals to the charge controller.

**Problem 6: Wire the system in figure D.15 to 12V.** Answer per figure D.27.

**Problem 7: Wire the system in figure D.16 to 12V.** Answer per figure D.28.

**Problem 8: Wire the system in figure D.17 to 24V.** Answer per figure D.29.

**Problem 9: Wire the system in figure D.18 to 24V.** Answer per figure D.30.

**Problem 10: Wire the system in figure D.19 to 48V.** Answer per figures D.31 and D.32.

Grid-tied systems rely on the grid as a power reserve. Batteries are included for power when the grid goes down or for backup grid storage. Grid-tied systems in the United States are limited to 600V on a string, which means PV modules wired in series cannot exceed 600V. Systems up to and including 600V are medium voltage generators, where systems above 600V are high voltage. Countries have different standards for the exact definition of high voltage based on AC or DC power; for the purposes of this book, series wiring should not exceed 600V DC. Pieces of equipment, such as controllers and inverters, have similar 600V thresholds for the energy profiles they can pass through.

Figure D.20 is an array of 9 modules (48V, 9A). Because they are in series, the volts add and the amps stay the same: W = V × A = 432V × 9A = 3,888W. Systems are rated by their DC kW: 3,888W / 1000kW/W = 3.88kW. Therefore, this is a 3.88kW system. All PV systems are rated this way no matter what their voltage is.

**Problem 11: Wire the system in figure D.21 to create a medium voltage array with a 24V battery backup system.** Answer per figure D.33.

Answers to problems are given in figures D.22–D.33.

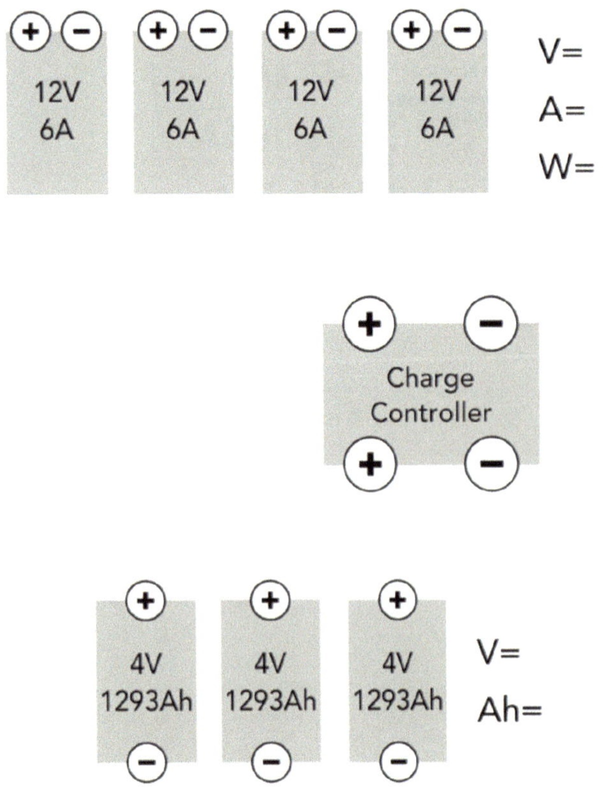


Figure D.15 Problem 6: Wire this system to 12V: 4×PV (12V, 6A), 3×batteries (4V, 1293Ah), charge controller (Answer figure D.27).

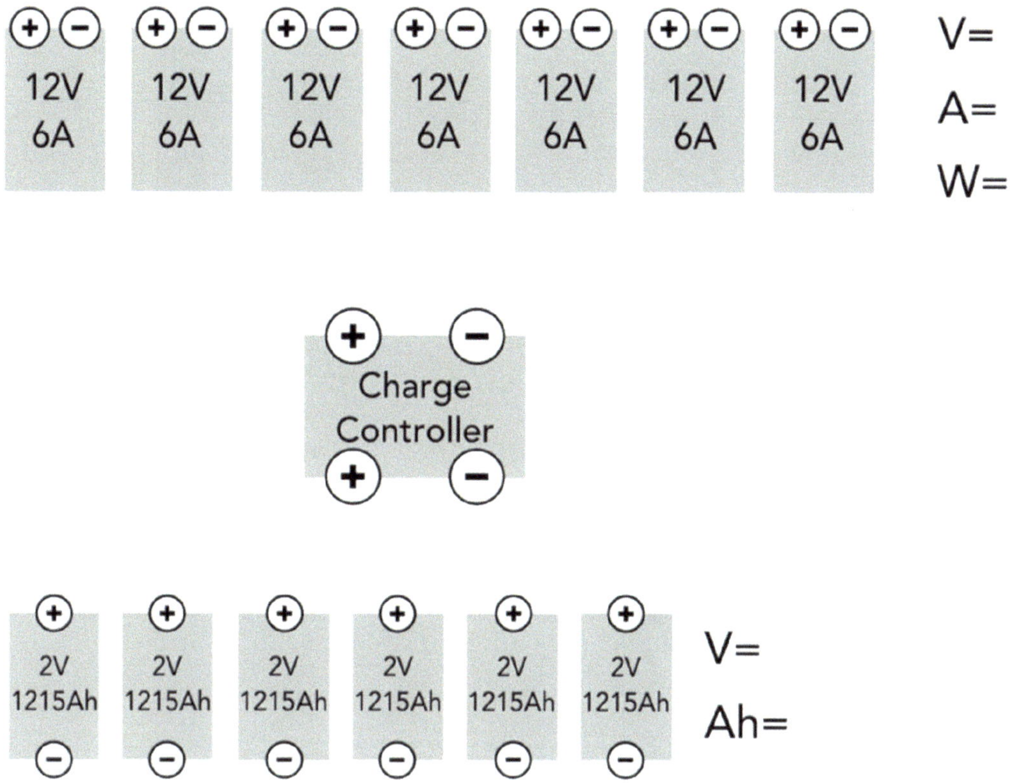


Figure D.16 Problem 7: Wire system to 12V: 7×PV (12V, 6A), 6×batteries (2V, 1215Ah), charge controller (Answer figure D.28).

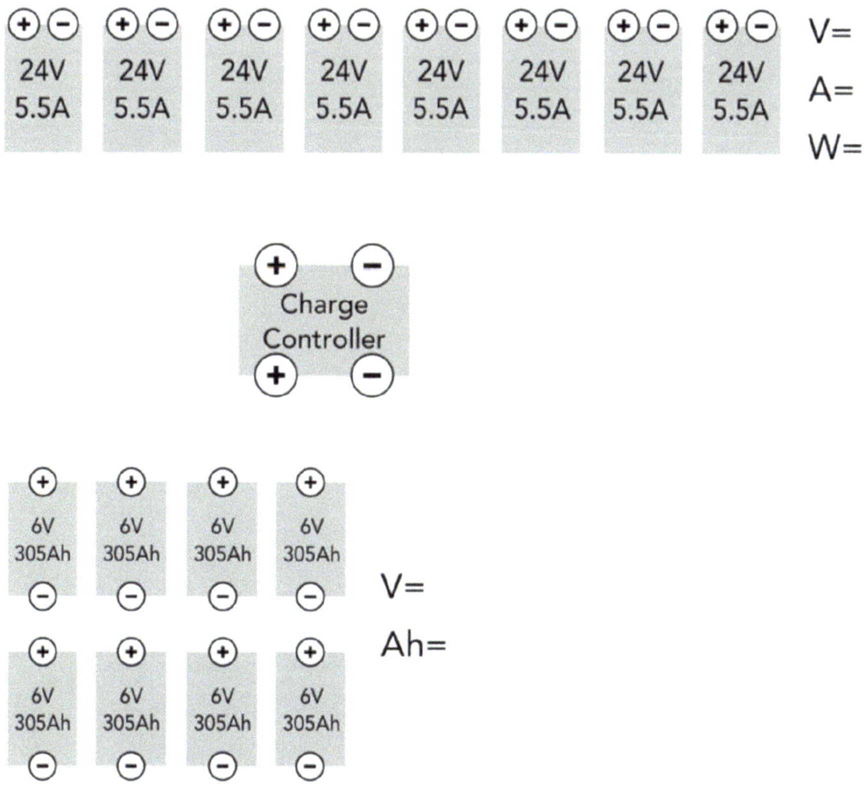


Figure D.17 Problem 8: Wire system to 24V: 8×PV (24V, 5.5A), 8×batteries (6V, 305Ah), charge controller (Answer figure D.29).

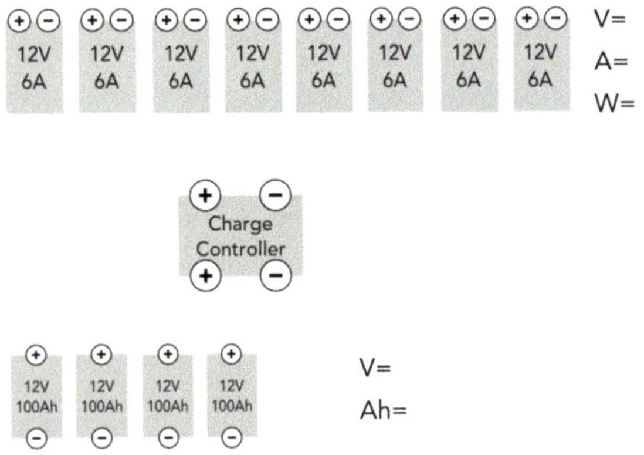


Figure D.18 Problem 9: Wire system to 24V: 8×PV (12V, 6A), 4×batteries (12V, 100Ah), charge controller (Answer figure D.30).

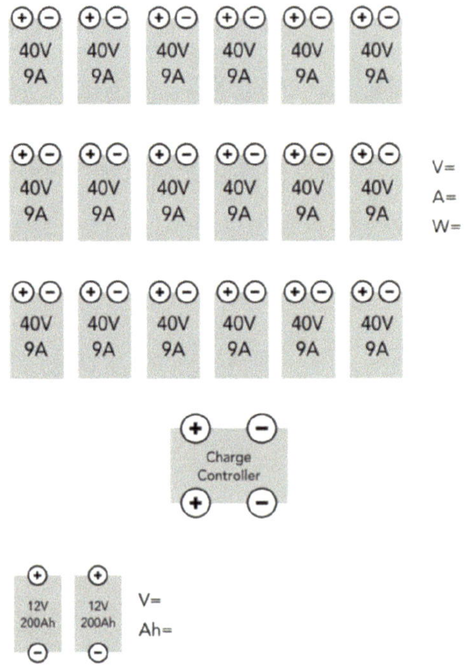


Figure D.21 Problem 11: Medium voltage array with 24V battery backup. 18×PV (40V, 9A), 2×batteries (12V, 200Ah), charge controller (Answer figure D.33).

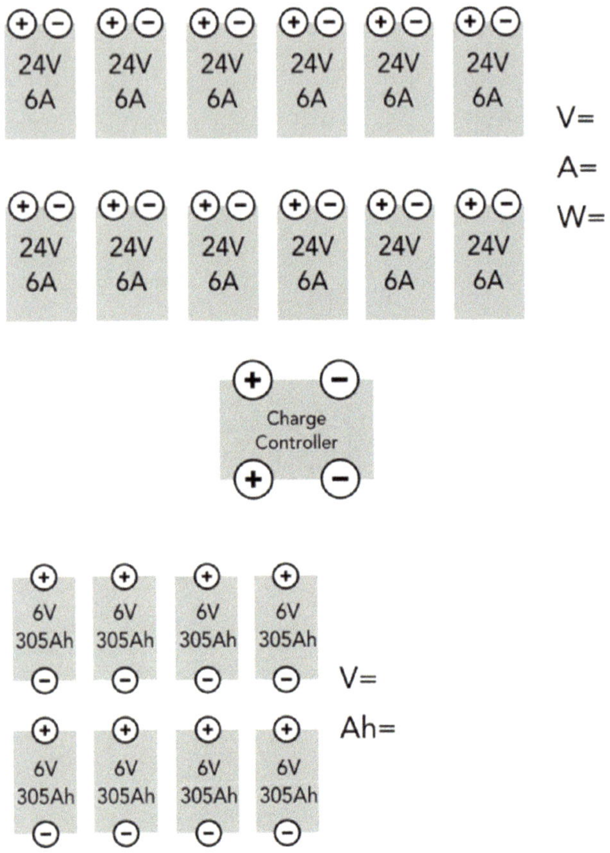


Figure D.19 Problem 10: Wire system to 48V: 12×PV (24V, 6A), 8×batteries (6V, 305Ah), charge controller (Answer figures D.31 and D.32).

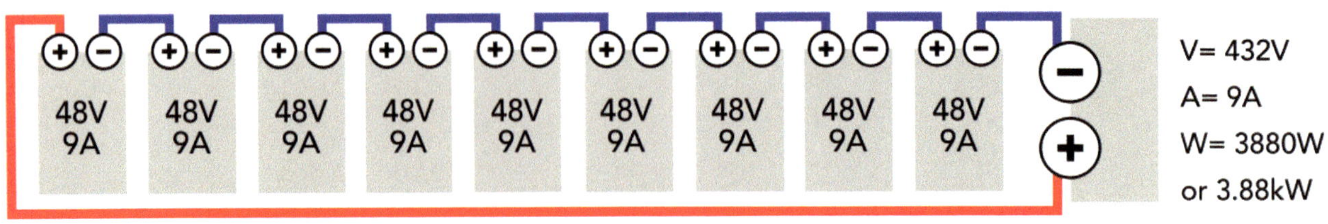


Figure D.20 Medium Voltage array of 9×PV (48V, 9A) in series. Array specifications: 432V, 9A, 3880W, 3.88kW.

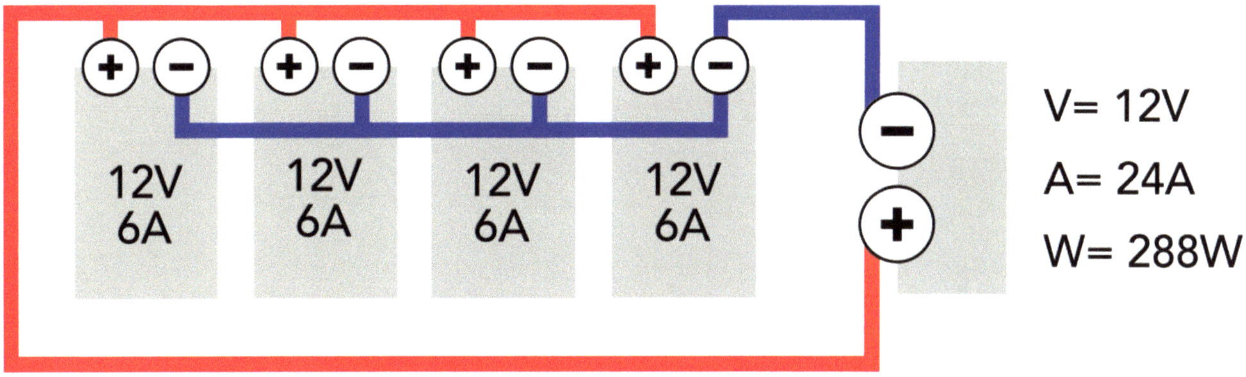


Figure D.22 Solution to problem 1: all items wired in parallel. System is 12V, 24A, 288W. The system is 12V and the modules are 12V so they are wired in parallel (+ to + and – to – ) as shown. They are also wired in parallel to the charge controller.

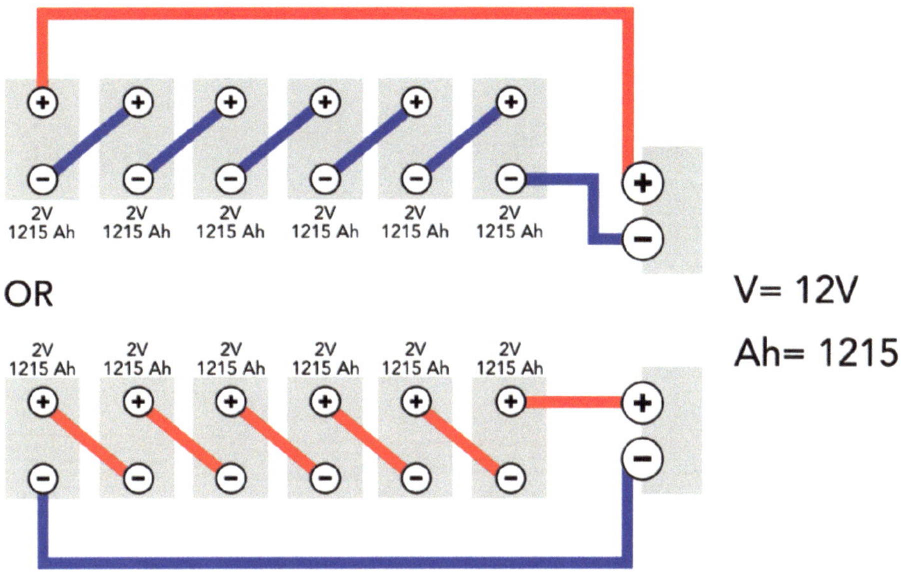


Figure D.23 Solution to problem 2: batteries wired in series to each other and in parallel to the charge controller. System is 12V, 1215Ah. The batteries are 2V and must be wired in series to get to 12V. The wiring can be either + to – or – to + as shown in the two possible solutions.

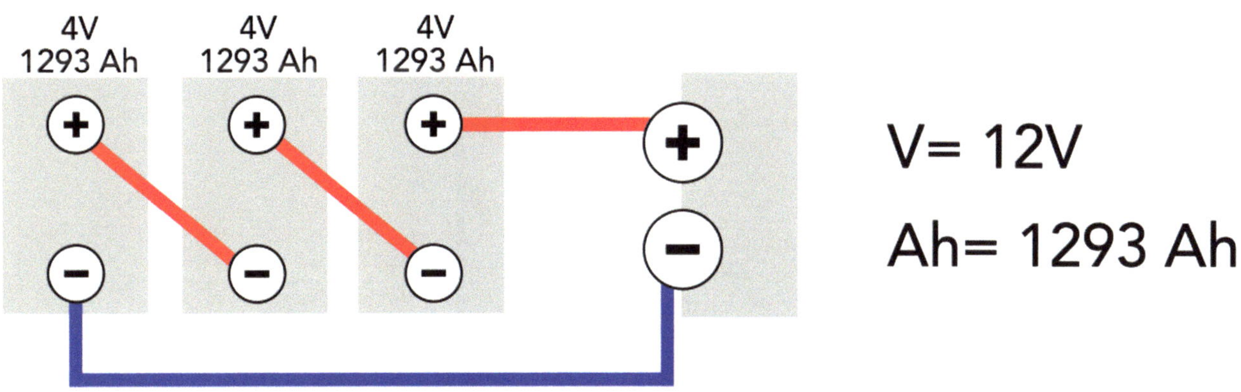


Figure D.24 Solution to problem 3: batteries are wired in series to each other and in parallel to the charge controller. System is 12V, 1293Ah. The batteries must be wired in series to each other to get to 12V. The battery bank is wired to the charge controller in parallel.

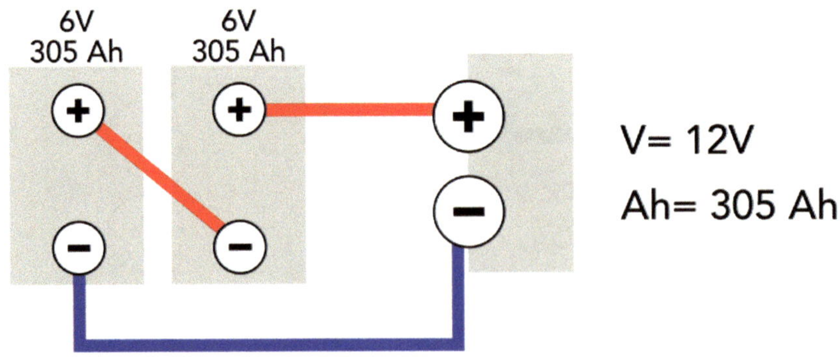


Figure D.25 Solution to problem 4: batteries are wired in series to each other and in parallel to the charge controller. System is 12V, 305Ah. The batteries must be wired in series to each other to get to 12V. The battery bank is wired to the charge controller in parallel.

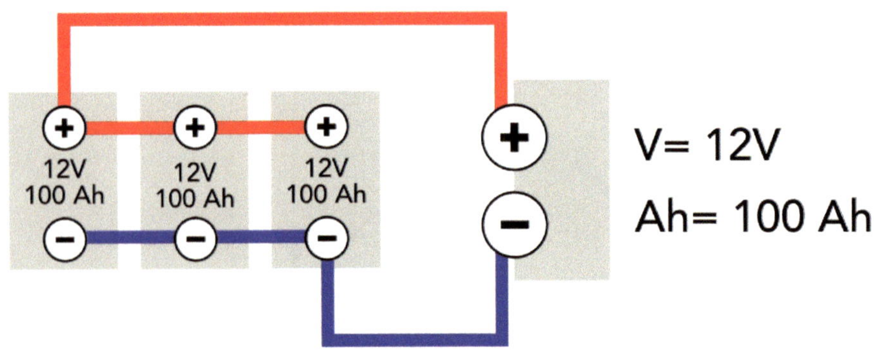


Figure D.26 Solution to problem 5: Batteries and charge controller are all wired in parallel. System is 12V, 100Ah. Batteries will be in parallel to each other and the charge controller.

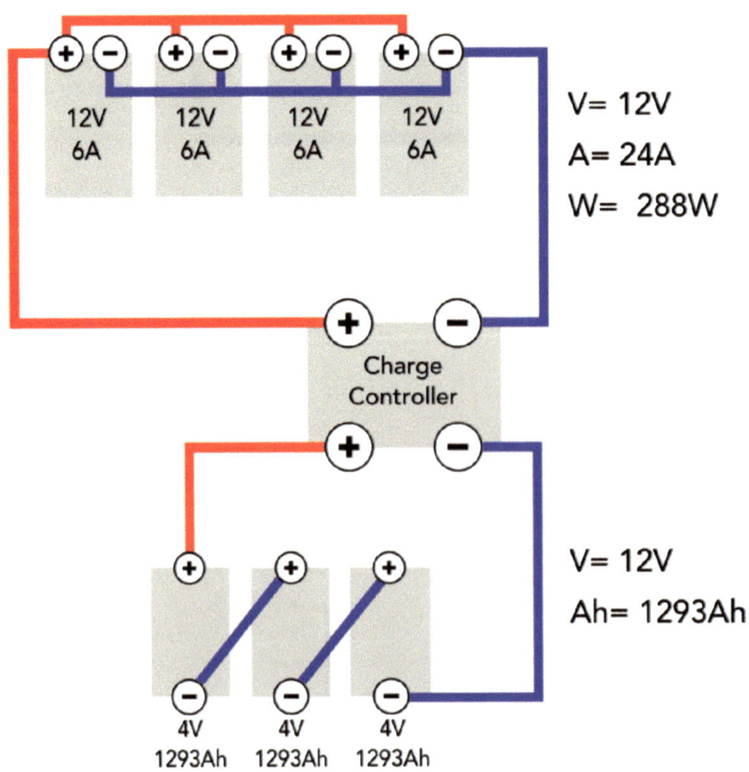


Figure D.27 Solution to problem 6: 4×PV (12V, 4A) are wired in parallel, 3×batteries (4V, 1293Ah) are wired in series, array and battery bank wired in parallel to charge controller. System is 12V, PV is 24A and 288W, batteries are 1293Ah. The PV (12V, 4A) are wired in parallel; the batteries (4V, 1293Ah) are wired in series. Everything is wired to the charge controller in parallel.

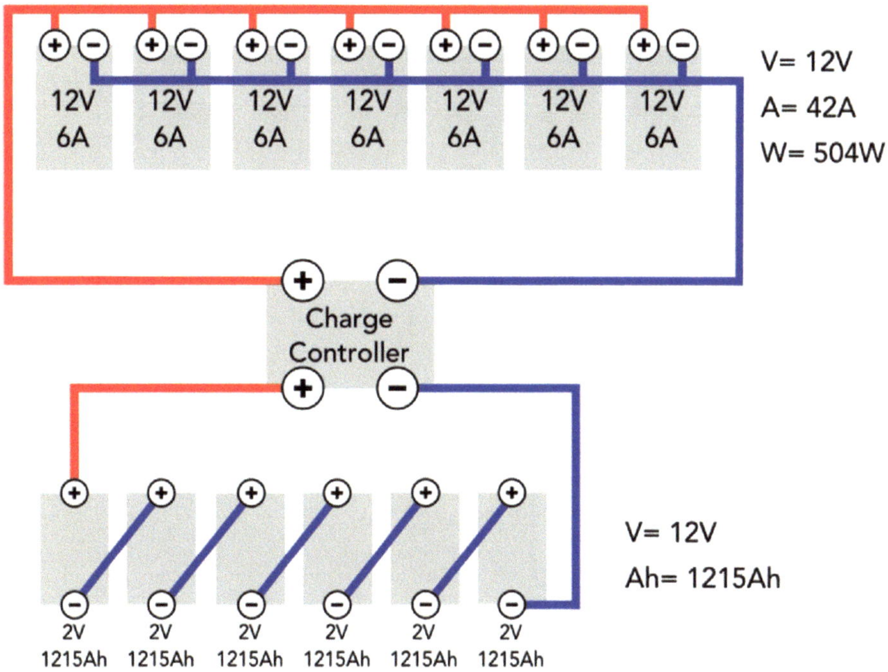


Figure D.28 Solution to problem 7: 7×PV (12V, 6A) are wired in parallel, 3×batteries (4V, 1293Ah) are wired in series, array and battery bank wired in parallel to charge controller. System is 12V, PV is 42A and 504W, batteries are 1215Ah. The PV (12V, 6A) are wired in parallel; the batteries (12V, 1215Ah) are wired in series. Everything is wired to the charge controller in parallel.

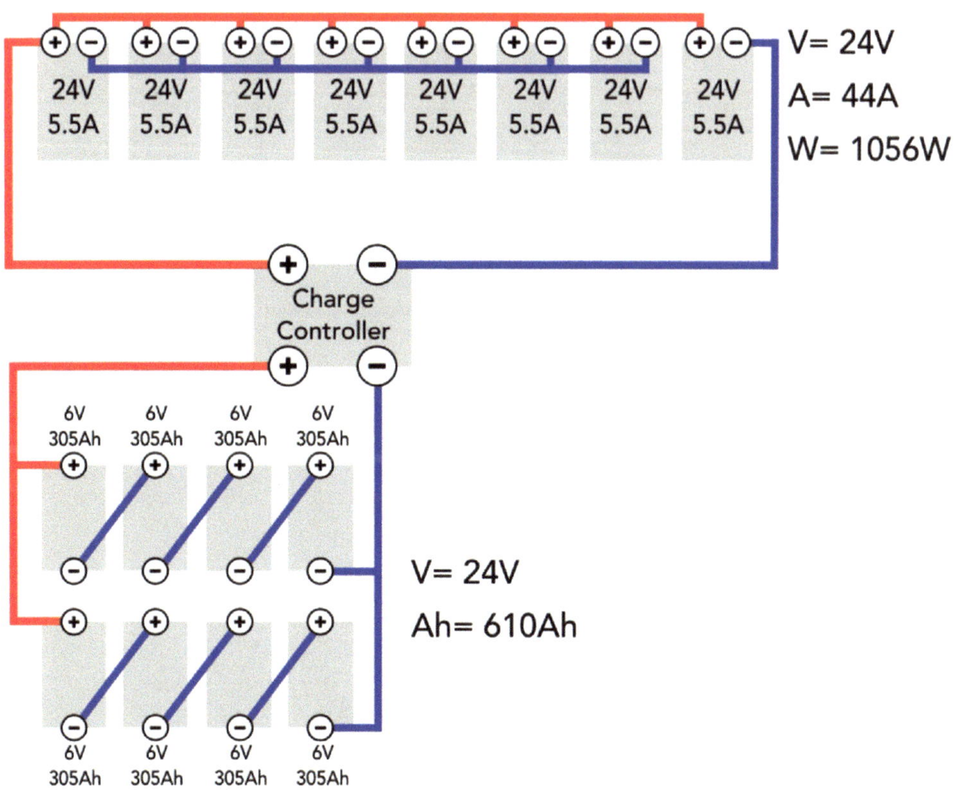


Figure D.29 Solution to problem 8: 8×PV (24V, 5.5A) are wired in parallel. 8×batteries (6V, 305A) require 4 batteries wired in series to reach 24V. The two sets of 4 are wired in parallel. The array and battery bank are wired in parallel to the charge controller. System is 24V, PV is 44A and 1056W, batteries are 610Ah. The PV (24V, 5.5A) are wired in parallel; batteries (6V, 305A) are wired 4 × series to reach 24V, then everything is wired in parallel.

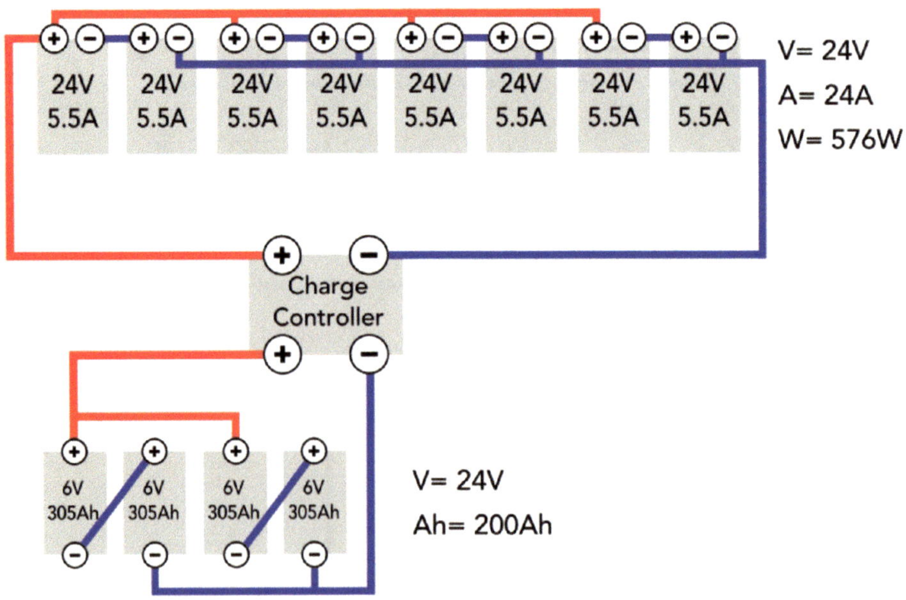


Figure D.30 Solution to problem 9: 8×PV (12V, 6A) are wired 2 in series with 4 strings in parallel. 4×batteries (12V, 100Ah) are wired 2 in series and the two groups in parallel. The array and the battery bank are wired in parallel to the charge controller. System is 24V, PV 24A and 576W, battery bank 200Ah.

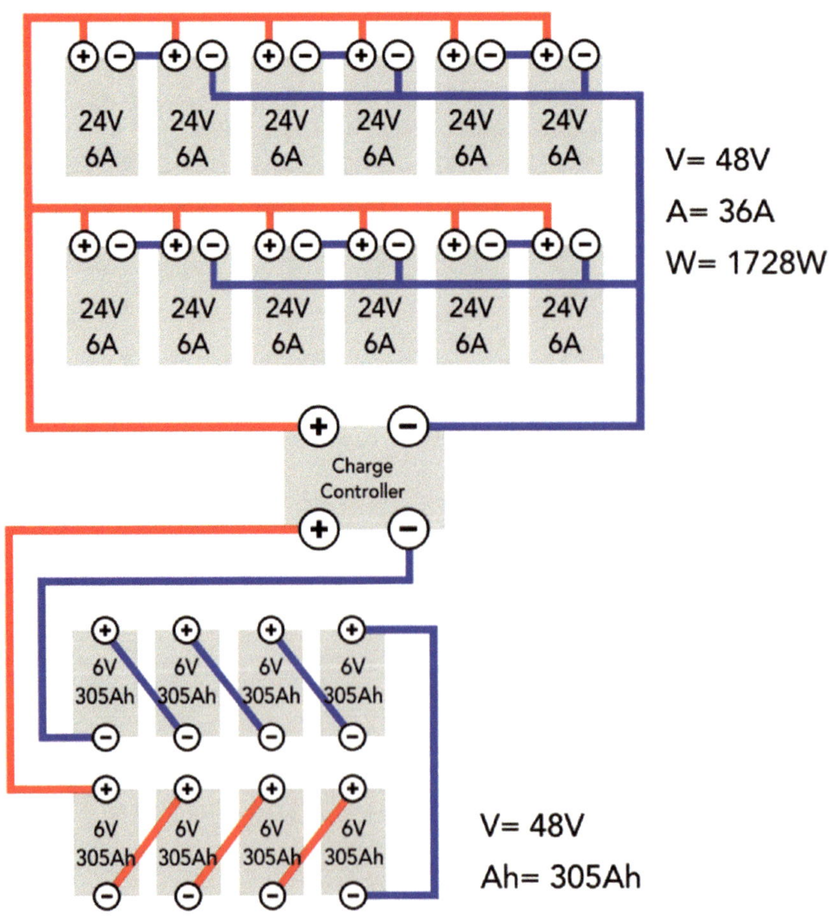


Figure D.31 Solution #1 to problem 10: 12×PV (24V, 6A) 2 modules need to be wired in series to get to 48V, creating 6 strings, series wiring to module to right. 8×batteries (6V, 305Ah) are all in series. Everything is wired in series to the charge controller. System is 48V, array is 36A and 1728W, batteries are 305Ah.

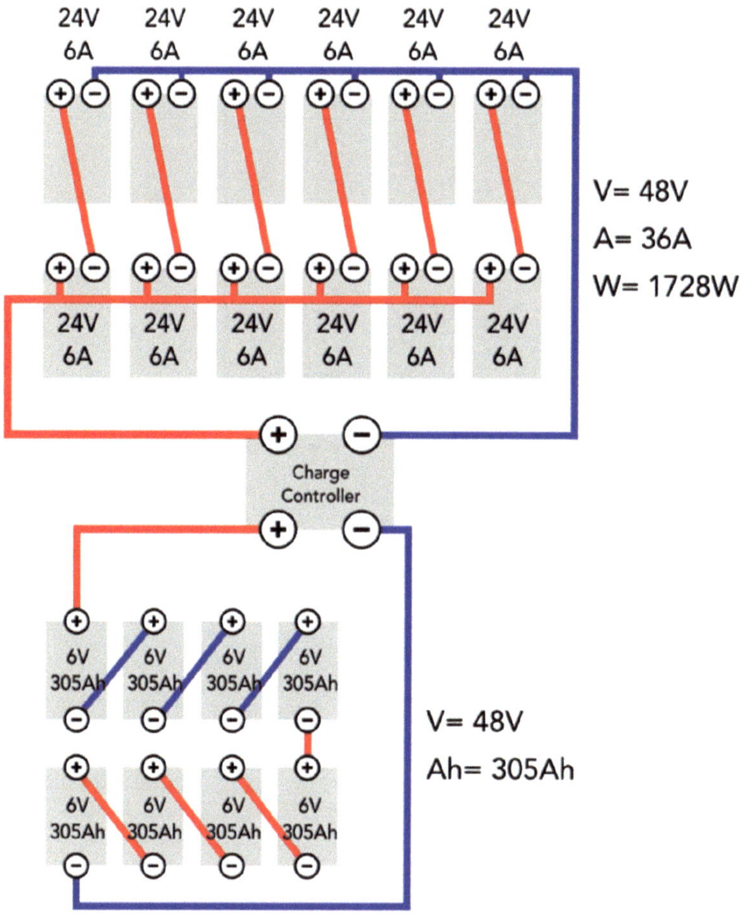


Figure D.32 Solution #2 to problem 10: 12×PV (24V, 6A) 2 modules need to be wired in series to get to 48V, creating 6 strings, series wiring to module underneath. 8×batteries (6V, 305Ah) are all in series. Everything is wired in parallel to the charge controller. System is 48V, array is 36A and 1728W, batteries are 305Ah.

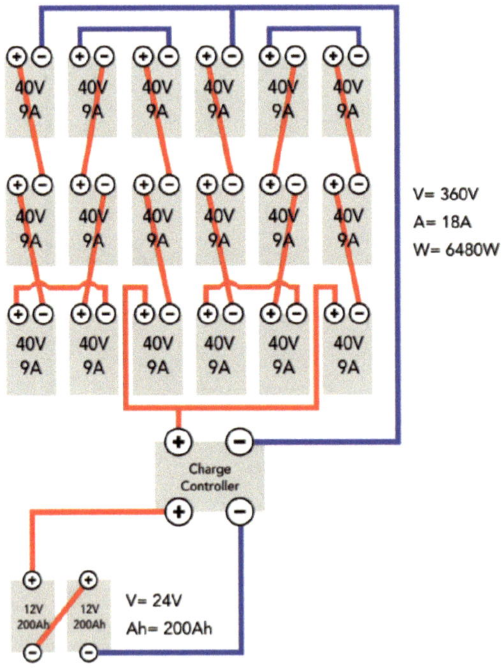


Figure D.33 Solution to problem 11: 18×PV (40V, 9A) create 2 equal strings (9×PV each). 2×batteries (12V, 200Ah) are wired in series. Everything is wired in parallel to the charge controller. Array is 360V, 18A, and 6480W (6.48kW). Battery bank 24V, 200Ah. To remain under 600V maximum, divide by the PV module voltage. 600V/40V per module = 15 modules maximum in a string. However, we have 18 and need equal strings. Therefore, 9 × PV are wired in series. Each string is 360V, 9A. This diagram would be less confusing looking if it were drawn with 9 modules in 2 rows, however, modules may be in varying configurations on a roof. If the diagram has wires that are not connected that cross over each other, one of the wires should be drawn with an arc where it intersects the other wire.

Appendix E

# Solar Thermal Sizing

## Solar Thermal Formulas

1kWh = 3,412Btu

1kWh = 3.6MJ, 1MJ = 1,000,000J

317.1 × kWh/$m^2$/day = 1Btu/$ft^2$/day

3.6 × kWh/$m^2$/day = 1MJ/$m^2$/day

1Btu = 1.055kJ

8.3Btu to heat 1gal water 1°F

4180J to heat 1L of water 1°C

Daily heating load = 8.3 × ΔG × total # gal/day = Btu/day

Daily heating load = 4180J × ΔG × total # L/day = J/day

Solar thermal collector efficiency = 73% of insolation levels

Solar fraction heating load ÷ Btu/$ft^2$/day collector production = area of collectors needed

Solar fraction heating load ÷ MJ/$m^2$/day collector production = area of collectors needed

## Steps of Solar Thermal Design

- Find the gradient: how much are you heating the water? ΔG = design temperature – groundwater temperature. The design temperature is what temperature the hot water is supposed to be for whatever you are using it for. The groundwater temperature is the temperature of the water being added to fill the storage tank as hot water is used.
- Determine how much energy it takes to heat the amount of water used every day.
    - Daily heating load = 8.3 × ΔG × total gal/day = Btu/day
    - Daily heating load = 4180J × ΔG × total L/day = J/day (note that you'll need to convert this to MJ/day so it matches units with the solar insolation values).
- Multiply the daily heating load by the solar fraction to determine how much heat energy you need your collectors to produce each day. Typical maximum solar fractions are 80% for DHW and 30% for heating. This will give you the solar fraction load.
- Look up the solar insolation values for annual average and peak (month when there is the highest amount of solar isolation). Remember to account for any shading that will occur. Convert these values to either Btu/$ft^2$/day or MJ/$m^2$/day. This is how much potential energy is falling on your site every day. Unfortunately, collectors are only about 73% efficient. So, you'll need to multiply these numbers by 73% to get the actual solar collector production value.
- Divide the solar fraction load by the solar collector production value of the peak month to get the required area of collectors. Always use peak month production if you are designing a system that will be working all year. NOTE: if you are designing for hot water to be used in a heating system, then you should use an average value for the months that heating is required. You'll need to determine what happens to the system in the months where no heating is required.
- Compare the area of collectors you want with what the manufacturer has available. If you sized the solar fraction at the highest percentage, round down. Remember that collectors are whatever size they are – you can't change the physical size or have a different tube configuration (i.e. you figured you need 13 tubes but the manufacturer only provides manifolds with 10 tubes or 15 tubes – you are going to have to decide on either 10 or 15 because you can't have 13). You have to use a whole sized collector.
- Compare the area needed for the collectors you've chosen to the roof size (or area on the ground). Remember that you have to keep the collectors vertical (the only exception is the direct flow

evacuated tube, which can be any angle, even horizontal) and the width and height are fixed. That means that you cannot divide the area of roof by the area of collectors! You can divide the width of the roof by the width of the collector. You can't fit part of a collector on a roof – it all has to fit. Most solar thermal collectors are fixed tilt. If your collectors are on stands, then the actual height will be different (because of being mounted at an angle on stands) from the height listed on the data sheet. If you have collectors that are in rows on stands, remember to space them apart so they are not shading each other. Remember that collectors cannot block roof access, see the discussion on both these topics in chapter 13 (Photovoltaics).

- Once you know what the area of collectors is, this will be the number based on what fits on the roof from the manufacturer's given size, then you can determine what the actual output of your collectors will be. The output will be the area of the collectors multiplied by the annual solar collector production value multiplied by 365 days in a year.
- The last steps are to size your tank and piping, figure out where all the balance of system parts fit, and decide on your racking system.

## Solar Thermal Example Problem

Let's say that we are designing a laundromat located in Boston, USA, that has 20 washing machines. The client wants the hot water temperature to be 120°F (48.9°C). This is going to be a very popular laundromat which is used 75% of the time. I used RETScreen[1] to determine that the facility would use 685.5gal/day (2595L/day) and the ground temperature is 52°F (11.1°C).

- Find the gradient: how much are you heating the water?

$\Delta G$ = design temperature – groundwater temperature

$\Delta G = 120°F - 52°F = 68°F$

$\Delta G = 48.9°C - 11.1°C = 37.8°C$

- Determine how much energy it takes to heat the amount of water used every day.

Daily heating load = 8.3 × $\Delta G$ × total # gal/day = Btu/day

Daily heating load = 8.3 × 68°F × 685.5gal/day = 386,896Btu/day

Daily heating load = 4180 × $\Delta G$ × total # L/day = J/day (Note that you'll need to convert this to MJ/day so it matches units with the solar insolation values.)

Daily heating load = 4180 × 37.8°C × 2595L/day = 410,020,380J/day or 410MJ/day

- Determine the solar fraction load.
- I'm going to use 80% solar fraction and then round everything down. This is a DHW use, so 80% is the maximum desirable solar fraction. The result here is how much heat energy that I want my collectors to produce.

Solar fraction load = daily heating load × solar fraction

Solar fraction load = 386,896Btu/day × 0.8 = 309,516Btu/day

Solar fraction load = 410MJ/day × 0.8 = 328MJ/day

- Insolation values from PVWatts for Boston at tilt angle latitude + 15° facing south.

Boston yearly average = 4.39 kWh/m$^2$/d
Boston August (peak) = 5.16 kWh/m$^2$/d

Annual = 4.39 kWh/m$^2$/d × 317.1)= 1,392Btu/ft$^2$/day

Annual = 4.39 kWh/m$^2$/d × 3.6 = 15.804MJ/m$^2$/day

Peak = 5.16 kWh/m$^2$/d × 317.1 = 1,636 Btu/ft$^2$/day

Peak = 5.16 kWh/m$^2$/d × 3.6 = 18.576MJ/m$^2$/day

Collector efficiency is 73%

Collector Annual = 1,392Btu/ft$^2$/day × 0.73 = 1016Btu/ft$^2$/day

Collector Annual = 15.804MJ/m$^2$/day × 0.73 = 11.536MJ/m$^2$/day

Collector Peak = 1,636 Btu/ft$^2$/day × 0.73 = 1,194Btu/ft$^2$/day

Collector Peak = 18.576MJ/m$^2$/day × 0.73 = 13.56MJ/m$^2$/day

- Collector area

Collector area = solar fraction load ÷ collector peak

Collector area = 309,516Btu/day ÷ 1,194Btu/ft$^2$/day = 259ft$^2$

Collector area = 328MJ/day ÷ 13.56MJ/m$^2$/day = 24m$^2$

Remember, this is the collector area to meet 80% of the demand. From here I need to compare this to what the manufacturer has available. It probably won't match the collector area exactly. But I want to get close. Once I have the number of collectors and the size space they'll take up, I can compare that to my roof.

## Resource


1 RETScreen Innovation Lab. (2021, September 21). *RETScreen* (Version 8.1) [Computer software]. Government of Canada. www.nrcan.gc.ca/maps-tools-

# Index